I0066336

Algebra: Advanced Mathematics

Algebra: Advanced Mathematics

Edited by
Derek Portman

www.willfordpress.com

Published by Willford Press,
118-35 Queens Blvd., Suite 400,
Forest Hills, NY 11375, USA

ISBN: 978-1-64728-346-9

Cataloging-in-Publication Data

Algebra : advanced mathematics / edited by Derek Portman.
p. cm.
Includes bibliographical references and index.
ISBN 978-1-64728-346-9
1. Algebra. 2. Mathematics. 3. Mathematical analysis. I. Portman, Derek.
QA155 .A44 2022
512--dc23

For information on all Willford Press publications
visit our website at www.willfordpress.com

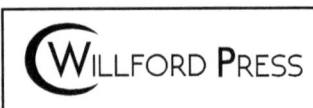

Contents

Preface ... VII

Chapter 1 **Real Zeros of a Class of Hyperbolic Polynomials with Random Coefficients** ... 1
Mina Ketan Mahanti, Amandeep Singh and Lokanath Sahoo

Chapter 2 **Hom-Lie Triple System and Hom-Bol Algebra Structures on Hom-Maltsev and Right Hom-Alternative Algebras** ... 8
Sylvain Attan and A. Nourou Issa

Chapter 3 **PS-Modules over Ore Extensions and Skew Generalized Power Series Rings** ... 20
Refaat M. Salem, Mohamed A. Farahat and Hanan Abd-Elmalk

Chapter 4 **Natural Partial Orders on Transformation Semigroups with Fixed Sets** ... 26
Yanisa Chaiya, Preeyanuch Honyam and Jintana Sanwong

Chapter 5 **Symmetric Integer Matrices Having Integer Eigenvalues** ... 33
Lei Cao and Selcuk Koyuncu

Chapter 6 **Green's Function for a Slice of the Korányi Ball in the Heisenberg Group $\mathbb{H}_n$** ... 39
Shivani Dubey, Ajay Kumar and Mukund Madhav Mishra

Chapter 7 **δ-Primary Hyperideals on Commutative Hyperrings** ... 46
Elif Ozel Ay, Gürsel Yesilot and Deniz Sonmez

Chapter 8 **On Degrees of Modular Common Divisors and the Big Prime** gcd **Algorithm** ... 50
Vahagn Mikaelian

Chapter 9 **New Branch of Intuitionistic Fuzzification in Algebras with their Applications** ... 63
Samaher Adnan Abdul-Ghani, Shuker Mahmood Khalil, Mayadah Abd Ulrazaq and Abu Firas Muhammad Jawad Al-Musawi

Chapter 10 **General Quadratic-Additive Type Functional Equation and its Stability** ... 69
Yang-Hi Lee and Soon-Mo Jung

Chapter 11 **Introduction to Neutrosophic BCI/BCK-Algebras** ... 79
A. A. A. Agboola and B. Davvaz

Chapter 12 **Annular Bounds for the Zeros of a Polynomial** ... 85
Le Gao and N. K. Govil

Chapter 13 **The Regular Part of a Semigroup of Full Transformations with Restricted Range: Maximal Inverse Subsemigroups and Maximal Regular Subsemigroups of its Ideals** ... 92
Worachead Sommanee

Chapter 14 **Explicit Formulas for Meixner Polynomials** 101
Dmitry V. Kruchinin and Yuriy V. Shablya

Chapter 15 **Characterizations of Regular Ordered Semirings by Ordered Quasi-Ideals** 106
Pakorn Palakawong na Ayutthaya and Bundit Pibaljommee

Chapter 16 **On Generalized Semiderivations of Prime Near Rings** 114
Abdelkarim Boua, A. Raji, Asma Ali and Farhat Ali

Chapter 17 **Characterization and Enumeration of Good Punctured Polynomials over Finite Fields** 121
Somphong Jitman, Aunyarut Bunyawat, Supanut Meesawat, Arithat Thanakulitthirat and Napat Thumwanit

Chapter 18 **Generalized Bell Numbers and Peirce Matrix via Pascal Matrix** 128
Eunmi Choi

Chapter 19 **On the Commutative Rings with At Most Two Proper Subrings** 136
David E. Dobbs

Chapter 20 **Structure of n-Lie Algebras with Involutive Derivations** 149
Ruipu Bai, Shuai Hou and Yansha Gao

Chapter 21 **On Killing Forms and Invariant Forms of Lie-Yamaguti Superalgebras** 158
Patricia L. Zoungrana and A. Nourou Issa

Chapter 22 **The Noncentral Version of the Whitney Numbers: A Comprehensive Study** 167
Mahid M. Mangontarum, Omar I. Cauntongan and Amila P. Macodi-Ringia

Chapter 23 **f_q-Derivations of G-Algebra** 183
Deena Al-Kadi

Chapter 24 **h-Adic Polynomials and Partial Fraction Decomposition of Proper Rational Functions over $\mathbb{R}$ or $\mathbb{C}$** 188
Kwang Hyun Kim and Xin Zhang

Chapter 25 **Vector Spaces of New Special Magic Squares: Reflective Magic Squares, Corner Magic Squares, and Skew-Regular Magic Squares** 194
Thitarie Rungratgasame, Pattharapham Amornpornthum, Phuwanat Boonmee, Busrun Cheko and Nattaphon Fuangfung

Chapter 26 **A Note on Primitivity of Ideals in Skew Polynomial Rings of Automorphism Type** 201
Edilson Soares Miranda

Permissions

List of Contributors

Index

Preface

This book has been a concerted effort by a group of academicians, researchers and scientists, who have contributed their research works for the realization of the book. This book has materialized in the wake of emerging advancements and innovations in this field. Therefore, the need of the hour was to compile all the required researches and disseminate the knowledge to a broad spectrum of people comprising of students, researchers and specialists of the field.

Algebra is one of the major branches of mathematics which studies mathematical symbols and the rules used for handling these symbols. It deals with numerous concepts such as elementary equations and abstractions such as groups, rings and fields. There are two subdomains within algebra, namely, elementary algebra and abstract algebra. Elementary algebra is the basic part of algebra and it is used in different branches of mathematics, science, engineering, medicine and economics. Advanced mathematics is a major area in abstract algebra. This book provides comprehensive insights into the field of algebra. The various advancements in this discipline are glanced at and their applications as well as ramifications are looked at in detail. The extensive content of this book provides the readers with a thorough understanding of the subject.

At the end of the preface, I would like to thank the authors for their brilliant chapters and the publisher for guiding us all-through the making of the book till its final stage. Also, I would like to thank my family for providing the support and encouragement throughout my academic career and research projects.

Editor

1

Real Zeros of a Class of Hyperbolic Polynomials with Random Coefficients

Mina Ketan Mahanti,[1] Amandeep Singh,[2] and Lokanath Sahoo[3]

[1]*Department of Mathematics, College of Basic Science and Humanities, OUAT, Bhubaneswar, India*
[2]*DPS Kalinga, Bhubaneswar, India*
[3]*Gopabandhu Science College, Athagad, India*

Correspondence should be addressed to Mina Ketan Mahanti; minaketan_mahanti@yahoo.com

Academic Editor: Niansheng Tang

We have proved here that the expected number of real zeros of a random hyperbolic polynomial of the form $y = P_n(t) = \sqrt{\binom{n}{1}}a_1 \cosh t + \sqrt{\binom{n}{2}}a_2 \cosh 2t + \cdots + \sqrt{\binom{n}{n}}a_n \cosh nt$, where $a_1, \ldots, a_n$ is a sequence of standard Gaussian random variables, is $\sqrt{n}/2 + o_p(1)$. It is shown that the asymptotic value of expected number of times the polynomial crosses the level $y = K$ is also $\sqrt{n}/2$ as long as K does not exceed $\sqrt{2^n e^{\mu(n)}}$, where $\mu(n) = o(n)$. The number of oscillations of $P_n(t)$ about $y = K$ will be less than $\sqrt{n}/2$ asymptotically only if $K = \sqrt{2^n e^{\mu(n)}}$, where $\mu(n) = O(n)$ or $n^{-1}\mu(n) \to \infty$. In the former case the number of oscillations continues to be a fraction of $\sqrt{n}$ and decreases with the increase in value of $\mu(n)$. In the latter case, the number of oscillations reduces to $o_p(\sqrt{n})$ and almost no trace of the curve is expected to be present above the level $y = K$ if $\mu(n)/(n \log n) \to \infty$.

1. Introduction

Let $(\Omega, A, \Pr)$ be a fixed probability space and let $\{a_k(\omega)\}_{k=1}^{j=n}$ be a sequence of independent random variables defined on Ω. The sum $a_0(\omega)f_1(t) + a_2(\omega)f_2(t) + \cdots + a_n(\omega)f_n(t)$ is traditionally known as a random algebraic polynomial if $f_i(t) = t^i$, a random trigonometric polynomial if $f_i(t) = \cos(it)$ or $\sin(it)$, and a random hyperbolic polynomial if $f_i(t) = \cosh(it)$ or $\sinh(it)$. One can have useful information about the behaviour of these ensembles of polynomials if the average number of times these polynomials oscillate about the line $y = K$ is known. The reader is referred to the book by Farahmand [1] where an exhaustive account of progress made in study of random polynomials has been presented. It is to be noted that there is significantly more published literature on random algebraic and random trigonometric polynomials than that of random hyperbolic polynomials. Let

$$y = Q_n(t) = \sum_{k=1}^{n} a_k(\omega) \cosh kt, \qquad (1)$$

where $a_k(\omega)$ are normally distributed random variables with mean zero and variance one. One knows that Das [2] first calculated the expected number of real zeros of $Q_n(t)$. Farahmand [3] calculated the asymptotic estimate of oscillations of $Q_n(t)$ about $y = K$ if $K = o(\sqrt{n})$. Some of the other works in this direction are due to Mahanti [4–6]. Wilkins [7] determined real zeros of $Q_n(t)$ when $\mathrm{var}(a_k(\omega)) = k^p$, $p \geq 0$. We observe that the asymptotic value of the oscillations of random hyperbolic polynomials is $(2/\pi)\log n$ in each of these cases.

One is tempted to ask whether $Q_n(t)$ has more than $(2/\pi)\log n$ oscillations under certain conditions. In this context, we are reminded of a recent work of Edelman and Kostlan [8] where it has been found out that the expected number of real zeros of random algebraic polynomials increases significantly if the variance of the coefficients changes from unity to $\sqrt{\binom{n}{k}}$. Therefore, we examine what effect this new assumption on variance of the coefficients has on number of oscillations of $Q_n(t)$. In other words, we calculate the number of oscillations of the polynomial

$$y = P_n(t) = \sum_{k=1}^{n} a_k(\omega)\sqrt{\binom{n}{k}}\cosh kt, \tag{2}$$

where $a_k(\omega)$ are normally distributed random variables with mean zero and variance one.

In Theorem 1 we have shown that the number of real zeros of $P_n(t)$ is substantially larger than that of $Q_n(t)$. Moreover, there is a significant difference in the way real zeros of $P_n(t)$ and $Q_n(t)$ lie on the t-axis. Most of the real zeros of $Q_n(t)$ are confined to the interval $[-1,1]$ and there are negligible numbers of them if $|t| > 1$ (see [4]). But there are large numbers of real zeros of $P_n(t)$ outside $[-1,1]$. This phenomenon can be deduced from the formula given in (26) and Lemma 10. In fact, the number of real zeros of $P_n(t)$ in the region $|t| > \alpha$, $\alpha > 1/\sqrt{n}$, is dependent on α. The real zeros decrease in number with increase in α in the region $|t| > \alpha$ and there are negligible numbers of them only when $\alpha \to \infty$.

Let $\mu(n)$ be any function of n such that $\mu(n)/n \to 0$. In Theorem 2 we have shown that the number of oscillations of $P_n(t)$ about the line $y = K$, where $K \le \sqrt{2^n e^{\mu(n)}}$, is equal to its axis crossings. Thus, for all these values of K one can say that most of the oscillations of $P_n(t)$ that cross the t-axis reach up to the level $y = K$. If $\mu(n) = O(n)$ or $n^{-1}\mu(n) \to \infty$, we have proved in the theorem that the asymptotic value of number of oscillations of $P_n(t)$ about $y = K = \sqrt{2^n e^{\mu(n)}}$ is less than $\sqrt{n}/2$. If $\mu(n) = O(n)$ the number of oscillations continues to be a fraction of $\sqrt{n}$ and decreases with the increase in value of $\mu(n)$. If $n^{-1}\mu(n) \to \infty$ the number of oscillations is reduced to $o_p(\sqrt{n})$. Inequality (35) together with Lemma 14 provides a glimpse of the manner in which the number of oscillations decreases with increase in value of K. Inequality (35) also shows that there is hardly any trace of the curve $P_n(t)$ above the level $y = K = \sqrt{2^n e^{\mu(n)}}$ if $\mu(n)/(n\log n) \to \infty$.

Let the coefficients $\{a_k(\omega)\}_{k=1}^{k=n}$ of $P_n(t)$ be standard normal random variables. The expected number of oscillations of $P_n(t)$ about the line $y = K$, $t \in (\alpha, \beta)$, has been denoted by us as $\mathrm{EN}_{n,K}(\alpha,\beta)$ in the following two theorems.

Theorem 1. *For sufficiently large n,*

$$\mathrm{EN}_{n,0}(-\infty,\infty) = \frac{\sqrt{n}}{2} + O_p\left(n^{-1/2}\log n\right). \tag{3}$$

Theorem 2. *For sufficiently large n,*

(i) $\mathrm{EN}_{n,K}(-\infty,\infty) = \sqrt{n}/2 + O_p(n^{-1/2}\log n)$ *if* $K^2/2^n \to o$,

(ii) $\mathrm{EN}_{n,K}(-\infty,\infty) = \sqrt{n}/2 + (2/\sqrt{\pi})\mathrm{erf}(K/\sqrt{2^n}) + O_p(n^{-1/2}\log n)$ *if* $K^2 = O(2^n)$,

(iii) $\mathrm{EN}_{n,K}(-\infty,\infty) = \sqrt{n}/2 - n^{-1/2}\mu(n)/(2\pi) + 2/\sqrt{\pi} + O_p(n^{-1/2}\log n)$ *if* $\mu(n)/n \to 0$, $\mu(n) \to \infty$, *and* $K = \sqrt{2^n e^{\mu(n)}}$,

(iv) $\mathrm{EN}_{n,K}(-\infty,\infty) = (\sqrt{n}/\pi)\sin^{-1}(e^{-n^{-1}\mu(n)}/\sqrt{2}) + 2/\sqrt{\pi} + O_p(n^{-1/2}\log n)$ *if* $\mu(n) = O(n)$ *and* $K = \sqrt{2^n e^{\mu(n)}}$,

(v) $\mathrm{EN}_{n,K}(-\infty,\infty) = \sqrt{n}e^{-n^{-1}\mu(n)}/(2\sqrt{2}) + 2/\sqrt{\pi} + O_p(n^{-1/2}\log n)$ *if* $\mu(n)/n \to \infty$ *and* $K = \sqrt{2^n e^{\mu(n)}}$.

Two more differences in behaviour of $P_n(t)$ and $Q_n(t)$ are noteworthy. In what follows we will find out that most of the axis crossings of $P_n(t)$ reach the level $\sqrt{2^n e^{\mu(n)}}$. However, the branches of $Q_n(t)$ that cross the axis do not travel beyond $y = K$, where $K = O(\sqrt{n})$ (see Mahanti and Sahoo [6]). Almost all of the polynomial $Q_n(t)$ lie below the level K, where $n^{-1}\log(K/\sqrt{n}) \to \infty$ (Mahanti and Sahoo [6]). However, a large part of $P_n(t)$ stretches above this level.

2. Formula for the Proof of the Theorems

The proof of the theorem is based on the formula for expected number of level crossings given by Crammer and Leadbetter [9, page 285]. Using it for $P_n(t) - K = 0$ in the interval (α,β) we can show that

$$\mathrm{EN}_{n,K}(\alpha,\beta) = \int_{\alpha}^{\beta}\frac{\sqrt{A_nC_n - B_n^2}}{A_n}\phi\left(-\frac{K}{\sqrt{A_n}}\right) \cdot \left[2\phi(\eta) + \eta\{2\Phi(\eta) - 1\}\right]dt, \tag{4}$$

where $A_n = \mathrm{var}\{P_n(t) - K\}$, $B_n = \mathrm{cov}[\{P_n(t) - K\}, P_n'(t)]$, $C_n = \mathrm{var}\{P_n'(t)\}$, $\Phi(t) = (1/\sqrt{2\pi})\int_{-\infty}^{t}\exp(-x^2/2)dx$, $\phi(t) = (1/\sqrt{2\pi})\exp(-x^2/2)$, and $\eta = -B_nK/\sqrt{A_n}\sqrt{A_nC_n - B_n^2}$. Since the coefficients of $P_n(t)$ are independent and $\mathrm{var}(a_j) = \sqrt{\binom{n}{k}}$, it is easy to derive using little algebra that

$$A_n = \sum_{k=1}^{n}\sqrt{\binom{n}{k}}\cosh^2 kt = 2^{n-1}\left(\cosh nt\cosh^n t + 1\right), \tag{5}$$

$$B_n = \frac{1}{2}\sum_{k=1}^{n}k\sqrt{\binom{n}{k}}\sinh 2kt = 2^{n-2}n\sinh(n+1)t \cdot \cosh^{n-1}t, \tag{6}$$

$$C_n = \sum_{k=1}^{n}k^2\sqrt{\binom{n}{k}}\sinh^2 k = 2^{n-3}\left[\left(n^2\cosh(n+2)t + n\cosh nt\right)\cosh^{n-2}t - n^2 - n\right]. \tag{7}$$

Let

$$D_n = A_nC_n - B_n^2 = 2^{2n-4}\left(n^2\left(\cosh^{2n}t - 1\right) + \left(n^2 - n\right)\cosh^{n-2}t\cosh nt\sinh^2 t + 2n^2\cosh^{n-1}t\sinh nt\sinh t + n\left(\cosh^{2n-2}t\cosh^2 nt - 1\right)\right). \tag{8}$$

Formula (4) now can be written as

$$\mathrm{EN}_{n,K}(\alpha,\beta) = I_1(\alpha,\beta) + I_2(\alpha,\beta), \tag{9}$$

where

$$I_1(\alpha,\beta) = \frac{1}{\pi}\int_\alpha^\beta \phi_n(t)\exp\left\{-\frac{K^2C_n}{2D_n}\right\}dt,$$

$$I_2(\alpha,\beta) = \int_\alpha^\beta \frac{\sqrt{2}KB_n}{\pi A_n^{3/2}}\exp\left\{-\frac{K^2}{2A_n}\right\}\operatorname{erf}\left(\frac{KB_n}{\sqrt{2A_nD_n}}\right)dt, \tag{10}$$

$$\phi_n(t) = \frac{\sqrt{D_n}}{A_n}.$$

As a special case, we can also obtain the famous Kac-Rice formula [10] for expected number of zeros of $P_n(t)$ by putting $K = 0$ in (9). Thus we have

$$\mathrm{EN}_{n,0}(\alpha,\beta) = \frac{1}{\pi}\int_\alpha^\beta \phi_n(t)\,dt. \tag{11}$$

3. Preliminary Analysis

To evaluate the integrals in (9) and (11) we need to find out the dominant terms of A_n, B_n, and C_n. The inequalities mentioned in Lemmas 4–7 will be helpful for the purpose. We first mention the following form of L'Hôpital's Rule which we use to derive some of the inequalities.

Lemma 3 (the monotonic form of L'Hôpital's Rule [11]). *For $-\infty < a < b < \infty$, let $g, h : [a,b] \to R$ be continuous on $[a,b]$ and differentiable on (a,b) with $h'(t) \neq 0$ on (a,b). If $g'(t)/h'(t)$ is increasing or decreasing on (a,b), then so are $(g(t)-g(a))/(h(t)-h(a))$ and $(g(t)-g(b))/(h(t)-h(b))$. If $g'(t)/h'(t)$ is strictly monotone, then the monotonicity of the above two quotients is also strict.*

Lemma 4. *If $0 \le t < \pi/2$, then $e^{nt^2/2-nt^4/8} \le (\cosh t)^n \le e^{nt^2/2}$. In particular in the interval $(0, n^{-1/2})$*

$$1 + \frac{n\sinh^2 t}{2} + \frac{n(n-2)\sinh^4 t}{8} < (\cosh t)^n < 1 + \frac{n\sinh^2 t}{2} + \frac{n^2\sinh^4 t}{4}. \tag{12}$$

Thus $(\cosh t)^n = 1 + O(nt^2)$ if $t = o(n^{-1/2})$.

Proof. The series representation of $\tanh t$ is given by [12, page 42] $\tanh t = \sum_{k=1}^\infty s_k t^{2k-1}$, where $s_k = 2^{2k}(4^k-1)B_{2k}/(2k)!$. Using the fact that $\zeta(2k) = 2^{2k-1}\pi^{2k}|B_{2k}|/(2k)!$ [12, page 1038], we have $|s_k|t^{2k-1} = \pi^{-2k}(2t)^{2k-1}(1-2^{-2k})\zeta(2k)$.

Now, $(1-2^{-n})\zeta(n) = \sum_{j=0}^\infty (2j+1)^{-n}$ is a monotonically decreasing function of n. Therefore, the power series of $\tanh t$ converges absolutely and uniformly for $0 \le t < \pi/2$. Integrating term by term we have the alternative series $\log\cosh t = \sum_{k=1}^\infty (2k)^{-1}s_k t^{2k}$ for $0 \le t < \pi/2$. Since the coefficients of t^{2k} are monotonically decreasing, we find that the first part of the lemma is true.

Let $h_1(t) = r_1(t)/r_2(t)$, where $r_1(t) = (\cosh t)^n - 1$ and $r_2(t) = n\sinh^2 t/2 + n^2\sinh^4 t/4$, with $r_1(0) = r_2(0) = 0$. Then

$$\left(\frac{r_1'(t)}{r_2'(t)}\right)' = \left[(n-2)\left(n^2-4n\right)(\cosh t)^{n-3}\sinh t \cdot \left\{\sinh^2 t - \frac{(n+2)}{(n^2-4n)}\right\}\right]\left\{2\left(1+n\sinh^2 t\right)^2\right\}^{-1}. \tag{13}$$

It is easy to verify that $t + t^3/5 < (n+2)/(n^2-4n)$ if $t \le n^{-1/2}$. Using the series representation of $\sinh t$ we find that

$$t + \frac{t^3}{5} - \sinh t = \frac{t^3}{30} - \sum_{k=2}^\infty \frac{t^{2k+1}}{(2k+1)!} \ge t^3\left(\frac{1}{30} - \sum_{k=2}^\infty \frac{1}{(2k+1)!}\right) \ge 0. \tag{14}$$

Therefore $\sinh^2 t - (n+2)/(n^2-4n) < 0$ if $t \in (0, n^{-1/2})$. As a consequence, $r_1'(t)/r_2'(t)$ is strictly decreasing in the interval $(0, n^{-1/2})$. By Lemma 3, $h_1(t)$ is decreasing in $(0, n^{-1/2})$ and by L'Hôpital's Rule $h_1(0+) = 1$. Therefore, $r_1(t) < r_2(t)$. Let $h_2(t) = r_3(t)/r_4(t)$, where $r_3(t) = (\cosh t)^n - 1 - (n\sinh^2 t)/2$ and $r_4(t) = (n^2\sinh^4 t)/4$ with $r_3(0) = r_4(0) = 0$. Then $r_3'(t)/r_4'(t) = (\cosh^{n-2}t - 1)(n\sinh^2 t)^{-1} = r_5(t)/r_6(t)$, where $r_5(t) = (\cosh t)^{n-2} - 1$ and $r_6(t) = n\sinh^2 t$ with $r_5(0) = r_6(0) = 0$. Now $r_5'(t)/r_6'(t) = (n-2)(\cosh t)^{n-4}/(2n)$, which is increasing. So, by Lemma 3, $h_2(t)$ is strictly increasing. By L'Hôpital's Rule we find that $h_2(0+) = (n-2)/(2n)$. Therefore,

$$1 + \frac{n\sinh^2 t}{2} + \frac{n(n-2)\sinh^4 t}{8} < (\cosh t)^n. \tag{15}$$

□

Lemma 5. *If $t \to 0$, then $\pi/4 + t/2 - t^3/12 < \tan^{-1}e^t < \pi/4 + t/2 - t^3/12 + t^5/48$.*

Proof. Let $r_7(t) = \tanh^{-1}e^t - \pi/4$, $r_8(t) = t/2 - t^3/12$, and $r_9(t) = t/2 - t^3/12 + t^5/48$. Observe that $r_7(0) = r_8(0) = r_9(0) = 0$. Let $h_3(t) = r_7(t)/r_8(t)$ and $h_4(t) = r_7(t)/r_9(t)$. Then $r_7'(t)/r_8'(t) = \operatorname{sec}ht$ is a monotonically decreasing and $r_7'(t)/r_9'(t) = 2\operatorname{sec}ht/(2-t^2)$ is monotonically increasing function of t. By L'Hôpital Rule $h_3(0+) = h_4(0+) = 1$. Therefore, by Lemma 3, $r_7(t) < r_8(t)$ and $r_9(t) < r_7(t)$ for $t > 0$. □

Lemma 6. *Let $m \ge 0$, $p \ge 1$. Then $\sinh^p t(\sinh nt)^{-1}(\cosh t)^{-m}$ is a monotonically decreasing function of t in $(0,\infty)$ if $p = 1$ and is a monotonically decreasing function of t in $(p/n, \infty)$ if $p \ge 2$ and $n > \sqrt{p}\{3(1-\tanh p)\}^{-1/2}$.*

Proof. Since

$$\left(\sinh^p t\,(\sinh nt)^{-1}(\coth t)^{-m}\right)' = \sinh^{p-1} t \cdot (\cosh t)^{m+1}\cosh nt \cdot \left(p\tanh nt - n\tanh t - m\tanh nt\tanh^2 t\right), \quad (16)$$

to prove the lemma we show that $p\tanh nt - n\tanh t < 0$ under the conditions mentioned in the lemma. We first note that $(\tanh nt - n\tanh t)' = n(\sec h^2 nt - \sec h^2 t) < 0$. Therefore, if $p = 1$ and $t > 0$, $p\tanh nt - n\tanh t < 0$ since its value is zero at $t = 0$.

Now let $p \geq 2$. Let $q(t) = r_1(t)/r_2(t)$, where $r_1(t) = \tanh t$ and $r_2(t) = t$ with $r_1(0) = 0$, $r_2(0) = 0$. As $r_1'(t)/r_2'(t) = \sec h^2 t$ is a decreasing function of t, it follows from Lemma 3 that $t^{-1}\tanh t$ is monotonically decreasing. Therefore, for a fixed p, $(p/n)^{-1}\tanh(p/n)$ increases with n. As $lt_{n\to\infty}(p/n)^{-1}\tanh(p/n) = 1$ and $\tanh p < 1$, there exists an integer n_0 such that, for $n > n_0$, $(p/n)^{-1}\tanh(p/n) > \tanh p$. In other words, for $n > n_0$, $p\tanh nt - n\tanh t < 0$ at $t = p/n$. Since $(\tanh nt - n\tanh t)' < 0$, we conclude that $p\tanh nt - n\tanh t < 0$ if $t \geq p/n$ and $n > n_0$. We can find out n_0 in the following manner.

We note that $t^{-1}\tanh t - 1 + t^2/3 > 0$ as $(\tanh t - t + t^3/3)' = t^2\sec h^2 t(\cosh^2 t - t^{-2}\sinh^2 t) > 0$ and $\tanh t - t + t^3/3$ is zero at $t = 0$. So $(p/n)^{-1}\tanh(p/n) > \tanh p$ if $1 - p^2/(3n^2) > \tanh p$; that is, $n > n_0 = \sqrt{p}\{3(1 - \tanh p)\}^{-1/2}$. □

Lemma 7. *If* $0 \leq x \leq 1/\sqrt{2}$, *then*

$$\frac{(\pi/2)\,x\sqrt{1-x^2}}{\left(1 + 2^{-1}(1-2x^2)\right)^{1.27\ldots}} < \sin^{-1} x < \frac{(\pi/2)\,x\sqrt{1-x^2}}{\left(1 + 2^{-1}(1-2x^2)\right)^{1.1137\ldots}}. \quad (17)$$

Proof. Let $x = \sin\theta$ and $\alpha = 2\theta$. Let $r_{10}(\theta) = \log(1 + 2^{-1}\cos\alpha)$ and $r_{11}(\theta) = \log(\pi\sin\alpha/(2\alpha))$. We observe that $r_{10}(\pi/2) = r_{11}(\pi/2) = 0$. After differentiation we have

$$\left(\frac{r_{11}'(\theta)}{r_{10}'(\theta)}\right)' = \left(\frac{(2+\cos\alpha)\left(\alpha^{-1} - \cot\alpha\right)}{\sin\alpha}\right)' = \frac{2\left[(2+\cos\alpha)\left(\cos ec^2\alpha - \alpha^{-2}\right)\sin\alpha + \left(\cot\alpha - \alpha^{-1}\right)(1 + 2\cos\alpha)\right]}{\sin^2\alpha}. \quad (18)$$

We now use the following power series representation (see [12]) convergent for $|t| < \pi^2$, $t^{-1} - \cot t = \sum_{k=1}^{\infty}(2^{2k}|B_{2k}|/(2k)!)t^{2k-1}$, where B_{2k} is the Bernoulli number of degree $2k$.

Then,

$$\sin^2\alpha\left(\frac{r_{11}'(\theta)}{r_{10}'(\theta)}\right)' = 2\sum_{k=1}^{\infty}\frac{2^{2k}|B_{2k}|}{(2k)!}(\alpha)^{2k-2} \cdot \left((2k-1)(2+\cos\alpha)\sin\alpha - \alpha(1+2\cos\alpha)\right). \quad (19)$$

We observe that $(2+\cos\alpha)\sin 2\theta - \alpha(1+2\cos\alpha)$ is nonnegative in $(0, \pi/4)$ since it vanishes at zero and its derivative, that is, $2(\alpha\sin\alpha - \sin^2\alpha)$ is positive.

It follows from (19) that $(r_{11}'(\theta)/r_{10}'(\theta)$ is a nondecreasing function in $(0, \pi/4)$ since

$$(2k-1)(2+\cos\alpha)\sin\alpha - \alpha(1+2\cos\alpha) \geq (2+\cos\alpha)\sin 2\theta - \alpha(1+2\cos\alpha). \quad (20)$$

$lt_{\theta\to 0}r_{11}(\theta)/r_{10}(\theta) = \ln(\pi/2)/\ln(3/2) = 1.1137\ldots$ and by L'Hôpital's Rule $lt_{\theta\to\pi/4}r_{11}(\theta)/r_{10}(\theta) = 4/\pi = 1.27\ldots$, we obtain the proof of Lemma 7 using Lemma 3. □

Lemma 8. *Consider*

$$b_0 - b_1 < \frac{\sqrt{\pi}}{2}\int_0^v e^{-q^2x^2}\operatorname{erf}(x)\,dx < b_0, \quad (21)$$

where

$$b_k = \frac{\left(1 - e^{-(qv)^2}\sum_{j=0}^{k}\left((qv)^{2j}/j!\right)\right)}{\left(2(2k+1)q^{2k+2}\right)}, \quad q > 1. \quad (22)$$

Proof. Consider the following power series representation of $\operatorname{erf}(x)$ [12, eq. 3.321.1] $\operatorname{erf}(x) = (2/\sqrt{\pi})\sum_{k=0}^{k=\infty}(-1)^k x^{2k+1}/(k!(2k+1))$. The power series converges absolutely and uniformly for $x \in R$. Hence [12, page 346]

$$\begin{aligned}\frac{\sqrt{\pi}}{2}\int_0^v e^{-q^2x^2}\operatorname{erf}(x)\,dx &= \sum_{k=0}^{k=\infty}(-1)^k\frac{\int_0^v x^{2k+1}e^{-q^2x^2}dx}{k!(2k+1)} \\ &= \sum_{k=0}^{k=\infty}(-1)^k\frac{\gamma\left(k+1, q^2v^2\right)}{2q^{2k+2}k!(2k+1)} = \sum_{k=0}^{k=\infty}(-1)^k b_k,\end{aligned} \quad (23)$$

where $b_k = (1 - e^{-(qv)^2}\sum_{j=0}^{k}((qv)^{2j}/j!))/(2(2k+1)q^{2k+2})$. It is easy to see that b_k is nonincreasing and $b_k \to 0$ as $k \to \infty$. Thus, the statement of the lemma is true. □

Lemma 9. *Let* $u_n = o(n^{-1/2})$; *then*

$$\int_0^{u_n}\phi_n(t)\,dt = O\left(u_n\sqrt{n}\right). \quad (24)$$

Proof. By Lemma 4, D_n defined in (8) can be written as

$$D_n = 2^{2n-4} n \sinh^2 nt \cdot \left\{(n-1)\sinh^2 t \operatorname{cos}ech^2 nt \left(\cosh^2 nt - \cosh nt\right) + (n \sinh t \operatorname{cos}echnt + 1)^2\right\} \left(1 + O\left(nt\right)^2\right). \quad (25)$$

By Lemma 6 we find that $(n\sinh t \operatorname{cos}echnt + 1)^2 = O(1)$ in (0,1) and $(\sinh t \operatorname{cos}echnt)^2 \cosh^2 nt = O(n^{-1})$ if $nt = O(1)$. Also, $\sinh^2 t \coth^2 nt = O(t^2)$ if $nt \to \infty$ and $t \in (0,1)$. So, $(2^{2n-4} n \sinh^2 nt)^{-1} D_n \sim C_1$, $1 \le C_1 \le 2$. Also using Lemma 4 and (5) we find that $A_n/2^{n-1} \sim \cosh nt + 1$. By (11), we see that Lemma 9 is valid. □

Lemma 10. *If $\beta > \alpha \ge 2n^{-1}\log n$, then*

$$\int_\alpha^\beta \phi_n(t)\,dt = \sqrt{n}\left(\tan^{-1} e^\beta - \tan^{-1} e^\alpha\right) + O\left(n^{3/2} \sinh\alpha \exp(-n\alpha)\right). \quad (26)$$

Proof. Using Lemmas 6 and 4 we obtain from (8) that $D_n = 2^{2n-4} n \cosh^{2n-2} t \cosh^2 nt (1 + O(n \sinh u \exp(-nu)))$ and $A_n = 2^{n-1} \cosh nt \cosh^n t (1 + O(\exp(-nu)))$. The conclusion of the lemma follows now from (11). □

4. Proof of Theorem 1

If u_n is taken as $2n^{-1}\log n$ we can derive from Lemmas 9 and 10 the following relations:

$$\mathrm{EN}_{n,0}\left(0, 2n^{-1}\log n\right) = O_p\left(n^{-1/2}\log n\right),$$
$$\mathrm{EN}_{n,0}\left(2n^{-1}\log n, \infty\right) = \frac{\sqrt{n}}{2} - \frac{\sqrt{n}\tan^{-1} e^{2n^{-1}\log n}}{\pi} + O_p\left(n^{-1/2}\right). \quad (27)$$

Since the integrand in (11) is an even function of t, we have

$$\mathrm{EN}_{n,0}(-\infty, \infty) = 2\mathrm{EN}_{n,0}(0,\infty). \quad (28)$$

Using Lemma 5 to approximate $\tan^{-1} e^{2n^{-1}\log n}$, we see that Theorem 1 is true.

5. Proof of Theorem 2

To determine $\mathrm{EN}_{n,K}(-\infty,\infty)$ we need to calculate $I_1(-\infty,\infty)$ and $I_2(-\infty,\infty)$ for different ranges of values of K as both quantities depend on magnitude of K. In Lemma 14 we have calculated $I_2(-\infty,\infty)$. Value of $I_1(-\infty,\infty)$ for three different ranges of value of K has been calculated in Lemmas 11–13. The relations (32), (34), and (35) and Lemma 14 establish Theorem 2. Note that we only need to calculate $I_1(0,\infty)$ and $I_2(0,\infty)$ since $I_1(\alpha,\beta)$ and $I_2(\alpha,\beta)$ are even functions of t.

Since $I_1(\alpha,\beta) \le \mathrm{EN}_{n,0}(\alpha,\beta)$, by Lemma 9 we have

$$I_1\left(0, 2n^{-1}\log n\right) = O_p\left(n^{-1/2}\log n\right). \quad (29)$$

In order to calculate $I_1(2n^{-1}\log n, \infty)$, we need the dominant term of C_n/D_n, which can be calculated with the help of Lemmas 4 and 6 as

$$\frac{C_n}{D_n} = n\left(2^{n-1}\cosh nt \cosh^n t\right)^{-1}\left(1 + O\left(n^{-1}\right)\right). \quad (30)$$

For brevity, we have written $K' = K^2/2^n$ in the following lemmas.

Lemma 11. *As long as K' does not exceed $e^{\mu(n)}$, where $\mu(n) = o(\sqrt{n})$,*

$$I_1(0,\infty) = \frac{\sqrt{n}}{4} + O_p\left(n^{-1/2}\log n\right). \quad (31)$$

Proof. Let $K' \le n^\alpha$, $\alpha > 0$. Then by Lemma 9 we find that $I_1(0, n^{-1}(\alpha+2)\log n) = O_p(n^{-1/2}\log n)$. Using (30) and Lemma 10, we obtain that $I_1(n^{-1}(\alpha+2)\log n, \infty) = \sqrt{n}/4 + O_p(n^{-1/2}\log n)$. Hence $I_1(0,\infty) = \sqrt{n}/4 + O_p(n^{-1/2}\log n)$.

Now let $K' \ge n^\alpha$, $\alpha \ge 2$, and $n^{-1/2}\log(K') = o(1)$. It follows from (30) that the maximum value of $e^{-K' C_n D_n^{-1}}$ in $(2n^{-1}\log n, n^{-1}\log(nK'^2/\log n))$ is $O(n^{-1})$.

Hence $I_1((2n^{-1}\log n, n^{-1}\log(nK'^2/\log n))) = O_p(n^{-1/2})$. As the integrand of $I_1(a,b)$ is bounded, we have $I_1(n^{-1}\log(nK'/\log n), n^{-1}\log(K'^2 n^2)) = O_p(n^{-1/2})$. By Lemma 10 and (30) we find that $I_1(n^{-1}\log(K'n^2), \infty) = \sqrt{n}/4 + O_p(n^{-1/2})$. Taking into account (29) we see that (31) is true. □

Lemma 12. *Let $n^{-1/2}\log K' = \mu(n)$, where either $\mu(n) = O(1)$ or $\mu(n) \to \infty$, but $\mu(n) = o(\sqrt{n})$; then*

$$I_1(0,\infty) = \frac{\sqrt{n}}{2} - \frac{\mu(n)}{2\pi} + O_p\left(n^{-1/2}\log n\right). \quad (32)$$

Proof. By Lemma 10 and (30) we find that the following relations are true:

$$I_1\left(2n^{-1}\log n, (2n)^{-1}\log n + \mu(n)\, n^{-1/2} - \frac{\mu^2(n)}{2}\right) = O_p\left(\sqrt{n}e^{-\sqrt{n}}\right),$$
$$I_1\left(2n^{-1}\log n + \mu(n)\, n^{-1/2} - \frac{\mu^2(n)}{2}, \infty\right) = \frac{\sqrt{n}}{4} - \frac{\mu(n)}{\pi} + O_p\left(n^{-1/2}\right),$$
$$I_1\left((2n)^{-1}\log n + \mu(n)\, n^{-1/2} - \frac{\mu^2(n)}{2}, 2n^{-1}\log n + \mu(n)\, n^{-1/2} - \frac{\mu^2(n)}{2}\right) = O_p\left(n^{-1/2}\log n\right). \quad (33)$$

These relations and (29) yield (32). □

Lemma 13. *Let* $\log K' = n\mu(n)$, *where* $\mu(n) = O(1)$ *or* $\mu(n) \to \infty$. *Then*

$$I_1(0,\infty) = \left(\frac{2\sqrt{n}}{\pi}\right)\sin^{-1}\left(\frac{e^{-\mu(n)}}{\sqrt{2}}\right) + O_p\left(n^{-1/2}\log n\right). \quad (34)$$

For large values of n

$$\frac{\sqrt{n}x\sqrt{1-x^2}}{4\left(1+\sqrt{1-x^2+x^4}\right)^{1.27\ldots}} < I_1(0,\infty) < \frac{\sqrt{n}x\sqrt{1-x^2}}{4\left(1+\sqrt{1-x^2+x^4}\right)^{1.1137\ldots}}, \quad (35)$$

where $x = e^{-\mu(n)}/\sqrt{2}$.

Moreover, $I_1(0,\infty) = O_p(1)$ *if* $\mu(n) = O(n\log n)$ *and* $I_1(0,\infty) = o_p(1)$ *if* $\mu(n)/(n\log n) \to \infty$.

Proof. Let t_1 and t_2 be the points in R where $(\exp(2t)+1)^n/2^n$ assumes the values $(n/\log n)\exp(n\mu(n))$ and $n^2\exp(n\mu(n))$, respectively. It follows from (30) and Lemma 10 that

$$I_1\left(2n^{-1}\log n, t_1\right) = O_p\left(n^{-1/2}\right). \quad (36)$$

Let $z = \ln((e^{2t}+1)/2)$ and $s = 2e^z$.

Then $\int \operatorname{sech} t\, dt = \int (dz/\sqrt{2e^z - 1}) = \int (ds/s\sqrt{s-1}) = -2\sin^{-1}(s^{-1/2})$. Therefore, by (30) and the definition of $I_1(\alpha,\beta)$, we have $I_1(t_1,t_2) = O_p(n^{-1/2}\log n)$ and $I_1(t_2,\infty) = \sqrt{n}\sin^{-1}(e^{-\mu(n)}/\sqrt{2}) + O_p(n^{-1/2)})$. We obtain (34) if (29) is also considered. Inequality (35) follows immediately from Lemma 7. From this inequality the other two estimates of $I_1(0,\infty)$ valid for $\mu(n) = O_p(n\log n)$ and $\mu(n)$ such that $\mu(n)/(n\log n) \to \infty$ follow. □

Lemma 14. *The dependence of* $I_2(-\infty,\infty)$ *on* K *is given by the following relations:*

(i) *if* $K^2 = o(2^n)$, *then* $I_2(-\infty,\infty) = o_p(1)$,

(ii) *if* $K^2 = O(2^n)$, *then* $I_2(-\infty,\infty) = (2/\sqrt{\pi})\mathrm{erf}(u) + O_p(e^{-n})$,

(iii) *if* $K^2/2^n \to \infty$, *then* $I_2(-\infty,\infty) = (2/\sqrt{\pi}) + O_p(e^{-K^2/2^n})$,

where $u = \sqrt{K'}$.

Proof. Using Lemma 4 we find that in $(0, u_n)$

$$D_n = 2^{2n-4} n \sinh^2 nt\,(1 + 2n\sinh t \cos echnt + o(1)), \quad B_n = 2^{n-2} n \sinh^2 nt\,(1 + o(1)). \quad (37)$$

Hence, $B_n/\sqrt{D_n} \sim \sqrt{n}/c$, where $1 \le c \le \sqrt{3}$.

On the other hand, using Lemma 6 and the definitions of B_n and D_n in (u_n,∞), where $u_n > 2n^{-1}\log n$, we have $B_n/\sqrt{D_n} \sim \sqrt{n}e^t$.

Let $s = K/\sqrt{2A_n}$. It is not difficult to see that $(K/s)^{1/n} \sim e^t$ in the interval (u_n,∞)

$$\begin{aligned} I_2(-\infty,\infty) &= 2I_2(0,\infty) \\ &= \left(\frac{4}{\pi}\right)\int_0^u \exp\left(-s^2\right)\mathrm{erf}(sp)\,ds \\ &= \left(\frac{2}{\sqrt{\pi}}\right)\mathrm{erf}(up)\,\mathrm{erf}(u) \\ &\quad - 4p\int_0^u \exp\left(-p^2s^2\right)\mathrm{erf}(s)\,ds, \end{aligned} \quad (38)$$

where $\sqrt{n}/\sqrt{3} \le p \le \sqrt{n}e^{\mu(n)}$.

Clearly $I_2(-\infty,\infty) = o_p(1)$ if $K^2 = o(2^n)$. The other two parts of the lemma are found to be true by virtue of Lemma 8. □

Conflict of Interests

The authors declare that there is no conflict of interests regarding the publication of this paper.

References

[1] K. Farahmand, *Topics in Random Polynomials*, vol. 393 of *Pitman Research Notes in Mathematics Series*, Longman, Harlow, UK, 1998.

[2] M. Das, *On real zeros of random polynomial with hyperbolic elements [Ph.D. thesis]*, Utkal University, Bhubaneswar, India, 1971.

[3] K. Farahmand, "Level crossings of a random polynomial with hyperbolic elements," *Proceedings of the American Mathematical Society*, vol. 123, no. 6, pp. 1887–1892, 1995.

[4] M. K. Mahanti, "Expected number of real zeros of random hyperbolic polynomial," *Statistics & Probability Letters*, vol. 70, no. 1, pp. 11–18, 2004.

[5] M. K. Mahanti, "On expected number of real zeros of a random hyperbolic polynomial with dependent coefficients," *Applied Mathematics Letters*, vol. 22, no. 8, pp. 1276–1280, 2009.

[6] M. K. Mahanti and L. Sahoo, "On expected number of level crossings of random hyperbolic polynomial," *Rocky Mountain Journal of Mathematics*. In press.

[7] J. R. Wilkins, "Mean number of real zeros of random hyperbolic polynomial," *International Journal of Mathematics and Mathematical Sciences*, vol. 23, no. 5, pp. 335–342, 2000.

[8] A. Edelman and E. Kostlan, "How many zeros of a random polynomial are real?" *Bulletin of the American Mathematical Society*, vol. 32, no. 1, pp. 1–37, 1995.

[9] H. Crammer and A. R. Leadbetter, *Stationary and Related Stochastic Process*, Wiley, New York, NY, USA, 1967.

[10] M. Kac and S. O. Rice, "Mathematical theory of random noise," *Bell System Technical Journal*, 1945, Reprinted in: Selected Papers Noise and Stochastic Processes, pp. 133–294, Dover, NY, USA, edited by: N. Wax, 1954.

[11] G. D. Anderson, M. K. Vamanamurthy, and M. K. Vuorinen, *Conformal Invariants, Inequalities and Quasiconformal Maps*, Canadian Mathematical Society Series of Monographs and Advanced Texts, John Wiley & Sons, New York, NY, USA, 1997.

[12] I. S. Gradshteyn and I. M. Ryzhik, *Tables of Integrals, Series and Products*, Academic Press, New York, NY, USA, 7th edition, 2007.

2

Hom-Lie Triple System and Hom-Bol Algebra Structures on Hom-Maltsev and Right Hom-Alternative Algebras

Sylvain Attan and A. Nourou Issa

Département de Mathématiques, Université d'Abomey-Calavi, 01 BP 4521 Cotonou, Benin

Correspondence should be addressed to A. Nourou Issa; woraniss@yahoo.fr

Academic Editor: Kaiming Zhao

Every multiplicative Hom-Maltsev algebra has a natural multiplicative Hom-Lie triple system structure. Moreover, there is a natural Hom-Bol algebra structure on every multiplicative Hom-Maltsev algebra and on every multiplicative right (or left) Hom-alternative algebra.

1. Introduction

The study of Lie triple systems (Lts) on their own as algebraic objects started from Jacobson's work [1] and developed further by, for example, Lister [2], Yamaguti [3], and other mathematicians. Lts constitute examples of ternary algebras. If $(g, [,])$ is a Lie algebra, then $(g, [,,])$ is a Lts, where $[x, y, z] := [[x, y], z]$ (see [1, 4, 5]). Another construction of Lts from binary algebras is the one from Maltsev algebras found by Loos [6].

Maltsev algebras were introduced by Maltsev [7] in a study of commutator algebras of alternative algebras and also as a study of tangent algebras to local smooth Moufang loops. Maltsev used the name "Moufang-Lie algebras" for these nonassociative algebras while Sagle [8] introduced the term "Malcev algebras." Equivalent defining identities of Maltsev algebras are pointed out in [8].

Alternative algebras, Maltsev algebras, and Lts (among other algebras) received a twisted generalization in the development of the theory of Hom-algebras during these latest years. The forerunner of the theory of Hom-algebras is the Hom-Lie algebra introduced by Hartwig et al. in [9] in order to describe the structure of some deformation of the Witt algebra and the Virasoro algebra. It is well-known that Lie algebras are related to associative algebras via the commutator bracket construction. In the search of a similar construction for Hom-Lie algebras, the notion of a Hom-associative algebra is introduced by Makhlouf and Silvestrov in [10], where it is proved that a Hom-associative algebra gives rise to a Hom-Lie algebra via the commutator bracket construction. Since then, various Hom-type structures are considered (see, e.g., [11–23]). Roughly speaking, Hom-algebraic structures are corresponding ordinary algebraic structures whose defining identities are twisted by a linear self-map. A general method for constructing a Hom-type algebra from the ordinary type of algebra with a linear self-map is given by Yau in [24].

In [11, 21], n-ary Hom-algebra structures generalizing n-ary algebras of Lie type or associative type were considered. In particular, generalizations of n-ary Nambu or Nambu-Lie algebras, called n-ary Hom-Nambu and Hom-Nambu-Lie algebras, respectively, were introduced in [11] while Hom-Jordan algebras were defined in [18] and Hom-Lie triple systems (Hom-Lts) were introduced in [21] (another definition of a Hom-Jordan algebra is given in [20]). It is shown [21] that Hom-Lts are ternary Hom-Nambu algebras with additional properties and that Hom-Lts arise also from Hom-Jordan triple systems or from other Hom-type algebras.

Motivated by the relationships between some classes of binary algebras and some classes of binary-ternary algebras, a study of Hom-type generalization of binary-ternary algebras is initiated in [16] with the definition of Hom-Akivis algebras.

Further, Hom-Lie-Yamaguti algebras are considered in [14] and Hom-Bol algebras [12] are defined as a twisted generalization of Bol algebras which are introduced and studied in [25–27] as infinitesimal structures tangent to smooth Bol loops (some aspects of the theory of Bol algebras are discussed in [28–30]).

In this paper, we will be concerned with right (or left) Hom-alternative algebras, Hom-Maltsev algebras, Hom-Lts, and Hom-Bol algebras. We extend Loos' construction of Lts from Maltsev algebras ([6], Satz 1) to the Hom-algebra setting (Section 3). Specifically, we prove (Theorem 14) that every multiplicative Hom-Maltsev algebra is naturally a multiplicative Hom-Lts by a suitable definition of the ternary operation. As a tool in the proof of this fact, we point out a kind of compatibility relation between the original binary operation of a given Hom-Maltsev algebra and the ternary operation mentioned above (Lemma 13). Moreover, we obtain that every multiplicative Hom-Maltsev algebra has a natural Hom-Bol algebra structure (Theorem 17). In [31] Mikheev proved that every right alternative algebra has a natural (left) Bol algebra structure. In [29] Hentzel and Peresi proved that not only a right alternative algebra but also a left alternative algebra has left Bol algebra structure. In Section 4 we prove that the Hom-analogue of these results holds. Specifically, every multiplicative right (or left) Hom-alternative algebra is a Hom-Bol algebra (Theorem 23). It could be observed that the methods used in the proof of results in [6, 29, 31] cannot be reported in the Hom-algebra setting at the present stage of the theory of Hom-algebras. In Section 5 we specify Theorem 23 to recover the construction of left Bol algebras from right alternative algebras (Theorem 26; one observes that, in our proof, we use essentially some fundamental properties of right alternative algebras). In Section 2 we recall some basic definitions and facts about Hom-algebras. We define the Hom-Jordan associator of a given Hom-algebra and point out that every Hom-algebra is a Hom-triple system with respect to the Hom-Jordan associator. This observation is used in the proof of Theorem 23.

All vector spaces and algebras are meant over an algebraically closed ground field $\mathbb{K}$ of characteristic 0.

2. Some Basics on Hom-Algebras

We first recall some relevant definitions about binary and ternary Hom-algebras. In particular, we recall the notion of a Hom-Maltsev algebra as well as some of its equivalent defining identities. Although various types of n-ary Hom-algebras are introduced and discussed in [11, 21], for our purpose, we will consider ternary Hom-algebras (ternary Hom-Nambu algebras and Hom-Lts) and Hom-Bol algebras. For fundamentals on Hom-algebras, one may refer, for example, to [9–11, 13, 17, 24, 32]. Some aspects of the theory of binary Hom-algebras are considered in [33], while some classes of binary-ternary Hom-algebras are defined and discussed in [12, 14, 16].

Definition 1. (i) A *Hom-algebra* is a triple $(A, *, \alpha)$ in which A is a $\mathbb{K}$-vector space, $* : A \times A \to A$ a bilinear map (the binary operation), and $\alpha : A \to A$ a linear map (the twisting map). The Hom-algebra A is said to be *multiplicative* if $\alpha(x * y) = \alpha(x) * \alpha(y)$ for all $x, y \in A$.

(ii) The *Hom-Jacobian* in $(A, *, \alpha)$ is the trilinear map $J_\alpha : A \times A \times A \to A$ defined as $J_\alpha(x, y, z) := \circlearrowleft_{x,y,z}(x * y) * \alpha(z)$, where $\circlearrowleft_{x,y,z}$ denotes the sum over cyclic permutation of x, y, z.

(iii) The *Hom-associator* of a Hom-algebra $(A, *, \alpha)$ is the trilinear map as : $A^{\otimes 3} \to A$ defined as $\mathrm{as}(x, y, z) = (x * y) * \alpha(z) - \alpha(x) * (y * z)$. If $\mathrm{as}(x, y, z) = 0$ for all $x, y, z \in A$, then $(A, *, \alpha)$ is said to be *Hom-associative*.

Remark 2. If $\alpha = \mathrm{id}$ (the identity map), then a Hom-algebra $(A, *, \alpha)$ reduces to an ordinary algebra $(A, *)$, the Hom-Jacobian J_α is the ordinary Jacobian J, and the Hom-associator is the usual associator for the algebra $(A, *)$. One observes that, in general, the map α is not always injective nor surjective (see [13, 15] for discussions on the subject). So, for example, a given algebra can be twisted into zero algebra and some properties of Hom-algebras may not be valid for corresponding ordinary algebras.

As for ordinary algebras, to each Hom-algebra $\mathscr{A} := (A, *, \alpha)$ are attached two Hom-algebras: the *commutator Hom-algebra* $\mathscr{A}^- := (A, [,], \alpha)$, where $[x, y] := x * y - y * x$ (the commutator of x and y), and the *plus Hom-algebra* $\mathscr{A}^+ := (A, \circ, \alpha)$, where $x \circ y := x * y + y * x$ (the *Jordan product*) for all $x, y \in A$.

For our purpose, we provide the following.

Definition 3. The *Hom-Jordan associator* of a Hom-algebra $\mathscr{A} := (A, *, \alpha)$ is the trilinear map $\mathrm{as}^J : A^{\otimes 3} \to A$ defined as $\mathrm{as}^J(x, y, z) = (x \circ y) \circ \alpha(z) - \alpha(x) \circ (y \circ z)$, where "$\circ$" is the Jordan product on A.

If $\alpha = \mathrm{id}$, the Hom-Jordan associator reduces to the usual Jordan associator.

Definition 4. (i) A *Hom-Lie algebra* is a Hom-algebra $(A, *, \alpha)$ such that the binary operation "$*$" is anticommutative and the *Hom-Jacobi identity*

$$J_\alpha(x, y, z) = 0 \tag{1}$$

holds for all x, y, and z in A ([9]).

(ii) A *Hom-Maltsev algebra* is a Hom-algebra $(A, *, \alpha)$ such that the binary operation "$*$" is anticommutative and that the *Hom-Maltsev identity*

$$J_\alpha(\alpha(x), \alpha(y), x * z) = J_\alpha(x, y, z) * \alpha^2(x) \tag{2}$$

holds for all x, y, z in A ([20]).

(iii) A *Hom-Jordan algebra* is a Hom-algebra $(A, *, \alpha)$ such that $(A, *)$ is a commutative algebra and the *Hom-Jordan identity*

$$\mathrm{as}(x * x, \alpha(y), \alpha(x)) = 0 \tag{3}$$

is satisfied for all x, y in A ([20]).

(iv) A Hom-algebra $(A, *, \alpha)$ is called a *right Hom-alternative algebra* if

$$\mathrm{as}(x, y, y) = 0 \tag{4}$$

for all x, y in A. A Hom-algebra $(A, *, \alpha)$ is called a *left Hom-alternative algebra* if

$$\mathrm{as}(x, x, y) = 0 \tag{5}$$

for all x, y in A. A Hom-algebra $(A, *, \alpha)$ is called a *Hom-alternative algebra* if it is both right and left Hom-alternative [18].

Remark 5. When $\alpha = \mathrm{id}$, the Hom-Jacobi identity (1) is the usual *Jacobi identity* $J(x, y, z) = 0$. Likewise, for $\alpha = \mathrm{id}$, the Hom-Maltsev identity (2) reduces to the *Maltsev identity* $J(x, y, x * z) = J(x, y, z) * x$. Therefore a Lie (resp., Maltsev) algebra $(A, *)$ may be seen as a Hom-Lie (resp., Hom-Maltsev) algebra with the identity map as the twisting map. Also Hom-Maltsev algebras generalize Hom-Lie algebras in the same way that Maltsev algebras generalize Lie algebras. For $\alpha = \mathrm{id}$ in the Hom-Jordan identity, we recover the usual Jordan identity. Observe that the definition of the Hom-Jordan identity in [20] is slightly different from the one formerly given in [18].

Hom-Maltsev algebras are introduced in [20] in connection with a study of Hom-alternative algebras introduced in [18]. In fact it is proved ([20], Theorem 3.8) that every Hom-alternative algebra is *Hom-Maltsev admissible;* that is, the commutator Hom-algebra of any Hom-alternative algebra is a Hom-Maltsev algebra (this is the Hom-analogue of Maltsev's construction of Maltsev algebras as commutator algebras of alternative algebras [7]). This result is also mentioned in [16], Section 4, using an approach via Hom-Akivis algebras (this approach is close to the one of Maltsev in [7]). Also, every Hom-alternative algebra is *Hom-Jordan admissible;* that is, its plus Hom-algebra is a Hom-Jordan algebra ([20]). Examples of Hom-alternative algebras and Hom-Jordan algebras could be found in [18, 20]. An example of a right Hom-alternative algebra that is not left Hom-alternative is given in [23].

Equivalent to (2) defining identities of Hom-Maltsev algebras are found in [20] where, in particular, it is shown that the identity

$$\begin{aligned} &J_\alpha(\alpha(x), \alpha(y), w * z) + J_\alpha(\alpha(w), \alpha(y), x * z) \\ &\quad = J_\alpha(x, y, z) * \alpha^2(w) + J_\alpha(w, y, z) * \alpha^2(x) \end{aligned} \tag{6}$$

is equivalent to (2) in any anticommutative Hom-algebra $(A, *, \alpha)$ ([20], Proposition 2.7). In [34], it is proved that, in any anticommutative Hom-algebra $(A, *, \alpha)$, the Hom-Maltsev identity (2) is equivalent to

$$\begin{aligned} &J_\alpha(\alpha(x), \alpha(y), u * v) \\ &\quad = \alpha^2(u) * J_\alpha(x, y, v) + J_\alpha(x, y, u) * \alpha^2(v) \\ &\quad\quad - 2J_\alpha(\alpha(u), \alpha(v), x * y). \end{aligned} \tag{7}$$

Therefore, apart from (2), identities (6) and (7) may be taken as defining identities of a Hom-Maltsev algebra.

The Hom-algebras mentioned above are *binary* Hom-algebras. The first generalization of binary algebras was the ternary algebras introduced in [1]. Ternary algebraic structures also appeared in various domains of theoretical and mathematical physics (see, e.g., [35]). Likewise, binary Hom-algebras are generalized to n-ary Hom-algebra structures in [11] (see also [21]).

Definition 6 (see [11]). A *ternary Hom-Nambu algebra* is a triple $(A, [,,], \alpha)$ in which A is a $\mathbb{K}$-vector space, $[,,] : A \times A \times A \to A$ is a trilinear map, and $\alpha = (\alpha_1, \alpha_2)$ is a pair of linear maps (the twisting maps) such that the identity

$$\begin{aligned} &[\alpha_1(x), \alpha_2(y), [u, v, w]] \\ &\quad = [[x, y, u], \alpha_1(v), \alpha_2(w)] \\ &\quad\quad + [\alpha_1(u), [x, y, v], \alpha_2(w)] \\ &\quad\quad + [\alpha_1(u), \alpha_2(v), [x, y, w]] \end{aligned} \tag{8}$$

holds for all u, v, w, x, and y in A. Identity (8) is called the *ternary Hom-Nambu identity.*

Remark 7. When $\alpha_1 = \mathrm{id} = \alpha_2$ one recovers the usual ternary Nambu algebra. One may refer to [35] for the origins of Nambu algebras. In [11], examples of n-ary Hom-Nambu algebras that are not Nambu algebras are provided.

Definition 8 (see [21]). A *Hom-Lie triple system* (Hom-Lts) is a ternary Hom-algebra $(A, [,,], \alpha = (\alpha_1, \alpha_2))$ such that

$$[x, y, z] = -[y, x, z], \tag{9}$$

$$\circlearrowleft_{x,y,z} [x, y, z] = 0, \tag{10}$$

and the ternary Hom-Nambu identity (8) holds in $(A, [,,], \alpha = (\alpha_1, \alpha_2))$.

One notes that when the twisting maps α_1, α_2 are both equal to the identity map id, then we recover the usual notion of a Lie triple system [2, 3]. Examples of Hom-Lts could be found in [21].

A particular situation, interesting for our setting, occurs when the twisting maps α_i are all equal, $\alpha_1 = \alpha_2 = \alpha$ and $\alpha([x, y, z]) = [\alpha(x), \alpha(y), \alpha(z)]$ for all x, y, and z in A. The Hom-Lie triple system $(A, [,,], \alpha)$ is then said to be *multiplicative* [21]. In case of multiplicativity, the ternary Hom-Nambu identity (8) then reads

$$\begin{aligned} &[\alpha(x), \alpha(y), [u, v, w]] \\ &\quad = [[x, y, u], \alpha(v), \alpha(w)] \\ &\quad\quad + [\alpha(u), [x, y, v], \alpha(w)] \\ &\quad\quad + [\alpha(u), \alpha(v), [x, y, w]]. \end{aligned} \tag{11}$$

In [14] a (multiplicative) *Hom-triple system* is defined as a (multiplicative) ternary Hom-algebra $(A, [,,], \alpha)$ such that (9) and (10) are satisfied (thus a multiplicative Hom-Lts is seen as a Hom-triple system in which identity (11) holds; observe that this definition of a Hom-triple system is different from the one formerly given in [21], where a

Hom-triple system is just the Hom-algebra $(A, [,,], \alpha)$). With this vision of a Hom-triple system, it is shown [14] that every multiplicative non-Hom-associative algebra (i.e., not necessarily Hom-associative algebra) has a natural Hom-triple system structure if defining $[x, y, z] := [[x, y], \alpha(z)] - \mathrm{as}(x, y, z) + \mathrm{as}(y, x, z)$. We note here that we get the same result if defining another ternary operation on a given Hom-algebra. Specifically, we have the following result.

Proposition 9. *Let $\mathcal{A} = (A, *, \alpha)$ be a multiplicative Hom-algebra. Define on $\mathcal{A}$ the ternary operation*

$$(x, y, z) := as^J(y, z, x) \tag{12}$$

for all $x, y,$ and $z \in A$. Then $(A, (,,), \alpha)$ is a multiplicative Hom-triple system.

Proof. A proof follows from the straightforward checking of identities (9) and (10) for "$(,,)$" using the commutativity of the Jordan product "$\circ$." □

Since our results here depend on multiplicativity, in the rest of this paper we assume that all Hom-algebras (binary or ternary) are multiplicative and while dealing with the binary operation "$*$" and where there is no danger of confusion, we will use juxtaposition in order to reduce the number of braces; that is, for example, $xy * \alpha(z)$ means $(x * y) * \alpha(z)$.

Various results and constructions related to Hom-Lts are given in [21]. In particular, it is shown that every Lts L can be twisted along any self-morphism of L into a multiplicative Hom-Lts. For our purpose we just mention the following result.

Proposition 10 (see [21]). *Let $(A, *)$ be a Maltsev algebra and $\alpha : A \to A$ an algebra morphism. Then $A_\alpha := (A, [,,]_\alpha, \alpha)$ is a multiplicative Hom-Lts, where $[x, y, z]_\alpha = \alpha(2xy * z - yz * x - zx * y)$, for all $x, y,$ and z in A.*

One observes that the product $[x, y, z] = 2xy * z - yz * x - zx * y$ is the one defined in [6] providing a Maltsev algebra $(A, *)$ with a Lts structure. A construction describing another view of Proposition 10 above will be given in Section 3 (see Proposition 16) via Hom-Maltsev algebras. For the time being, we point out the following slight generalization of the result above, producing a sequence of multiplicative Hom-Lts from a given Maltsev algebra.

Proposition 11. *Let $(A, *)$ be a Maltsev algebra and $\alpha : A \to A$ an algebra morphism. Let $\alpha^0 = id$ and, for any integer $n \geq 1$, $\alpha^n = \alpha \circ \alpha^{n-1}$. If one defines on A a trilinear operation $[,,]_{\alpha^n}$ by*

$$[x, y, z]_{\alpha^n} = \alpha^n(2xy * z - yz * x - zx * y) \tag{13}$$

for all $x, y,$ and z in A, then $(A, [,,]_{\alpha^n}, \alpha^n)$ is a multiplicative Hom-Lts.

Proof. Let $[x, y, z] = 2xy * z - yz * x - zx * y$ and then $[x, y, z]_{\alpha^n} = \alpha^n([x, y, z])$. We shall use the fact that $(A, [,,])$ is a Lts [6]. Identities (9) and (10) for $[x, y, z]_{\alpha^n}$ are quite obvious. Next,

$$\begin{aligned}
&[\alpha^n(x), \alpha^n(y), [u, v, w]_{\alpha^n}]_{\alpha^n} \\
&= [\alpha^n(x), \alpha^n(y), \alpha^n([u, v, w])]_{\alpha^n} \\
&= \alpha^{2n}([x, y, [u, v, w]]) \\
&= \alpha^{2n}([[x, y, u], v, w]) + \alpha^{2n}([u, [x, y, v], w]) \\
&\quad + \alpha^{2n}([u, v, [x, y, w]]) \\
&= [\alpha^n([x, y, u]), \alpha^n(v), \alpha^n(w)]_{\alpha^n} \\
&\quad + [\alpha^n(u), \alpha^n([x, y, v]), \alpha^n(w)]_{\alpha^n} \\
&\quad + [\alpha^n(u), \alpha^n(v), \alpha^n([x, y, w])]_{\alpha^n} \\
&= [[x, y, u]_{\alpha^n}, \alpha^n(v), \alpha^n(w)]_{\alpha^n} \\
&\quad + [\alpha^n(u), [x, y, v]_{\alpha^n}, \alpha^n(w)]_{\alpha^n} \\
&\quad + [\alpha^n(u), \alpha^n(v), [x, y, w]_{\alpha^n}]_{\alpha^n}
\end{aligned} \tag{14}$$

and so (11) holds for $[,,]_{\alpha^n}$. Thus $(A, [,,]_{\alpha^n}, \alpha^n)$ is a multiplicative Hom-Lts. □

In [12] we defined a Hom-Bol algebra as a twisted generalization of a (left) Bol algebra. For the introduction and original studies of Bol algebras, we refer to [25–27] (the defining identities of left Bol algebras are recalled in Section 5 of the present paper). Bol algebras are further considered in, for example, [29, 30].

Definition 12 (see [12]). A *Hom-Bol algebra* is a quadruple $(A, [,], (,,), \alpha)$ in which A is a vector space, "$[,]$" a binary operation, "$(,,)$" a ternary operation on A, and $\alpha : A \to A$ a linear map such that

(HB1) $\alpha([x, y]) = [\alpha(x), \alpha(y)]$.

(HB2) $\alpha((x, y, z)) = (\alpha(x), \alpha(y), \alpha(z))$.

(HB3) $[x, y] = -[y, x]$.

(HB4) $(x, y, z) = -(y, x, z)$.

(HB5) $\circlearrowleft_{x,y,z}(x, y, z) = 0$.

(HB6) $(\alpha(x), \alpha(y), [u, v]) = [(x, y, u), \alpha^2(v)] + [\alpha^2(u), (x, y, v)] + (\alpha(u), \alpha(v), [x, y]) - [[\alpha(u), \alpha(v)], [\alpha(x), \alpha(y)]]$.

(HB7) $(\alpha^2(x), \alpha^2(y), (u, v, w)) = ((x, y, u), \alpha^2(v), \alpha^2(w)) + (\alpha^2(u), (x, y, v), \alpha^2(w)) + (\alpha^2(u), \alpha^2(v), (x, y, w))$

for all $u, v, w, x, y,$ and $z \in A$.

Identities (HB1) and (HB2) mean the multiplicativity of $(A, [,], (,,), \alpha)$. It is built into our definition for convenience.

One observes that for $\alpha = id$ identities (HB3)–(HB7) reduce to the defining identities of a (left) Bol algebra [25] (see also [29, 30]). If $[x, y] = 0$ for all $x, y \in A$, then $(A, [,], (,,), \alpha)$ becomes a (multiplicative) Hom-Lts $(A, (,,), \alpha^2)$.

Construction results and some examples of Hom-Bol algebras are given in [12]. In particular, Hom-Bol algebras can be constructed from Maltsev algebras. The Hom-analogues of the construction of Bol algebras from Maltsev algebras [25] or from right alternative algebras [31] (see also [29]) are considered in this paper.

3. Hom-Lts and Hom-Bol Algebras from Hom-Maltsev Algebras

In this section, we prove that every multiplicative Hom-Maltsev algebra has a natural multiplicative Hom-Lts structure (Theorem 14) and, moreover, a natural Hom-Bol algebra structure (Theorem 17). Theorem 14 could be seen as the Hom-analogue of Loos' result ([6], Satz 1) although his proof cannot be reproduced here. Besides identities (6) and (7), Lemma 13 below is a tool in the proof of this result. Theorem 17 could be seen as the Hom-analogue of a construction by Mikheev [25] of Bol algebras from Maltsev algebras. Proposition 16 is another view of a result in [21] (see Proposition 10 above).

In his work [6], Loos considered in a Maltsev algebra $(A, *)$ the following ternary operation:

$$\{x, y, z\} = 2xy * z - yz * x - zx * y. \tag{15}$$

Then $(A, \{,,\})$ turns out to be a Lts. This result, in the Hom-algebra setting, looks as in Theorem 14 below. Similarly as in the Loos construction, our investigations are based on the following ternary operation in a Hom-Maltsev algebra $(A, *, \alpha)$:

$$\{x, y, z\}_\alpha = 2xy * \alpha(z) - yz * \alpha(x) - zx * \alpha(y). \tag{16}$$

From (16) it clearly follows that $\{,,\}_\alpha$ can also be written as

$$\{x, y, z\}_\alpha = -J_\alpha(x, y, z) + 3xy * \alpha(z). \tag{17}$$

One observes that when $\alpha = \mathrm{id}$, we recover product (15). First, we prove the following.

Lemma 13. *Let $(A, *, \alpha)$ be a Hom-Maltsev algebra. If one defines on $(A, *, \alpha)$ a ternary operation "$\{,,\}_\alpha$" by (16), then*

$$\begin{aligned}\{\alpha(x), \alpha(y), u * v\}_\alpha &= \alpha^2(u) * \{x, y, v\}_\alpha \\ &\quad + \{x, y, u\}_\alpha * \alpha^2(v) \\ &\quad - J_\alpha(\alpha(u), \alpha(v), x * y)\end{aligned} \tag{18}$$

for all u, v, x, and y in A.

Proof. Let us write (7) as

$$\begin{aligned}&-J_\alpha(\alpha(x), \alpha(y), u * v) \\ &\quad = -J_\alpha(x, y, u) * \alpha^2(v) + \alpha^2(u) * (-J_\alpha(x, y, v)) \\ &\qquad + 3J_\alpha(\alpha(u), \alpha(v), x * y) \\ &\qquad - J_\alpha(\alpha(u), \alpha(v), x * y).\end{aligned} \tag{19}$$

That is,

$$\begin{aligned}&-J_\alpha(\alpha(x), \alpha(y), u * v) \\ &\quad = -J_\alpha(x, y, u) * \alpha^2(v) + \alpha^2(u) * (-J_\alpha(x, y, v)) \\ &\qquad + 3\alpha(u)\alpha(v) * \alpha(x * y) + 3(\alpha(v) * xy) \\ &\qquad * \alpha^2(u) + 3(xy * \alpha(u)) * \alpha^2(v) \\ &\qquad - J_\alpha(\alpha(u), \alpha(v), x * y).\end{aligned} \tag{20}$$

Therefore, by multiplicativity, we have

$$\begin{aligned}&-J_\alpha(\alpha(x), \alpha(y), u * v) + 3\alpha(x)\alpha(y) * \alpha(u * v) \\ &\quad = (-J_\alpha(x, y, u) + 3xy * \alpha(u)) * \alpha^2(v) + \alpha^2(u) \\ &\qquad * (-J_\alpha(x, y, v) + 3xy * \alpha(v)) \\ &\qquad - J_\alpha(\alpha(u), \alpha(v), x * y)\end{aligned} \tag{21}$$

and so, we get (18) by (17). □

We now prove the following.

Theorem 14. *Let $(A, *, \alpha)$ be a multiplicative Hom-Maltsev algebra. If one defines on $(A, *, \alpha)$ a ternary operation "$\{,,\}_\alpha$" by (16), then $(A, \{,,\}_\alpha, \alpha^2)$ is a multiplicative Hom-Lts.*

Proof. We must prove the validity of (9), (10), and (11) for operation (16) in the Hom-Maltsev algebra $(A, *, \alpha)$.

First observe that the multiplicativity of $(A, *, \alpha)$ implies that $\alpha^2(\{x, y, z\}_\alpha) = \{\alpha^2(x), \alpha^2(y), \alpha^2(z)\}_\alpha$, with x, y, and z in A.

From the skew-symmetry of "$*$" and $J_\alpha(x, y, z)$, it clearly follows from (17) that $\{x, y, z\}_\alpha = -\{y, x, z\}_\alpha$ which is (9) for "$\{,,\}_\alpha$."

Next, using (17) and the skew-symmetry of $J_\alpha(x, y, z)$ where applicable, we compute

$$\begin{aligned}&\{x, y, z\}_\alpha + \{y, z, x\}_\alpha + \{z, x, y\}_\alpha \\ &\quad = -J_\alpha(x, y, z) + 3xy * \alpha(z) - J_\alpha(y, z, x) + 3yz \\ &\qquad * \alpha(x) - J_\alpha(z, x, y) + 3zx * \alpha(y) \\ &\quad = -3J_\alpha(x, y, z) + 3J_\alpha(x, y, z) = 0\end{aligned} \tag{22}$$

and thus $\circlearrowleft_{x,y,z}\{x, y, z\}_\alpha = 0$, so we get (10) for "$\{,,\}_\alpha$."

Consider now $\{\alpha^2(x), \alpha^2(y), \{u, v, w\}_\alpha\}_\alpha$ in $(A, *, \alpha)$. Then

$$\begin{aligned}&\{\alpha^2(x), \alpha^2(y), \{u, v, w\}_\alpha\}_\alpha = \{\alpha^2(x), \alpha^2(y), 2uv \\ &\qquad * \alpha(w) - vw * \alpha(u) - wu * \alpha(v)\}_\alpha \quad (\text{by } (16)) \\ &\quad = \{\alpha^2(x), \alpha^2(y), 2uv * \alpha(w)\}_\alpha - \{\alpha^2(x), \alpha^2(y), \\ &\qquad vw * \alpha(u)\}_\alpha - \{\alpha^2(x), \alpha^2(y), wu * \alpha(v)\}_\alpha \\ &\quad = \{\alpha(x), \alpha(y), 2u * v\}_\alpha * \alpha^3(w) + \alpha^2(2u * v)\end{aligned}$$

$$\begin{aligned}
&* \{\alpha(x), \alpha(y), \alpha(w)\}_\alpha - J_\alpha\big(\alpha(2u * v), \alpha^2(w),\\
&\alpha(x * y)\big) - \{\alpha(x), \alpha(y), v * w\}_\alpha * \alpha^3(u)\\
&- \alpha^2(v * w) * \{\alpha(x), \alpha(y), \alpha(u)\}_\alpha + J_\alpha\big(\alpha(v\\
&* w), \alpha^2(u), \alpha(x * y)\big) - \{\alpha(x), \alpha(y), w * u\}_\alpha\\
&* \alpha^3(v) - \alpha^2(w * u) * \{\alpha(x), \alpha(y), \alpha(v)\}_\alpha\\
&+ J_\alpha\big(\alpha(w * u), \alpha^2(v), \alpha(x * y)\big) \quad (\text{by } (18))\\
&= \big(2\{x, y, u\}_\alpha * \alpha^2(v) + 2\alpha^2(u) * \{x, y, v\}_\alpha\\
&- 2J_\alpha(\alpha(u), \alpha(v), x * y)\big) * \alpha^3(w) + 2\alpha^2(u * v)\\
&* \{\alpha(x), \alpha(y), \alpha(w)\}_\alpha - J_\alpha\big(\alpha(2u * v), \alpha^2(w),\\
&\alpha(x * y)\big) - \big(\{x, y, v\}_\alpha * \alpha^2(w) + \alpha^2(v)\\
&* \{x, y, w\}_\alpha - J_\alpha(\alpha(v), \alpha(w), x * y)\big) * \alpha^3(u)\\
&- \alpha^2(v * w) * \{\alpha(x), \alpha(y), \alpha(u)\}_\alpha + J_\alpha\big(\alpha(v\\
&* w), \alpha^2(u), \alpha(x * y)\big) - \big(\{x, y, w\}_\alpha * \alpha^2(u)\\
&+ \alpha^2(w) * \{x, y, u\}_\alpha - J_\alpha(\alpha(w), \alpha(u), x * y)\big)\\
&* \alpha^3(v) - \alpha^2(w * u) * \{\alpha(x), \alpha(y), \alpha(v)\}_\alpha\\
&+ J_\alpha\big(\alpha(w * u), \alpha^2(v), \alpha(x * y)\big)\\
&\qquad\qquad\qquad\qquad (\text{again by } (18))\\
&= 2\{x, y, u\}_\alpha \alpha^2(v) * \alpha^3(w) + 2\alpha^2(u)\{x, y, v\}_\alpha\\
&* \alpha^3(w) - 2J_\alpha(\alpha(u), \alpha(v), x * y) * \alpha^3(w)\\
&+ 2\alpha^2(u * v) * \{\alpha(x), \alpha(y), \alpha(w)\}_\alpha - J_\alpha\big(\alpha(2u\\
&* v), \alpha^2(w), \alpha(x * y)\big) - \{x, y, v\}_\alpha \alpha^2(w) * \alpha^3(u)\\
&- \alpha^2(v)\{x, y, w\}_\alpha * \alpha^3(u) + J_\alpha(\alpha(v), \alpha(w), x\\
&* y) * \alpha^3(u) - \alpha^2(v * w) * \{\alpha(x), \alpha(y), \alpha(u)\}_\alpha\\
&+ J_\alpha\big(\alpha(v * w), \alpha^2(u), \alpha(x * y)\big) - \{x, y, w\}_\alpha\\
&\cdot \alpha^2(u) * \alpha^3(v) - \alpha^2(w)\{x, y, u\}_\alpha * \alpha^3(v)\\
&+ J_\alpha(\alpha(w), \alpha(u), x * y) * \alpha^3(v) - \alpha^2(w * u)\\
&* \{\alpha(x), \alpha(y), \alpha(v)\}_\alpha + J_\alpha\big(\alpha(w * u), \alpha^2(v), \alpha(x\\
&* y)\big) = 2\{x, y, u\}_\alpha \alpha^2(v) * \alpha^3(w) - \alpha^2(v * w)\\
&* \alpha(\{x, y, u\}_\alpha) - \alpha^2(w)\{x, y, u\}_\alpha * \alpha^3(v)\\
&+ 2\alpha^2(u)\{x, y, v\}_\alpha * \alpha^3(w) - \{x, y, v\}_\alpha \alpha^2(w)\\
&* \alpha^3(u) - \alpha^2(w * u) * \alpha(\{x, y, v\}_\alpha) + 2\alpha^2(u\\
&* v) * \alpha(\{x, y, w\}_\alpha) - \alpha^2(v)\{x, y, w\}_\alpha * \alpha^3(u)\\
&- \{x, y, w\}_\alpha \alpha^2(u) * \alpha^3(v) - 2J_\alpha(\alpha(u), \alpha(v), x\\
&* y) * \alpha^3(w) - J_\alpha\big(\alpha(2u * v), \alpha^2(w), \alpha(x * y)\big)\\
&+ J_\alpha(\alpha(v), \alpha(w), x * y) * \alpha^3(u) + J_\alpha\big(\alpha(v * w),\\
&\alpha^2(u), \alpha(x * y)\big) + J_\alpha(\alpha(w), \alpha(u), x * y)\\
&* \alpha^3(v) + J_\alpha\big(\alpha(w * u), \alpha^2(v), \alpha(x * y)\big)\\
&\qquad\qquad\qquad\qquad (\text{rearranging terms})\\
&= \big\{\{x, y, u\}_\alpha, \alpha^2(v), \alpha^2(w)\big\}_\alpha + \big\{\alpha^2(u), \{x, y, v\}_\alpha,\\
&\alpha^2(w)\big\}_\alpha + \big\{\alpha^2(u), \alpha^2(v), \{x, y, w\}_\alpha\big\}_\alpha\\
&+ \big[-2\big(J_\alpha(\alpha(u), \alpha(v), x * y) * \alpha^3(w)\\
&+ J_\alpha\big(\alpha(u * v), \alpha^2(w), \alpha(x * y)\big)\big)\\
&+ J_\alpha(\alpha(v), \alpha(w), x * y) * \alpha^3(u)\\
&+ J_\alpha\big(\alpha(v * w), \alpha^2(u), \alpha(x * y)\big)\\
&+ J_\alpha(\alpha(w), \alpha(u), x * y) * \alpha^3(v)\\
&+ J_\alpha\big(\alpha(w * u), \alpha^2(v), \alpha(x * y)\big)\big].
\end{aligned} \tag{23}$$

In this latest expression, denote by $N(u, v, w, x, y)$ the expression in "$[\cdots]$"; to conclude, we proceed to show that $N(u, v, w, x, y) = 0$.

Observe first that, by (6), we have

$$\begin{aligned}
&J_\alpha(\alpha(u), x * y, \alpha(w)) * \alpha^2(\alpha(v))\\
&\quad + J_\alpha(\alpha(v), x * y, \alpha(w)) * \alpha^2(\alpha(u))\\
&\quad = J_\alpha\big(\alpha^2(u), \alpha(x * y), \alpha(v) * \alpha(w)\big)\\
&\quad + J_\alpha\big(\alpha^2(v), \alpha(x * y), \alpha(u) * \alpha(w)\big).
\end{aligned} \tag{24}$$

That is,

$$\begin{aligned}
&J_\alpha(\alpha(w), \alpha(u), x * y) * \alpha^3(v)\\
&\quad + J_\alpha\big(\alpha(w * u), \alpha^2(v), \alpha(x * y)\big)\\
&\quad = J_\alpha\big(\alpha(v * w), \alpha^2(u), \alpha(x * y)\big)\\
&\quad + J_\alpha(\alpha(v), \alpha(w), x * y) * \alpha^3(u).
\end{aligned} \tag{25}$$

With this observation, the expression $N(u, v, w, x, y)$ is transformed as follows:

$$\begin{aligned}
N(u, v, w, x, y) = 2\big[&-J_\alpha(\alpha(u), \alpha(v), x * y)\\
&* \alpha^3(w) - J_\alpha\big(\alpha(u * v), \alpha^2(w), \alpha(x * y)\big)\big]
\end{aligned}$$

$$
\begin{aligned}
&+2\left[J_\alpha\left(\alpha(v*w),\alpha^2(u),\alpha(x*y)\right)\right.\\
&\left.+J_\alpha(\alpha(v),\alpha(w),x*y)*\alpha^3(u)\right]\\
&=2\left[-J_\alpha\left(\alpha^2(w),\alpha(x*y),\alpha(u)*\alpha(v)\right)\right.\\
&-J_\alpha\left(\alpha^2(u),\alpha(x*y),\alpha(w)*\alpha(v)\right)\\
&-J_\alpha(\alpha(u),\alpha(v),x*y)*\alpha^3(w)\\
&\left.+J_\alpha(\alpha(v),\alpha(w),x*y)*\alpha^3(u)\right]\\
&=2\left[-J_\alpha(\alpha(w),x*y,\alpha(v))*\alpha^3(u)\right.\\
&-J_\alpha(\alpha(u),x*y,\alpha(v))*\alpha^3(w)\\
&-J_\alpha(\alpha(u),\alpha(v),x*y)*\alpha^3(w)\\
&\left.+J_\alpha(\alpha(v),\alpha(w),x*y)*\alpha^3(u)\right]
\end{aligned}
$$

$\Big($applying (6) to

$$
\begin{aligned}
&-J_\alpha\left(\alpha^2(w),\alpha(x*y),\alpha(u)*\alpha(v)\right)\\
&-J_\alpha\left(\alpha^2(u),\alpha(x*y),\alpha(w)*\alpha(v)\right)\Big)\\
&=0\,(\text{by the skew-symmetry of } J_\alpha(x,y,z)).
\end{aligned}
\tag{26}
$$

Therefore, we obtain that (11) holds for "$\{,,\}_\alpha$" and we conclude that $(A,\{,,\}_\alpha,\alpha^2)$ is a Hom-Lts. □

Remark 15. In the proof of his result, Loos ([6], Satz 1) used essentially the fact that the left translations $L(x)$ in a Maltsev algebra $(A,*)$ are derivations with respect to the ternary operation "$\{,,\}$" defined by (15). Unfortunately, for Hom-Maltsev algebras such a tool is still not available at hand.

From [20] (Theorem 2.12) we know that any Maltsev algebra A can be twisted into a Hom-Maltsev algebra along any linear self-map of A. Consistent with this result, we recall the following method for constructing Hom-Lts which, in fact, is a result in [21] (see also Propositions 10 and 11 above) but using a Hom-Maltsev algebra construction in our proof (as a consequence of Theorem 14).

Proposition 16. *Let $(A,*)$ be a Maltsev algebra and α any self-morphism of $(A,*)$. If one defines on $(A,*)$ a ternary operation "$\{,,\}_\alpha$" by*

$$\{x,y,z\}_\alpha=\alpha^2(2xy*z-yz*x-zx*y), \tag{27}$$

then $(A,\{,,\}_\alpha,\alpha^2)$ is a multiplicative Hom-Lts.

Proof. One knows ([20], Theorem 2.12) that, from $(A,*)$ and any self-morphism α of $(A,*)$, we get a (multiplicative) Hom-Maltsev algebra $(A,\tilde{*},\alpha)$, where $x\tilde{*}y=\alpha(x*y)$ for all x,y in A. Next, if one defines on $(A,\tilde{*},\alpha)$ a ternary operation

$$
\begin{aligned}
\{x,y,z\}_\alpha:={}&2(x\tilde{*}y)\tilde{*}\alpha(z)-(y\tilde{*}z)\tilde{*}\alpha(x)\\
&-(z\tilde{*}x)\tilde{*}\alpha(y),
\end{aligned}
\tag{28}
$$

then, by Theorem 14, $(A,\{,,\}_\alpha,\alpha^2)$ is a Hom-Lts and "$\{,,\}_\alpha$" is expressed through "$*$" as

$$
\begin{aligned}
\{x,y,z\}_\alpha&=2\alpha(\alpha(x*y)*\alpha(z))\\
&\quad-\alpha(\alpha(y*z)*\alpha(x))\\
&\quad-\alpha(\alpha(z*x)*\alpha(y))\\
&=2\alpha^2(xy*z)-\alpha^2(yz*x)-\alpha^2(zx*y)\\
&=\alpha^2(2xy*z-yz*x-zx*y).
\end{aligned}
\tag{29}
$$

□

Observe that though constructed in quite a different way, the operation "$\{,,\}_\alpha$" in Proposition 16 above coincides with "$[,,]_{\alpha^n}$" in Proposition 11 for $n=2$.

Combining Lemma 13 and Theorem 14, we get the following result.

Theorem 17. *Let $(A,*,\alpha)$ be a multiplicative Hom-Maltsev algebra. If one defines on $(A,*,\alpha)$ a ternary operation $(,,)_\alpha$ by*

$$(x,y,z)_\alpha:=\frac{1}{3}\{x,y,z\}_\alpha, \tag{30}$$

where "$\{,,\}_\alpha$" is defined by (17), then $(A,*,(,,)_\alpha,\alpha)$ is a Hom-Bol algebra.

Proof. Definition (30) and Theorem 14 imply that $(A,(,,)_\alpha,\alpha^2)$ is a multiplicative Hom-Lts; that is, (HB4), (HB5), and (HB7) hold for $(A,*,(,,)_\alpha,\alpha)$. Now, (HB1), (HB2), and (HB3) are, respectively, the multiplicativity and skew-symmetry of "$*$"; next, we are done if we prove (HB6) for $(A,*,(,,)_\alpha,\alpha)$.

From (17) and multiplicativity we have

$$
\begin{aligned}
&-J_\alpha(\alpha(u),\alpha(v),x*y)\\
&=\{\alpha(u),\alpha(v),x*y\}_\alpha-3(\alpha(u)\alpha(v))\\
&\quad*(\alpha(x)\alpha(y))
\end{aligned}
\tag{31}
$$

and then (18) takes the form

$$
\begin{aligned}
&\{\alpha(x),\alpha(y),u*v\}_\alpha\\
&=\{x,y,u\}_\alpha*\alpha^2(v)+\alpha^2(u)*\{x,y,v\}_\alpha\\
&\quad+\{\alpha(u),\alpha(v),x*y\}_\alpha-3(\alpha(u)\alpha(v))\\
&\quad*(\alpha(x)\alpha(y)).
\end{aligned}
\tag{32}
$$

Multiplying by $1/3$ each member of this latter equality and using (30), we get

$$
\begin{aligned}
&(\alpha(x),\alpha(y),u*v)_\alpha\\
&=(x,y,u)_\alpha*\alpha^2(v)+\alpha^2(u)*(x,y,v)_\alpha\\
&\quad+(\alpha(u),\alpha(v),x*y)_\alpha-(\alpha(u)\alpha(v))\\
&\quad*(\alpha(x)\alpha(y))
\end{aligned}
\tag{33}
$$

which is (HB6) for $(A, *, (,,)_\alpha, \alpha)$. So $(A, *, (,,)_\alpha, \alpha)$ is a Hom-Bol algebra. □

Example 18. Let A be a vector space with basis $\{e_1, e_2, e_3, e_4\}$. From [20] (Example 2.13) we know that if one considers the linear map $\alpha : A \to A$ given by

$$\begin{aligned} \alpha(e_1) &= e_1 + e_3; \\ \alpha(e_2) &= 2e_2 + 2e_4; \\ \alpha(e_3) &= -e_3; \\ \alpha(e_4) &= -2e_4 \end{aligned} \tag{34}$$

and the multiplication table given by

$$\begin{aligned} e_1 * e_2 &= -2e_2 - 2e_4 \ (= -e_2 * e_1); \\ e_1 * e_3 &= e_3 \ (= -e_3 * e_1); \\ e_1 * e_4 &= -2e_4 \ (= -e_4 * e_1); \\ e_2 * e_3 &= -4e_4 \ (= -e_3 * e_2) \end{aligned} \tag{35}$$

(only nonzero products are specified), then $(A, *, \alpha)$ is a multiplicative Hom-Maltsev algebra. It is observed that $(A, *, \alpha)$ is not a Hom-Lie algebra nor a Maltsev algebra.

Now, by (17) and (30), one checks that the only nonzero ternary products $(x, y, z)_\alpha$ on A with respect to the basis elements are

$$\begin{aligned} (e_1, e_2, e_1)_\alpha &= -4e_2 \ (= -(e_2, e_1, e_1)_\alpha); \\ (e_1, e_3, e_1)_\alpha &= -e_3 \ (= -(e_3, e_1, e_1)_\alpha); \\ (e_1, e_4, e_1)_\alpha &= -4e_4 \ (= -(e_4, e_1, e_1)_\alpha). \end{aligned} \tag{36}$$

By Theorem 17 we get that $(A, *, (,,)_\alpha, \alpha)$ is a Hom-Bol algebra.

Since any Hom-alternative algebra is Hom-Maltsev admissible ([20], Theorem 3.8), from Theorem 17 we have the following.

Corollary 19. *Let $(A, *, \alpha)$ be a multiplicative Hom-alternative algebra. Then $(A, [,], (,,)_\alpha, \alpha)$ is a Hom-Bol algebra, where $(x, y, z)_\alpha := -(1/3)(2[[x, y], \alpha(z)] - [[y, z], \alpha(x)] - [[z, x], \alpha(y)])$, for all x, y, and $z \in A$.*

The aim of Section 4 is a generalization of Corollary 19 to multiplicative right (or left) Hom-alternative algebras.

Various constructions of Hom-Lts are offered in [21] starting from either Hom-associative algebras, Hom-Lie algebras, Hom-Jordan triple systems, ternary totally Hom-associative algebras, Maltsev algebras, or alternative algebras. In practice, it is easier to construct Hom-Lts or Hom-Bol algebras from well-known (binary) algebras such as alternative algebras or Maltsev algebras. From this point of view, our construction results (Theorem 14, Proposition 16, and Theorem 17) have rather a theoretical feature (the extension to Hom-algebra setting of Loos' result [6] and a result by Mikheev [25]) than a practical method for constructing Hom-Lts or Hom-Bol algebras. However, it could be of some interest to get a Hom-Lts or a Hom-Bol algebra from a given Hom-Maltsev algebra without resorting to the corresponding Maltsev algebra.

4. Hom-Lts and Hom-Bol Algebras from Right (or Left) Hom-Alternative Algebras

In this section we prove that every multiplicative right (or left) Hom-alternative algebra has a natural Hom-Bol algebra structure (and, subsequently, a natural Hom-Lts structure). This is the Hom-analogue of a result by Mikheev [31] and by Hentzel and Peresi [29] although with a different scheme of proof.

First we recall some few basic properties of right Hom-alternative algebras that could be found in [18, 23].

The linearized form of the right Hom-alternative identity $\mathrm{as}(x, y, y) = 0$ is given by the following result.

Lemma 20 (see [18]). *If $(A, *, \alpha)$ is a Hom-algebra, then the following statements are equivalent.*

(i) $(A, *, \alpha)$ is right Hom-alternative.

(ii) $(A, *, \alpha)$ satisfies

$$as(x, y, z) = -as(x, z, y) \tag{37}$$

for all x, y, and $z \in A$.

(iii) $(A, *, \alpha)$ satisfies

$$\alpha(x) * (yz + zy) = xy * \alpha(z) + xz * \alpha(y) \tag{38}$$

for all x, y, and $z \in A$.

Observe that if $(A, *, \alpha)$ is a right Hom-alternative algebra, then $(A, *^{\mathrm{op}}, \alpha)$ is a left Hom-alternative algebra, where $x *^{\mathrm{op}} y := y * x$. So the mirrors of (37) and (38) hold for $(A, *^{\mathrm{op}}, \alpha)$:

$$\mathrm{as}(x, y, z) = -\mathrm{as}(y, x, z), \tag{39}$$

$$\begin{aligned} &((x *^{\mathrm{op}} y) + (y *^{\mathrm{op}} x)) *^{\mathrm{op}} \alpha(z) \\ &\quad = \alpha(x) *^{\mathrm{op}} (y *^{\mathrm{op}} z) + \alpha(y) *^{\mathrm{op}} (x *^{\mathrm{op}} z). \end{aligned} \tag{40}$$

Now we have the following.

Lemma 21. *In any multiplicative right Hom-alternative algebra $(A, *, \alpha)$, the identity*

$$\begin{aligned} &as([u, v], \alpha(x), \alpha(y)) \\ &\quad = [as(u, x, y), \alpha^2(v)] + [\alpha^2(u), as(v, x, y)] \\ &\quad + as(\alpha(v), \alpha(u), [x, y]) \\ &\quad - as(\alpha(u), \alpha(v), [x, y]) \end{aligned} \tag{41}$$

holds for all x, y, and $z \in A$.

Proof. The identity

$$\begin{aligned}\operatorname{as}(uv,\alpha(x),\alpha(y)) &= \operatorname{as}(u,x,y)\alpha^2(v)\\ &\quad+\alpha^2(u)\operatorname{as}(v,x,y)\\ &\quad-\operatorname{as}(\alpha(u),\alpha(v),[x,y])\end{aligned} \tag{42}$$

holds in any right Hom-alternative algebra (see [23], Theorem 7.1 (7.1.1c)). Next, in this identity, switching u and v, we have

$$\begin{aligned}\operatorname{as}(vu,\alpha(x),\alpha(y)) &= \operatorname{as}(v,x,y)\alpha^2(u)\\ &\quad+\alpha^2(v)\operatorname{as}(u,x,y)\\ &\quad-\operatorname{as}(\alpha(v),\alpha(u),[x,y]).\end{aligned} \tag{43}$$

Then, subtracting memberwise this latter equality from the one above and using the linearity of as, we get (41). □

Note that in the case when $(A,*,\alpha)$ is a left Hom-alternative algebra, identity (41) reads as

$$\begin{aligned}&\operatorname{as}(\alpha(x),\alpha(y),[u,v])\\ &\quad=\left[\operatorname{as}(x,y,u),\alpha^2(v)\right]+\left[\alpha^2(u),\operatorname{as}(x,y,v)\right]\\ &\qquad+\operatorname{as}([x,y],\alpha(v),\alpha(u))\\ &\qquad-\operatorname{as}([x,y],\alpha(u),\alpha(v)).\end{aligned} \tag{44}$$

In any multiplicative right (or left) Hom-alternative algebra $(A,*,\alpha)$ we consider the ternary operation defined by (12); that is,

$$(x,y,z) := \operatorname{as}^J(y,z,x), \tag{45}$$

where as^J is the Hom-Jordan associator defined in Section 2. Observe that for $\alpha = \mathrm{id}$ the ternary operation "(, ,)" is precisely the one defined in [29] (see also [31], Remark 2) and that makes any right (or left) alternative algebra into a left Bol algebra. In [29], Hentzel and Peresi used the approach of Mikheev [31] who formerly proved that the commutator algebra of any right alternative algebra has a left Bol algebra structure.

Proposition 22. *(i) If $(A,*,\alpha)$ is a multiplicative right Hom-alternative algebra, then*

$$(x,y,z) = [[x,y],\alpha(z)] - 2as(z,x,y) \tag{46}$$

for all x, y, and $z \in A$.

(ii) If $(A,*,\alpha)$ is a multiplicative left Hom-alternative algebra, then

$$(x,y,z) = [[x,y],\alpha(z)] - 2as(x,y,z) \tag{47}$$

for all x, y, and $z \in A$.

Proof. (i) From (12) we have

$$\begin{aligned}(x,y,z) &= (y\circ z)\circ\alpha(x)-\alpha(y)\circ(z\circ x)\\ &= ((y*z)+(z*y))*\alpha(x)\\ &\quad+[\alpha(x)*((y*z)+(z*y))]\\ &\quad-[\alpha(y)*((z*x)+(x*z))]\\ &\quad-((z*x)+(x*z))*\alpha(y)\\ &= ((y*z)+(z*y))*\alpha(x)\\ &\quad+[(x*y)*\alpha(z)+(x*z)*\alpha(y)]\\ &\quad-[(y*z)*\alpha(x)+(y*x)*\alpha(z)]\\ &\quad-((z*x)+(x*z))*\alpha(y)\quad(\text{by }(38))\\ &= (z*y)*\alpha(x)+(x*y)*\alpha(z)\\ &\quad-(y*x)*\alpha(z)-(z*x)*\alpha(y)\\ &= (z*y)*\alpha(x)-(z*x)*\alpha(y)+[x,y]\\ &\quad*\alpha(z)\\ &= (z*y)*\alpha(x)-(z*x)*\alpha(y)\\ &\quad+[[x,y],\alpha(z)]+\alpha(z)*[x,y]\\ &= [[x,y],\alpha(z)]+(z*y)*\alpha(x)-\alpha(z)\\ &\quad*(y*x)-(z*x)*\alpha(y)+\alpha(z)\\ &\quad*(x*y)\\ &= [[x,y],\alpha(z)]+\operatorname{as}(z,y,x)-\operatorname{as}(z,x,y)\\ &= [[x,y],\alpha(z)]-2\operatorname{as}(z,x,y)\quad(\text{by }(37))\end{aligned} \tag{48}$$

and so we get (46).

(ii) Proceeding as above, but using (40) and then (39), one gets (47). □

We are now in a position to prove the main result of this section.

Theorem 23. *Let $(A,*,\alpha)$ be a multiplicative right (resp., left) Hom-alternative algebra. If one defines on A a ternary operation "(, ,)" by (46) (resp., (47)), then $(A,(,,),\alpha^2)$ is a Hom-Lts and $(A,[,],(,,),\alpha)$ is a Hom-Bol algebra.*

Proof. We prove the theorem for a multiplicative right Hom-alternative algebra $(A,*,\alpha)$ (the proof of the left case is the mirror of the right one).

Identities (HB1) and (HB2) follow from the multiplicativity of $(A,*,\alpha)$. Identities (HB3) and (HB4) are obvious from the definition of "[,]" and "(, ,)"; identity (HB5) follows from Proposition 9.

In [22] Yau showed that if, on a multiplicative Hom-Jordan algebra $(A,\circ,\alpha)$, define a ternary operation by

$$[x,y,z] := 2(\alpha(x)\circ(y\circ z)-\alpha(y)\circ(x\circ z)), \tag{49}$$

then $(A,[,,],\alpha^2)$ is a multiplicative Hom-Lts (see [22], Corollary 4.1). Now, observe that $[x,y,z] = 2\operatorname{as}^J(y,z,x)$; that is, $[x,y,z] = 2(x,y,z)$. Therefore, since every multiplicative right Hom-alternative algebra is Hom-Jordan admissible (see [23], Theorem 4.3), we conclude that $(A, (,,), \alpha^2)$ is a multiplicative Hom-Lts and so identity (HB7) holds for $(A, [,], (,,), \alpha)$.

Next, $(A, [,], (,,), \alpha)$ is a Hom-Bol algebra if we prove that (HB6) additionally holds.

Write (46) as

$$-2\mathrm{as}(z, x, y) = (x, y, z) - [[x, y], \alpha(z)]. \tag{50}$$

Multiplying each member of (41) by -2 and next using (50), we get

$$\begin{aligned}(\alpha(x), \alpha(y), [u, v]) &- [[\alpha(x), \alpha(y)], \alpha([u, v])] \\ &= [(x, y, u) - [[x, y], \alpha(u)], \alpha^2(v)] \\ &\quad + [\alpha^2(u), (x, y, v) - [[x, y], \alpha(v)]] \\ &\quad + (\alpha(u), [x, y], \alpha(v)) \\ &\quad - [[\alpha(u), [x, y]], \alpha^2(v)] \\ &\quad - (\alpha(v), [x, y], \alpha(u)) \\ &\quad + [[\alpha(v), [x, y]], \alpha^2(u)].\end{aligned} \tag{51}$$

That is,

$$\begin{aligned}(\alpha(x), \alpha(y), [u, v]) = {} & [(x, y, u), \alpha^2(v)] \\ & + [\alpha^2(u), (x, y, v)] \\ & - ([x, y], \alpha(u), \alpha(v)) \\ & + ([x, y], \alpha(v), \alpha(u)) \\ & + \alpha([[x, y], [u, v]]).\end{aligned} \tag{52}$$

Observe that

$$\begin{aligned}-([x, y], \alpha(u), \alpha(v)) &+ ([x, y], \alpha(v), \alpha(u)) \\ &= (\alpha(u), [x, y], \alpha(v)) + ([x, y], \alpha(v), \alpha(u)) \\ &= -(\alpha(v), \alpha(u), [x, y])\end{aligned} \tag{53}$$

$$(\text{since } \circlearrowleft_{a,b,c}(a, b, c) = 0 \text{ by (HB5)}) = (\alpha(u), \alpha(v), [x, y]).$$

Therefore, (52) now reads

$$\begin{aligned}(\alpha(x), \alpha(y), [u, v]) = {} & [(x, y, u), \alpha^2(v)] \\ & + [\alpha^2(u), (x, y, v)] \\ & + (\alpha(u), \alpha(v), [x, y]) \\ & - \alpha([[u, v], [x, y]])\end{aligned} \tag{54}$$

and so (HB6) holds for $(A, [,], (,,), \alpha)$. Thus we conclude that $(A, [,], (,,), \alpha)$ is a Hom-Bol algebra. One gets the same result in the case when $(A, *, \alpha)$ is a multiplicative left Hom-alternative algebra and essentially using (47) and (44). This finishes the proof. □

Example 24. Let A be a five-dimensional vector space with basis $\{e, u, v, w, z\}$ and let $\alpha : A \to A$ be a linear map given by

$$\begin{aligned}\alpha(e) &= e + u + v; \\ \alpha(u) &= -u; \\ \alpha(v) &= -v; \\ \alpha(w) &= -w; \\ \alpha(z) &= -z.\end{aligned} \tag{55}$$

Define on A a binary operation "$*$" by

$$\begin{aligned}e * e &= e + u + v; \\ e * u &= -v; \\ e * w &= -w + z; \\ e * z &= -z; \\ u * e &= -u; \\ z * e &= -z\end{aligned} \tag{56}$$

(again, only nonzero products are specified). Then $(A, *, \alpha)$ is a multiplicative right Hom-alternative algebra (see [23], Example 2.9). Then, using $[x, y] = x * y - y * x$ and (46), one could find (although the computation is somewhat lengthy) all the nonzero products "$[,]$" and "$(,,)$" with respect to the basis elements e, u, v, w, and z of A. We just point out that they are nonzero products; for example, $[e, u] = u - v$, $[e, w] = -w + z$, $(e, u, e) = -u - v$, and $(e, w, e) = -w - z$. Therefore, Theorem 23 implies that $(A, [,], (,,), \alpha)$ is a Hom-Bol algebra.

5. The Construction of Bol Algebras from Right Alternative Algebras Revisited

As already mentioned in Section 2, for $\alpha = \mathrm{id}$ in Definition 12 we get the definition of a left Bol algebra.

Definition 25 (see [25, 27]). A *left Bol algebra* is a triple $(A, [,], (,,))$ in which A is a vector space, "$[,]$" a binary operation, and "$(,,)$" a ternary operation on A such that

(B1) $[x, y] = -[y, x]$,

(B2) $(x, y, z) = -(y, x, z)$,

(B3) $\circlearrowleft_{x,y,z}(x, y, z) = 0$,

(B4) $(x, y, [u, v]) = [(x, y, u), v] + [u, (x, y, v)] + (u, v, [x, y]) - [[u, v], [x, y]]$,

(B5) $(x, y, (u, v, w)) = ((x, y, u), v, w) + (u, (x, y, v), w) + (u, v, (x, y, w))$,

for all $u, v, w, x, y, z \in A$.

In this section we show how the construction of Hom-Bol algebras from right or left Hom-alternative algebras described in Section 4 can be specified to the ordinary untwisted case of construction of (left) Bol algebras from right or left alternative algebras ([29, 31]). In fact, for $\alpha = \mathrm{id}$ in Theorem 23 and specifying the right alternative case, we get the following.

Theorem 26. *Let $(A, *)$ be a right alternative algebra. If one defines on A a ternary operation "$(,,)$" by*

$$(x, y, z) = [[x, y], z] - 2as(z, x, y), \quad (57)$$

where $as(u, v, w) = uv * w - u * vw$, then $(A, (,,))$ is a Lts and $(A, [,], (,,))$ is a left Bol algebra.

Proof. Identities (B1) and (B2) are obvious. For α = id, the Hom-Jordan associator (see Definition 3) reduces to the usual Jordan associator $as^J(u, v, w) := (u \circ v) \circ w - u \circ (v \circ w)$ in $(A, *)$. The fact that (B3) and (B5) hold in $(A, *)$ follows from the equality $(x, y, z) = as^J(y, z, x)$ that holds in right alternative algebras (the untwisted form of (46)) and from that right alternative algebras are Jordan admissible [36]. Therefore $(A, (,,))$ is a Lts since any Jordan algebra is a Lts with respect to the operation $(x, y, z) = as^J(y, z, x)$ (see [1]). So we get the untwisted version of (HB5) and (HB7).

In order to show that (B4) holds in $(A, *)$, we proceed as follows. First, recall that the identity

$$\begin{aligned} as(uv, y, x) &= u as(v, y, x) + as(u, v, y)x \\ &\quad + as(u, vy, x) - as(u, v, yx) \end{aligned} \quad (58)$$

holds in any algebra. Also, in a right alternative algebra $(A, *)$ (over a ground field of characteristic different from 2), the following identity holds [37]:

$$as(u, v, v * y) = as(u, v, y) * v; \quad (59)$$

that is, by linearization and right alternativity,

$$\begin{aligned} as(u, v * y, x) &= as(u, v, x * y) - as(u, v, y) * x \\ &\quad - as(u, x, y) * v. \end{aligned} \quad (60)$$

Putting (60) in (58), we get

$$\begin{aligned} as(u * v, y, x) &= u * as(v, y, x) - as(u, x, y) * v \\ &\quad + as(u, v, [x, y]); \end{aligned} \quad (61)$$

that is, by right alternativity,

$$\begin{aligned} as(u * v, x, y) &= as(u, x, y) * v + u * as(v, x, y) \\ &\quad - as(u, v, [x, y]). \end{aligned} \quad (62)$$

Now, in (62) switching u and v and then subtracting the obtained equality from (62), one gets

$$\begin{aligned} as([u, v], x, y) &= [as(u, x, y), v] + [u, as(v, x, y)] \\ &\quad + as(v, u, [x, y]) - as(u, v, [x, y]) \end{aligned} \quad (63)$$

(observe that (63) is the untwisted form of (41)). Next, write (57) as

$$-2as(z, x, y) = (x, y, z) - [[x, y], z]. \quad (64)$$

Then multiplying (63) by -2 and using the equality above, and next proceeding as in the proof of Theorem 23, one proves the validity of (B4) for $(A, [,], (,,))$. Thus we get that $(A, [,], (,,))$ is a left Bol algebra. □

Remark 27. (i) The process of constructing left Bol algebras from right alternative algebras described in Theorem 26 above is different from the ones given in [29, 31]. In our approach here, we rely essentially on fundamental properties of right alternative algebras (see, e.g., [36, 37]) without subsidiary constructions.

(ii) If $(A, *)$ is a left alternative algebra, it is also possible to get a natural left Bol algebra structure on $(A, *)$. Indeed, one needs to consider the counterparts of (x, y, z) and (63) that looks, respectively, as

$$(x, y, z) = [[x, y], z] - as(x, y, z) \quad (65)$$

(the untwisted version of (47)) and

$$\begin{aligned} as(x, y, [u, v]) &= [as(x, y, u), v] + [u, as(x, y, v)] \\ &\quad + as([x, y], v, u) \\ &\quad - as([x, y], u, v). \end{aligned} \quad (66)$$

Next one proceeds as in Theorem 26 observing that a left alternative algebra is also Jordan-admissible (see [36], Theorem 2, for right alternative algebras).

Conflicts of Interest

The authors declare that they have no conflicts of interests.

References

[1] N. Jacobson, "Lie and Jordan triple systems," *American Journal of Mathematics*, vol. 71, pp. 149–170, 1949.

[2] W. G. Lister, "A structure theory of Lie triple systems," *Transactions of the American Mathematical Society*, vol. 72, pp. 217–242, 1952.

[3] K. Yamaguti, "On algebras of totally geodesic spaces (Lie triple systems)," *Journal of Science of the Hiroshima University, Series A*, vol. 21, pp. 107–113, 1957/1958.

[4] S. Kobayashi and K. Nomizu, *Foundations of Differential Geometry*, Interscience Publishers, New York, NY, USA, 1963.

[5] O. Loos, *Symmetric Spaces*, vol. 1-2, W. A. Benjamin, New York, NY, USA, 1969.

[6] O. Loos, "Über eine beziehung zwischen malcev-algebren und lie-tripelsystemen," *Pacific Journal of Mathematics*, vol. 18, pp. 553–562, 1966.

[7] A. I. Maltsev, "Analytic loops," *Matematicheskii Sbornik*, vol. 78, pp. 569–578, 1955.

[8] A. A. Sagle, "Malcev algebras," *Transactions of the American Mathematical Society*, vol. 101, pp. 426–458, 1961.

[9] J. T. Hartwig, D. Larsson, and S. D. Silvestrov, "Deformations of Lie algebras using σ-derivations," *Journal of Algebra*, vol. 295, no. 2, pp. 314–361, 2006.

[10] A. Makhlouf and S. D. Silvestrov, "Hom-algebra structures," *Journal of Generalized Lie Theory and Applications*, vol. 2, no. 2, pp. 51–64, 2008.

[11] H. Ataguema, A. Makhlouf, and S. Silvestrov, "Generalization of n-ary Nambu algebras and beyond," *Journal of Mathematical Physics*, vol. 50, no. 8, Article ID 083501, 083501, 15 pages, 2009.

[12] S. Attan and A. N. Issa, "Hom-Bol algebras," *Quasigroups and Related Systems*, vol. 21, no. 2, pp. 131–146, 2013.

[13] Y. Frégier, A. Gohr, and S. D. Silvestrov, "Unital algebras of Hom-associative type and surjective or injective twistings," *Journal of Generalized Lie Theory and Applications*, vol. 3, no. 4, pp. 285–295, 2009.

[14] D. Gaparayi and A. N. Issa, "A twisted generalization of Lie-Yamaguti algebras," *International Journal of Algebra*, vol. 6, no. 5-8, pp. 339–352, 2012.

[15] A. Gohr, "On hom-algebras with surjective twisting," *Journal of Algebra*, vol. 324, no. 7, pp. 1483–1491, 2010.

[16] A. N. Issa, "Hom-Akivis algebras," *Commentationes Mathematicae*, vol. 52, no. 4, pp. 485–500, 2011.

[17] A. Makhlouf, "Paradigm of nonassociative Hom-algebras and Hom-superalgebras," in *Proceedings of the Jordan Structures in Algebra and Analysis Meeting*, pp. 143–177, Editorial Círculo Rojo, Almería, Spain, 2010.

[18] A. Makhlouf, "Hom-alternative algebras and Hom-Jordan algebras," *International Electronic Journal of Algebra*, vol. 8, pp. 177–190, 2010.

[19] A. Makhlouf and S. Silvestrov, "Hom-algebras and Hom-coalgebras," *Journal of Algebra and Its Applications*, vol. 9, no. 4, pp. 553–589, 2010.

[20] D. Yau, "Hom-Maltsev, Hom-alternative, and Hom-Jordan algebras," *International Electronic Journal of Algebra*, vol. 11, pp. 177–217, 2012.

[21] D. Yau, "On n-ary Hom-Nambu and Hom-Nambu-Lie algebras," *Journal of Geometry and Physics*, vol. 62, no. 2, pp. 506–522, 2012.

[22] D. Yau, *On n-ary Hom-Nambu and Hom-Maltsev Algebras*, Cornell University, 2010.

[23] D. Yau, *Right Hom-alternative algebras*, Cornell University, New York, NY, USA, 2010.

[24] D. Yau, "Hom-algebras and homology," *Journal of Lie Theory*, vol. 19, no. 2, pp. 409–421, 2009.

[25] P. O. Mikheev, *Geometry of smooth Bol loops [PhD Thesis]*, Friendship University, Moscow, Russia, 1986 (Russian).

[26] L. V. Sabinin and P. O. Mikheev, "Analytic Bol loops," in *Webs and quasigroups*, pp. 102–109, Kalinin Gos. University, Kalinin, Russia, 1982.

[27] L. V. Sabinin and P. O. Mikheev, "The geometry of smooth Bol loops," in *Webs and Quasigroups*, pp. 144–154, Kalinin Gos. University, Kalinin, Russia, 1984.

[28] T. B. Bouetou, "On Bol algebras," in *Webs and Quasigroups*, pp. 75–83, Tver State University, Tver, Russia, 1995.

[29] I. R. Hentzel and L. A. Peresi, "Special identities for Bol algebras," *Linear Algebra and its Applications*, vol. 436, no. 7, pp. 2315–2330, 2012.

[30] J. M. Pérez-Izquierdo, "An envelope for Bol algebras," *Journal of Algebra*, vol. 284, no. 2, pp. 480–493, 2005.

[31] P. O. Mikheev, "Commutator algebras of right-alternative algebras," *Matematicheskii Issledovania*, vol. 113, pp. 62–65, 1990.

[32] D. Yau, "Enveloping algebras of Hom-Lie algebras," *Journal of Generalized Lie Theory and Applications*, vol. 2, no. 2, pp. 95–108, 2008.

[33] Y. Sheng, "Representations of hom-Lie algebras," *Algebras and Representation Theory*, vol. 15, no. 6, pp. 1081–1098, 2012.

[34] A. N. Issa, "On identities in Hom-Malcev algebras," *International Electronic Journal of Algebra*, vol. 17, pp. 1–10, 2015.

[35] Y. Nambu, "Generalized Hamiltonian dynamics," *Physical Review D: Particles, Fields, Gravitation and Cosmology*, vol. 7, pp. 2405–2412, 1973.

[36] A. A. Albert, "On the right alternative algebras," *Annals of Mathematics: Second Series*, vol. 50, pp. 318–328, 1949.

[37] E. Kleinfeld, "Right alternative rings," *Proceedings of the American Mathematical Society*, vol. 4, pp. 939–944, 1953.

3

PS-Modules over Ore Extensions and Skew Generalized Power Series Rings

Refaat M. Salem,[1] Mohamed A. Farahat,[1,2] and Hanan Abd-Elmalk[3]

[1]*Mathematics Department, Faculty of Science, Al-Azhar University, P.O. Box 11884, Nasr City, Cairo, Egypt*
[2]*Department of Mathematics and Statistics, Faculty of Science, Taif University, P.O. Box 888, Al-Hawiyah, Taif 21974, Saudi Arabia*
[3]*Department of Mathematics, Faculty of Science, Ain Shams University, P.O. Box 11566, Abbasaya, Cairo, Egypt*

Correspondence should be addressed to Mohamed A. Farahat; refaat_salem@cic-cairo.com

Academic Editor: Ram N. Mohapatra

A right R-module M_R is called a PS-module if its socle, $\mathrm{Soc}(M_R)$, is projective. We investigate PS-modules over Ore extension and skew generalized power series extension. Let R be an associative ring with identity, M_R a unitary right R-module, $O = R[x;\alpha,\delta]$ Ore extension, $M[x]_O$ a right O-module, $(S,\leq)$ a strictly ordered additive monoid, $\omega : S \to \mathrm{End}(R)$ a monoid homomorphism, $A = [[R^{S,\leq},\omega]]$ the skew generalized power series ring, and $B_A = [[M^{S,\leq}]]_{[[R^{S,\leq},\omega]]}$ the skew generalized power series module. Then, under some certain conditions, we prove the following: (1) If M_R is a right PS-module, then $M[x]_O$ is a right PS-module. (2) If M_R is a right PS-module, then B_A is a right PS-module.

1. Introduction

Throughout this paper R denotes an associative ring with identity and M_R a unitary right R-module. According to Nicholson and Watters [1], M_R is called a *PS-module* if every simple submodule is projective and equivalently if its socle, $\mathrm{Soc}(M_R)$, is projective. Examples of PS-modules include nonsingular modules, regular modules in the sense of Zelmanowitz [2], and modules with zero socle. The class of PS-modules is closed under direct sums and submodules. In [3], Weimin proved that PS-modules are preserved by Morita equivalences and excellent extensions.

For any subset X of R, denote

$$l_M(X) = \{m \in M \mid mX = 0\}. \tag{1}$$

Theorem 1 (see [3]). *The following statements are equivalent for a right R-module M_R:*

(1) *M_R is a PS-module.*

(2) *If L is a maximal right ideal of R, then either $l_M(L) = 0$ or $L = eR$, where $e^2 = e \in R$.*

A left PS-module ${}_RM$ is defined analogously. A ring R is said to be a *left PS-ring* if ${}_RR$ is a PS-module. Every semiprime ring is a PS-ring. Every PP-ring is a PS-ring (where a ring R is called PP-ring if every principal left ideal is projective). In particular every Baer ring is a PS-ring (where a ring R is called Baer if every left (or right) annihilator is generated by an idempotent). A ring for which every simple singular module is injective is a PS-ring. If $l_R(J(R)) = 0$, then R is a PS-ring. In fact $J(R) \subset L$ for every maximal right ideal so $l_R(L) = 0$.

The notion of PS-rings is not left-right symmetric (cf. [1]). A ring R is *duo* if each one-sided ideal of R is a two-sided ideal. As a generalization of left duo rings, a ring R is called *weakly left duo* if for every $r \in R$ there is a natural number $n(r)$ such that $Rr^{n(r)}$ is a two-sided ideal of R. A ring R is weakly duo if it is weakly right and left duo. In [3], Weimin proved that a duo ring R is a PS-ring if and only if it is a right PS-ring. In [4], Dingguo generalized this result to weakly duo rings as follows: a weakly duo reduced ring R is a PS-ring if and only if R is a right PS-ring.

If R is a PS-ring so also are $R[x]$ and $R[[x]]$. The converse of this result is false by the following example.

Example 2 (see [1], Example 3.2). If $R = \mathbb{Z}_4$, then $R[x]$ and $R[[x]]$ are PS-rings but R is not PS-ring.

The motivation of this paper is to investigate the PS property of Ore extension modules and the skew generalized power series extension modules. These results generalize the corresponding results for polynomial rings, generalized power series rings, and modules [5, 6].

2. PS-Modules over Ore Extension Rings

This section is devoted to study the relationship between the PS property of a right R-module M_R and the PS property of the right Ore extension module $M[x]_O(M[x]_{R[x;\alpha,\delta]})$.

Let α be an endomorphism of R and $\delta : R \to R$ an α-derivation of R, that is, an additive map such that

$$\delta(ab) = \delta(a)b + \alpha(a)\delta(b), \quad \forall a, b \in R. \tag{2}$$

In case α is the identity map, δ is called just a derivation of R.

The Ore extension $O = R[x;\alpha,\delta]$ is the set of all polynomials $\sum_{i=0}^{n} a_i x^i$ with the usual sum and the following multiplication rule:

$$xa = \alpha(a)x + \delta(a). \tag{3}$$

We assume that 1 is the identity element of $O = R[x;\alpha,\delta]$. This means that $\alpha(1) = 1$ and $\delta(1) = 0$. This definition of noncommutative polynomial rings with identity was first introduced by Ore [7]. Ever since the appearance of Ore's fundamental paper [7], Ore extensions have played an important role in noncommutative ring theory and many noncommutative ring theorists have investigated Ore extensions from different points of view such as ideal theory, order theory, Galois theory, and homological algebras.

For integers i, j with $j \geq i \geq 0$, $f_i^j \in \mathrm{End}(R,+)$ will denote the map which is the sum of all possible words in α, δ built with i letters of α and $j-i$ letters of δ. For instance,

$$\begin{aligned} f_0^0 &= \mathrm{Id}_R, \\ f_j^j &= \alpha^j, \\ f_0^j &= \delta^j, \\ f_{j-1}^j &= \alpha^{j-1}\delta + \alpha^{j-2}\delta\alpha + \cdots + \delta\alpha^{j-1}. \end{aligned} \tag{4}$$

For any positive integer n and $r \in R$, we have

$$x^n r = \sum_{i=0}^{n} f_i^n(r)x^i \tag{5}$$

(see [8], Lemma 4.1). This formula uniquely determines a general product of (left) polynomials in O and will be used freely in what follows.

Given a right R-module M_R, $M[x]_O$ is a right O-module with the natural action of O on $M[x]$ applying the above twist whenever necessary. The verification that this defines a valid O-module structure on $M[x]$ is almost identical to the verification that $O = R[x;\alpha,\delta]$ is a ring and it is straightforward (see [9]).

Definition 3 (see [9]). Given a module M_R, an endomorphism $\alpha : R \to R$ and an α-derivation $\delta : R \to R$. One says that M_R is α-compatible if for each $m \in M_R$, $r \in R$, one has $mr = 0 \Leftrightarrow m\alpha(r) = 0$. Moreover, One says that M_R is δ-compatible if for each $m \in M_R$, $r \in R$, one has $mr = 0 \Rightarrow m\delta(r) = 0$. If M_R is both α-compatible and δ-compatible, one says that M_R is (α,δ)-compatible.

Note that if M_R is α-compatible (resp., δ-compatible), then M_R is α^i-compatible (resp., δ^i-compatible) for all $i \geq 1$. It is clear that M_R is α-compatible (resp., δ-compatible), then so is any submodule of M_R. A ring R is (α,δ)-compatible if and only if R_R is an (α,δ)-compatible module.

As an immediate consequence of Definition 3, we obtain the following.

Lemma 4. *Let M_R be an (α,δ)-compatible module. For each $m \in M$ and $a \in R$, one has the following:*

(1) *$ma = 0$ if and only if $m\alpha^n(a) = 0$ for any positive integer n.*

(2) *If $ma = 0$, then $mf_i^j(a) = 0$ for all $j \geq i \geq 0$.*

Lemma 5 (see [10], Lemma 2.5). *Let M_R be an (α,δ)-compatible module, $m(x) = m_0 + \cdots + m_k x^k \in M[x]$, and $r \in R$. If $m(x)r = 0$, then $m_i r = 0$ for each i.*

Definition 6 (see [10]). Given a module M_R, an endomorphism $\alpha : R \to R$ and an α-derivation $\delta : R \to R$. One says that M_R is (α,δ)-Armendariz if whenever $m(x) = \sum_{i=0}^{k} m_i x^i \in M[x]$ and $f(x) = \sum_{j=0}^{n} a_j x^j \in R[x;\alpha,\delta]$ satisfy $m(x)f(x) = 0$, one has $m_i x^i a_j x^j = 0$ for all i, j.

A ring R is called (α,δ)-Armendariz if R_R is an (α,δ)-Armendariz module.

Using Lemma 5 it is easy to deduce that if M_R is (α,δ)-compatible and (α,δ)-Armendariz, then for any $m(x) = \sum_{i=0}^{k} m_i x^i \in M[x]$ and $f(x) = \sum_{j=0}^{n} a_j x^j \in R[x;\alpha,\delta]$, $m(x)f(x) = 0$ if and only if $m_i a_j = 0$ for all i, j.

Theorem 7. *Let M_R be an (α,δ)-compatible and (α,δ)-Armendariz module. If M_R is a PS-module, then $M[x]_O$ is a PS-module.*

Proof. Let L be a maximal right ideal of O. We will show that either $\mathrm{l}_{M[x]}(L) = 0$ or $L = hO$, where $h^2 = h \in O$. Let I be the set of all coefficients of all polynomials in L and let J be the right ideal of R generated by I. If $J = R$, then there exist $a_1, \ldots, a_n \in I$ and $r_1, \ldots, r_n \in R$ such that

$$1 = a_1 r_1 + \cdots + a_n r_n. \tag{6}$$

Suppose that $\varphi(x) = \sum_{i=0}^{k} m_i x^i \in 1_{M[x]}(L)$ and $\varphi \neq 0$, then for every $g(x) = \sum_{j=0}^{n} a_j x^j \in L$, we have

$$\varphi(x) g(x) = \left(\sum_{i=0}^{k} m_i x^i\right)\left(\sum_{j=0}^{n} a_j x^j\right) = 0. \quad (7)$$

Since M_R is (α, δ)-compatible and (α, δ)-Armendariz, it follows that

$$m_i a_j = 0, \quad \forall 0 \le i \le k, \; 0 \le j \le n. \quad (8)$$

Consequently, for every $a \in I$, $m_i a = 0$, $0 \le i \le k$. Hence we get

$$\begin{aligned} m_i &= m_i 1 = m_i (a_1 r_1 + \cdots + a_n r_n) \\ &= (m_i a_1) r_1 + \cdots + (m_i a_n) r_n = 0, \end{aligned} \quad (9)$$

a contradiction. Then $1_{M[x]}(L) = 0$. Suppose that $J \neq R$. We will show that J is a maximal right ideal of R. Let $r \in R - J$. If $r \in L$, then $r \in I$ and so $r \in J$, a contradiction. Thus $r \notin L$. Since L is a maximal right ideal of O,

$$O = L + rO. \quad (10)$$

It follows that there exist $g(x) = \sum_{i=0}^{n} a_i x^i \in L$ and $h(x) = \sum_{j=0}^{m} b_j x^j \in O$ such that $1 = a_0 + rb_0$. If $a_0 = 0$, then $1 = rb_0 \in rR$ and so $R = J + rR$. If $a_0 \neq 0$, then $a_0 \in I \subseteq J$ which implies that $R = J + rR$. Hence J is a maximal right ideal of R. Since M_R is a PS-module, it follows that either $1_M(J) = 0$ or $J = eR$, where $e^2 = e \in R$. According to that we have the following two cases.

Case 1. Suppose that $1_M(J) = 0$. We will show that $1_{M[x]}(L) = 0$. Let $\varphi(x) = \sum_{i=0}^{k} m_i x^i \in 1_{M[x]}(L)$ and $\varphi \neq 0$; then for every $g(x) = \sum_{j=0}^{n} a_j x^j \in L$, we have

$$\varphi(x) g(x) = \left(\sum_{i=0}^{k} m_i x^i\right)\left(\sum_{j=0}^{n} a_j x^j\right) = 0. \quad (11)$$

Since M_R is (α, δ)-compatible and (α, δ)-Armendariz, it follows that $m_i a_j = 0$ for all $0 \le i \le k$ and $0 \le j \le n$. Consequently, for every $a \in I$, $m_i a = 0$, $0 \le i \le k$. For any $r \in J$, there exist $a_1, \ldots, a_n \in I$ and $r_1, \ldots, r_n \in R$ such that $r = a_1 r_1 + \cdots + a_n r_n$. Hence

$$\begin{aligned} m_i r &= m_i (a_1 r_1 + \cdots + a_n r_n) \\ &= (m_i a_1) r_1 + \cdots + (m_i a_n) r_n = 0, \end{aligned} \quad (12)$$

which implies that $m_i \in 1_M(J) = 0$. Thus $\varphi(x) = 0$, a contradiction. Hence $1_{M[x]}(L) = 0$.

Case 2. Suppose that $J = eR$, where $e^2 = e \in R$. We will show that $L = eO$, where $e^2 = e \in O$. To show that $eO \subseteq L$, we need to prove that $e \in L$. If $e \notin L$, then $O = L + eO$. Thus there exist $g(x) = \sum_{i=0}^{n} a_i x^i \in L$ and $h(x) = \sum_{j=0}^{m} b_j x^j \in O$ such that $1 = a_0 + eb_0$. If $a_0 = 0$, then $1 = eb_0 \in eR = J$, a contradiction. If $a_0 \neq 0$, then $a_0 \in I \subseteq J$ which implies that $1 = a_0 + eb_0 \in J + eR = J$, a contradiction. Therefore $e \in L$ which implies that $eO \subseteq L$. Now we show that $L \subseteq eO$. Suppose that $g(x) = \sum_{i=0}^{n} a_i x^i \in L$; then, for all $0 \le i \le n$, $a_i \in I \subseteq J = eR$ and so $a_i = ea_i$. We have

$$eg(x) = e\sum_{i=0}^{n} a_i x^i = \sum_{i=0}^{n} (ea_i) x^i = \sum_{i=0}^{n} a_i x^i = g(x); \quad (13)$$

it follows that $g(x) \in eO$. Thus $L \subseteq eO$. □

If $M_R = R_R$, we get the following.

Corollary 8. *Let R be an (α, δ)-compatible and (α, δ)-Armendariz ring. If R is a right PS-ring, then $R[x; \alpha, \delta]$ is a right PS-ring.*

3. PS-Modules over Skew Generalized Power Series Rings

Let $(S, \le)$ be an ordered commutative monoid. Unless stated otherwise, the operation of S will be denoted additively, and the identity by 0. Recall that $(S, \le)$ is artinian if every strictly decreasing sequence of elements of S is finite and that $(S, \le)$ is narrow if every subset of pairwise order-incomparable elements of S is finite. The following construction is due to Zhongkui [11].

Let $(S, \le)$ be a strictly ordered monoid (i.e., if $s, s', t \in S$ and $s < s'$, then $s + t < s' + t$), R a ring, and $\omega : S \to \mathrm{End}(R)$ a monoid homomorphism. Consider the set $A = [[R^{S,\le}, \omega]]$ of all maps $f : S \to R$ whose support ($\mathrm{supp}(f) = \{s \in S \mid f(s) \neq 0\}$) is artinian and narrow.

For every $s \in S$ and $f, g \in A$, let

$$\begin{aligned} &X_s(f, g) \\ &= \{(u, v) \in S \times S \mid u + v = s; f(u) \neq 0, g(v) \neq 0\}. \end{aligned} \quad (14)$$

It follows from ([12], 4.1) that $X_s(f, g)$ is a finite set.

This fact allows defining the operation of multiplication (*convolution*) as follows:

$$(fg)(s) = \sum_{(u,v)\in X_s(f,g)} f(u)\, \omega_u(g(v)), \quad (15)$$

and $(fg)(s) = 0$ if $X_s(f, g) = \phi$. With this operation and pointwise addition $A = [[R^{S,\le}, \omega]]$ becomes a ring, which is called the *ring of skew generalized power series* with coefficients in R and exponents in S.

In [13], Zhao and Jiao generalized this construction to obtain the skew generalized power series modules over skew generalized power series rings as follows.

Let M_R be a right R-module; let B be the set of all maps $\varphi : S \to M$ such that $\mathrm{supp}(\varphi) = \{s \in S \mid \varphi(s) \neq 0\}$ is artinian and narrow. With pointwise addition, $B = [[M^{S,\le}]]$ is an abelian additive group. For each $f \in A = [[R^{S,\le}, \omega]]$ and $\varphi \in B$, the set

$$\begin{aligned} &X_s(\varphi, f) \\ &= \{(u, v) \in S \times S \mid u + v = s; \varphi(u) \neq 0, f(v) \neq 0\} \end{aligned} \quad (16)$$

is finite (see [14], Lemma 1). This allows defining the scalar multiplication of the elements of B by scalars from A as follows:

$$(\varphi f)(s) = \sum_{(u,v)\in X_s(\varphi,f)} \varphi(u)\,\omega_u(f(v)), \tag{17}$$

and $(\varphi f)(s) = 0$ if $X_s(\varphi, f) = \phi$. With this operation and pointwise addition, one can easily show that B is a right A-module, which is called the *module of skew generalized power series* with coefficients in M and exponents in S.

For every $s \in S$ if we set $\omega(s) = \mathrm{Id}_R \in \mathrm{Aut}(R) \subset \mathrm{End}(R)$, the identity map of R, then $A = [[R^{S,\le}, \omega]] = [[R^{S,\le}]]$ is the ring of generalized power series in the sense of Ribenboim [12] and $B = [[M^{S,\le}]]$ is the untwisted module of generalized power series in the sense of [15].

For any $r \in R$ we associated the map $c_r \in A$ defined by

$$c_r(x) = \begin{cases} r, & \text{if } x = 0, \\ 0, & \text{if } x \neq 0. \end{cases} \tag{18}$$

For any $m \in M$ and $s \in S$, we define a map $d^s_m \in B$ by

$$d^s_m(x) = \begin{cases} m, & \text{if } x = s, \\ 0, & \text{if } x \neq s. \end{cases} \tag{19}$$

Definition 9 (see [13]). A right R-module M_R is called ω-compatible whenever $ma = 0$ if and only if $m\omega_s(a) = 0$ for any $s \in S$, $m \in M$, and $a \in R$.

Clearly, R is an ω-compatible ring if and only if R_R is an ω-compatible R-module.

Theorem 10. *Let $(S, \le)$ be a strictly totally ordered monoid which satisfies the condition $0 \le s$ for every $s \in S$ and let M_R be an ω-compatible module. If M_R is a PS-module, then B_A is a PS-module.*

Proof. Let L be a maximal right ideal of A. We will show that either $\mathrm{l}_B(L) = 0$ or $L = hA$, where $h^2 = h \in A$. Since $(S, \le)$ is a strictly totally ordered monoid, $\mathrm{supp}(f)$ is a nonempty well-ordered subset of S, for every $0 \neq f \in A$. We denote by $\pi(f)$ the smallest element of support f.

For any $s \in S$, set

$$\begin{aligned} I_s &= \{f(s) \mid f \in L, \pi(f) = s\} \subset R, \\ I &= \bigcup_{s\in S} I_s. \end{aligned} \tag{20}$$

Let J be the right ideal of R generated by I. If $J = R$, then there exist $s_1, \dots, s_n \in S$, $f_1, \dots, f_n \in L$, and $r_1, \dots, r_n \in R$ such that

$$1 = f_1(s_1) r_1 + \cdots + f_n(s_n) r_n, \tag{21}$$

where $f_i(s_i) \in I_{s_i}$ and $\pi(f_i) = s_i$, for every $1 \le i \le n$. We will show that $\mathrm{l}_B(L) = 0$. Suppose that $\varphi \in \mathrm{l}_B(L)$ and $\varphi \neq 0$. Then $\mathrm{supp}(\varphi)$ is a nonempty well-ordered subset of S. Let $t = \pi(\varphi)$; if

$$\varphi(t)\,\omega_t(f_i(s_i)) \neq 0, \quad \text{for some } 1 \le i \le n, \tag{22}$$

then

$$\begin{aligned} (\varphi f_i)(t + s_i) &= \sum_{(u,v)\in X_{t+s_i}(\varphi,f_i)} \varphi(u)\,\omega_u(f_i(v)) \\ &= \varphi(t)\,\omega_t(f_i(s_i)) \neq 0. \end{aligned} \tag{23}$$

This means that $\varphi f_i \neq 0$ for some $1 \le i \le n$, a contradiction. Thus

$$\varphi(t)\,\omega_t(f_i(s_i)) = 0, \quad \forall 1 \le i \le n. \tag{24}$$

Since M_R is an ω-compatible module, we get

$$\varphi(t)\, f_i(s_i) = 0, \quad \forall 1 \le i \le n. \tag{25}$$

Consequently

$$\begin{aligned} \varphi(t) &= \varphi(t)\,1 = \varphi(t)\,(f_1(s_1) r_1 + \cdots + f_n(s_n) r_n) \\ &= (\varphi(t) f_1(s_1)) r_1 + \cdots + (\varphi(t) f_n(s_n)) r_n = 0, \end{aligned} \tag{26}$$

a contradiction. Thus $\mathrm{l}_B(L) = 0$. Suppose that $J \neq R$. We will show that J is a maximal right ideal of R. Let $r \in R - J$. If $c_r \in L$, then $r = c_r(0) \in I_0 \subset I$ and so $r \in J$, a contradiction. Therefore $c_r \notin L$. Since L is a maximal right ideal of A,

$$A = L + c_r A. \tag{27}$$

It follows that there exist $f \in L$ and $g \in A$ such that $c_1 = f + c_r g$. Thus

$$\begin{aligned} 1 &= c_1(0) = f(0) + (c_r g)(0) = f(0) + r\omega_0(g(0)) \\ &= f(0) + rg(0). \end{aligned} \tag{28}$$

If $f(0) = 0$, then $1 = rg(0) \in rR$. So, $R = J + rR$.

If $f(0) \neq 0$, then $0 \in \mathrm{supp}(f)$. Since $0 \le s$ for every $s \in S$, $\pi(f) = 0$. Thus $f(0) \in I_0 \subset I \subset J$, which implies that $R = J + rR$.

Hence J is a maximal right ideal of R. Since M_R is a PS-module, it follows that either $\mathrm{l}_M(J) = 0$ or $J = eR$, where $e^2 = e \in R$. According to that we have the following two cases.

Case 1. Suppose that $\mathrm{l}_M(J) = 0$. We will show that $\mathrm{l}_B(L) = 0$. Let $\varphi \in \mathrm{l}_B(L)$ and $\varphi \neq 0$. Then $\mathrm{supp}(\varphi)$ is a nonempty well-ordered subset of S. Let $s = \pi(\varphi)$. For any $r \in J$, there exist $s_1, \dots, s_n \in S$, $f_1, \dots, f_n \in L$, and $r_1, \dots, r_n \in R$ such that

$$r = f_1(s_1) r_1 + \cdots + f_n(s_n) r_n, \tag{29}$$

where $f_i(s_i) \in I_{s_i}$ and $\pi(f_i) = s_i$, for every $1 \le i \le n$. Since $\varphi \in \mathrm{l}_B(L)$, $f_1, \dots, f_n \in L$, we get $\varphi f_i = 0$ for every $1 \le i \le n$. If

$$\varphi(s)\,\omega_s(f_i(s_i)) \neq 0, \quad \text{for some } 1 \le i \le n, \tag{30}$$

then

$$(\varphi f_i)(s+s_i) = \sum_{(u,v)\in X_{s+s_i}(\varphi,f_i)} \varphi(u)\,\omega_u(f_i(v)) = \varphi(s)\,\omega_s(f_i(s_i)) \neq 0. \tag{31}$$

This means that $\varphi f_i \neq 0$ for some $1 \leq i \leq n$, a contradiction. Thus

$$\varphi(s)\,\omega_s(f_i(s_i)) = 0, \quad \forall 1 \leq i \leq n. \tag{32}$$

Since M_R is an ω-compatible module, we get

$$\varphi(s)\,f_i(s_i) = 0, \quad \forall 1 \leq i \leq n. \tag{33}$$

Consequently

$$\begin{aligned}\varphi(s)\,r &= \varphi(s)\,(f_1(s_1)\,r_1 + \cdots + f_n(s_n)\,r_n)\\ &= (\varphi(s)\,f_1(s_1))\,r_1 + \cdots + (\varphi(s)\,f_n(s_n))\,r_n\\ &= 0.\end{aligned} \tag{34}$$

Therefore, $\varphi(s) \in \mathrm{l}_M(J) = 0$ and $\pi(\varphi) = s$. Thus $\varphi = 0$, a contradiction. Hence $\mathrm{l}_B(L) = 0$.

Case 2. Suppose that $J = eR$, where $e^2 = e \in R$. We will show that $L = c_eA$, where $(c_e)^2 = c_e \in A$. To show that $c_eA \subseteq L$, we need to prove that $c_e \in L$. If $c_e \notin L$, then $A = L + c_eA$. Thus there exist $f \in L$ and $g \in A$ such that $c_1 = f + c_eg$. Thus

$$\begin{aligned}1 = c_1(0) &= f(0) + (c_eg)(0) = f(0) + e\omega_0(g(0))\\ &= f(0) + eg(0).\end{aligned} \tag{35}$$

If $f(0) = 0$, then $1 = eg(0) \in eR = J$, a contradiction.

If $f(0) \neq 0$, then $0 \in \mathrm{supp}(f)$. Since $0 \leq s$ for every $s \in S$, $\pi(f) = 0$. Thus $f(0) \in I_0 \subset I \subset J$, which implies that $f(0) \in J$ and $J = eR$. Hence $1 = f(0) + eg(0) \in J + eR = J$, a contradiction. Therefore $c_e \in L$ which implies that $c_eA \subseteq L$.

Conversely, suppose that $f \in L$ and $\pi(f) = s$; then $f(s) \in I_s \subset I \subset J = eR$ and so $f(s) = ef(s)$. We claim that $f(u) = ef(u)$ for any $u \in \mathrm{supp}(f)$.

Suppose that $f(v) = ef(v)$ for each $v < u$. Consider the following element $f_u \in A$ defined by

$$f_u(x) = \begin{cases} f(x), & x < u,\\ 0, & x \geq u.\end{cases} \tag{36}$$

Thus $\pi(f - f_u) = u$. By hypothesis it is easy to see that $f_u = c_ef_u \in c_eA \subset L$. Thus $f - f_u \in L$. By analogy with the proof above, it follows that

$$(f - f_u)(u) = e(f - f_u)(u), \tag{37}$$

which implies that $f(u) = ef(u)$. Thus our claim holds. Therefore

$$(c_ef)(t) = \sum_{(u,v)\in X_t(c_e,f)} c_e(u)\,\omega_u(f(v)) = c_e(0)\,\omega_0(f(t)) = ef(t) = f(t). \tag{38}$$

Hence $f = c_ef \in c_eA$. Thus $L = c_eA$ and the result follows since c_e is an idempotent of A. □

As a special case of the last result if we set $M_R = R_R$ we get the following.

Corollary 11. *Let $(S, \leq)$ be a strictly totally ordered monoid which satisfies the condition that $0 \leq s$ for every $s \in S$ and let R be an ω-compatible ring. If R is a right PS-ring, then $A = [[R^{S,\leq}, \omega]]$ is a right PS-ring.*

If we set $\omega(s) = \mathrm{Id}_R$, for every $s \in S$, we get the following as a corollary.

Corollary 12 (see [6], Theorem 1). *Let $(S, \leq)$ be a strictly totally ordered monoid which satisfies the condition that $0 \leq s$ for every $s \in S$. If M_R is a PS-module, then $[[M^{S,\leq}]]_{[[R^{S,\leq}]]}$ is a PS-module.*

If $M_R = R_R$, we get the following as a corollary.

Corollary 13 (see [5], Theorem 4). *Let $(S, \leq)$ be a strictly totally ordered monoid which satisfies the condition that $0 \leq s$ for every $s \in S$. If R is a right PS-ring, then $[[R^{S,\leq}]]$ is a right PS-ring.*

Conflict of Interests

The authors declare that there is no conflict of interests regarding the publication of this paper.

References

[1] W. K. Nicholson and J. F. Watters, "Rings with projective socle," *Proceedings of the American Mathematical Society*, vol. 102, no. 3, pp. 443–450, 1988.

[2] J. Zelmanowitz, "Regular modules," *Transactions of the American Mathematical Society*, vol. 163, pp. 341–355, 1972.

[3] X. Weimin, "Modules with projective socles," *Rivista di Matematica della Università di Parma*, vol. 1, no. 5, pp. 311–315, 1992.

[4] W. Dingguo, "Modules with flat socles and almost excellent extensions," *Acta Mathematica Vietnamica*, vol. 21, no. 2, pp. 295–301, 1996.

[5] Z. Liu and F. Li, "PS-rings of generalized power series," *Communications in Algebra*, vol. 26, no. 7, pp. 2283–2291, 1998.

[6] Z. Liu, "PS-modules over rings of generalized power series," *Northeastern Mathematical Journal*, vol. 18, no. 3, pp. 254–260, 2002.

[7] Ø. Ore, "Theory of non-commutative polynomials," *The Annals of Mathematics*, vol. 34, no. 3, pp. 480–508, 1933.

[8] T. Y. Lam, A. Leroy, and J. Matczuk, "Primeness, semiprimeness and prime radical of Ore extensions," *Communications in Algebra*, vol. 25, no. 8, pp. 2459–2506, 1997.

[9] S. Annin, "Associated primes over Ore extension rings," *Journal of Algebra and its Applications*, vol. 3, no. 2, pp. 193–205, 2004.

[10] E. Hashemi, "Extensions of BAEr and quasi-BAEr modules," *Iranian Mathematical Society—Bulletin*, vol. 37, no. 1, pp. 1–13, 2011.

[11] L. Zhongkui, "Triangular matrix representations of rings of generalized power series," *Acta Mathematica Sinica (English Series)*, vol. 22, no. 4, pp. 989–998, 2006.

[12] P. Ribenboim, "Semisimple rings and von Neumann regular rings of generalized power series," *Journal of Algebra*, vol. 198, no. 2, pp. 327–338, 1997.

[13] R. Zhao and Y. Jiao, "Principal quasi-Baerness of modules of generalized power series," *Taiwanese Journal of Mathematics*, vol. 15, no. 2, pp. 711–722, 2011.

[14] L. Zhongkui, "A note on Hopfian modules," *Communications in Algebra*, vol. 28, no. 6, pp. 3031–3040, 2000.

[15] K. Varadarajan, "Generalized power series modules," *Communications in Algebra*, vol. 29, no. 3, pp. 1281–1294, 2001.

4

Natural Partial Orders on Transformation Semigroups with Fixed Sets

Yanisa Chaiya, Preeyanuch Honyam, and Jintana Sanwong

Department of Mathematics, Chiang Mai University, Chiang Mai 50200, Thailand

Correspondence should be addressed to Jintana Sanwong; jintana.s@cmu.ac.th

Academic Editor: Pentti Haukkanen

Let X be a nonempty set. For a fixed subset Y of X, let $\text{Fix}(X,Y)$ be the set of all self-maps on X which fix all elements in Y. Then $\text{Fix}(X,Y)$ is a regular monoid under the composition of maps. In this paper, we characterize the natural partial order on $\text{Fix}(X,Y)$ and this result extends the result due to Kowol and Mitsch. Further, we find elements which are compatible and describe minimal and maximal elements.

1. Introduction

For any semigroup S, the natural partial order on $E(S)$, the set of all idempotents on S, is defined by

$$e \le f \quad \text{iff } e = ef = fe. \tag{1}$$

In 1980, Hartwig [1] and Nambooripad [2] proved that if S is a regular semigroup, then the relation

$$a \le b \quad \text{iff } a = eb = bf \text{ for some } e, f \in E(S) \tag{$*$}$$

is a partial order on S which extends the usual ordering of the set $E(S)$.

Later in 1986, the natural partial order on a regular semigroup was further extended to any semigroup S by Mitsch [3] as follows:

$$\begin{aligned} a \le b \quad & \text{iff } a = xb = by, \\ xa = a \quad & \text{for some } x, y \in S^1. \end{aligned} \tag{2}$$

Let X be a set and $B(X)$ denote the semigroup of binary relations on the set X under the composition of relations. A *partial transformation semigroup* is the collection of functions from a subset of X into X with composition which is denoted by $P(X)$. Let $T(X)$ be the set of all transformations from X into itself and it is called the *full transformation semigroup* on X. Then $P(X)$ and $T(X)$ are subsemigroups of $B(X)$. It is well known that $P(X)$ and $T(X)$ are regular semigroups.

In 1986, Kowol and Mitsch [4] characterized the natural partial order on $T(X)$ in terms of images and kernels. They also proved that an element $\alpha \in T(X)$ is maximal with respect to the natural order if and only if α is surjective or injective; α is minimal if and only if α is a constant map. Moreover, they described lower and upper bounds for two transformations and gave necessary and sufficient conditions for their existence.

Later in 2006, Namnak and Preechasilp [5] studied two natural partial orders on $B(X)$ and characterized when two elements of $B(X)$ are related under these orders. They also described the minimality, maximality, left compatibility, and right compatibility of elements with respect to each order.

Let Y be a subset of X. Recently, Fernandes and Sanwong [6] defined

$$PT(X,Y) = \{\alpha \in P(X) : X\alpha \subseteq Y\}, \tag{3}$$

where $X\alpha$ denotes the image of α. Moreover, they defined $I(X,Y)$ to be the set of all injective transformations in $PT(X,Y)$. Hence $PT(X,Y)$ and $I(X,Y)$ are subsemigroups of $P(X)$.

In [7], Sangkhanan and Sanwong described natural partial order $\leq$ on $PT(X,Y)$ and $I(X,Y)$ in terms of domains, images, and kernels. They also compared $\leq$ with the subset order and characterized the meet and join of these two orders. Furthermore, they found elements of $PT(X,Y)$ and $I(X,Y)$ which are compatible and determined the minimal and maximal elements.

Let Y be a fixed subset of X and

$$\operatorname{Fix}(X,Y) = \{\alpha \in T(X) : y\alpha = y \ \forall y \in Y\}. \quad (4)$$

In 2013, Honyam and Sanwong [8] proved that $\operatorname{Fix}(X,Y)$ is a regular semigroup and they also determined its Green's relations and ideals. Moreover, they proved that $\operatorname{Fix}(X,Y)$ is never isomorphic to $T(Z)$ for any set Z when $\emptyset \neq Y \subsetneq X$, and every semigroup S is isomorphic to a subsemigroup of $\operatorname{Fix}(X',Y')$ for some appropriate sets X' and Y' with $Y' \subseteq X'$. Note that this also follows trivially from the fact that $T(X)$ embeds in $\operatorname{Fix}(X \cup Z, Z)$ for any set Z with $X \cap Z = \emptyset$. Recently, the authors in [9] proved that there are only three types of maximal subsemigroups of $\operatorname{Fix}(X,Y)$ and these maximal subsemigroups coincide with the maximal regular subsemigroups when $X\setminus Y$ is a finite set with $|X\setminus Y| \geq 2$. They also gave necessary and sufficient conditions for $\operatorname{Fix}(X,Y)$ to be factorizable, unit-regular, and directly finite.

In this paper, we characterize the natural partial order on $\operatorname{Fix}(X,Y)$ and find elements which are compatible under this order in Section 3. In Section 4, we describe the minimal elements, the maximal elements, and the covering elements. Moreover, we find the number of upper covers of minimal elements and the number of lower covers of maximal elements.

2. Preliminaries and Notations

In [8], the authors proved that $\operatorname{Fix}(X,Y)$ is a regular subsemigroup of $T(X)$. Note that $\operatorname{Fix}(X,Y)$ contains 1_X, the identity map on X. If $Y = \emptyset$, then $\operatorname{Fix}(X,Y) = T(X)$; and if $|X| = 1$ or $X = Y$, then $\operatorname{Fix}(X,Y)$ consists of one element, 1_X. So, throughout this paper we will consider the case $Y \subsetneq X$ and $|X| > 1$.

For any $\alpha \in T(X)$, the symbol π_α denotes the partition of X induced by the map α, namely,

$$\pi_\alpha = \{x\alpha^{-1} : x \in X\alpha\}. \quad (5)$$

For $\alpha, \beta \in T(X)$, $\mathscr{A} \subseteq \pi_\alpha$, and $\mathscr{B} \subseteq \pi_\beta$, we say that $\mathscr{A}$ refines $\mathscr{B}$ if for each $A \in \mathscr{A}$ there exists $B \in \mathscr{B}$ such that $A \subseteq B$.

Throughout this paper, unless otherwise stated, let $Y = \{y_i : i \in I\}$.

For each $\alpha \in \operatorname{Fix}(X,Y)$, we have $y_i\alpha = y_i$ for all $i \in I$. So $Y = Y\alpha \subseteq X\alpha$. If $\alpha \in \operatorname{Fix}(X,Y)$, then we write

$$\alpha = \begin{pmatrix} A_i & B_j \\ y_i & b_j \end{pmatrix} \quad (6)$$

and take as understood that the subscripts i and j belong to the index sets I and J, respectively, such that $X\alpha = \{y_i : i \in I\} \cup \{b_j : j \in J\}$, $y_i\alpha^{-1} = A_i$, and $b_j\alpha^{-1} = B_j$. Thus $A_i \cap Y = \{y_i\}$ for all $i \in I$, $B_j \subseteq X\setminus Y$ for all $j \in J$ and $\{b_j : j \in J\} \subseteq X\setminus Y$. Here J can be an empty set.

An idempotent e in a semigroup S is said to be minimal if e has the property $f \in E(S)$ and $f \leq e$ implies $f = e$.

In [8] the authors showed that

$$E_m = \left\{\begin{pmatrix} A_i \\ y_i \end{pmatrix} : \{A_i : i \in I\} \text{ is a partition of } X \text{ with } y_i \in A_i\right\} \quad (7)$$

is the set of all minimal idempotents in $\operatorname{Fix}(X,Y)$ and it is an ideal of $\operatorname{Fix}(X,Y)$. We note that E_m is simply the set $\{\alpha \in \operatorname{Fix}(X,Y) : X\alpha = Y\}$ and α is an idempotent in $\operatorname{Fix}(X,Y)$ if and only if $x\alpha = x$ for all $x \in X\alpha\setminus Y$.

3. Natural Partial Order on Fix(X,Y)

Kowol and Mitsch [4] gave a characterization of the natural partial order on $T(X)$. Later in 1994, Higgins [10] showed that if T is a regular subsemigroup of a semigroup S, then the natural partial order on T is the restriction to T of the natural partial order on S. Here we describe the natural partial order on $\operatorname{Fix}(X,Y)$ which is a regular subsemigroup of $T(X)$ without making use of Higgins' result and when we take $Y = \emptyset$, we recapture the result above by Kowol and Mitsch.

We note that if $\alpha, \beta \in \operatorname{Fix}(X,Y)$ and $\alpha = \beta\gamma$ for some $\gamma \in \operatorname{Fix}(X,Y)$, then π_β refines π_α.

Since $\operatorname{Fix}(X,Y)$ is regular, we use $(*)$ to study the natural partial order on this semigroup.

Theorem 1. *Let $\alpha, \beta \in \operatorname{Fix}(X,Y)$. Then $\alpha \leq \beta$ if and only if the following statements hold:*

(1) *$X\alpha \subseteq X\beta$;*

(2) *π_β refines π_α;*

(3) *if $x\beta \in X\alpha$, then $x\beta = x\alpha$.*

Proof. Suppose that $\alpha \leq \beta$. Then, by $(*)$, we have

$$\alpha = \lambda\beta = \beta\gamma \quad (8)$$

for some $\lambda, \gamma \in E(\operatorname{Fix}(X,Y))$. Thus $X\alpha = (X\lambda)\beta \subseteq X\beta$. Since $\alpha = \beta\gamma$, we get that π_β refines π_α. Now, let $x\beta \in X\alpha$. Then $x\beta = x'\alpha$ for some $x' \in X$ and thus $x\beta = x'\alpha = x'\beta\gamma = (x'\beta)\gamma$. Hence $x\beta \in X\gamma$ and then $x\alpha = x\beta\gamma = x\beta$ since γ is an idempotent.

Conversely, assume that conditions (1)–(3) hold. By condition (1), we can write

$$\alpha = \begin{pmatrix} A_i & B_j \\ y_i & b_j \end{pmatrix}, \quad \beta = \begin{pmatrix} A'_i & C_j & C_k \\ y_i & b_j & b_k \end{pmatrix}, \quad (9)$$

where $y_i \in A_i \cap A'_i$, $b_j, b_k \in X\setminus Y$, and $B_j, C_j, C_k \subseteq X\setminus Y$. Since $y_i \in A_i \cap A'_i$ and π_β refines π_α, we obtain $A'_i \subseteq A_i$ for all $i \in I$.

If $J = \emptyset$, then define $\lambda = \alpha$ and thus $\alpha = \lambda\beta$. If $J \neq \emptyset$, then, for each $j \in J$, let $c_j \in C_j$. So $c_j\beta = b_j \in X\alpha$. By condition (3), $c_j\alpha = c_j\beta = b_j$; that is, $c_j \in B_j$ and hence $C_j \subseteq B_j$. Define

$$\lambda = \begin{pmatrix} A_i & B_j \\ y_i & c_j \end{pmatrix}. \quad (10)$$

We get $\lambda \in E(\mathrm{Fix}(X,Y))$ and $\alpha = \lambda\beta$.

If $K = \emptyset$, then $\alpha = \beta 1_X$. If $K \neq \emptyset$, then, for each $k \in K$, we choose $c_k \in C_k$.

Case 1. Consider $X\beta = X$. Then $X \setminus X\alpha = \{b_k : k \in K\}$. We define $\gamma \in \mathrm{Fix}(X,Y)$ by

$$x\gamma = \begin{cases} x, & x \in X\alpha, \\ c_k\alpha, & x = b_k \in X \setminus X\alpha. \end{cases} \quad (11)$$

To prove that $\alpha = \beta\gamma$, let $x \in X$. If $x \in A_i'$ for some i or $x \in C_j$ for some j, then it is clear that $x\alpha = x\beta\gamma$. Now, if $x \in C_k$ for some k, then $x\beta = c_k\beta$ and thus $x\alpha = c_k\alpha$ since π_β refines π_α. So $(x\beta)\gamma = b_k\gamma = c_k\alpha = x\alpha$. Hence $\alpha = \beta\gamma$. It remains to show that γ is an idempotent. Let $x\gamma \in X\gamma \setminus X\alpha$. Then $x\gamma = c_k\alpha$ for some k. Thus $(x\gamma)\gamma = (c_k\alpha)\gamma = c_k\alpha = x\gamma$ since $c_k\alpha \in X\alpha$.

Case 2. Consider $X\beta \subsetneq X$. We choose $c_0 \in X \setminus X\beta$ and define $\gamma' \in \mathrm{Fix}(X,Y)$ by

$$x\gamma' = \begin{cases} x, & x \in X\alpha, \\ c_k\alpha, & x = b_k \in X\beta \setminus X\alpha, \\ c_0, & x \in X \setminus X\beta. \end{cases} \quad (12)$$

By the same prove as given in Case 1, we get $\alpha = \beta\gamma'$ and $(x\gamma')\gamma' = x\gamma'$ for all $x\gamma' \in X\beta$. If $x\gamma' = c_0$, then $(x\gamma')\gamma' = c_0\gamma' = c_0 = x\gamma'$. So γ' is an idempotent. Therefore, $\alpha \leq \beta$ by $(*)$. □

Remark 2. If $Y = \emptyset$, then $\mathrm{Fix}(X,Y) = T(X)$, and we have the characterization of $\leq$ on $T(X)$ which first appeared in [4, Proposition 2.3].

As a direct consequence of Theorem 1, we get the following corollary.

Corollary 3. *Let $\alpha, \beta \in \mathrm{Fix}(X,Y)$ with $\alpha \leq \beta$. If $X\alpha \setminus Y = X\beta \setminus Y$, then $\alpha = \beta$.*

Let S be a semigroup. An element $a \in S$ is said to be *left (right) compatible* with respect to the partial order $\leq$ if $ab \leq ac$ $(ba \leq ca)$ whenever $b \leq c$.

The following results describe all the left compatible and right compatible elements in $\mathrm{Fix}(X,Y)$ when $\emptyset \neq Y \subsetneq X$. We also write $\alpha < \beta$ instead of $\alpha \leq \beta$ and $\alpha \neq \beta$ for $\alpha, \beta \in \mathrm{Fix}(X,Y)$.

Theorem 4. *Assume that $\emptyset \neq Y \subsetneq X$ and let $\lambda \in \mathrm{Fix}(X,Y)$. Then λ is left compatible if and only if λ is a minimal idempotent or λ is surjective.*

Proof. Suppose that λ is left compatible. Assume by contrary that λ is not a minimal idempotent and λ is not surjective. So there are $a \in X\lambda \setminus Y$ and $b \in X \setminus X\lambda$. Define

$$\alpha = \begin{pmatrix} y_i & X \setminus Y \\ y_i & a \end{pmatrix}, \quad \beta = \begin{pmatrix} y_i & b & X \setminus (Y \cup \{b\}) \\ y_i & a & b \end{pmatrix}. \quad (13)$$

Then $\alpha, \beta \in \mathrm{Fix}(X,Y)$ with $\alpha < \beta$ and thus $\lambda\alpha \leq \lambda\beta$ since λ is left compatible. However, $X\lambda\alpha \nsubseteq X\lambda\beta$ since $a \in X\lambda\alpha$ but $a \notin X\lambda\beta$, a contradiction.

Conversely, let $\alpha \leq \beta$. If λ is a minimal idempotent, then $\lambda\alpha = \lambda = \lambda\beta$. Now, assume that λ is surjective. So $X\lambda\alpha = X\alpha \subseteq X\beta = X\lambda\beta$. Let $A \in \pi_{\lambda\beta}$. So $A = x(\lambda\beta)^{-1} = (x\beta^{-1})\lambda^{-1}$ for some $x \in X\lambda\beta$. Since $\alpha \leq \beta$, we have that π_β refines π_α and hence $x\beta^{-1} \subseteq x'\alpha^{-1}$ for some $x' \in X\alpha$. Since $x' \in X\alpha$, we get $x' = u\alpha$ for some $u \in X$ and $u = v\lambda$ for some $v \in X$ because λ is surjective. Hence $v\lambda\alpha = u\alpha = x'$; that is, $x' \in X\lambda\alpha$. Further, $A = (x\beta^{-1})\lambda^{-1} \subseteq (x'\alpha^{-1})\lambda^{-1} = x'(\lambda\alpha)^{-1} \in \pi_{\lambda\alpha}$, thus $\pi_{\lambda\beta}$ refines $\pi_{\lambda\alpha}$. Let $a\lambda\beta \in X\lambda\alpha$. So $(a\lambda)\beta \in X\alpha$ and then $a\lambda\beta = a\lambda\alpha$. By Theorem 1, we have $\lambda\alpha \leq \lambda\beta$ which implies that λ is left compatible. □

Theorem 5. *The following statements hold.*

(1) *If $|Y| = 1$, then $\lambda \in \mathrm{Fix}(X,Y)$ is right compatible if and only if λ is a minimal idempotent or λ is injective.*

(2) *If $|Y| \geq 2$, then $\lambda \in \mathrm{Fix}(X,Y)$ is right compatible if and only if λ is injective.*

Proof. (1) Assume that $Y = \{y\}$ and λ is right compatible. Suppose in the contrary that λ is not a minimal idempotent and λ is not injective. So we can write

$$\lambda = \begin{pmatrix} A & B_j \\ y & b_j \end{pmatrix}, \quad (14)$$

where $y \in A$ and $J \neq \emptyset$. Since λ is not injective, two cases arise.

Case 1. Consider $|A| \geq 2$. Choose $a \in A \setminus \{y\}$ and $c \in B_{j_0}$ for some $j_0 \in J$. Let $X \setminus \{a,c\} = \{x_k : k \in K\}$ and define $\alpha \in \mathrm{Fix}(X,Y)$ by

$$\alpha = \begin{pmatrix} \{a,c\} & x_k \\ c & x_k \end{pmatrix}; \quad (15)$$

we get $\alpha < 1_X$. Moreover, we have $a(1_X\lambda) = a\lambda = y = y(1_X\lambda)$, hence there is $B \in \pi_{1_X\lambda}$ such that $\{a,y\} \subseteq B$. However, $\{a,y\} \nsubseteq C$ for all $C \in \pi_{\alpha\lambda}$ since $a(\alpha\lambda) = c\lambda = b_{j_0} \neq y = y(\alpha\lambda)$. This means that $\pi_{1_X\lambda}$ does not refine $\pi_{\alpha\lambda}$. By Theorem 1, we get $\alpha\lambda \nleq 1_X\lambda$, a contradiction.

Case 2. Consider $|B_{j_0}| \geq 2$ for some $j_0 \in J$. Choose $a, b \in B_{j_0}$ such that $a \neq b$. Let $X \setminus \{a, b, y\} = \{x_k : k \in K\}$. Define $\alpha, \beta \in \mathrm{Fix}(X, Y)$ by

$$\alpha = \begin{pmatrix} \{y, a\} & b & x_k \\ y & a & x_k \end{pmatrix}, \qquad \beta = \begin{pmatrix} a & b & y & x_k \\ b & a & y & x_k \end{pmatrix}, \tag{16}$$

we get $\alpha < \beta$. Since $a(\beta\lambda) = b\lambda = a\lambda = b(\beta\lambda)$, there is $B \in \pi_{\beta\lambda}$ such that $\{a, b\} \subseteq B$. However, $\{a, b\} \nsubseteq C$ for all $C \in \pi_{\alpha\lambda}$ since $a(\alpha\lambda) = y\lambda = y \neq b_{j_0} = a\lambda = b(\alpha\lambda)$. So $\pi_{\beta\lambda}$ does not refine $\pi_{\alpha\lambda}$. By Theorem 1, we get $\alpha\lambda \nleq \beta\lambda$, a contradiction.

Conversely, let $\alpha, \beta \in \mathrm{Fix}(X, Y)$ be such that $\alpha \leq \beta$. If λ is a minimal idempotent, then $\lambda = \binom{X}{y}$ and $\alpha\lambda = \lambda = \beta\lambda$; that is, λ is right compatible. Now, assume that λ is injective. Since $X\alpha \subseteq X\beta$, we get $X\alpha\lambda \subseteq X\beta\lambda$. Let $A \in \pi_{\beta\lambda}$. So $A = x(\beta\lambda)^{-1} = (x\lambda^{-1})\beta^{-1}$ for some $x \in X\beta\lambda$ and hence $(x\lambda^{-1})\beta^{-1} \subseteq x'\alpha^{-1}$ for some $x' \in X\alpha$. So $x' = u\alpha$ for some $u \in X$. Since λ is injective, $\{x'\} = v\lambda^{-1}$ for some $v \in X\lambda$ and $u\alpha\lambda = x'\lambda = v$; that is, $v \in X\alpha\lambda$. Thus $x'\alpha^{-1} = (v\lambda^{-1})\alpha^{-1} = v(\alpha\lambda)^{-1} \in \pi_{\alpha\lambda}$ which implies that $\pi_{\beta\lambda}$ refines $\pi_{\alpha\lambda}$. Let $a\beta\lambda \in X\alpha\lambda$. So $a\beta\lambda = b\alpha\lambda$ for some $b \in X$. Since λ is injective, $a\beta = b\alpha$ and then $a\beta \in X\alpha$. Thus $a\beta = a\alpha$ since $\alpha \leq \beta$ and that $a\beta\lambda = a\alpha\lambda$. Therefore, $\alpha\lambda \leq \beta\lambda$, and we conclude that λ is right compatible.

(2) Suppose that λ is right compatible and λ is not injective. Write

$$\lambda = \begin{pmatrix} A_i & B_j \\ y_i & b_j \end{pmatrix}, \tag{17}$$

where $y_i \in A_i$ and $|I| \geq 2$. Since λ is not injective, two cases arise.

Case 1. $|A_{i_0}| \geq 2$ for some $i_0 \in I$. Choose $a \in A_{i_0} \setminus \{y_{i_0}\}$ and $y_{i_1} \in Y \setminus \{y_{i_0}\}$. Let $X \setminus \{y_{i_1}, a\} = \{x_k : k \in K\}$ and define

$$\alpha = \begin{pmatrix} \{y_{i_1}, a\} & x_k \\ y_{i_1} & x_k \end{pmatrix}. \tag{18}$$

Then $\alpha < 1_X$ and hence $\alpha\lambda \leq 1_X\lambda$. We can see that $\{y_{i_0}, a\} \subseteq A_{i_0} \in \pi_\lambda = \pi_{1_X\lambda}$, but $\{y_{i_0}, a\} \nsubseteq B$ for all $B \in \pi_{\alpha\lambda}$ since $a\alpha\lambda = y_{i_1}\lambda = y_{i_1} \neq y_{i_0} = y_{i_0}\alpha\lambda$. This means that $\pi_{1_X\lambda}$ does not refine $\pi_{\alpha\lambda}$, a contradiction.

Case 2. $|B_{j_0}| \geq 2$ for some $j_0 \in J$. This is virtually identical to Case 2 of (1) above. □

4. Minimal and Maximal Elements

Let S be a semigroup together with the partial order $\leq$. S is said to be *directed downward* if every pair of elements has a lower bound. In other words, for any a and b in S, there exists c in S with $c \leq a$ and $c \leq b$. A directed upward semigroup is defined dually.

If $Y = \emptyset$, then $\mathrm{Fix}(X, Y) = T(X)$ and it has neither minimum nor maximum elements under the natural order (see [4]). So, in Lemmas 6 and 7 we assume that $\emptyset \neq Y \subsetneq X$.

Lemma 6. *Assume that $\emptyset \neq Y \subsetneq X$. Then the following statements are equivalent.*

(1) $\mathrm{Fix}(X, Y)$ *has a minimum element.*

(2) $\mathrm{Fix}(X, Y)$ *is directed downward.*

(3) $|Y| = 1$.

Proof. (1)⇒(2) This is clear.

(2)⇒(3) Assume that $\mathrm{Fix}(X, Y)$ is directed downward. Let $y_{i_1}, y_{i_2} \in Y$ and $J = I \setminus \{i_1, i_2\}$. Consider

$$\alpha = \begin{pmatrix} y_j & y_{i_2} & (X \setminus Y) \cup \{y_{i_1}\} \\ y_j & y_{i_2} & y_{i_1} \end{pmatrix}, \qquad \beta = \begin{pmatrix} y_j & y_{i_1} & (X \setminus Y) \cup \{y_{i_2}\} \\ y_j & y_{i_1} & y_{i_2} \end{pmatrix}. \tag{19}$$

We have $\alpha, \beta \in \mathrm{Fix}(X, Y)$ and there is $\gamma \in \mathrm{Fix}(X, Y)$ such that $\gamma \leq \alpha$ and $\gamma \leq \beta$. By Theorem 1, π_α refines π_γ and π_β refines π_γ. Then there is $A \in \pi_\gamma$ such that $(X \setminus Y) \cup \{y_{i_1}\} \subseteq A$ and $(X \setminus Y) \cup \{y_{i_2}\} \subseteq A$. Thus $y_{i_1}, y_{i_2} \in A$ and hence $y_{i_1} = y_{i_2}$. Since y_{i_1}, y_{i_2} are arbitrary elements in Y, we obtain that $|Y| = 1$.

(3)⇒(1) Assume that $Y = \{y\}$. It is easy to see that $\theta = \binom{X}{y}$ is the minimum element in $\mathrm{Fix}(X, Y)$. □

Lemma 7. *Assume that $\emptyset \neq Y \subsetneq X$. Then the following statements are equivalent.*

(1) $\mathrm{Fix}(X, Y)$ *has a maximum element.*

(2) $\mathrm{Fix}(X, Y)$ *is directed upward.*

(3) $|X \setminus Y| = 1$.

Proof. (1)⇒(2) This is clear.

(2)⇒(3) Assume that $\mathrm{Fix}(X, Y)$ is directed upward. Let $a, b \in X \setminus Y$ and $X \setminus \{a, b\} = \{x_k : k \in K\}$. Define

$$\alpha = \begin{pmatrix} a & b & x_k \\ b & a & x_k \end{pmatrix} \in \mathrm{Fix}(X, Y). \tag{20}$$

Then there is $\gamma \in \mathrm{Fix}(X, Y)$ such that $\alpha \leq \gamma$ and $1_X \leq \gamma$. Since α and 1_X are bijective, γ is also bijective and thus $b\gamma \in (X\alpha \setminus Y) \cap (X1_X \setminus Y)$. So $a = b\alpha = b\gamma = b1_X = b$. Since a, b are arbitrary elements in $X \setminus Y$, we get $|X \setminus Y| = 1$.

(3)⇒(1) Assume that $|X \setminus Y| = 1$. It is easy to see that 1_X is the maximum element in $\mathrm{Fix}(X, Y)$. □

We now describe minimal and maximal elements in $\mathrm{Fix}(X, Y)$ when $\emptyset \neq Y \subsetneq X$. If $|Y| = 1$, then $\mathrm{Fix}(X, Y)$ has a minimum element by Lemma 6 and it is minimal. In the same way, if $|X \setminus Y| = 1$, then $\mathrm{Fix}(X, Y)$ has a maximum element by Lemma 7 and it is maximal.

Theorem 8. *Assume that $\emptyset \neq Y \subsetneq X$ and let $\alpha \in \mathrm{Fix}(X, Y)$. Then α is minimal if and only if α is a minimal idempotent.*

Proof. Assume that α is minimal but α is not a minimal idempotent. So we can write

$$\alpha = \begin{pmatrix} A_i & B_j \\ y_i & b_j \end{pmatrix}, \quad (21)$$

where $J \neq \emptyset$. Choose $i_0 \in I$ and $j_0 \in J$. Let $I' = I \setminus \{i_0\}$, $J' = J \setminus \{j_0\}$ and define $\beta \in \mathrm{Fix}(X, Y)$ by

$$\beta = \begin{pmatrix} A_{i_0} \cup B_{j_0} & A_{i'} & B_{j'} \\ y_{i_0} & y_{i'} & b_{j'} \end{pmatrix}. \quad (22)$$

Hence $\beta < \alpha$, which contradicts the minimality of α.

Conversely, assume that α is a minimal idempotent and $\beta \leq \alpha$. Since $Y \subseteq X\beta \subseteq X\alpha = Y$, we get $X\beta = X\alpha$ and hence $X\beta \setminus Y \subseteq X\alpha \setminus Y$. By Corollary 3, we obtain $\beta = \alpha$. □

Theorem 9. *Assume that $\emptyset \neq Y \subsetneq X$ and let $\alpha \in \mathrm{Fix}(X, Y)$. Then α is maximal if and only if α is injective or α is surjective.*

Proof. Let α be maximal. Assume that α is not injective and surjective. So there are $a, b, c \in X$ such that $a\alpha = b\alpha$ with $a \neq b$ and $c \in X \setminus X\alpha$. Write

$$\alpha = \begin{pmatrix} A_i & B_j \\ y_i & b_j \end{pmatrix}. \quad (23)$$

Case 1. $a, b \in A_{i_0}$ for some $i_0 \in I$. We may assume that $a \neq y_{i_0}$. Let $I' = I \setminus \{i_0\}$ and define

$$\beta = \begin{pmatrix} A_{i_0} \setminus \{a\} & a & A_{i'} & B_j \\ y_{i_0} & c & y_{i'} & b_j \end{pmatrix}. \quad (24)$$

Then $\beta \in \mathrm{Fix}(X, Y)$ and $\alpha < \beta$ which contradicts the maximality of α.

Case 2. $a, b \in B_{j_0}$ for some $j_0 \in J$. Then we let $J' = J \setminus \{j_0\}$ and define

$$\gamma = \begin{pmatrix} A_i & B_{j_0} \setminus \{a\} & a & B_{j'} \\ y_i & b_{j_0} & c & b_{j'} \end{pmatrix}. \quad (25)$$

Then $\gamma \in \mathrm{Fix}(X, Y)$ and $\alpha < \gamma$ which contradicts the maximality of α.

Conversely, assume that α is injective or α is surjective and $\alpha \leq \beta$ for some $\beta \in \mathrm{Fix}(X, Y)$. Then $X\alpha \subseteq X\beta$ and $X\alpha \setminus Y \subseteq X\beta \setminus Y$. Consider the case where α is injective, by letting $z \in X\beta \setminus Y$. Then $z = x\beta$ for some $x \in X \setminus Y$ and $x\alpha \in X\alpha \setminus Y \subseteq X\beta \setminus Y$; that is, $x\alpha = x'\beta$ for some $x' \in X \setminus Y$. So $x'\beta \in X\alpha$ and $x'\beta = x'\alpha$ by Theorem 1. Since α is injective, we get $x = x'$ and thus $z = x\beta = x'\beta = x\alpha \in X\alpha \setminus Y$, whence $X\beta \setminus Y \subseteq X\alpha \setminus Y$. Hence, in this case, $X\alpha \setminus Y = X\beta \setminus Y$ and by Corollary 3 we obtain $\alpha = \beta$. In the case α is surjective, we get $X \setminus Y = X\alpha \setminus Y \subseteq X\beta \setminus Y \subseteq X \setminus Y$; that is, $X\alpha \setminus Y = X\beta \setminus Y$. Again by Corollary 3, we have that $\alpha = \beta$. Therefore, α is maximal. □

Figure 1 shows the diagram of $\mathrm{Fix}(X, Y)$ when $X = \{1, 2, 3, 4\}$ and $Y = \{1, 2\}$. The notation $(abcd)$ for $\alpha \in \mathrm{Fix}(X, Y)$ means that $1\alpha = a$, $2\alpha = b$, $3\alpha = c$, and $4\alpha = d$.

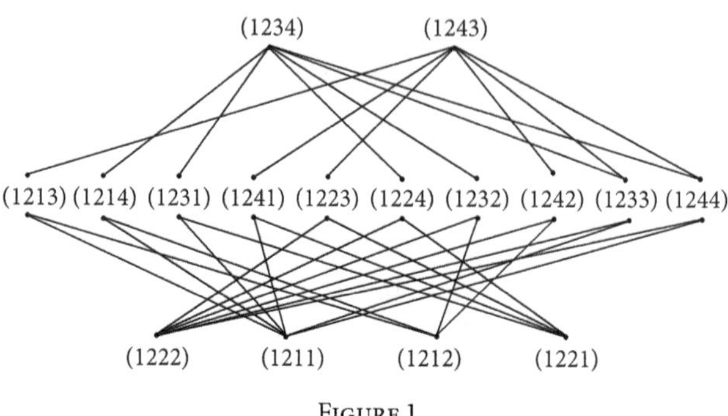

FIGURE 1

An element $\beta \in \mathrm{Fix}(X, Y)$ is called an *upper cover* for $\alpha \in \mathrm{Fix}(X, Y)$ if $\alpha < \beta$ and there is no $\gamma \in \mathrm{Fix}(X, Y)$ such that $\alpha < \gamma < \beta$; lower covers are defined dually.

Lemma 10. *Assume that $\emptyset \neq Y \subsetneq X$ and let $\alpha \in \mathrm{Fix}(X, Y)$. Then the following statements hold.*

(1) *If α is not minimal in $\mathrm{Fix}(X, Y)$, then there is some lower cover of α in $\mathrm{Fix}(X, Y)$.*

(2) *If α is not maximal in $\mathrm{Fix}(X, Y)$, then there is some upper cover of α in $\mathrm{Fix}(X, Y)$.*

Proof. (1) Let $\alpha \in \mathrm{Fix}(X, Y)$ be not minimal. By Theorem 8, α is not a minimal idempotent. So we can write

$$\alpha = \begin{pmatrix} A_i & B_j \\ y_i & b_j \end{pmatrix}, \quad (26)$$

where $J \neq \emptyset$. Define β as in the proof of Theorem 8, we get $\beta < \alpha$. Suppose that there is $\lambda \in \mathrm{Fix}(X, Y)$ such that $\beta \leq \lambda \leq \alpha$. Then by Theorem 1, $X\beta \subseteq X\lambda \subseteq X\alpha$ and thus $X\beta \setminus Y \subseteq X\lambda \setminus Y \subseteq X\alpha \setminus Y$. Since $X\alpha \setminus Y = (X\beta \setminus Y) \cup \{b_{j_0}\}$ which implies $X\lambda \setminus Y = X\beta \setminus Y$ or $X\lambda \setminus Y = X\alpha \setminus Y$, thus $\lambda = \beta$ or $\lambda = \alpha$ by Corollary 3. Therefore, β is a lower cover of α.

(2) The proof is similar to (1), using β or γ from the proof of Theorem 9 as appropriate. □

Now, we aim to find the number of upper covers of minimal elements and the number of lower covers of maximal elements when X is a finite set. The following lemma is needed in finding such numbers.

Lemma 11. *Assume that $\emptyset \neq Y \subsetneq X$ and let $\alpha, \beta \in \mathrm{Fix}(X, Y)$ with $\alpha < \beta$. Then β is an upper cover of α if and only if $|X\beta \setminus X\alpha| = 1$.*

Proof. Write

$$\alpha = \begin{pmatrix} A_i & B_j \\ y_i & b_j \end{pmatrix}. \quad (27)$$

Since $\alpha < \beta$, we can write

$$\beta = \begin{pmatrix} A'_i & C_j & C_k \\ y_i & b_j & b_k \end{pmatrix}, \quad (28)$$

where $y_i \in A_i' \subseteq A_i$, $C_j \subseteq B_j$, and C_k is contained in either A_i for some i or B_j for some j. We get $|K| = |X\beta \setminus X\alpha|$.

Assume that β is an upper cover of α. If $|X\beta \setminus X\alpha| = 0$, then $X\beta = X\alpha$ which implies that $\beta = \alpha$, a contradiction. For the case $|X\beta \setminus X\alpha| > 1$, we choose $k_0 \in K$ and hence $C_{k_0} \subseteq A_{i_0}$ for some $i_0 \in I$ or $C_{k_0} \subseteq B_{j_0}$ for some $j_0 \in J$. Assume that $C_{k_0} \subseteq A_{i_0}$ (the other case being similar). Let $I' = I \setminus \{i_0\}$ and $K' = K \setminus \{k_0\}$. Define

$$\gamma = \begin{pmatrix} A_{i_0}' \cup C_{k_0} & A_{i'}' & C_j & C_{k'} \\ y_{i_0} & y_{i'} & b_j & b_{k'} \end{pmatrix} \in \operatorname{Fix}(X, Y). \quad (29)$$

Since $K' \neq \emptyset$, we get $\alpha < \gamma < \beta$, a contradiction. Therefore, $|X\beta \setminus X\alpha| = 1$.

The converse is proved in similar fashion to Lemma 10 (1). □

Let X be a finite set with n elements and Y a nonempty proper subset of X with r elements. If $|Y| = 1$, then $\operatorname{Fix}(X, Y)$ has unique minimal element, say $\alpha = \begin{pmatrix} X \\ y \end{pmatrix}$. By Lemma 11, each of upper covers of α is of the form $\begin{pmatrix} X\setminus B & B \\ y & b \end{pmatrix}$, where $\emptyset \neq B \subseteq X \setminus \{y\}$ and $b \in X \setminus Y$. Since there are $(2^{n-1} - 1)$ ways to choose B and $n - 1$ choice of b, in this case there are in total $(2^{n-1} - 1)(n - 1)$ upper covers of α.

If $|X \setminus Y| = 1$, then $\operatorname{Fix}(X, Y)$ has unique maximal element, the identity map. Let $I = \{1, 2, \ldots, n-1\}$, $Y = \{y_i : i \in I\}$, and $X \setminus Y = \{b\}$. Then each of lower covers of 1_X is of the form $\begin{pmatrix} \{y_{i_0}, b\} & y_{i'} \\ y_{i_0} & y_{i'} \end{pmatrix}$, where $I' = I \setminus \{i_0\}$. Since i_0 can be chosen from I, there are in total $n - 1$ lower covers of 1_X.

Theorem 12. *Assume that $\emptyset \neq Y \subsetneq X$ and let $\alpha \in \operatorname{Fix}(X, Y)$. Then the following statements hold.*

(1) *If $\alpha = \begin{pmatrix} A_i \\ y_i \end{pmatrix}$ is minimal, then there are*

$$\sum_{i=1}^{r} \left(2^{|A_i|-1} - 1\right)(n - r) \quad (30)$$

upper covers of α.

(2) *If α is maximal, then there are $(n-r)(n-1)$ lower covers of α.*

Proof. Since Y is a finite set with r elements, $Y = \{y_1, \ldots, y_r\}$ and $I = \{1, \ldots, r\}$.

(1) Let $\alpha = \begin{pmatrix} A_i \\ y_i \end{pmatrix}$ be minimal in $\operatorname{Fix}(X, Y)$ and β an upper cover of α. Then $|X\beta \setminus X\alpha| = 1$ by Lemma 11; that is, $X\beta = Y \cup \{b\}$ for some $b \in X \setminus Y$. Since π_β must refine π_α, we can write

$$\beta = \begin{pmatrix} A_i' & B \\ y_i & b \end{pmatrix}, \quad (31)$$

where $A_i' \subseteq A_i$ and $\emptyset \neq B \subseteq A_{i_0} \setminus \{y_{i_0}\}$ for some $i_0 \in I$. We claim that $A_i' = A_i$ for all $i \in I \setminus \{i_0\}$. Assume by contrary that there is $i_1 \in I \setminus \{i_0\}$ such that $A_{i_1}' \subsetneq A_{i_1}$. Let $B_1 = A_{i_1} \setminus A_{i_1}'$. So $\emptyset \neq B_1 \cap A_i' \subseteq A_i$ for some $i \neq i_1$, but $B_1 \subseteq A_{i_1}$; that is $A_i \cap A_{i_1} \neq \emptyset$, a contradiction. So we can write

$$\beta = \begin{pmatrix} A_{i'} & A_{i_0} \setminus B & B \\ y_{i'} & y_{i_0} & b \end{pmatrix}, \quad (32)$$

where $I' = I \setminus \{i_0\}$. Since there are $2^{|A_{i_0}|-1} - 1$ ways to choose B and $n-r$ choices of b, in this case β can have $(2^{|A_{i_0}|-1} - 1)(n-r)$ forms, but i_0 can be chosen from $I = \{1, \ldots, r\}$, so that there are in total $\sum_{i=1}^{r} (2^{|A_i|-1} - 1)(n - r)$ upper covers of α.

(2) Assume that α is maximal. Then α is a bijection and we can write

$$\alpha = \begin{pmatrix} y_i & b_j \\ y_i & c_j \end{pmatrix}, \quad (33)$$

where $J = \{1, \ldots, n - r\}$ and $\{b_j : j \in J\} = X \setminus Y = \{c_j : j \in J\}$. Let β be a lower cover of α. Then $|X\alpha \setminus X\beta| = 1$; that is, $X\beta = X\alpha \setminus \{c_{j_0}\}$ for some $j_0 \in J$. Let $J' = J \setminus \{j_0\}$ and $b_{j'} \in \{b_j : j \in J'\}$. So $b_{j'}\alpha = c_{j'} \in X\beta \setminus Y$, then $b_{j'}\alpha = b_{j'}\beta$ since $\beta < \alpha$. Hence $x\alpha = x\beta$ for all $x \in X \setminus \{b_{j_0}\}$ and $b_{j_0}\beta = y_{i_0}$ for some $i_0 \in I$ or $b_{j_0}\beta = b_{j_1}\beta = c_{j_1}$ for some $j_1 \in J'$. Thus

$$\beta = \begin{pmatrix} \{y_{i_0}, b_{j_0}\} & y_{i'} & b_{j'} \\ y_{i_0} & y_{i'} & c_{j'} \end{pmatrix}, \quad (34)$$

where $I' = I \setminus \{i_0\}$, or

$$\beta = \begin{pmatrix} y_i & \{b_{j_1}, b_{j_0}\} & b_k \\ y_i & c_{j_1} & c_k \end{pmatrix}, \quad (35)$$

where $K = J \setminus \{j_0, j_1\}$. For the first form and the second form, the numbers of ways of placing b_{j_0} is r and $n - r - 1$, respectively. So the total number of ways of placing b_{j_0} is $n - 1$. But j_0 varies in the index set J; hence there are in total $(n - 1)(n - r)$ lower covers of α. □

Competing Interests

The authors declare that there is no conflict of interests regarding the publication of this paper.

Acknowledgments

This research was supported by Chiang Mai University.

References

[1] R. Hartwig, "How to partially order regular elements," *Mathematica Japonica*, vol. 35, pp. 1-13, 1980.

[2] K. S. Nambooripad, "The natural partial order on a regular semigroup," *Proceedings of the Edinburgh Mathematical Society (Series 2)*, vol. 23, no. 3, pp. 249–260, 1980.

[3] H. Mitsch, "A natural partial order for semigroups," *Proceedings of the American Mathematical Society*, vol. 97, no. 3, pp. 384–388, 1986.

[4] G. Kowol and H. Mitsch, "Naturally ordered transformation semigroups," *Monatshefte für Mathematik*, vol. 102, no. 2, pp. 115–138, 1986.

[5] C. Namnak and P. Preechasilp, "Natural partial orders on the semigroup of binary relations," *Thai Journal of Mathematics*, vol. 4, no. 3, pp. 39–50, 2006.

[6] V. H. Fernandes and J. Sanwong, "On the ranks of semigroups of transformations on a finite set with restricted range," *Algebra Colloquium*, vol. 21, no. 3, pp. 497–510, 2014.

[7] K. Sangkhanan and J. Sanwong, "Partial orders on semigroups of partial transformations with restricted range," *Bulletin of the Australian Mathematical Society*, vol. 86, no. 1, pp. 100–118, 2012.

[8] P. Honyam and J. Sanwong, "Semigroups of transformations with fixed sets," *Quaestiones Mathematicae. Journal of the South African Mathematical Society*, vol. 36, no. 1, pp. 79–92, 2013.

[9] Y. Chaiya, P. Honyam, and J. Sanwong, "Maximal subsemigroups and finiteness conditions on transformation semigroups with fixed sets," *Turkish Journal of Mathematics*, In press.

[10] P. M. Higgins, "The Mitsch order on a semigroup," *Semigroup Forum*, vol. 49, no. 2, pp. 261–266, 1994.

5

Symmetric Integer Matrices Having Integer Eigenvalues

Lei Cao[1] and Selcuk Koyuncu[2]

[1]*Department of Mathematics, Georgian Court University, Lakewood, NJ 08701, USA*
[2]*Department of Mathematics, University of North Georgia, Gainesville, GA 30566, USA*

Correspondence should be addressed to Selcuk Koyuncu; skoyuncu@ung.edu

Academic Editor: Niansheng Tang

We provide characterization of symmetric integer matrices for rank at most 2 that have integer spectrum and give some constructions for such matrices of rank 3. We also make some connection between Hanlon's conjecture and integer eigenvalue problem.

1. Introduction

The study of matrices with integer entries combines linear algebra, number theory, and group theory (the study of arithmetic groups). It was shown that the eigenvalues of symmetric matrices over the integers $\mathbb{Z}$ stem from as to what algebraic integers occur as eigenvalues for the incidence matrix of a graph (see [1]). Integer eigenvalues of a nonsymmetric matrix with entries as certain simple functions are presented in [2]. A graph is Laplacian integral if the spectrum of its Laplacian matrix consists entirely of integers. A number of papers on Laplacian matrices investigate the class of Laplacian integral graphs (see [3–5]). Integer matrices that arise from Laplacians are connected to the three-dimensional Heisenberg Lie algebra and the eigenvalues and eigenvectors were explicitly given for the subclass of these matrices (see [6]). An interesting class of matrices called B_n was introduced in [7]; the most interesting property of the B_n-class is that the spectra of the matrices consist of the consecutive integers $\{0, 1, \dots, n-1\}$; that is, the eigenvalues do not depend on the values of the elements of $B \in B_n$. In this paper, we characterize all symmetric integer matrices for rank at most 2 that have integer spectrum and give some constructions for such matrices of rank 3. We also open a discussion on the fact that integer eigenvalue problem has strong connection with Hanlon's conjecture (see [6]). We provide some examples and conjectures that relate these two problems.

We start with some basic definitions from linear algebra. Let A be a square matrix of size n and let λ be a scalar quantity. Then $P_A(\lambda) = \det(A - \lambda I)$ is called the characteristic polynomial of A. It is clear that the characteristic polynomial is an nth degree polynomial in λ and $\det(A - \lambda I) = 0$ will have n (not necessarily distinct) solutions for λ. The values of λ that satisfy $\det(A - \lambda I) = 0$ are the characteristic roots or eigenvalues of A. An $n \times n$ matrix A is called real symmetric if A^T, the transpose of A, coincide with A. If $A = [a_{ij}]$ is an $m \times n$ matrix and $B = [b_{ij}]$ is an $p \times q$ matrix, then the tensor product of A and B, denoted by $A \otimes B$, is the $mp \times nq$ matrix and is defined as

$$A \otimes B = \begin{pmatrix} a_{11}B & \cdots & a_{1n}B \\ \vdots & \ddots & \vdots \\ a_{m1}B & \cdots & a_{mn}B \end{pmatrix}. \quad (1)$$

If A is $n \times n$ and B is $m \times m$, then the Kronecker sum (or tensor sum) of A and B, denoted by $A \oplus B$, is the $mn \times mn$ matrix of the form $(I_m \otimes A) + (B \otimes I_n)$. Let $M_n(\mathbb{Z})$ be the set of all $n \times n$ symmetric matrices with integer entries.

Let A be an $n \times n$ matrix and $A_{[k]}$ be the sum of all kth order principal minors of A. Then all coefficients in the characteristic polynomial of A can be expressed by $A_{[k]}$ for $k = 1, 2, \dots, n$. In particular, $A_{[1]} = \text{trace}(A)$ and $A_{[n]} = \det(A)$.

Lemma 1. *Let $A \in M_n(\mathbb{R})$ with eigenvalues $\lambda_1, \lambda_2, \ldots, \lambda_n$. The characteristic polynomial of A is given by*

$$\begin{aligned} p_A(\lambda) &= \det(\lambda I - A) \\ &= \lambda^n + c_{n-1}\lambda^{n-1} + c_{n-2}\lambda^{n-2} + \cdots + c_1\lambda + c_0. \end{aligned} \tag{2}$$

Then

$$c_i = A_{[n-i]}, \quad i = 0, \ldots, n-1. \tag{3}$$

Lemma 2. *Let $A \in M_n(\mathbb{Z})$ with rank 1. Then A has integer eigenvalues.*

We now present the characterization of all symmetric integer matrices for rank at most 2 that have integer spectrum.

2. The Rank 2 Case

Theorem 3. *Let $A \in M_n(\mathbb{Z})$ with rank 2. Then A has integer eigenvalues if and only if there exist two integers m and n such that* $\text{trace}(A) = m + n$ *and* $A_{[2]} = mn$ *where* $A_{[2]}$ *is the sum of determinants of all 2nd order principal minors of A.*

Proof. Since A has rank 2, the characteristic polynomial of A has the form

$$\lambda^n - \text{trace}(A)\lambda^{n-1} + A_{[2]}\lambda^{n-2}. \tag{4}$$

It is clear that the two nonzero eigenvalues of A are

$$\begin{aligned} \lambda_1 &= \frac{\text{trace}(A) + \sqrt{\text{trace}(A)^2 - 4A_{[2]}}}{2}, \\ \lambda_2 &= \frac{\text{trace}(A) - \sqrt{\text{trace}(A)^2 - 4A_{[2]}}}{2}. \end{aligned} \tag{5}$$

"$\Leftarrow$" Suppose there exist integers m and n such that $\text{trace}(A) = m + n$ and $A_{[2]} = mn$. Then it follows that

$$\begin{aligned} \lambda_1 &= m, \\ \lambda_2 &= n. \end{aligned} \tag{6}$$

"$\Rightarrow$" If all eigenvalues of A are integers, then there exists an integer k such that

$$\text{trace}(A)^2 - 4A_{[2]} = k^2. \tag{7}$$

Letting

$$\begin{aligned} m &= \frac{\text{trace}(A) + k}{2}, \\ m &= \frac{\text{trace}(A) - k}{2} \end{aligned} \tag{8}$$

and using (7), the difference of $\text{trace}(A)^2$ and k^2 is $4A_{[2]}$ which is even, so either both $\text{trace}(A)$ and k are even or both $\text{trace}(A)$ and k are odd, and hence both m and n are integers. □

Lemma 4. *Let* $A = \begin{pmatrix} B & C \\ D & E \end{pmatrix}$. *If E is invertible and DE = ED, then* $\det(A) = \det(BE - CD)$.

Theorem 5. *Let* $A = \begin{pmatrix} B & C \\ C & B \end{pmatrix}$. *Suppose both B and C are integer matrices of order n with integer eigenvalues. If B and C commute, then A has integer eigenvalues.*

Proof. According to Lemma 4,

$$\begin{aligned} \det(\lambda I_{2n} - A) &= \det\left((\lambda I_{2n} - B)^2 - C^2\right) \\ &= \det\left((\lambda I_{2n} - B - C)(\lambda I_{2n} - B + C)\right). \end{aligned} \tag{9}$$

Since B and C commute, they can be diagonalized simultaneously and hence all eigenvalues of both $B + C$ and $B - C$ are integers. □

Lemma 6. *Let*

$$A = \begin{pmatrix} B & I & O & \cdots & O \\ O & B & I & O & \vdots \\ \vdots & \ddots & \ddots & \ddots & O \\ O & \cdots & O & B & I \\ O & \cdots & \cdots & O & B \end{pmatrix} \in M_{2n}(\mathbb{Z}). \tag{10}$$

If $B = \begin{pmatrix} a & b \\ b & a \end{pmatrix} \in M_2(\mathbb{Z})$, *then A has integer eigenvalues.*

Proof. Define

$$J = \begin{pmatrix} 0 & 1 & 0 & \cdots & 0 \\ 0 & 0 & 1 & 0 & \vdots \\ \vdots & \ddots & \ddots & \ddots & 0 \\ 0 & \cdots & 0 & 0 & 1 \\ 0 & \cdots & \cdots & 0 & 0 \end{pmatrix} \in M_n(\mathbb{Z}). \tag{11}$$

Then we can write A as

$$A = (I_n \otimes B) + (J \otimes I_2) = B \oplus J. \tag{12}$$

Since B and J have integer eigenvalues, then $A = B \oplus J$ has also integer eigenvalues. □

We now give some constructions for all symmetric matrices of rank 3 that has integer spectrum.

3. The Rank 3 Case

Theorem 7. *Let $A \in M_n(\mathbb{Z})$ be symmetric integer matrix with rank 3. If one of the following cases holds, then A has integer eigenvalues.*

(i) *One of the eigenvalues of A is 1 or −1 and there exists a positive integer k such that*

$$\left[A_{[3]} - A_{[2]}\right]^2 + 4A_{[3]} = k^2. \tag{13}$$

(ii) *All nonzero eigenvalues of A are the same and*

$$A_{[2]} = \frac{(\text{trace}(A))^2}{3}, \quad A_{[3]} = \frac{(\text{trace}(A))^3}{27}. \tag{14}$$

(iii) *One of the nonzero eigenvalues of A has multiplicity two and there exists a positive integer such that*

$$(\text{trace}(A))^2 - 3A_{[2]} = k^2. \tag{15}$$

(iv) *The trace of A is equal to zero and there exists a positive integer k and integers m, n such that*

$$k = \sqrt{\frac{(A_{[3]})^2}{4} + \frac{(A_{[2]})^3}{27}}, \quad m^3 = \frac{A_{[3]}}{2} + k, \quad n^3 = \frac{A_{[3]}}{2} - k. \tag{16}$$

In fact, one of eigenvalues is $m+n$.

Proof. Since the rank of A is 3, the characteristics polynomial of A can be written as

$$\begin{aligned} P_A(\lambda) &= \lambda^n - \text{trace}(A)\lambda^{n-1} + A_{[2]}\lambda^{n-2} - A_{[3]}\lambda^{n-3} \\ &= \lambda^{n-3}\left(\lambda^3 - \text{trace}(A)\lambda^2 + A_{[2]}\lambda - A_{[3]}\right). \end{aligned} \tag{17}$$

(i) Suppose that one of the eigenvalues of A is $\lambda = 1$. By substituting this eigenvalue in (17), one obtains

$$A_{[3]} - A_{[2]} = 1 - \text{trace}(A). \tag{18}$$

In addition, (17) can be factored as

$$P_A(\lambda) = \lambda^{n-3}(\lambda - 1)\left[\lambda^2 + (1 - \text{trace}(A))\lambda + A_{[3]}\right]. \tag{19}$$

By Theorem 3, the quadratic factor has integer roots if and only if there exists a positive integer k, such that

$$(1 - \text{trace}(A))^2 + 4A_{[3]} = k^2. \tag{20}$$

Now combining (18) and (20) yields

$$\left(A_{[3]} - A_{[2]}\right)^2 + 4A_{[3]} = k^2. \tag{21}$$

And in fact, the other eigenvalues are

$$\lambda_2 = \frac{1 - \text{trace}(A) + k}{2}, \quad \lambda_3 = \frac{-1 + \text{trace}(A) - k}{2} \tag{22}$$

which are integers because either both $1 - \text{trace}A$ and k are even or both of them are odd by (20).

(ii) Let $\tilde{\lambda}$ be the only nonzero eigenvalue of A. Then

$$P_A(\lambda) = \lambda^{n-3}\left(\lambda - \tilde{\lambda}\right)^3, \tag{23}$$

$$\left(\lambda - \tilde{\lambda}\right)^3 = \lambda^3 - \text{trace}(A)\lambda^2 + A_{[2]}\lambda - A_{[3]}. \tag{24}$$

By comparing the coefficients on both sides of (24), one obtains

$$\text{trace}(A) = 3\tilde{\lambda}, \quad A_{[2]} = 3\tilde{\lambda}^2, \quad A_{[3]} = \tilde{\lambda}^3. \tag{25}$$

Thus

$$A_{[2]} = \frac{(\text{trace}(A))^2}{3}, \quad A_{[3]} = \frac{(\text{trace}(A))^3}{27}. \tag{26}$$

(iii) Suppose that A has two nonzero eigenvalues λ_1 and λ_2 with multiplicity one and two, respectively. Then the characteristic polynomial of A can be written as

$$P_A(\lambda) = \lambda^{n-3}(\lambda - \lambda_1)^2(\lambda - \lambda_2), \tag{27}$$

and hence

$$(\lambda - \lambda_1)^2(\lambda - \lambda_2) = \lambda^3 - \text{trace}(A)\lambda^2 + A_{[2]}\lambda - A_{[3]}. \tag{28}$$

By comparing the coefficients on both sides of (28), one obtains

$$\text{trace}(A) = 2\lambda_1 + \lambda_2, \quad A_{[2]} = \lambda_1^2 + 2\lambda_1\lambda_2, \quad A_{[3]} = \lambda_1^2\lambda_2. \tag{29}$$

In addition, since λ_1 has multiplicity two, both $P_A(\lambda_1)$ and its derivative $P_A'(\lambda_1)$ are equal to zero and hence

$$3\lambda_1^2 - 2\,\text{trace}(A)\lambda_1 + A_{[2]} = 0. \tag{30}$$

Now taking the derivative of both sides of (28), we get

$$\begin{aligned} &3\lambda^2 - 2\,\text{trace}(A)\lambda + A_{[2]} \\ &\quad = (\lambda - \lambda_1)^2 + 2(\lambda - \lambda_1)(\lambda - \lambda_2). \end{aligned} \tag{31}$$

Since (31) is quadratic and has one integer root λ_1, then the other root must be rational. Thus there exists a positive integer k such that

$$4\,(\text{trace}\,(A))^2 - 12A_{[2]} = (2k)^2 \tag{32}$$

which yields that

$$(\text{trace}\,(A))^2 - 3A_{[2]} = k^2. \tag{33}$$

(iv) Denote the nonzero eigenvalues of A by λ_1, λ_2, and λ_3. We have $\lambda_3 = -(\lambda_1 + \lambda_2)$ due to zero trace. Also, we have the following equations

$$\lambda_1^2 + \lambda_1\lambda_2 + \lambda_2^2 = -A_{[2]}, \tag{34}$$

$$\lambda_1^2\lambda_2 + \lambda_1\lambda_2^2 = -A_{[3]}. \tag{35}$$

Multiplying (34) by λ_1 yields that

$$\lambda_1^3 + \lambda_1^2\lambda_2 + \lambda_1\lambda_2^2 = -A_{[2]}\lambda_1. \tag{36}$$

Subtracting (35) from (36), one obtains

$$\lambda_1^3 + A_{[2]}\lambda_1 - A_{[3]} = 0. \tag{37}$$

Note that the following

$$\begin{aligned}&\sqrt[3]{\frac{A_{[3]}}{2} + \sqrt{\frac{(A_{[3]})^2}{4} + \frac{(A_{[3]})^3}{27}}}\\ &+ \sqrt[3]{\frac{A_{[3]}}{2} - \sqrt{\frac{(A_{[3]})^2}{4} + \frac{(A_{[2]})^3}{27}}} = \sqrt[3]{\frac{A_{[3]}}{2} + k}\\ &+ \sqrt[3]{\frac{A_{[3]}}{2} - k} = m + n\end{aligned} \tag{38}$$

is always a solution of (36). To see this, first note that

$$\begin{aligned} m\cdot n &= \sqrt[3]{\frac{A_{[3]}}{2} + k}\cdot\sqrt[3]{\frac{A_{[3]}}{2} - k}\\ &= \sqrt[3]{\frac{(A_{[3]})^2}{4} - \frac{(A_{[3]})^2}{4} - \frac{(A_{[2]})^3}{27}} = -\frac{A_{[3]}}{3},\\ (m+n)^3 &= m^3 + 3m^2n + 3mn^2 + n^3\\ &= \left(\frac{A_{[3]}}{2} + k\right) + 3mn\,(m+n) + \left(\frac{A_{[3]}}{2} - k\right)\\ &= \left(\frac{A_{[3]}}{2} + k\right) - \frac{A_{[2]}}{3}\cdot(m+n)\\ &\quad + \left(\frac{A_{[3]}}{2} - k\right) = -\frac{A_{[2]}}{3}\cdot(m+n) + A_{[3]}.\end{aligned} \tag{39}$$

□

Lemma 8. *Let*

$$F = \begin{pmatrix} 1 & 1 & \cdots & 1\\ 1 & 1 & \cdots & 1\\ \vdots & \vdots & \ddots & \vdots\\ 1 & 1 & \cdots & 1\end{pmatrix} \in M_m \tag{40}$$

and let $A = (F \otimes B) + (C \otimes I_m)$. Then A has integer eigenvalues if $BC = CB$ and both B and C have integer eigenvalues.

Proof. Since B and C commute, they can be diagonalized simultaneously. Without loss of generality, suppose

$$B = \begin{pmatrix} b_1 & & & \\ & b_2 & & \\ & & \ddots & \\ & & & b_n\end{pmatrix},\qquad C = \begin{pmatrix} c_1 & & & \\ & c_2 & & \\ & & \ddots & \\ & & & c_n\end{pmatrix}. \tag{41}$$

Note that A has eigenvalues $b_i + (n-1)c_i$ with multiplicity 1 for $i = 1, 2, \ldots, n$ and A has eigenvalue $b_i - c_i$ with multiplicities $m-1$ for $i = 1, 2, \ldots, n$ since $(b_i - c_i)I - A$ has rank $(n-1)m$. □

Lemma 9. *Let $A \in M_n(\mathbb{Z})$. If $A = A_1 \otimes A_2 \otimes \cdots \otimes A_k$ where each A_i has integer eigenvalues and $A_i \in M_i(\mathbb{Z})$, $i = 1, \ldots, k$, then A has integer eigenvalues.*

Proposition 10. *Let $A^T = A \in M_3(\mathbb{Z})$ and $B = \left(\begin{smallmatrix} a & b\\ b & a\end{smallmatrix}\right) \in M_2(\mathbb{Z})$. Suppose both A and B have integer eigenvalues. Then $A \oplus B$ has integer eigenvalues.*

In this section, we open a discussion on possible connection between integer eigenvalue problem and Hanlon's conjecture. We support our approach with some examples and adopt the notation used in [6].

4. Connection to Hanlon's Conjecture

Definition 11. Let a, b, and k be nonnegative integers with $a \le k+1$ and $b \le k+1$. Let $\Omega_k(a, b)$ be the set of pairs (U, V) such that U is an A-subset of k_0 and V is a B-subset of k_0, where $k_0 = \{0, 1, \ldots, k\}$ and k is a nonnegative integer. Define the weight of a pair (U, V) to be

$$\mathbb{W}(U, V) = \sum_{u\in U} u + \sum_{v\in V} v. \tag{42}$$

Let $\Omega_k(a, b, w)$ be the set of pairs (U, V) such that $\mathbb{W}(U, V) = w$.

Example 12. Let $a = 2$, $b = 1$, and $w = 4$. If $k = 2$, then the ordered basis is $\Omega_2(2, 1, 4) = \{(12, 1), (02, 2)\}$.

If $k = 3$, the ordered basis is $\Omega_2(2,1,4) = \{(13,0),(03,1),(12,1),(02,2),(01,3)\}$. In this case, if $k = 0$ then $\Omega_2(2,1,4) = \emptyset$.

We define a matrix $T_k(a,b,w)$ with respect to corresponding basis $\Omega_k(a,b,w)$. We need the following definition in order to define the matrix $T_k(a,b,w)$.

Definition 13. Let (U,V) and (X,Y) be elements of $\Omega_k(a,b,w)$, and let (u,v,z) be a triple with $u \in U$, $v \in V$ and z is any integer. we say that (U,V) and (X,Y) are (u,v,z)-neighbors if

(1) $X = (U \setminus \{u\}) \cup \{u+z\}$,

(2) $Y = (Y \setminus \{v\}) \cup \{v-z\}$,

(3) $u + v \leq k$.

In [6] (Conjecture 1.12), it was conjectured that the eigenvalues $T_k(a,b,w)$ are nonnegative integers. Let us present some examples.

Example 14. $T_2(2,1,4)$ with respect to the ordered basis $\{(12,1),(02,2)\}$ is

$$\begin{pmatrix} 2 & 1 \\ 1 & 2 \end{pmatrix}. \tag{43}$$

Theorem 3 guarantees that both eigenvalues of $T_2(2,1,4)$ are integers.

Example 15. $T_3(2,1,4)$ with respect to the ordered basis

$$\{(13,0),(03,1),(12,1),(02,2),(01,3)\} \tag{44}$$

is

$$\begin{pmatrix} 2 & 1 & 1 & 0 & -1 \\ 1 & 2 & 0 & 1 & 1 \\ 1 & 0 & 2 & 1 & -1 \\ 0 & 1 & 1 & 2 & 1 \\ -1 & 1 & -1 & 1 & 2 \end{pmatrix}. \tag{45}$$

If the last row and column are deleted, the resulting matrix is in the form of block matrix given in Theorem 5. The eigenvalues of $(1,1)$-block are 1, 3 and the eigenvalues of $(2,2)$-block are both 1 with multiplicity 2. The eigenvalues of $T_2(2,1,4)$ are 0, 0, 2, 4, 4. Notice that these eigenvalues can also be obtained from the sum and difference of $(1,1)$ and $(1,2)$ blocks with given multiplicities.

Example 16. $T_4(2,1,4)$ with respect to the ordered basis

$$\{(04,0),(13,0),(03,1),(12,1),(02,2),(01,3)\} \tag{46}$$

is

$$\begin{pmatrix} 2 & 0 & 1 & 0 & 1 & 1 \\ 0 & 2 & 1 & 1 & 0 & -1 \\ 1 & 1 & 2 & 0 & 1 & 1 \\ 0 & 1 & 0 & 2 & 1 & -1 \\ 1 & 0 & 1 & 1 & 2 & 1 \\ 1 & -1 & 1 & -1 & 1 & 2 \end{pmatrix}. \tag{47}$$

The eigenvalues of $T_4(2,1,4)$ are 0, 0, 1, 2, 4, and 5. We do not observe an obvious connection in this example.

Example 17. $T_3(3,2,5)$ with respect to the ordered basis

$$\{(012,02)\} \tag{48}$$

is

$$(4). \tag{49}$$

Example 18. $T_3(2,2,4)$ with respect to the ordered basis

$$\{(13,01),(12,02),(02,12),(01,13)\} \tag{50}$$

is

$$\begin{pmatrix} 3 & 1 & 0 & 1 \\ 1 & 3 & 1 & 0 \\ 0 & 1 & 3 & 1 \\ 1 & 0 & 1 & 3 \end{pmatrix}. \tag{51}$$

Notice that the matrix $T_3(2,2,4)$ is in block form given in Theorem 5 and its eigenvalues are the eigenvalues of the sum and difference of $(1,1)$ and $(2,2)$-blocks with given multiplicities. In this case the eigenvalues are 1, 3, 3, and 5.

Example 19. $T_3(2,2,6)$ with respect to the ordered basis

$$\{(23,01),(13,02),(03,12),(03,03),(12,12),(12,03),(02,13),(01,23)\} \tag{52}$$

is

$$\begin{pmatrix} 3 & 1 & -1 & 1 & 1 & -1 & 1 & 0 \\ 1 & 3 & 1 & 1 & 1 & 1 & 0 & -1 \\ -1 & 1 & 2 & 0 & 0 & 0 & 1 & -1 \\ 1 & 1 & 0 & 3 & 0 & 0 & 1 & 1 \\ 1 & 1 & 0 & 0 & 3 & 0 & 1 & 1 \\ -1 & 1 & 0 & 0 & 0 & 2 & 1 & -1 \\ 1 & 0 & 1 & 1 & 1 & 1 & 3 & 1 \\ 0 & 1 & -1 & 1 & 1 & -1 & 1 & 3 \end{pmatrix}. \tag{53}$$

Note that some examples given above follow directly from the theory we have provided and some do not. We aim to search more general connections in a followup paper. In [6], the author search for algebraic expression for the eigenvalues. We use the same notation and state the conjecture. For each a, b, w, k, and each nonnegative integer r, let $\mu_k(a,b,w;r)$ denote the multiplicity of r as an eigenvalue of $T_k(a,b,w)$. Let

$$\mu_k(a,b;r) = \sum_{w} \mu_k(a,b,w;r). \quad (54)$$

Let $M_k(x,y,\lambda)$ be the following generating function for the numbers $\mu_k(a,b;r)$:

$$M_k(x,y,\lambda) = \sum_{a,b,r} \mu_k(a,b;r)\, x^a y^b \lambda^r. \quad (55)$$

It was conjectured in [6] (Conjecture 1.14) that $M_k(x, y,\lambda) = \prod_{i=0}^{k}(1+x+y+\lambda^{i+1}xy)$. This conjecture is still open in general. We hope to find more obvious connections and use them to solve given conjectures in this paper.

Competing Interests

Lei Cao and Selcuk Koyuncu declare that there is no conflict of interests regarding the publication of this paper.

References

[1] D. R. Estes, "Eigenvalues of symmetric integer matrices," *Journal of Number Theory*, vol. 42, no. 3, pp. 292–296, 1992.

[2] L. Bondesson and I. Traat, "A nonsymmetric matrix with integer eigenvalues," *Linear and Multilinear Algebra*, vol. 55, no. 3, pp. 239–247, 2007.

[3] F. Harary and A. J. Schwenk, "Which graphs have integral spectra?" in *Graphs and Combinatorics: Proceedings of the Capital Conference on Graph Theory and Combinatorics at the George Washington University June 18–22, 1973*, R. A. Bari and F. Harary, Eds., vol. 406 of *Lecture Notes in Mathematics*, pp. 45–51, Springer, Berlin, Germany, 1974.

[4] S. Kirkland, "Constructably Laplacian integral graphs," *Linear Algebra and Its Applications*, vol. 423, no. 1, pp. 3–21, 2007.

[5] W. So, "Rank one perturbation and its application to the Laplacian spectrum of a graph," *Linear and Multilinear Algebra*, vol. 46, no. 3, pp. 193–198, 1999.

[6] P. Hanlon, "Some remarkable combinatorial matrices," *Journal of Combinatorial Theory, Series A*, vol. 59, no. 2, pp. 218–239, 1992.

[7] T. von Rosen and D. von Rosen, "On a class of singular nonsymmetric matrices with nonnegative integer spectra," in *Algebraic Methods in Statistics and Probability II*, M. A. G. Viana and H. P. Wynn, Eds., vol. 516, pp. 319–325, American Mathematical Society, Providence, RI, USA, 2010.

6

Green's Function for a Slice of the Korányi Ball in the Heisenberg Group $\mathbb{H}_n$

Shivani Dubey,[1] Ajay Kumar,[1] and Mukund Madhav Mishra[2]

[1] *Department of Mathematics, University of Delhi, Delhi 110007, India*
[2] *Department of Mathematics, Hans Raj College, University of Delhi, Delhi 110007, India*

Correspondence should be addressed to Ajay Kumar; akumar@maths.du.ac.in

Academic Editor: Heinrich Begehr

We give a representation formula for solution of the inhomogeneous Dirichlet problem on the upper half Korányi ball and for the slice of the Korányi ball in the Heisenberg group $\mathbb{H}_n$ by obtaining explicit expressions of Green-like kernel when the given data has certain radial symmetry.

1. Introduction

The Heisenberg groups, in discrete and continuous versions, appear in several streams of mathematics, including Fourier analysis, several complex variables, geometry, and topology. The well known concepts of Green, Neumann, and Robin functions are important for representing solutions to certain boundary value problems for elliptic equations and to realize their smoothness properties. While the existence of these particular fundamental solutions for admissible domains are presented in all textbooks (see, e.g., [1–3]), explicit expressions for these functions are rare. In case of the Laplace operator mostly just the unit ball serves as an example.

The objective of this paper is to continue the search for explicit Green's functions for domains other than the Korányi ball in the Heisenberg group. There is little hope to get explicit kernels that work for arbitrary continuous boundary data, but it is possible to find some if one is restricted to boundary data having certain symmetry properties. This line of investigation was started by Gaveau et al. in [4] where they dealt with the case of the unit ball in the 3-dimensional Heisenberg group $\mathbb{H}_1$ and functions invariant under a circle action. This result was extended in [5] to the general Heisenberg group $\mathbb{H}_n$ with its natural metric, for functions invariant under the unitary group $U(n)$. Further, in the case of $\mathbb{H}_n$, it has been shown that the method of [5] works for the much larger class of circular functions, that is, functions invariant under a circle action [6]. The Dirichlet problem on the Heisenberg group and the existence of unique solution was discussed in [7]. Green's function for circular data in the Heisenberg group has been studied for various domains, for example, for half space in [6], for quarter space in [8], and for annulus in [9].

In following sections, we obtain the circular Green's function for the upper half Korányi ball and a slice of the Korányi ball by two parallel planes. We apply the method of infinitely many reflections along the boundaries of the domain, which was introduced by Courant and Hilbert in [2], generalized for the annulus in the Heisenberg group [10].

2. Analysis on the Heisenberg Group

We begin by recalling some notions from [11], which have laid the foundation for harmonic analysis on the Heisenberg group.

Definition 1. The Heisenberg group $\mathbb{H}_n$ (of degree n) is the Lie group structure on $\mathbb{C}^n \times \mathbb{R}$ whose group law is given by

$$[z,t] \cdot [z',t'] = [z+z', t+t' + 2\mathrm{Im}(z \cdot \overline{z}')], \qquad z \cdot \overline{z}' = \sum_{j=1}^{n} z_j \cdot \overline{z}'_j. \tag{1}$$

We write $z_j = x_j + iy_j$, $j = 1, \ldots, n$, and we define on $\mathbb{H}_n$ the vector fields

$$\begin{aligned} X_j &= \partial_{x_j} + 2y_j\partial_t, \\ Y_j &= \partial_{y_j} - 2x_j\partial_t, \\ T &= \partial_t, \\ Z_j &= \frac{1}{2}\left(X_j - iY_j\right) = \partial_{z_j} + i\overline{z}_j\partial_t, \\ \overline{Z}_j &= \frac{1}{2}\left(X_j + iY_j\right) = \partial_{\overline{z}_j} - iz_j\partial_t. \end{aligned} \tag{2}$$

One may note the following:

$$\begin{aligned} &\left[X_j, Y_k\right] = -4\delta_{jk}T, \\ &\left[X_j, X_k\right] = \left[Y_j, Y_k\right] = \left[X_j, T\right] = \left[Y_j, T\right] = 0, \\ &\qquad \forall 1 \le j,\ k \le n. \end{aligned} \tag{3}$$

Equivalently,

$$\begin{aligned} &\left[Z_j, \overline{Z}_k\right] = -2i\delta_{jk}T, \\ &\left[Z_j, Z_k\right] = \left[\overline{Z}_j, \overline{Z}_k\right] = \left[Z_j, T\right] = \left[\overline{Z}_j, T\right] = 0, \\ &\qquad \forall 1 \le j,\ k \le n. \end{aligned} \tag{4}$$

Here [,] is the Lie bracket.

Observe that Z_j, T are left invariant vector fields generated by the tangent vectors ∂_{z_j}, ∂_t at e.

If $\xi = [z,t] \in \mathbb{H}_n$, the natural gauge on $\mathbb{H}_n$ is given by

$$N(z,t) = \left(|z|^4 + t^2\right)^{1/4}. \tag{5}$$

On the Heisenberg group we have an analogue of the Laplacian which was first studied by Folland and Stein [11]. The sub-Laplacian L on $\mathbb{H}_n$ is explicitly given by

$$L = -\sum_{j=1}^{n}\left(X_j^2 + Y_j^2\right). \tag{6}$$

Let L_0 denote the slightly modified subelliptic operator $(-1/4)L$.

The fundamental solution for the sub-Laplacian L_0 on the Heisenberg group was first given by Folland in [12] with pole at identity by

$$g_e(\xi) = g_e([z,t]) = a_0\left(|z|^4 + t^2\right)^{-n/2}, \tag{7}$$

where

$$a_0 = 2^{n-2}\frac{\left(\Gamma(n/2)\right)^2}{\pi^{n+1}}, \tag{8}$$

is the normalization constant.

The fundamental solution of L_0 with pole at η is given by

$$g_\eta(\xi) = g_e\left(\eta^{-1}\xi\right). \tag{9}$$

From [6], for $\eta = [z', t']$ and $\xi = [z,t]$,

$$g_\eta(\xi) = a_0\left|C_{[z',t']}(z,t) - P_{[z',t']}(z,t)\right|^{-n}, \tag{10}$$

where

$$\begin{aligned} C_{[z',t']}(z,t) &= |z|^2 + \left|z'\right|^2 + i\left(t - t'\right), \\ P_{[z',t']}(z,t) &= 2z\cdot\overline{z}'. \end{aligned} \tag{11}$$

For an integrable function f on $\mathbb{H}_n$, we denote the average of f by

$$\overline{f}([z,t]) = \frac{1}{2\pi}\int_0^{2\pi} f\left(\left[e^{i\theta}z, t\right]\right)d\theta. \tag{12}$$

A function f is said to be circular, that is, invariant under circle action if $f([z,t]) = \overline{f}([z,t])$, for $[z,t] \in \mathbb{H}_n$.

The average of the fundamental solution with pole at η is given in [6] as

$$\begin{aligned} &\overline{g}_\eta(\xi) \\ &= a_0\left|C_{[z',t']}(z,t)\right|^{-n} F\left(\frac{n}{2}, \frac{n}{2}; n; \frac{\left|P_{[z',t']}(z,t)\right|^2}{\left|C_{[z',t']}(z,t)\right|^2}\right), \end{aligned} \tag{13}$$

where F is the Gaussian hypergeometric function [13]. We can easily see that the average of the fundamental solution is also L_0-harmonic away from η.

3. Green's Function for the Upper Half Korányi Ball

In this section, the domain D is defined as $D = B \cap H$, where B denotes the Korányi ball in $\mathbb{H}_n$; that is, $B = \{\xi = [z,t] \in \mathbb{H}_n : N(\xi) < 1\}$ and $H = \{\xi = [z,t] \in \mathbb{H}_n : t > 0\}$ denotes the half space in $\mathbb{H}_n$.

W. Thomson (Lord Kelvin) proved in 1847 the following fact: if U is harmonic on $\mathbb{R}^3$ then the function V defined by

$$V(x,y,z) := ar^{-1}U\left(a^2r^{-2}x, a^2r^{-2}y, a^2r^{-2}z\right), \tag{14}$$

($r = \sqrt{x^2+y^2+z^2}$, $a > 0$) is harmonic on $\mathbb{R}^3 \setminus \{0\}$ [14, p. 232]. For this reason the transformation $U \mapsto V$ is called the Kelvin transform. An analogue of the classical Kelvin transform is defined for functions on the Heisenberg group and it is shown that this transform preserves the class of functions annihilated by the left invariant sub-Laplacian [15]. For a function f on $\mathbb{H}_n$, the Kelvin transform is defined by

$$Kf = N^{-2n} f o h, \tag{15}$$

where h is the inversion defined as

$$h([z,t]) = \left[\frac{-z}{|z|^2 - it}, \frac{-t}{|z|^4 + t^2}\right], \tag{16}$$

for $[z,t] \in \mathbb{H}_n \backslash \{e\}$. The Kelvin transform sends a L_0-harmonic function on $\mathbb{H}_n \setminus \{e\}$ to a L_0-harmonic function. It was shown in [5] that, for a circular function f on $\mathbb{H}_n \setminus \{e\}$, we have

$$K(f)\left(\xi^{-1}\right) = f(\xi), \tag{17}$$

for all ξ with $N(\xi) = 1$.

From [5, (3.3)] we have, for $\eta \neq e \in \mathbb{H}_n$

$$K\left(g_\eta\right) = N\left(\eta\right)^{-2n} g_{\eta^\dagger}, \tag{18}$$

where we wrote $\eta^\dagger$ for $h(\eta)$.

For D, let

$$\begin{aligned} G_D\left(\eta,\xi\right) &= \overline{g}_\eta\left(\xi\right) - K\left(\overline{g}_\eta\right)\left(\xi^{-1}\right) - \overline{g}_{\eta*}\left(\xi\right) \\ &\quad + K\left(\overline{g}_{\eta*}\right)\left(\xi^{-1}\right), \end{aligned} \tag{19}$$

η^* being the reflection of the point η with respect to the boundary $t = 0$; that is, $\eta^* = [z', -t']$.

In the next result, we take the differential operator L_0 with respect to the variable ξ.

Theorem 2. *For each $\eta \in D$, the function $G_D(\eta,\xi)$ as defined in (19) is a smooth function on $D \setminus \{\eta\}$ which extends continuously up to ∂D and satisfies the following:*

(i) $L_0 G_D(\eta,\xi) = \delta_\eta$.

(ii) *Limits of the function $G_D(\eta,\xi)$ vanish at the boundaries of the upper half Korányi ball, that is, at $N(\xi) = 1$ and $t = 0$.*

Proof. (i) For $\eta = [z', t'] \in D$, we have

$$K\left(\overline{g}_\eta\right)\left(\xi^{-1}\right) = \left(N\left(\eta\right)\right)^{-2n} \overline{g}_{\eta^\dagger}\left(\xi^{-1}\right), \tag{20}$$

since $\eta^\dagger = h(\eta)$ lies outside D and we know that $\overline{g}_\eta(\xi)$ is L_0-harmonic away from η therefore $K(\overline{g}_\eta)(\xi^{-1})$ is L_0-harmonic on D.

Similarly $\overline{g}_{\eta*}(\xi)$ and $K(\overline{g}_{\eta*})(\xi^{-1})$ are L_0-harmonic on D because $\eta^* = [z', -t']$ and $h(\eta^*) = h([z', -t'])$ lie outside D.

Hence, $L_0(G_D(\eta,\xi)) = L_0(\overline{g}_\eta(\xi)) = \delta_\eta$.

(ii) We have

$$\begin{aligned} &\overline{g}_{\eta*}\left(\xi\right) \\ &= a_0 \left|C_{[z',-t']}\left(z,t\right)\right|^{-n} F\left(\frac{n}{2}, \frac{n}{2}; n; \frac{\left|P_{[z',t']}\left(z,t\right)\right|^2}{\left|C_{[z',-t']}\left(z,t\right)\right|^2}\right), \\ &\overline{g}_\eta\left(\xi\right) \\ &= a_0 \left|C_{[z',t']}\left(z,t\right)\right|^{-n} F\left(\frac{n}{2}, \frac{n}{2}; n; \frac{\left|P_{[z',t']}\left(z,t\right)\right|^2}{\left|C_{[z',t']}\left(z,t\right)\right|^2}\right), \end{aligned} \tag{21}$$

where $C_{[z',\pm t']}(z,t) = |z|^2 + |z'|^2 + i(t \mp t')$ and $P_{[z',t']}(z,t) = 2z \cdot \overline{z}'$.

At $t = 0$, $C_{[z',t']}(z,t) = C_{[z',-t']}(z,t)$, therefore,

$$\begin{aligned} \overline{g}_{\eta*}\left(\xi\right) &= \overline{g}_\eta\left(\xi\right), \\ K\left(\overline{g}_{\eta*}\right)\left(\xi^{-1}\right) &= K\left(\overline{g}_\eta\right)\left(\xi^{-1}\right). \end{aligned} \tag{22}$$

Hence $G_D(\eta,\xi) = 0$ at $t = 0$.

Since $\overline{g}_\eta(\xi)$ is circular, so, by using (17) we get

$$\begin{aligned} \overline{g}_{\eta*}\left(\xi\right) &= K\left(\overline{g}_{\eta*}\right)\left(\xi^{-1}\right), \\ \overline{g}_\eta\left(\xi\right) &= K\left(\overline{g}_\eta\right)\left(\xi^{-1}\right) \\ &\qquad \text{for } N\left(\xi\right) = 1. \end{aligned} \tag{23}$$

Hence $G_D(\eta,\xi) = 0$ at $N(\xi) = 1$. □

By using the circular Green's function, we can easily solve the inhomogeneous Dirichlet problem

$$\begin{aligned} L_0 u &= f \quad \text{in D}, \\ u &= h \quad \text{on } \partial D, \end{aligned} \tag{24}$$

where f and h are continuous circular functions and the solution is given by the representation formula

$$\begin{aligned} u\left(\eta\right) &= \int_D G_D\left(\eta,\xi\right) f\left(\xi\right) dv\left(\xi\right) \\ &\quad + \int_{\partial D} P_D\left(\eta,\xi\right) h\left(\xi\right) d\sigma\left(\xi\right), \end{aligned} \tag{25}$$

where

$$P_D\left(\eta,\xi\right) = -\frac{1}{4}\frac{\partial}{\partial n_0} G_D\left(\eta,\xi\right), \quad \xi \in \partial D \tag{26}$$

is the Poisson kernel and $d\sigma$ denote the Riemannian surface element, dv being the corresponding Riemannian volume element on $\mathbb{H}_n$; for detailed study refer [5].

4. Green's Function for a Slice of the Korányi Ball

In this section S denotes a slice of the Korányi ball in $\mathbb{H}_n$ which is defined as follows:

$$\begin{aligned} S = \left\{\xi = [z,t] \in \mathbb{H}_n : N\left(\xi\right) < 1, c_1 < t < c_2\right\}, \\ -1 < c_1 < c_2 < 1. \end{aligned} \tag{27}$$

For $\xi = [z,t]$ and $\eta = [z', t'] \in \mathbb{H}_n$ we define

$$\begin{aligned} &H_\xi\left(z', t'\right) \\ &= a_0 \left|C_{[z',t']}\left(z,t\right)\right|^{-n} F\left(\frac{n}{2}, \frac{n}{2}; n; \frac{\left|P_{[z',t']}\left(z,t\right)\right|^2}{\left|C_{[z',t']}\left(z,t\right)\right|^2}\right), \end{aligned} \tag{28}$$

where $C_{[z',t']}(z,t)$, $P_{[z',t']}(z,t)$ are as in (11). A differential operator, whenever applied to function $H_\xi(z', t')$, will be with respect to the variable ξ. Now, we assert that the Green function for the domain S is given by

$$\begin{aligned} G_S(\eta,\xi) = \sum_{m=0}^{\infty} \Big[& H_\xi\left(z', 2m(c_2-c_1)+t'\right) \\ & - K\left(H_{\xi^{-1}}\left(z', 2m(c_2-c_1)+t'\right)\right)\Big] \\ + \sum_{m=1}^{\infty} \Big[& H_\xi\left(z', 2m(c_1-c_2)+t'\right) \\ & - K\left(H_{\xi^{-1}}\left(z', 2m(c_1-c_2)+t'\right)\right)\Big] \\ - \sum_{m=1}^{\infty} \Big[& H_\xi\left(z', (2mc_2-(2m-2)c_1)-t'\right) \\ & - K\left(H_{\xi^{-1}}\left(z', (2mc_2-(2m-2)c_1)-t'\right)\right)\Big] \\ - \sum_{m=1}^{\infty} \Big[& H_\xi\left(z', (2mc_1-(2m-2)c_2)-t'\right) \\ & - K\left(H_{\xi^{-1}}\left(z', (2mc_1-(2m-2)c_2)-t'\right)\right)\Big], \end{aligned} \tag{29}$$

where K denotes the Kelvin transform in $\mathbb{H}_n$. In the following lemma, we prove that $G_S(\eta,\xi)$ is well defined for all η, ξ with $|z| \neq |z'|$ or $t \neq t'$.

Lemma 3. *For a fixed $\eta = [z',t']$ and $\xi = [z,t]$ such that $|z| \neq |z'|$ or $t \neq t'$, the series on RHS of (29) are absolutely and uniformly convergent on a compact neighbourhood of ξ.*

Proof.

Case 1. Suppose that $t = t'$, we choose a relatively compact neighbourhood U of ξ such that $||z|-|z'|| > \delta$, $\delta > 0$. Then,

$$\begin{aligned} & \frac{|P_{[z',t']}(z,t)|^2}{\left|C_{[z',2m(c_2-c_1)+t']}(z,t)\right|^2} \\ & = \frac{4|z|^2|z'|^2}{|z|^4+|z'|^4+2|z|^2|z'|^2+(t-(2m(c_2-c_1)+t'))^2} \\ & \leq \frac{M}{4m^2(c_2-c_1)^2}, \quad \text{where } M = \sup_U |z|^2 \text{ for } m > 0, \end{aligned} \tag{30}$$

and therefore, for sufficiently large m,

$$\frac{|P_{[z',t']}(z,t)|^2}{\left|C_{[z',2m(c_2-c_1)+t']}(z,t)\right|} < 1-\epsilon \quad \text{for some } \epsilon > 0. \tag{31}$$

On the set $[0, 1-\epsilon]$, the function $F(n/2, n/2; n; |P_{[z',t']}(z,t)|^2/|C_{[z',2m(c_2-c_1)+t']}(z,t)|^2)$ is bounded and so, for sufficiently large m, $H_\xi(z', 2m(c_2-c_1)+t')$ are uniformly bounded. Also, for smaller positive values of m

$$\begin{aligned} \frac{|P_{[z',t']}(z,t)|^2}{\left|C_{[z',2m(c_2-c_1)+t']}(z,t)\right|^2} & \leq \frac{2|z|^2|z'|^2}{2|z|^2|z'|^2+4m^2(c_2-c_1)^2} \\ & \leq \frac{1}{1+2m^2(c_2-c_1)^2/|z|^2|z'|^2} \\ & \leq \frac{1}{1+2m^2(c_2-c_1)^2/M^2|z'|^2} < 1. \end{aligned} \tag{32}$$

Hence $H_\xi(z', 2m(c_2-c_1)+t')$ are uniformly bounded for all $m > 0$. For $m = 0$, the term is

$$\frac{|P_{[z',t']}(z,t)|^2}{\left|C_{[z',2m(c_2-c_1)+t']}(z,t)\right|^2} = \frac{4|z|^2|z'|^2}{|z|^4+|z'|^4+2|z|^2|z'|^2}. \tag{33}$$

In case $|z| = 0$ or $|z'| = 0$, above expression is equal to zero. For $|z| \neq 0$ and $|z'| \neq 0$,

$$\begin{aligned} 1 - & \frac{|P_{[z',t']}(z,t)|^2}{\left|C_{[z',2m(c_2-c_1)+t']}(z,t)\right|^2} \\ & = \frac{\left(|z|^2-|z'|\right)^2}{|z|^4+|z'|^4+2|z|^2|z'|^2} \\ & \geq \frac{\delta^2}{|z|^4+|z'|^4+2|z|^2|z'|^2}, \end{aligned} \tag{34}$$

on a neighbourhood U of ξ

and therefore, $F(n/2, n/2; n; |P_{[z',t']}(z,t)|^2/|C_{[z',2m(c_2-c_1)+t']}(z,t)|^2)$ is uniformly bounded on U. Now,

$$\begin{aligned} & \left|C_{[z',2m(c_2-c_1)+t']}(z,t)\right|^{-n} \\ & = \frac{1}{\left||z|^2+|z'|^2+i(t-2m(c_2-c_1)-t')\right|^n} \\ & = \frac{1}{\left||z|^2+|z'|^2-2m(c_2-c_1)i\right|^n} \\ & \leq \frac{1}{(2m(c_2-c_1))^n}. \end{aligned} \tag{35}$$

Thus,

$$\sum_{m=0}^{\infty} H_\xi\left(z', 2m(c_2-c_1)+t'\right) < C\sum_{m=0}^{\infty} \frac{1}{(2m(c_2-c_1))^n}. \tag{36}$$

Case 2. Suppose that $|z| = |z'|$, we choose a compact neighbourhood O of ξ such that $|t - t'| > \delta_1$, $\delta_1 > 0$, for all $t' \in O$.

$$\frac{|P_{[z',t']}(z,t)|^2}{|C_{[z',2m(c_2-c_1)+t'](z,t)}|^2} = \frac{4|z|^4}{4|z|^4 + (t - t' - 2m(c_2 - c_1))^2}$$

$$= \frac{1}{1 + ((t - t' - 2m(c_2 - c_1))/2|z|^2)^2} < 1 - \epsilon_1 \quad \text{for some } \epsilon_1 > 0. \tag{37}$$

So, the functions $F(n/2, n/2; n; |P_{[z',t']}(z,t)|^2/|C_{[z',2m(c_2-c_1)+t']}(z,t)|^2)$ are uniformly bounded on O. Now

$$\begin{aligned}|C_{[z',2m(c_2-c_1)+t']}(z,t)|^{-n} &= \frac{1}{||z|^2 + |z'|^2 + i(t - 2m(c_2 - c_1) - t')|^n} = \frac{1}{(4|z|^4 + (t - 2m(c_2 - c_1) - t')^2)^{n/2}} \\ &= \frac{1}{(4m^2(c_2 - c_1)^2)^{n/2}(|z|^4/m^2(c_2 - c_1)^2 + ((t - t')/2m(c_2 - c_1))^2)^{n/2}} < \frac{1}{(2m(c_2 - c_1))^n}.\end{aligned} \tag{38}$$

Case 3. Now, suppose that $|z| \neq |z'|$, $t \neq t'$, and consider

$$\begin{aligned}\frac{|P_{[z',t']}(z,t)|^2}{|C_{[z',2m(c_2-c_1)+t']}(z,t)|^2} &= \frac{4|z|^2|z'|^2}{|z|^4 + |z'|^4 + 2|z|^2|z'|^2 + (t - (2m(c_2 - c_1) + t'))^2} \\ &\leq \frac{4|z|^2|z'|^2}{|z|^4 + |z'|^4 + 2|z|^2|z'|^2}.\end{aligned} \tag{39}$$

Therefore,

$$1 - \frac{|P_{[z',t']}(z,t)|^2}{|C_{[z',2m(c_2-c_1)+t']}(z,t)|^2} \geq \frac{(|z|^2 - |z'|^2)^2}{|z|^4 + |z'|^4 + 2|z|^2|z'|^2}. \tag{40}$$

We can choose a suitable compact neighbourhood of ξ such that $|z|^2 - |z'|^2 > \epsilon$ for some $\epsilon > 0$. We have

$$1 - \frac{|P_{[z',t']}(z,t)|^2}{|C_{[z',2m(c_2-c_1)+t']}(z,t)|^2} > \frac{\epsilon^2}{|z|^4 + |z'|^4 + 2|z|^2|z'|^2} > \epsilon_1. \tag{41}$$

This implies that $|P_{[z',t']}(z,t)|^2/|C_{[2m(c_2-c_1)+t']}(z,t)|^2 < 1 - \epsilon_1$. Clearly,

$$|C_{[z',2m(c_2-c_1)+t']}(z,t)|^{-n} \leq (2m(c_2 - c_1))^{-n} \quad \text{for sufficiently large } m. \tag{42}$$

Hence, in all the three cases

$$\sum_{m=0}^{\infty} H_\xi(z', 2m(c_2 - c_1) + t') < C\sum_{m=0}^{\infty} \frac{1}{(2m(c_2 - c_1))^n}, \tag{43}$$

for some positive constant C. Thus, $\sum_{m=0}^{\infty} H_\xi(z', 2m(c_2 - c_1) + t')$ is absolutely and uniformly convergent for $\eta \neq \xi$. Similarly, one can easily show that the other series in the RHS of (29) are uniformly and absolutely convergent for $\eta \neq \xi$.

Therefore, $G_S(\eta, \xi)$ is well defined for all $\eta \neq \xi$. □

In the next lemma, we show that $G_S(\eta, \xi) = 0$ on the boundaries of the domain S.

Lemma 4. *For $m \geq 0$,*

$$\begin{aligned}&\lim_{t \to c_1} H_\xi(z', 2m(c_2 - c_1) + t') = \lim_{t \to c_1} H_\xi(z', (2(m+1)c_1 - (2(m+1) - 2)c_2) - t'), \\ &\lim_{t \to c_1} K[H_{\xi^{-1}}(z', 2m(c_2 - c_1) + t')] = \lim_{t \to c_1} K[H_{\xi^{-1}}(z', (2(m+1)c_1 - (2(m+1) - 2)c_2) - t')],\end{aligned} \tag{44}$$

and for $m \geq 1$

$$\begin{aligned}&\lim_{t \to c_1} H_\xi(z', (2mc_2 - (2m - 2)c_1) - t') = \lim_{t \to c_1} H_\xi(z', 2m(c_1 - c_2) + t'), \\ &\lim_{t \to c_1} K[H_{\xi^{-1}}(z', (2mc_2 - (2m - 2)c_1) - t')] = \lim_{t \to c_1} K[H_{\xi^{-1}}(z', 2m(c_1 - c_2) + t')].\end{aligned} \tag{45}$$

Similarly, for $m \geq 0$,

$$\lim_{t \to c_2} H_\xi(z', 2m(c_1 - c_2) + t') = \lim_{t \to c_2} H_\xi(z', (2(m+1)c_2 - (2(m+1) - 2)c_1) - t'),$$

$$\lim_{t\to c_2} K\left[H_{\xi^{-1}}\left(z', 2m\left(c_1-c_2\right)+t'\right)\right]$$
$$= \lim_{t\to c_2} K\left[H_{\xi^{-1}}\left(z',\right.\right.$$
$$\left.\left.\left(2\left(m+1\right)c_2-\left(2\left(m+1\right)-2\right)c_1\right)-t'\right)\right], \tag{46}$$

and for $m \geq 1$

$$\lim_{t\to c_2} H_\xi\left(z', \left(2mc_1-\left(2m-2\right)c_2\right)-t'\right)$$
$$= \lim_{t\to c_2} H_\xi\left(z', 2m\left(c_2-c_1\right)+t'\right),$$
$$\lim_{t\to c_2} K\left[H_{\xi^{-1}}\left(z', \left(2mc_1-\left(2m-2\right)c_2\right)-t'\right)\right] \tag{47}$$
$$= \lim_{t\to c_2} K\left[H_{\xi^{-1}}\left(z', 2m\left(c_2-c_1\right)+t'\right)\right].$$

Finally, for $m \geq 0$

$$\lim_{N(\xi)\to 1} H_\xi\left(z', 2m\left(c_2-c_1\right)+t'\right)$$
$$= \lim_{N(\xi)\to 1} K\left[H_{\xi^{-1}}\left(z', 2m\left(c_2-c_1\right)+t'\right)\right], \tag{48}$$

and for $m \geq 1$

$$\lim_{N(\xi)\to 1} H_\xi\left(z', 2m\left(c_1-c_2\right)+t'\right)$$
$$= \lim_{N(\xi)\to 1} K\left[H_{\xi^{-1}}\left(z', 2m\left(c_1-c_2\right)+t'\right)\right],$$
$$\lim_{N(\xi)\to 1} H_\xi\left(z', \left(2mc_2-\left(2m-2\right)c_1\right)-t'\right)$$
$$= \lim_{N(\xi)\to 1} K\left[H_{\xi^{-1}}\left(z', \left(2mc_2-\left(2m-2\right)c_1\right)-t'\right)\right], \tag{49}$$
$$\lim_{N(\xi)\to 1} H_\xi\left(z', \left(2mc_1-\left(2m-2\right)c_2\right)-t'\right)$$
$$= \lim_{N(\xi)\to 1} K\left[H_{\xi^{-1}}\left(z', \left(2mc_1-\left(2m-2\right)c_2\right)-t'\right)\right].$$

Proof. Consider

$$\left|C_{[z',2m(c_2-c_1)+t']}(z,t)\right| = \Big|\|z\|^2+\left|z'\right|^2$$
$$+i\left(t-\left(2m\left(c_2-c_1\right)+t'\right)\right)\Big| = \Big|\|z\|^2+\left|z'\right|^2$$
$$+i\left(c_1-2mc_2+2mc_1-t'\right)\Big| \quad \text{at } t=c_1 \tag{50}$$
$$= \Big|\|z\|^2+\left|z'\right|^2$$
$$+i\left(\left(2\left(m+1\right)c_1-\left(2\left(m+1\right)-2\right)c_2\right)-t'\right)\Big|$$
$$= \left|C_{[z',(2(m+1)c_1-(2(m+1)-2)c_2)-t']}(z,t)\right| \quad \text{for } m \geq 0.$$

Therefore,

$$\lim_{t\to c_1} H_\xi\left(z', 2m\left(c_2-c_1\right)+t'\right)$$
$$= \lim_{t\to c_1} H_\xi\left(z', \left(2\left(m+1\right)c_1-\left(2\left(m+1\right)-2\right)c_2\right)\right. \tag{51}$$
$$\left.-t'\right).$$

On the same lines, one can easily prove the limits agree over the other boundaries also. □

Now, we show that $G_S(\eta,\xi)$ works as Green's function for domain S when applied to circular data.

Theorem 5. *For each $\eta \in S$, the function $G_S(\eta,\xi)$ as defined in (29) is a smooth function on $S\setminus\{\eta\}$ which extends continuously up to ∂S and satisfies the following:*

(i) $L_0 G_S(\eta,\xi) = \delta_\eta(\xi)$.

(ii) *Limits of the function $G_S(\eta,\xi)$ vanish at the boundaries of Korányi ball slice, that is, at $t = c_1$, $t = c_2$, and $N(\xi) = 1$.*

Proof. (i) For $\eta = [z',t'] \in S$, $m \geq 1$, the functions $H_\xi(z',2m(c_2-c_1)+t')$, $H_\xi(z',2m(c_1-c_2)+t')$, $H_\xi(z',(2mc_2-(2m-2)c_1)-t')$, $H_\xi(z',(2mc_1-(2m-2)c_2)-t')$, $K(H_{\xi^{-1}}(z',2m(c_2-c_1)+t'))$, $K(H_{\xi^{-1}}(z',2m(c_1-c_2)+t'))$, $K(H_{\xi^{-1}}(z',(2mc_2-(2m-2)c_1)-t'))$, and $K(H_{\xi^{-1}}(z',(2mc_1-(2m-2)c_2)-t'))$ are all harmonic at $\xi = [z,t]$ because poles of these functions lie outside the domain S. Also we have $K(H_{\xi^{-1}}(z',t'))$ is harmonic and from the definition of $H_\xi(z',t')$ it follows that

$$L_0 H_\xi\left(z',t'\right) = \delta_\eta\left(\xi\right). \tag{52}$$

Therefore, by using Lemma 3

$$L_0 G_S\left(\eta,\xi\right) = \sum_{m=0}^{\infty} L_0\left[H_\xi\left(z', 2m\left(c_2-c_1\right)+t'\right)\right.$$
$$\left.-K\left(H_{\xi^{-1}}\left(z', 2m\left(c_2-c_1\right)+t'\right)\right)\right]$$
$$+\sum_{m=1}^{\infty} L_0\left[H_\xi\left(z', 2m\left(c_1-c_2\right)+t'\right)\right.$$
$$\left.-K\left(H_{\xi^{-1}}\left(z', 2m\left(c_1-c_2\right)+t'\right)\right)\right]$$
$$-\sum_{m=1}^{\infty} L_0\left[H_\xi\left(z', \left(2mc_2-\left(2m-2\right)c_1\right)-t'\right)\right. \tag{53}$$
$$\left.-K\left(H_{\xi^{-1}}\left(z', \left(2mc_2-\left(2m-2\right)c_1\right)-t'\right)\right)\right]$$
$$-\sum_{m=1}^{\infty} L_0\left[H_\xi\left(z', \left(2mc_1-\left(2m-2\right)c_2\right)-t'\right)\right.$$
$$\left.-K\left(H_{\xi^{-1}}\left(z', \left(2mc_1-\left(2m-2\right)c_2\right)-t'\right)\right)\right]$$
$$= \delta_\eta\left(\xi\right).$$

(ii) This part easily follows from Lemma 4.

Hence the theorem is established. □

Now, we have circular Green's function for a slice of the Korányi ball so we can obtain explicit expression of the Poisson kernel for this domain to solve inhomogeneous Dirichlet boundary value problem. The Poisson kernel is the normal derivative of Green's function and, from [5, 6], is given by

$$P(\eta,\xi) = -\frac{1}{4}\frac{\partial}{\partial n_0}G(\eta,\xi), \quad \xi \in \partial S, \tag{54}$$

where

$$\frac{\partial}{\partial n_0} = \begin{cases} \frac{1}{|z|}\left(\overline{A}E + A\overline{E}\right) & \text{at } \{\xi \in \mathbb{H}_n : N(\xi) = 1\} \\ \frac{i}{|z|}\left(E - \overline{E}\right) & \text{at } \{\xi \in \mathbb{H}_n : t = c_1, t = c_2\}. \end{cases} \tag{55}$$

Theorem 6. *Green's function and the Poisson kernel which we have obtained solve the inhomogeneous Dirichlet boundary value problem for S and the solution for BVP*

$$\begin{aligned} L_0 u &= f \quad \text{in } S, \\ u &= h \quad \text{on } \partial S \end{aligned} \tag{56}$$

is given by

$$\begin{aligned} u(\eta) = {} & \int_S G(\eta,\xi) f(\xi)\, dv(\xi) \\ & + \int_{t=c_1} P(\eta,\xi) h(\xi)\, d\sigma(\xi) \\ & + \int_{t=c_2} P(\eta,\xi) h(\xi)\, d\sigma(\xi) \\ & + \int_{N(\xi)=1} P(\eta,\xi) h(\xi)\, d\sigma(\xi), \end{aligned} \tag{57}$$

where f and h are continuous circular functions.

Conflict of Interests

The authors declare that there is no conflict of interests regarding the publication of this paper.

Acknowledgments

The first author is supported by the Senior Research Fellowship of Council of Scientific and Industrial Research, India (Grant no. 09/045(1152)/2012-EMR-I), and the second author is supported by R&D grant from University of Delhi, Delhi, India. The authors would like to thank the referees for their careful reading and substantial comments on an earlier version of this paper.

References

[1] H. G. Begehr, *Complex Analytic Methods for Partial Differential Equations*, World Scientific Publishing, Singapore, 1994.

[2] R. Courant and D. Hilbert, *Methods of Mathematical Physics*, vol. 1, Wiley-VCH, Weinheim, Germany, 2008.

[3] G. C. Wen and H. Begehr, *Boundary Value Problems for Elliptic Equations and Systems*, Longman, Harlow, UK, 1990.

[4] B. Gaveau, P. Greiner, and J. Vauthier, "Polynôes harmoniques et problème de Dirichlet de la boule du groupe de Heisenberg en présence de symétrie radiale," *Bulletin des Sciences Mathématiques*, vol. 108, no. 4, pp. 337–354, 1984.

[5] A. Korányi and H. M. Riemann, "Horizontal normal vectors and conformal capacity of spehrical rings in the Heisenberg group," *Bulletin des Sciences Mathématiques*, vol. 111, no. 2, pp. 3–21, 1987.

[6] A. Korányi, "Poisson formulas for circular functions and some groups of type H," *Science in China Series A: Mathematics*, vol. 49, no. 11, pp. 1683–1695, 2006.

[7] B. Gaveau, "Principe de moindre action, propagation de la chaleur et estimées sous elliptiques sur certains groupes nilpotents," *Acta Mathematica*, vol. 139, no. 1-2, pp. 95–153, 1977.

[8] A. Kumar and M. M. Mishra, "Green's functions on the Heisenberg group," *Analysis*, vol. 30, no. 2, pp. 147–155, 2010.

[9] A. Kumar and M. M. Mishra, "Green functions and related boundary value problems on the Heisenberg group," *Complex Variables and Elliptic Equations*, vol. 58, no. 4, pp. 547–556, 2013.

[10] M. M. Mishra, A. Kumar, and S. Dubey, "Green's function for certain domains in the Heisenberg group Hn," *Boundary Value Problems*, vol. 2014, article 182, 2014.

[11] G. B. Folland and E. M. Stein, "Estimates for the $\overline{\partial}_b$ complex and analysis on the Heisenberg group," *Communications on Pure and Applied Mathematics*, vol. 27, pp. 429–522, 1974.

[12] G. B. Folland, "A fundamental solution for a subelliptic operator," *Bulletin of the American Mathematical Society*, vol. 79, pp. 373–376, 1973.

[13] E. D. Rainville, *Special Functions*, The MacMillan Company, New York, NY, USA, 1960.

[14] O. D. Kellogg, *Foundations of Potential Theory*, Ungar, New York, NY, USA, 1929.

[15] A. Korányi, "Kelvin transforms and harmonic polynomials on the Heisenberg group," *Journal of Functional Analysis*, vol. 49, no. 2, pp. 177–185, 1982.

7

δ-Primary Hyperideals on Commutative Hyperrings

Elif Ozel Ay, Gürsel Yesilot, and Deniz Sonmez

Department of Mathematics, Yildiz Technical University, Davutpasa, Istanbul, Turkey

Correspondence should be addressed to Elif Ozel Ay; elifozel314@gmail.com

Academic Editor: Aloys Krieg

The purpose of this paper is to define the hyperideal expansion. Hyperideal expansion is associated with prime hyperideals and primary hyperideals. Then, we define some of their properties. Prime and primary hyperideals' numerous results can be extended into expansions.

1. Introduction

The hyperstructure theory was introduced by Marty (1934). Hyperstructures have many applications to several sectors of both pure and applied mathematics. A hypergroup in the sense of Marty is a nonempty set H endowed by a hyperoperation $* : H \times H \rightarrow P^*(H)$ [1], the set of the entire nonempty set H, which satisfies the associative law and reproduction axiom. Canonical hypergroups are a special class of the hypergroup of Marty. The more general structure that satisfies the ring-like axioms is the hyperring in the general sense: $(R, +, \cdot)$ is a hyperring if $+$ and $\cdot$ are two hyperoperations such that $(R, +)$ is a hypergroup and $\cdot$ is an associative hyperoperation, which is distributive with respect to $+$. There are different notions of hyperrings [1]. If only the addition $+$ is hyperoperation and the multiplication $\cdot$ is usual operation, then we say that R is an additive hyperring. A special case of this type is the hyperring introduced by Krasner (1957) [2]. Also, Krasner (1983) introduced a class of hyperring and hyperfields and the quotient hyperrings and hyperfields. If only $\cdot$ is a hyperoperation, we shall say that R is a multiplicative hyperring [2]. Rota (1982) introduced the multiplicative hyperrings; subsequently, many authors worked on this field (Nakassis, 1988; Olson and Ward, 1997; Procesi and Rota, 1999; Rota, 1996) [2]. Algebraic hyperstructures have been studied in the following decades and nowadays by many mathematicians.

Although the δ-primary ideals have been investigated by Dongsheng [3], the concept of δ-primary hyperideals which unify prime hyperideals and primary hyperideals has not been studied yet. So, this work shows some elementary properties of the hyperideal expansion; then we show some new results of δ-primary hyperideals. After this introductory section, Section 2 is devoted to some definitions and properties related to δ primary ideals and hyperideals that will be needed later. In Section 3, the definitions of hyperideal expansion and δ primary hyperideals will be given and some basic properties of these concepts will be studied.

2. Preliminaries

Throughout this paper $(R, +, \cdot)$ denotes the Krasner hyperring.

Definition 1 (see [4]). A Krasner hyperring is an algebraic structure $(R, +, \cdot)$ which satisfies the following axioms:

(1) $(R, +)$ is a canonical hypergroup; that is,

- (i) for every $x, y, z \in R$, $x + (y + z) = (x + y) + z$,
- (ii) for every $x, y \in R$, $x + y = y + x$,
- (iii) there exists $0 \in R$ such that $0 + x = \{x\}$ for every $x \in R$,
- (iv) for every $x \in R$ there exists a unique element $x' \in R$ such that $0 \in x + x'$,
- (v) $z \in x + y$ implies $y \in -x + z$ and $x \in z - y$,

(2) $(R, \cdot)$ is a semigroup having zero as a bilaterally absorbing element; that is, $x \cdot 0 = 0 \cdot x = 0$.

(3) The multiplication is distributive with respect to the hyperoperation +.

Definition 2 (see [2]). Let $(R, +, \cdot)$ be a hyperring and A be a nonempty subset of R. Then A is said to be a subhyperring of R if $(A, +, \cdot)$ is itself a hyperring.

Definition 3 (see [1]). A subhyperring A of a hyperring R is a left (right) hyperideal of R if $ra \in A$ ($ar \in A$) for all $r \in R$ and $a \in A$. A is called a hyperideal if A is both a left and a right hyperideal.

Lemma 4 (see [2]). *A nonempty subset A of a hyperring R is a hyperideal if and only if*

(1) $a, b \in A$ *implies* $a - b \subseteq A$,

(2) $a \in A$ *and* $r \in R$ *imply* $ra \in A$.

Definition 5 (see [2]). Let R_1 and R_2 be hyperrings. A mapping φ from R_1 into R_2 is said to be a good (strong) homomorphism if, for all $a, b \in R_1$,

$$\varphi(a+b) = \varphi(a) + \varphi(b),$$
$$\varphi(a \cdot b) = \varphi(a) \cdot \varphi(b), \quad (1)$$
$$\varphi(0) = 0.$$

Definition 6 (see [1]). Let $f : R \rightarrow S$ be a hyperring homomorphism. The kernel of f, denoted $\ker(f)$, is the set of elements of R that map to 0 in S; that is, $\ker(f) = \{x \in R \mid f(x) = 0\}$.

Definition 7 (see [2]). A hyperideal P of a hyperring R is called a prime hyperideal if whenever $a \cdot b \in P$, either $a \in P$ or $b \in P$.

Definition 8 (see [2]). Let I be a hyperideal of the hyperring R. Then the radical of I, denoted by $\sqrt{I}$, is defined as $\sqrt{I} = \{x \mid x^n \in I \text{ for some } n \in N\}$.

Definition 9 (see [2]). A hyperideal I of a hyperring R is called a primary hyperideal if whenever $a \cdot b \in P$, either $a \in P$ or $b \in \sqrt{P}$.

3. Hyperideal Expansion and δ Primary Hyperideals

Definition 10. An expansion of hyperideals, or briefly hyperideal expansion, is a function δ which assigns to each hyperideal I of a hyperring R another hyperideal $\delta(I)$ of the same ring such that the following conditions are satisfied:

(i) $I \subseteq \delta(I)$.

(ii) $P \subseteq Q$ implies $\delta(P) \subseteq \delta(Q)$ for P, Q hyperideals of R.

Example 11. Let $\mathrm{Id}(R)$ denote the set of all hyperideals of the hyperring R. The identity function δ_0, where $\delta(I) = I$ for every $I \in \mathrm{Id}(R)$, is an expansion of hyperideals.

For each I hyperideal define $\delta_1(I) = \sqrt{I}$, the radical of I. Then δ_1 is an expansion of hyperideals.

Definition 12. Given an expansion δ of hyperideals, a hyperideal I of R is called δ-primary if $ab \in I$ and $a \notin I$ imply $b \in \delta(I)$ for all $a, b \in R$.

Obviously the definition of δ-primary hyperideals can be also stated as $ab \in I$ and $a \notin \delta(I)$ implies $b \in I$ for all $a, b \in R$.

Example 13.

(1) A Hyperideal I Is δ_0-Primary If and Only If It Is Prime. Let I be δ_0-primary hyperideal. We show that I is prime. Assume that $ab \in I$ and that I is δ_0-primary $a \in I$ or $b \in \delta_0(I) = I$ so I is a prime hyperideal.

Conversely, let I be a prime hyperideal. Assume that $ab \in I$. Since I is prime $a \in I$ or $b \in I = \delta_0(I)$ and I is δ_0-primary.

(2) A Hyperideal I is δ_1-Primary If and Only If It Is Primary. Let I be δ_1-primary. We show that I is primary. Assume that $ab \in I$. Since I is δ_1-primary then we can say that $a \in I$ or $b \in \delta_1(I)$. That is $a \in I$ or $b \in \sqrt{I}$. So I is primary.

Conversely let I be a primary hyperideal. Assume that $ab \in I$. Since I primary hyperideal $a \in I$ or $b \in \sqrt{I}$, thus $a \in I$ or $b \in \sqrt{I} = \delta_1(I)$.

Remark 14. (1) If δ and γ are two hyperideal expansions and $\delta(I) \subseteq \gamma(I)$ for each hyperideal I, then every δ-primary hyperideal is also γ-primary. Thus, in particular, a prime hyperideal is δ-primary for every δ hyperideal expansion. Let I be δ-primary. Assume that, for all $a, b \in R$, $ab \in I$ and $a \notin I$. Since I is δ-primary and $\delta(I) \subseteq \gamma(I)$ and $b \in \delta(I) \subseteq \gamma(I)$, thus $b \in \gamma(I)$.

(2) Given two hyperideal expansions δ_1 and δ_2, define $\delta(I) = \delta_1(I) \cap \delta_2(I)$. Then δ is also a hyperideal expansion. Since δ_1 and δ_2 are hyperideal expansions $I \subseteq \delta_1(I)$ and $I \subseteq \delta_2(I)$, then $I \subseteq \delta_1 \cap \delta_2(I) = \delta(I)$ and $I \subseteq \delta(I)$.

Let P and Q be any hyperideals of R and $P \subseteq Q$. Thus $\delta_1(P) \subseteq \delta_1(Q)$ and $\delta_2(P) \subseteq \delta_2(Q)$. Finally we find $\delta_1(P) \cap \delta_2(P) \subseteq \delta_1(Q) \cap \delta_2(Q)$.

(3) Let δ be a hyperideal expansion. Define $E_\delta(P) = \bigcap\{J \in \mathrm{Id}(R) \mid P \subseteq J,\ J \text{ is } \delta\text{-primary}\}$. Then E_δ is still a hyperideal expansion.

For all $P \in \mathrm{Id}(R)$, we show that $P \subseteq E_\delta(P)$ for any $K, L \in \mathrm{Id}(R)$, if $K \subseteq L$; then $E_\delta(K) \subseteq E_\delta(L)$. By the definition of $E_\delta(P)$, we conclude that $P \subseteq E_\delta(P)$. For any $K, L \in \mathrm{Id}(R)$, if $K \subseteq L$, then the δ-primary hyperideals which contain L contain also K. In addition, there may be δ-primary hyperideals which contained L but did not contain K. Hence, we conclude that $E_\delta(K) \subseteq E_\delta(L)$.

Lemma 15. *A hyperideal P is δ-primary if and only if for any two hyperideals I and J, if $IJ \subseteq P$ and $I \nsubseteq P$ then $J \subseteq \delta(P)$.*

Proof. Let P be δ-primary. Suppose $IJ \subseteq P$ and $I \nsubseteq P$, but $J \nsubseteq \delta(P)$, and then we can choose $a \in I - P$ and $b \in J - \delta(P)$. Then $ab \in IJ \subseteq P$ but $a \notin P$ and $b \notin \delta(P)$. This contradicts the assumption that P is δ-primary.

Conversely, if the condition is satisfied, for any two elements a and b, suppose $ab \in P$ and $a \notin P$. Then $(a)(b) \subseteq P$ and $(a) \nsubseteq P$. So $(b) \subseteq \delta(P)$. Hence $b \in (b) \subseteq \delta(P)$ implies $b \in \delta(P)$. Thus P is δ-primary. □

Recall that if I and J are ideals of a commutative ring R, then their ideal quotient denotes $(I : J)$ defined by $(I : J) = \{r \in R \mid rJ \subset I\}$. We recall also ideal quotient $(I : J)$ is itself an ideal in R.

Theorem 16. *Let δ be a hyperideal expansion. Then*

(1) *if P is a δ-primary hyperideal and I is a hyperideal with $I \nsubseteq \delta(P)$, then $(P : I) = P$,*

(2) *for any δ-primary hyperideal P and any subset N of the R, $(P : N)$ is also δ-primary.*

Proof. (1) From the definition of $(P : I)$, for all $x \in I \cdot (P : I)$, $x \in \sum_{i=1}^{n} a_i p_i$, $a_i \in I$ and $p_i \in (P : I)$. Since P is a hyperideal $x \in \sum_{i=1}^{n} a_i p_i \subseteq P$, then we get $x \in P$. In other words $I \cdot (P : I) \subseteq P$. Since P is δ-primary, if $I \nsubseteq \delta(P)$ then $(P : I) \subseteq P$.

Conversely, since $P \subseteq (P : I)$ then $(P : I) = P$.

(2) For all $a, b \in R$, assume that $ab \in (P : N)$ and $a \notin (P : N)$. Then there exists a $n \in N$ such that $an \notin P$. But $anb = abn \in P$. Thus $b \in \delta(P)$. Since $\delta(P) \subseteq \delta(P : N)$, $b \in \delta(P : N)$. By this way we get that $(P : N)$ is δ-primary. □

Theorem 17. *If δ is a hyperideal expansion such that $\delta(I) \subseteq \delta_1(I)$ for every hyperideal I, then, for any δ-primary hyperideal P, $\delta(P) = \delta_1(P)$.*

Proof. For all I hyperideals, since $\delta(I) \subseteq \delta_1(I)$, $\delta(P) \subseteq \delta_1(P)$.

Conversely, let $a \in \delta_1(P)$. We show that $a \in \delta(P)$.

Then there exists k which is the least positive integer k with $a^k \in P$. If $k = 1$ then $a \in P \subseteq \delta(P)$. If $k > 1$ then $a^{k-1}a \in P$. But $a^{k-1} \notin P$, so $a \in \delta(P)$. Hence $\delta_1(P) \subseteq \delta(P)$ and $\delta(P) = \delta_1(P)$. □

4. Expansions with Extra Properties

In this section we investigate δ-primary hyperideals where δ satisfy additional conditions and prove more results with respect to such expansions.

Definition 18. A hyperideal expansion δ is intersection preserving if it satisfies

$$\delta(I \cap J) = \delta(I) \cap \delta(J) \quad \text{for any } I, J \in \mathrm{Id}(R). \quad (2)$$

An expansion is said to be global if for any hyperring homomorphism $f : R \to S$

$$\delta\left(f^{-1}(I)\right) = f^{-1}(\delta(I)) \quad \forall I \in \mathrm{Id}(S). \quad (3)$$

The expansions δ_0 and δ_1 are both intersection preserving and global.

For any $I, J \in \mathrm{Id}(R)$, $\delta_0(I \cap J) = I \cap J = \delta_0(I) \cap \delta_0(J) = I \cap J$:

$$\delta_1(I \cap J) = \sqrt{I \cap J} = \delta_1(I) \cap \delta_1(J) = \sqrt{I} \cap \sqrt{J}. \quad (4)$$

And $\delta_0(f^{-1}(I)) = \delta(I) = I = f^{-1}(I)$.

$\delta_1(f^{-1}(I)) = \delta_1(I) = \sqrt{I} = f^{-1}(I)$. Thus δ_0 and δ_1 are both intersection preserving and global.

Theorem 19. *Let δ be an intersection preserving hyperideal expansion. If $Q_1, Q_2, \ldots, Q_n$ are δ-primary hyperideals of R and $P = \delta(Q_i)$ for all i, then $Q = \bigcap_{i=1}^{n} Q_i$ is δ-primary.*

Proof. Let δ be an intersection preserving hyperideal expansion. If $x, y \in Q$ and $x \notin Q$, then $x \notin Q_k$ for some k. But $xy \in Q \subseteq Q_k$ and Q_k is δ-primary, so $y \in \delta(Q_i)$. But $\delta(Q) = \delta(\bigcap_{i=1}^{n} Q_i) = \bigcap_{i=1}^{n} \delta(Q_i) = P = \delta(Q_k)$. Thus $y \in \delta(Q)$. So Q is δ-primary. □

Definition 20. Let R be a hyperring and δ be a hyperideal expansion. If for an $a \in R$, $a \in \delta\{0\}$ then a is called δ nilpotent.

Note that δ_0 nilpotent element of a ring is the zero element of the ring. Also δ_1 nilpotent elements are exactly the ordinary nilpotent elements.

Theorem 21. *Let δ be a global expansion. Let a hyperideal I of R be δ-primary and then every zero divisor of the quotient hyperring R/I is δ nilpotent.*

Proof. Let I be δ-primary. If $\tilde{r} = r + I$ is a zero divisor of R/I, then there is a $\tilde{s} = s + I \neq I$ with $\tilde{r}\tilde{s} = rs + I = I$. This means that $rs \in I$ and $s \notin I$. Since I is δ-primary so $r \in \delta(I)$; that is, $\tilde{r} \in \delta(I)/I$.

Let $q : R \to R/I$ be the natural quotient hyperring homomorphism. As δ is global, we have $\delta(I) = \delta(q^{-1}(0_{R/I})) = q^{-1}(\delta(0_{R/I}))$.

Since q is onto, so $\delta(I)/I = q(\delta)(I) = \delta(0_{R/I})$. Hence we get $\tilde{r} \in \delta(0_{R/I})$, so $\tilde{r}$ is δ nilpotent. □

Theorem 22. *If δ is global and $f : R \to S$ is a hyperring homomorphism, then, for any δ-primary hyperideal I of S, f^{-1} is a δ-primary hyperideal of R.*

Proof. Let $a, b \in R$ with $ab \in f^{-1}(I)$. If $a \notin f^{-1}(I)$ then $f(a)f(b) \in I$ but $f(a) \notin I$. So, as I is δ-primary, $f(b) \in \delta(I)$. So $b \in f^{-1}(\delta(I)) = \delta(f^{-1}(I))$. Hence $f^{-1}(I)$ is δ-primary. □

Theorem 23. *Let $f : R \to S$ be a surjective hyperring homomorphism. Then a hyperideal I of R that contains* $\ker(f)$ *is δ-primary hyperideal of S.*

Proof. If $f(I)$ is δ-primary, then by $I = f^{-1}(f(I))$ and Theorem 22, I is δ-primary. Now suppose $f(I)$ is δ-primary. If $a, b \in S$ and $ab \in f(\mathrm{I})$ and $a \notin f(I)$, then there are $x, y \in R$ with $f(x) = a$, $f(y) = b$. Then $f(xy) = f(x)f(y) = ab \in f(I)$ implies $xy \in f^{-1}(f(I)) = I$ and $f(x) = a \notin f(I)$ implies $xy \in f^{-1}(f(I)) = I$ and $f(x) = a \notin f(I)$ implies $x \notin I$.

So $y \in \delta(I)$, and hence $b = f(y) \in f(\delta(I))$. Now one only needs to prove $f(\delta(I)) = \delta(f(I))$. But this follows directly from $\delta(I) = \delta(f^{-1}(f(I))) = f^{-1}(\delta(f(I)))$ and that f is surjective. □

The following theorem does not need a proof because it is a consequence of Theorems 22 and 23.

Correspondence Theorem for δ-Primary Hyperideals. *Let f be a hyperring homomorphism of a hyperring R onto a* *hyperring S and let δ be global hyperring expansion. Then f induces a one-one inclusion preserving correspondence between δ-primary hyperideals of R containing $\ker f$ and the δ-primary hyperideals of S in such a way that if I is a δ-primary hyperideal of R that I contains $\ker f$, then $f(I)$ is the corresponding δ-primary hyperideal of S, and if J is a δ-primary hyperideal of S, then $f^{-1}(J)$ is the corresponding δ-primary hyperideal of R.*

Conflicts of Interest

The authors declare that they have no conflicts of interest.

References

[1] R. Ameri and M. Norouzi, "On Commutative Hyperrings," *International Journal of Algebraic Hyperstructures and its Applications*, vol. 1, no. 1, pp. 45–58, 2014.

[2] B. Davvaz and V. Leoreanu, *Hyperring Theory and Applications*, International Academic Press, Palm Harbor, Fla, USA, 2007.

[3] Z. Dongsheng, "δ-primary ideals of commutative rings," *Kyungpook Mathematical Journal*, vol. 41, pp. 17–22, 2001.

[4] B. Davvaz and A. Salasi, "A realization of hyperrings," *Communications in Algebra*, vol. 34, no. 12, pp. 4389–4400, 2006.

8

On Degrees of Modular Common Divisors and the Big Prime gcd Algorithm

Vahagn Mikaelian[1,2]

[1] *Yerevan State University, Alex Manoogian 1, 0025 Yerevan, Armenia*
[2] *American University of Armenia, 40 Marshal Baghramyan Ave., 0019 Yerevan, Armenia*

Correspondence should be addressed to Vahagn Mikaelian; vmikaelian@ysu.am

Academic Editor: Nawab Hussain

We consider a few modifications of the Big prime modular gcd algorithm for polynomials in $\mathbb{Z}[x]$. Our modifications are based on bounds of degrees of modular common divisors of polynomials, on estimates of the number of prime divisors of a resultant, and on finding preliminary bounds on degrees of common divisors using auxiliary primes. These modifications are used to suggest improved algorithms for gcd calculation and for coprime polynomials detection. To illustrate the ideas we apply the constructed algorithms on certain polynomials, in particular on polynomials from Knuth's example of intermediate expression swell.

1. Introduction

This work is one of the articles in which we would like to present parts from new introduction to computer algebra [1] that currently is under preparation. In [1] we try to give a "more algebraic" and detailed view on some of the areas of computer algebra, such as algorithms on the Euclidean rings, extensions of fields, operators in spaces on finite fields, and factorization in UFDs.

The Big prime modular gcd algorithm is one of the first and most popular algorithms of computer algebra. In its classical form, it allows calculating the greatest common divisor $\gcd(f(x), g(x))$ for any nonzero polynomials $f(x), g(x) \in \mathbb{Z}[x]$. There are a few modifications of this algorithm for other UFDs, such as multivariate polynomial rings. Attention to the gcd calculation is partially explained by the first examples that were built to explain importance of application of algebraic methods to computer science. In particular, Knuth's well-known example of intermediate expression swell discusses the polynomials

$$f(x) = x^8 + x^6 - 3x^4 - 3x^3 + 8x^2 + 2x - 5,$$
$$g(x) = 3x^6 + 5x^4 - 4x^2 - 9x + 21, \tag{1}$$

and it shows that calculation of $\gcd(f(x), g(x))$ by traditional Euclidean algorithm on rational numbers generates very large integers to deal with, whereas consideration of these polynomials modulo p, that is, consideration of their images under ring homomorphism $\varphi_p : \mathbb{Z}[x] \to \mathbb{Z}_p[x]$ (where $\mathbb{Z}_p[x]$ is the polynomial ring over the residue ring $\mathbb{Z}_p \cong \mathbb{Z}/p\mathbb{Z}$) very easily shows that $\gcd(f(x), g(x)) = 1$ (see [2] and also [3–6]). We are going to use polynomials (1) as examples below to apply the algorithms below (see Examples 16, 18, 22, 25, 27, and 29).

The main idea of the Big prime modular gcd algorithm is that for the given polynomials $f(x), g(x) \in \mathbb{Z}[x]$ one may first consider their images $f_p(x) = \varphi_p(f(x)), g_p(x) = \varphi_p(g(x)) \in \mathbb{Z}_p[x]$ under φ_p. Unlike $\mathbb{Z}[x]$, the ring $\mathbb{Z}_p[x]$ is an Euclidean domain, since it is a polynomial ring over a field, so $\gcd(f_p(x), g_p(x))$ can be computed in it by the well-known Euclidean algorithm. There remains "to lift" a certain fold $t \cdot \gcd(f_p(x), g_p(x))$ of it to the ring $\mathbb{Z}[x]$ to reconstruct the preimage $\gcd(f(x), g(x))$. The "lifting" procedure consists of selecting the suitable value for prime p, then finding in $\mathbb{Z}[x]$ an appropriate preimage for $\gcd(f_p(x), g_p(x))$, and then checking if that preimage divides both $f(x)$ and $g(x)$. If yes, it is $\gcd(f(x), g(x))$ we are looking for. If not, then a new p need be selected to repeat the process. Arguments based on resultants and on Landau-Mignotte bounds show that we can effectively choose p such that the number of required repetitions is "small."

The first aim of this work is to present in Sections 2–5 a slightly modified argumentation of the algorithm, based on comparison of the degrees of common divisors of $f(x)$ and $g(x)$ in $\mathbb{Z}[x]$ and of $f_p(x)$ and $g_p(x)$ in $\mathbb{Z}_p[x]$ (see Algorithm 2). This approach allows some simplification of a step of the algorithm: for some primes p, we need not reconstruct the preimage of $t \cdot \gcd(f_p(x), g_p(x))$, but we immediately get an indication that this prime is not suitable, and we should proceed to a new p (see Remark 15).

Then in Section 6 we discuss the problem if the Big prime modular gcd algorithm could output the correct answer using just one prime p or not. The answer is positive, but for some reasons it should not be used to improve the algorithm (to make it work with one p) because it evolves a very large prime (see Remark 19). Instead, we show that we can estimate the maximal number of p (repetitions of steps) that may be used in traditional Big prime modular gcd algorithm. For example, for polynomials (1) of Knuth's example, this number is at most 31. Estimates of this type can be found in literature elsewhere. We just make the bound considerably smaller (see Remark 23).

The obtained bounds on the number of primes p are especially effective when we are interested not in gcd but just in detection if the polynomials $f(x), g(x) \in \mathbb{Z}[x]$ are coprime or not. We consider this in Section 7 (see Algorithm 3).

In Section 8 we consider four other ideas to modify the Big prime modular gcd algorithm. Two first ideas are based on checking the number of primes p. The third idea is based on using an auxiliary prime q to estimate the degree of $\gcd(f(x), g(x))$ by means of the degree of $\gcd(f_q(x), g_q(x))$ (see Algorithm 4). Example 27 shows how much better results we may get by this modification. The fourth idea combines both approaches: it uses a set of auxiliary primes $q_1, \ldots, q_{k+1}$ to correctly find the degree of $\gcd(f(x), g(x))$, and then we use a modified version of Landau-Mignotte bound to find a single big prime p by which we can calculate the $\gcd(f_p(x), g_p(x))$.

The arguments used here can be generalized for the case of polynomials on general UFDs. From the unique factorization in a UFD, it easily follows that gcd always exists, and it is easy to detect if the given common divisor of maximal degree is gcd or not. The less simple part is to find ways to compute gcd (without having the prime-power factorization). That can be done for some classes of UFDs, such as multivariate polynomials on fields.

2. gcd **in Polynomial Rings and the Degrees of Common Divisors**

The problem of finding the greatest common divisor $\gcd(a, b)$ of any nonzero elements a, b in a ring R can be separated to two tasks:

(1) Finding out if $\gcd(a, b)$, in general, *exists* for $a, b \in R$.

(2) Finding an *effective* way to calculate $\gcd(a, b)$.

The Euclidean algorithm gives an easy answer to both of these tasks in any Euclidean domain, that is, an integrity domain R possessing *Euclidean norm* $\delta : R\backslash 0 \to \mathbb{N}\cup\{0\}$, such that $\delta(ab) \geq \delta(a)$ hold for any nonzero elements $a, b \in R$; and for any $a, b \in R$, where $b \neq 0$, there exist elements $q, r \in R$, such that $a = qb + r$, where either $r = 0$ or $r \neq 0$ and $\delta(r) < \delta(b)$ [4, 5, 7–10]. The Euclidean algorithm works for any polynomial ring $K[x]$ over a field K, such as $\mathbb{Q}[x]$, $\mathbb{R}[x]$, $\mathbb{C}[x]$, and $\mathbb{Z}_p[x]$ because these rings can easily be turned to an Euclidean domain by defining $\delta(f(x)) = \deg f(x)$ for any nonzero $f(x) \in K[x]$.

The situation is less simple in non-Euclidean domains, even in such a widely used ring as the ring $\mathbb{Z}[x]$ of polynomials with integer coefficients. That $\mathbb{Z}[x]$ is not an Euclidean domain easy to show by elements $x, 2 \in \mathbb{Z}[x]$. If $\mathbb{Z}[x]$ were an Euclidean domain, it would contain elements $u(x), v(x)$ such that $x \cdot u(x) + 2 \cdot v(x) = \gcd(x, 2) = \pm 1$, which is not possible.

The first of two tasks mentioned above, namely, *existence* of gcd, can be accomplished for $\mathbb{Z}[x]$ by proving that $\mathbb{Z}[x]$ is a UFD, that is, an integrity domain in which every nonzero element a has a factorization $a = \epsilon\, p_1 \cdots p_k$, where $\epsilon \in R^*$ is a unit (invertible) element in R, the elements p_i are prime for all $i = 1, \ldots, k$, and the factorization above is unique in the sense that if a has another factorization of that type $\theta\, q_1 \cdots q_s$, where $\theta \in R^*$ and the elements q_i are prime, then $k = s$ and (perhaps after some reordering of the prime factors) the respective prime elements are associated: $p_i \approx q_i$ for all $i = 1, \ldots, k$. For briefness, in the sequel we will often omit the phrase "perhaps after some reordering of the prime factors" and this will cause no confusion.

After merging the associated prime elements together, we get a unique factorization into prime-power elements:

$$a = \nu\, p_1^{\alpha_1} \cdots p_n^{\alpha_n}, \quad \nu \in R^*, \; \alpha_i \in \mathbb{N}, \; p_i \not\approx p_j \text{ for any } i \neq j; \; i, j = 1, \ldots, n \tag{2}$$

(in some arguments below we may admit that some of the factors $p_i^{\alpha_i}$ participate with degrees $\alpha_i = 0$; this makes some notations simpler). From this, it is easy to see that, in a UFD R, $\gcd(a, b)$ exists for any nonzero elements $a, b \in R$. Assume $b \in R$ has the factorization

$$b = \kappa\, p_1^{\alpha'_1} \cdots p_n^{\alpha'_n}, \quad \kappa \in R^* \tag{3}$$

(we use the same primes p_i in both factorizations because if, e.g., p_i is not actually participating in one of those factorizations, we can add it as $p_i^{\alpha_i}$ with $\alpha_i = 0$). Then

$$\gcd(a, b) = d = p_1^{\gamma_1} \cdots p_n^{\gamma_n}, \tag{4}$$

where $\gamma_i = \min\{\alpha_i, \alpha'_i\}$. This follows from uniqueness of factorization in UFD. For, if h is a common divisor of a, b and if p_i is a prime divisor of h, then it also is a prime divisor of a and of b. The elements p_i cannot participate in factorization of h by a power greater than $\min\{\alpha_i, \alpha'_i\}$, because then a (or b) would have an alternative factorization in which p_i occurs more than α_i (or α'_i) times.

The shortest way to see that $\mathbb{Z}[x]$ is a UFD is to apply Gauss's theorem: if the ring R is a UFD, then the polynomial ring $R[x]$ also is a UFD [4, 7, 8, 10, 11]. Since $\mathbb{Z}$ is a UFD (that fact is known as "the fundamental theorem of arithmetic"), $\mathbb{Z}[x]$ also is a UFD.

Clearly, $\gcd(a,b)$ is defined up to a unit multiplier from R^*. For integers from $R = \mathbb{Z}$ or for polynomials from $R = \mathbb{Z}[x]$, this unit multiplier can be just -1 or 1. So to say, $\gcd(a,b)$ is defined "up to the sign ± 1" because $\mathbb{Z}^* = \mathbb{Z}[x]^* = \{-1,1\}$. And for polynomials from $R = \mathbb{Z}_p[x]$, $\gcd(a,b)$ is defined up to any nonzero multiplier $t \in \mathbb{Z}_p^* = \{1,\ldots,p-1\}$. Taking this into account, we can use $\gcd(a,b) = 1$ and $\gcd(a,b) \approx 1$ as equivalent notations, since associated elements are defined up to a unit multiplier. Notice that in some sources they prefer to additionally introduce a normal form of gcd to distinguish one fixed instance of gcd. Instead of using that extra term, we will just in a few places refer to the "positive gcd," meaning that we take, for example, $2 = \gcd(6,8)$ and not -2.

Furthermore, since the content $\text{cont}(f(x))$ of a polynomial $f(x)$ is gcd for some elements (coefficients of the polynomials), the constant and the primitive part $\text{pp}(f(x)) = f(x)/\text{cont}(f(x))$ can also be considered up to a unit multiplier. For a nonzero polynomial $f(x) \in \mathbb{Z}[x]$, we can choose $\text{cont}(f(x))$ so that $\text{sgn}\,\text{cont}(f(x)) = \text{sgn}\,\text{lc}(f(x))$; that is, the $\text{cont}(f(x))$ has the same sign as the leading coefficient of $f(x)$. Then the leading coefficient $\text{lc}(\text{pp}(f(x)))$ of the primitive part $\text{pp}(f(x)) = f(x)/\text{cont}(f(x))$ will be positive. We will use this below without special notification.

Now we would like to little restrict the algebraic background we use. Two main algebraic systems, used in the Big prime modular gcd algorithm, are the Euclidean domains and the UFDs. However, their usage is "asymmetric" in the sense that the Euclidean domains and Euclidean algorithm are used in many parts of the Big prime modular gcd algorithm, whereas the UFDs are used just to prove that gcd does exist. Moreover, it is easy to understand that (2) and (4) may hardly be effective tools to calculate gcd, since they are using factorization of elements to primes, while finding such a factorization is a more complicated task than finding just gcd. Thus, it is reasonable to drop the UFDs from consideration and to obtain (2) directly using Gauss's lemma on primitive polynomials in $\mathbb{Z}[x]$ (a polynomial $f(x) \in \mathbb{Z}[x]$ is primitive if $\text{cont}(f(x)) \approx 1$, that is, $\text{pp}(f(x)) = f(x)/\text{cont}(f(x)) \approx f(x)$).

By Gauss's lemma, a product of two primitive polynomials is primitive in $\mathbb{Z}[x]$ [4, 7, 8, 10, 11]. So if

$$\begin{aligned} f(x) &= \text{cont}(f(x)) \cdot \text{pp}(f(x)), \\ g(x) &= \text{cont}(g(x)) \cdot \text{pp}(g(x)), \end{aligned} \tag{5}$$

then

$$\begin{aligned} \text{cont}(f(x) \cdot g(x)) &= \text{cont}(f(x)) \cdot \text{cont}(g(x)), \\ \text{pp}(f(x) \cdot g(x)) &= \text{pp}(f(x)) \cdot \text{pp}(g(x)). \end{aligned} \tag{6}$$

The following is easy to deduce from Gauss's lemma.

Lemma 1. *If $f(x), t(x) \in \mathbb{Z}[x]$ and $t(x)$ is primitive, then if $t(x)$ divides $f(x)$ in the ring $\mathbb{Q}[x]$ and then $t(x)$ also divides $f(x)$ in $\mathbb{Z}[x]$.*

The unique factorization of any nonzero $f(x) \in \mathbb{Z}[x]$ is easy to obtain from decompositions (6) above and from Lemma 1. Let us just outline it; the details can be found in [1, 4, 5, 7, 8]. By the fundamental theorem of arithmetic, $\text{cont}(f(x))$ can in a unique way be presented as a product of powers of primes: $\text{cont}(f(x)) = \nu\, p_1^{\alpha_1} \cdots p_n^{\alpha_n}$. So, if $\deg f(x) = 0$, then we are done.

Assume $\deg f(x) > 0$. If $f(x)$ is not prime, then, by repeatedly splitting it to products of factors of lower degree as many times as needed, we will eventually get a presentation of $f(x)$ as a product of $\text{cont}(f(x))$ and of some finitely many primitive prime polynomials $q_i(x)$ of degrees greater than 0. We do not yet have the uniqueness of this decomposition, but we can still group the associated elements together to get the presentation

$$\begin{aligned} f(x) &= \text{cont}(f(x)) \cdot \text{pp}(f(x)) \\ &= \nu\, p_1^{\alpha_1} \cdots p_n^{\alpha_n} \cdot q_1^{\beta_1}(x) \cdots q_m^{\beta_m}(x). \end{aligned} \tag{7}$$

If $f(x)$ has another, alternative presentation of this sort and if $t(x)$ is one of the primitive prime factors (of degree greater than 0) of that presentation, then product (7) is divisible by $t(x)$. By Lemma 1, $t(x)$ divides $f(x)$ also in $\mathbb{Z}[x]$. Since $t(x)$ is prime, it is associated with one of $q_i(x)$. Eliminate one instance of this $q_i(x)$ in (7) and consider $f(x)/q_i(x)$. If $f(x)/q_i(x)$ also is divisible by $q_i(x)$, we repeat the process. If not, we turn to other primitive prime polynomials (of degree greater than 0) dividing what remains from (7) after elimination. After finitely many steps, (7) will become $\nu\, p_1^{\alpha_1} \cdots p_n^{\alpha_n}$, and also from the other alternative presentation a constant should be left only. So we apply the fundamental theorem of arithmetic one more time to get that (7) is the unique factorization.

We see that (7) is a particular case of (2). The proof above avoided usage of Gauss's theorem and the formal definitions of the UFDs. And we see that the prime elements of $\mathbb{Z}[x]$ are of two types: *prime numbers* and *primitive prime polynomials* of degrees greater than 0.

Existence of $\gcd(f(x), g(x))$ for any two nonzero polynomials in $f(x), g(x) \in \mathbb{Z}[x]$ can be deduced from (7) in analogy with (4). If

$$g(x) = \nu'\, p_1^{\alpha_1'} \cdots p_n^{\alpha_n'} \cdot q_1^{\beta_1'}(x) \cdots q_m^{\beta_m'}(x), \tag{8}$$

then

$$\gcd(f(x), g(x)) = \kappa\, p_1^{\gamma_1} \cdots p_n^{\gamma_n} \cdot q_1^{\delta_1}(x) \cdots q_m^{\delta_m}(x), \tag{9}$$

where $\kappa = \pm 1$, $\gamma_i = \min\{\alpha_i, \alpha_i'\}$, and $\delta_j = \min\{\beta_j, \beta_j'\}$ ($i = 1,\ldots,n$; $j = 1,\ldots,m$). However, like we admitted earlier, (4) and (9) are no effective tools to calculate gcd. We will turn to gcd calculation algorithm in the next sections (see Algorithms 4 and 5).

Equations (4) and (9) allow us to get some information that will be essential later. Observe that the following definition of gcd, often used in elementary mathematics, is no longer true for general polynomial rings: "*$d(x)$ is the greatest common divisor of $f(x)$ and $g(x)$ if it is their common divisor of maximal degree.*" For example, for $f(x) = 12x^2 + 24x + 12$ and $g(x) = 8x + 8$, the maximum of degree of their common divisors is 1. Nevertheless, $h(x) = x+1$ is not $\gcd(f(x), g(x))$, although $h(x) \mid f(x), h(x) \mid g(x)$, and $\deg h(x) = 1$. For, $x+1$ is not divisible by the common divisor $2x + 2$. We can detect the cases when the divisor of highest degree is gcd.

Lemma 2. *For polynomials $f(x), g(x) \in \mathbb{Z}[x]$, their common divisor of maximal degree $h(x)$ is their* gcd *if and only if* $\text{cont}(h(x)) \approx \gcd(\text{cont}(f(x)), \text{cont}(g(x)))$.

The lemma easily follows from (7), (8), and (9). We see that in example above the condition was missing: $\text{cont}(x + 1) = 1$ but $\gcd(\text{cont}(f(x)), \text{cont}(g(x))) = \gcd(12, 8) \approx 4 \not\approx 1$. In fact, $\gcd(f(x), g(x)) \approx 4x + 4$.

Corollary 3. *For primitive polynomials $f(x), g(x) \in \mathbb{Z}[x]$, their common divisor of maximal degree $h(x)$ is their* gcd *if and only if $h(x)$ is primitive.*

In the case if polynomials are over a field, the situation is simpler. For any field K, the polynomial ring $K[x]$ is a UFD (and even an Euclidean domain). Any nonzero $f(x) \in K[x]$ has a factorization

$$f(x) = \theta \cdot q_1^{\beta_1}(x) \cdots q_m^{\beta_m}(x), \quad \theta \in K^*, \ \deg q_i(x) > 0, \ i = 1, \ldots, m, \tag{10}$$

which is unique in the sense mentioned above. Since all nonzero scalars in K are units, what we, in (7) above, had as a product of some prime numbers actually "merges" in K into a unit:

$$\nu \cdot p_1^{\alpha_1} \cdots p_n^{\alpha_n} = \theta \in K^* = K \setminus \{0\}. \tag{11}$$

Comparing factorizations of type (10) for any nonzero polynomials $f(x), g(x) \in K[x]$, we easily get the following.

Lemma 4. *For any nonzero polynomials $f(x), g(x) \in K[x]$ over a field K, their common divisor of maximal degree $h(x)$ is their* gcd.

This, in particular, is true for rings mentioned above: $\mathbb{Q}[x]$, $\mathbb{R}[x]$, $\mathbb{C}[x]$, and $\mathbb{Z}_p[x]$. We will use this fact later to construct the Big prime modular gcd algorithm and its modifications.

The analog of Lemma 4 was not true for $\mathbb{Z}[x]$ because in factorization (9) we have the nonunit prime-power factors $p_i^{\gamma_i}$ which do participate in factorization of $d(x) = \gcd(f(x), g(x))$ but which add *nothing* to the degree of $d(x)$. This is why maximality of the degree is no longer the only criterion in $\mathbb{Z}[x]$ to detect if the given $h(x)$ is gcd or not.

3. Some Notations for Modular Reductions

The following notations, adopted from [1], are to make our arguments shorter and more uniform when we deal with numerals, polynomials, and matrices. As above, let $\mathbb{Z}_p$ be the residue ring (finite Galois field $\mathbb{Z}_p = \mathbb{F}_p = \{0, \ldots, p-1\}$) and let

$$\varphi_p : \mathbb{Z} \longrightarrow \mathbb{Z}_p \tag{12}$$

be the rings homomorphism mapping each $z \in \mathbb{Z}$ to the remainder after division of z by p. That is, $\varphi_p(z) \equiv z(\text{mod } p)$, and $\varphi_p(z) \in \mathbb{Z}_p$.

We use the same symbol φ_p to denote the homomorphism

$$\varphi_p : \mathbb{Z}[x] \longrightarrow \mathbb{Z}_p[x], \tag{13}$$

where $\mathbb{Z}_p[x]$ is the ring of polynomials over $\mathbb{Z}_p$ and φ_p is mapping each of the coefficients a_i of $f(x) \in \mathbb{Z}[x]$ to the remainder after division of a_i by p.

Similarly, we define the homomorphism of matrix rings

$$\varphi_p : M_{m,n}(\mathbb{Z}) \longrightarrow M_{m,n}(\mathbb{Z}_p), \tag{14}$$

which maps each of the elements a_{ij} of a matrix $A \in M_{m,n}(\mathbb{Z})$ to the remainder after division of a_{ij} by p.

Using the same symbol φ_p for numeric, polynomial, and matrix homomorphisms causes no misunderstanding below, and it is more comfortable for some reasons. These homomorphisms are called *"modular reductions"* or just *"reductions."* We can also specify these homomorphisms as *"numeric modular reduction," "polynomial modular reduction,"* or *"matrix modular reduction"* where needed [1].

For $a \in \mathbb{Z}$, denote $\varphi_p(a) = a_p$. For $f(x) \in \mathbb{Z}[x]$, denote $\varphi_p(f(x)) = f_p(x) \in \mathbb{Z}_p[x]$. So if

$$f(x) = a_0 x^n + \cdots + a_n \tag{15}$$

then

$$f_p(x) = \varphi_p(f(x)) = \varphi_p(a_0) x^n + \cdots + \varphi_p(a_n) = a_{0,p} x^n + \cdots + a_{n,p} \in \mathbb{Z}_p[x]. \tag{16}$$

And for a matrix $A \in M_{m,n}(\mathbb{Z})$ denote $\varphi_p(A) = A_p \in M_{m,n}(\mathbb{Z}_p)$. If $A = \|a_{i,j}\|_{m\times n}$, then $A_p = \|\varphi_p(a_{i,j})\|_{m\times n} = \|a_{i,j,p}\|_{m\times n}$.

4. Problems at Lifting the Modular gcd to $\mathbb{Z}[x]$

Now we turn to the second task mentioned earlier: *effective calculation* of the actual $\gcd(f(x), g(x))$ for the given nonzero polynomials $f(x), g(x) \in \mathbb{Z}[x]$.

The ring $\mathbb{Z}_p[x]$ is an Euclidean domain, unlike the ring $\mathbb{Z}[x]$. So we can use the Euclidean algorithm to calculate gcd for any nonzero polynomials in $\mathbb{Z}_p[x]$, including the modular images $f_p(x)$ and $g_p(x)$. Since the notation $\gcd(f_p(x), g_p(x))$ is going to be used repeatedly, for briefness denote by $e_p(x)$ gcd calculated by Euclidean algorithm for $f_p(x), g_p(x)$. Let us stress that $\gcd(f_p(x), g_p(x))$ is not determined uniquely, since for any nonzero $t \in \mathbb{Z}_p$ the product $t \cdot \gcd(f_p(x), g_p(x))$ also is gcd for $f_p(x), g_p(x)$. We are denoting just *one* of these gcd's (namely, that computed by the Euclidean algorithm) by $e_p(x)$. This $e_p(x)$ is unique, since at each step of the Euclidean algorithm we have a unique action to take (to see this, just consider the steps of "long division" used to divide $f_p(x)$ by $g_p(x)$ on field $\mathbb{Z}_p$).

The main idea of the algorithm is to calculate $e_p(x) \approx \gcd(f_p(x), g_p(x))$ for some suitable p and to reconstruct $d(x) = \gcd(f(x), g(x))$ by it. We separate the process to four main problems that may occur and show how to overcome each one to arrive to a correctly working algorithm.

4.1. Problem 1: Avoiding the Eliminating Coefficients. After reduction φ_p, some of the coefficients of $f(x)$ and $g(x)$ may change or even be eliminated. So their images $f_p(x) = \varphi_p(f(x))$ and $g_p(x) = \varphi_p(g(x))$ may keep very little information to reconstruct $d(x)$ based on $e_p(x)$.

Example 5. If $f(x) = 7x^2 + 22$ and $g(x) = 49x^3 + 154x$, then for $p = 7$ we get $f_p(x) = 1$ and $g_p(x) = 0$. So these values contain no reliable information to reconstruct $\gcd(f(x), g(x))$.

The first simple idea to avoid such elimination is to take p larger than the absolute value of all coefficients of $f(x)$ and $g(x)$. This, however, is not enough since a divisor $h(x)$ of a polynomial $f(x)$ may have coefficients, larger than those of $f(x)$. Moreover, using the cyclotomic polynomials for large enough n,

$$\phi_n(x) = \prod_{k=1,\dots,n;(k,n)=1} \left(x - e^{2i\pi k/n}\right), \tag{17}$$

one can get divisors of $f(x) = x^n - 1$ which have a coefficient larger than any pregiven number [1, 4, 8]. Since we do not know the divisors of $f(x)$ and $g(x)$, we cannot be sure if the abovementioned large p will be large enough to prevent elimination of coefficients of $h(x)$. To overcome this, one can use the Landau-Mignotte bounds (in different sources, the bounds on coefficients of the divisors are called differently, associating them with names of L. Landau or M. Mignotte or with both of them; these authors have different roles in development of the formulas, which in turn are consequence of a formula by A. L. Cauchy), as done in [4–6]. For a polynomial $f(x)$ given by (15), denote its norm by $\|f(x)\| = \sqrt{\sum_{i=0}^{n} a_i^2}$.

Theorem 6 (L. Landau, M. Mignotte). *Let $f(x) = a_0x^n + \cdots + a_n$ and $h(x) = c_0x^k + \cdots + c_k$ be nonzero polynomials in $\mathbb{Z}[x]$. If $h(x)$ is a divisor of $f(x)$, then*

$$\sum_{i=0}^{n} |c_i| \leq 2^k \cdot \left|\frac{c_0}{a_0}\right| \cdot \|f(x)\|. \tag{18}$$

The proof is based on calculations on complex numbers, and it can be found, for example, in [1, 4]. We are going to use the Landau-Mignotte bounds in the following two shapes.

Corollary 7. *In notations of Theorem 6, there is the following upper bound for the coefficients of $h(x)$:*

$$|c_i| \leq N_f = 2^{n-1} \|f(x)\|. \tag{19}$$

Proof. To obtain this from (18), first notice that $|c_0/a_0| \leq 1$.

Next, if $k = \deg h(x) = \deg f(x) = n$, then $f(x) = r \cdot h(x)$, where r is a nonzero integer. Then $|c_i| \leq \max\{|c_i| \mid i = 0, \dots, n\} \leq \max\{|a_i| \mid i = 0, \dots, n\} \leq \|f(x)\|$.

Finally, if $k = \deg h(x) \leq n - 1$ (k is unknown to us), then we can simply replace in (18) the value 2^k by 2^{n-1}. □

Remark 8. In literature, they use the rather less accurate bound $|c_i| \leq 2^n \|f(x)\|$, but the second paragraph of our proof above allows replacing 2^n by 2^{n-1}. See also Remark 23.

Corollary 9. *In notations of Theorem 6, if $h(x)$ also is a divisor of the polynomial $g(x) = b_0x^m + \cdots + b_m$, then there is the following upper bound for the coefficients of $h(x)$:*

$$\begin{aligned} |c_i| &\leq N_{f,g} \\ &= 2^{\min\{n,m\}} \cdot \gcd(a_0, b_0) \\ &\quad \cdot \min\left\{\frac{\|f(x)\|}{|a_0|}, \frac{\|g(x)\|}{|b_0|}\right\}. \end{aligned} \tag{20}$$

Proof. To obtain this from (18), just notice that if $h(x)$ is a common divisor for $f(x)$ and $g(x)$, then its leading coefficient c_0 divides both a_0 and b_0. □

Formula (20) provides the hint to overcome Problem 1 about eliminating coefficients, mentioned at the start of this subsection. Although the divisors $h(x)$ of $f(x)$ and $g(x)$ are yet unknown, we can compute $N_{f,g}$ and take $p > N_{f,g}$. If we apply the reduction φ_p for this p, we can be sure that none of the coefficients of $h(x)$ has changed "much" under that homomorphism, for φ_p does not alter the nonnegative coefficients of $h(x)$, and it just adds p to all negative coefficients of $h(x)$. The same holds true for $d(x) = \gcd(f(x), g(x))$.

4.2. Problem 2: Negative Coefficients and Reconstruction of the Preimage. The reduction φ_p is not a bijection, and $d_p(x)$ has infinitely many preimages in $\mathbb{Z}[x]$. But the relatively uncomplicated relationship between coefficients of $d(x)$ and $d_p(x)$, obtained in previous subsection, may allow us to reconstruct $d(x)$ if we know $d_p(x)$. The condition $p > N_{f,g}$ puts a restriction on the preimage $d(x)$: the coefficients of $d(x)$ either are equal to respective coefficients of $d_p(x)$ (if they are nonnegative) or are the respective coefficients of $d_p(x)$ minus p (if they are negative). Reconstruction may cause problems connected with negative coefficients.

Example 10. If for some polynomials $f(x), g(x)$ we have $N_{f,g} = 15$, we can take the prime, say $p = 17 > N_{f,g}$. Assume we have somehow calculated $d_{17}(x) = 12x^3 + 3x + 10$; we can be sure that $d(x)$ is not the preimage $29x^3 - 17x^2 + 20x + 27$ because $d(x)$ cannot have coefficients greater than 15 by absolute value. But we still cannot be sure if the preimage $d(x)$ is $12x^3 + 3x + 10$, or $-5x^3 + 3x + 10$, or maybe $-5x^3 - 14x - 7$.

It is easy to overcome this by just taking a larger value:

$$p > 2 \cdot N_{f,g}. \tag{21}$$

If the coefficient c_i of $d(x)$ is nonnegative, then $\varphi_p(c_i) = c_i < p/2$, and if it is negative, then $\varphi_p(c_i) = c_i + p > p/2$. This provides us with the very simple method as shown in Algorithm 1 to reconstruct $d(x)$ if we have already computed $d_p(x)$ for sufficiently large prime p.

Input: For an unknown polynomial $d(x) \in \mathbb{Z}[x]$ *we know the upper bound* N *of absolute values of its coefficients, and for arbitrarily large prime number* p *we have the modular image* $d_p(x) = b_{0,p}x^k + \cdots + b_{k,p} \in \mathbb{Z}_p[x]$.
Reconstruct the polynomial $d(x)$.

Choose any prime $p > 2 \cdot N$.
Set $k = \deg d_p(x)$.
Set $i = 0$.
While $i \le k$
    if $b_{i,p} < p/2$
        set $b_i = b_{i,p}$;
    else
        set $b_i = b_{i,p} - p$;
    set $i = i + 1$.
Output $d(x) = b_0 x^k + \cdots + b_k$.

Algorithm 1: The polynomial reconstruction by modular image.

4.3. Problem 3: Finding the Correct Fold of the Modular gcd *of Right Degree.* Now additionally assume the polynomials $f(x), g(x) \in \mathbb{Z}[x]$ to be primitive. Since $\mathrm{cont}(f(x))$ and $\mathrm{cont}(g(x))$ are defined up to the sign ± 1, we can without loss of generality admit the leading coefficients of $f(x), g(x)$ to be positive.

Below, in Problem 4, we will see that for some p the polynomial $e_p(x)$, computed by the Euclidean algorithm in $\mathbb{Z}_p[x]$, may not be the image of $d(x)$ and, moreover, its degree may be different from that of $d(x)$. This means that by applying Algorithm 1 to $e_p(x)$ we may not obtain $d(x)$. Assume, however, we have p, which meets the condition $p > 2 \cdot N_{f,g}$ and for which

$$\deg d(x) = \deg e_p(x). \tag{22}$$

By Corollary 3, a common divisor of $f(x), g(x)$ is $d(x) = \gcd(f(x), g(x))$ if and only if it is primitive and if its degree is the maximum of degrees of all common divisors. Since φ_p does not change the degree of $d(x)$, we get by Lemma 4 (applied for the field $K = \mathbb{Z}_p$) that $d_p(x)$ is gcd of $f_p(x), g_p(x)$ in $\mathbb{Z}_p[x]$. This correspondence surely is not one-to-one, because in $\mathbb{Z}[x]$ gcd is calculated up to the unit element of $\mathbb{Z}[x]$, which is ± 1, whereas in $\mathbb{Z}_p[x]$ gcd is calculated up to the unit element of $\mathbb{Z}_p[x]$, which can be any nonzero number $t \in \mathbb{Z}_p^* = \{1, \ldots, p-1\}$. So the polynomial $e_p(x)$ calculated by the Euclidean algorithm may not be the image $d_p(x)$ of $d(x)$.

Example 11. For $f(x) = x^2 + 4x + 3$ and $g(x) = x^2 + 2x + 1$, whichever prime $p > 4$ we take, we will get by the Euclidean algorithm

$$e_p(x) = \gcd(f_p(x), g_p(x)) = 2x + 2 \in \mathbb{Z}_p[x]. \tag{23}$$

But in $\mathbb{Z}[x]$ we have $d(x) = x + 1$. So regardless how large p we choose, we will never get $\varphi_p(x + 1) = 2x + 2$.

In other words, we are aware that the image $d_p(x)$ is one of the folds $t \cdot e_p(x)$ of $e_p(x)$ for some $t \in \{1, \ldots, p-1\}$, but we are not aware which t is that.

The leading coefficient $c_0 = \mathrm{lc}(d(x))$ of $d(x)$ can also be assumed to be positive. Denote by w the positive $\gcd(a_0, b_0)$. Since both c_0 and w are not altered by φ_p, their fraction w/c_0 also is not altered. Take such t that

$$\mathrm{lc}(t \cdot e_p(x)) = w. \tag{24}$$

Even if $t \cdot e_p(x)$ is not the image $d_p(x)$, it is the image of $l \cdot d(x)$, where l divides w/c_0. If we calculate the preimage $k(x) \in \mathbb{Z}[x]$ of $t \cdot e_p(x)$ by Algorithm 1, we will get a polynomial, which is either $d(x)$ or some fold of $d(x)$. Since $f(x), g(x)$ are primitive, it remains to go to the primitive part $d(x) = \mathrm{pp}(k(x))$.

The general case, when $f(x), g(x)$ may not be primitive, can easily be reduced to the following: for arbitrary $f(x), g(x)$, take their decompositions by formula (5) and set

$$r = \gcd(\mathrm{cont}(f(x)), \mathrm{cont}(g(x))) \in \mathbb{Z}. \tag{25}$$

Then assign $f(x) = \mathrm{pp}(f(x))$, $g(x) = \mathrm{pp}(g(x))$ and do the steps above for these new polynomials. After $d(x) = \mathrm{pp}(k(x))$ is computed, we get the final answer as $r \cdot d(x) = r \cdot \mathrm{pp}(k(x))$.

Notice that for Algorithm 1 we need p to be greater than any coefficient $|c_i|$ of the polynomial we reconstruct. The bound $p > 2 \cdot N_{f,g}$ assures that p meets this condition for $d(x)$. We, however, reconstruct not $d(x)$ but $l \cdot d(x)$, which may have larger coefficients. One could overcome this point by taking $p > w \cdot 2 \cdot N_{f,g}$, but this is not necessary because as we see later, while the Big prime modular gcd algorithm works, the value of p will grow and this issue will be covered.

4.4. Problem 4: Finding the Right Degree for the Modular gcd. As we saw, one can reconstruct $d(x)$ if we find $p > 2 \cdot N_{f,g}$ such that condition (22) holds. Consider an example to see that (22) may actually not hold for some p even if φ_p is not altering the coefficients of $f(x)$ and $g(x)$!

Example 12. For $f(x) = x^2 + 1$ and $g(x) = x + 1$, we have $d(x) = \gcd(f(x), g(x)) = 1$. Taking $p = 2$ we get $f_2(x) = x^2 + 1$ and $g_2(x) = x + 1$. In $\mathbb{Z}_2[x]$ we have $f_2(x) = x^2 + 1 = x^2 + 1^2 = (x+1)(x+1)$; thus, $e_2(x) = \gcd(f_2(x), g_2(x)) = x + 1$. We get that $1 = \deg(x + 1) > \deg(x) = 0$. In particular, whatever t we take, $t \cdot (x + 1)$ is not the image of $d(x) = 1$ under φ_2.

The idea to overcome this problem is to show that the number of primes p, for which (22) falsifies, is "small." So if the selected p is not suitable, we take another p and do the calculation again by the new prime. And we will not have to repeat these steps for many times (we will turn to this point in Section 6).

The proof of the following theorem and the definition of the resultant $\mathrm{res}(f(x), g(x))$ (i.e., of the determinant of the Sylvester matrix $S_{f,g}$ of polynomials $f(x), g(x)$) can be found, for example, in [1, 4, 7, 10]. The resultant is a comfortable tool to detect if the given polynomials are coprime.

Theorem 13. *Let* R *be an integrity domain. The polynomials* $f(x), g(x) \in R[x]$ *are coprime if and only if* $\mathrm{res}(f(x), g(x)) \neq 0$.

Input: non-zero polynomials $f(x), g(x) \in \mathbb{Z}[x]$.
Calculate their greatest common divisor $\gcd(f(x), g(x)) \in \mathbb{Z}[x]$.
(01) Calculate $\text{cont}(f(x))$, $\text{cont}(g(x))$ in the Euclidean domain $\mathbb{Z}$, choose their signs so that $\text{sgn cont}(f(x)) = \text{sgn lc}(f(x))$ and $\text{sgn cont}(g(x)) = \text{sgn lc}(g(x))$.
(02) Set $f(x) = \text{pp}(f(x))$ and $g(x) = \text{pp}(g(x))$.
(03) Calculate r in the Euclidean domain $\mathbb{Z}$ by (25).
(04) Set $a_0 = \text{lc}(f(x))$ and $b_0 = \text{lc}(g(x))$ (they are positive by our selection of signs for $\text{cont}(f(x))$ and $\text{cont}(g(x))$).
(05) Calculate the positive $w = \gcd(a_0, b_0)$ in the Euclidean domain $\mathbb{Z}$.
(06) Set $D = \min\{\deg f(x), \deg g(x)\} + 1$.
(07) Compute the Landau-Mignotte bound $N_{f,g}$ by (20).
(08) Choose a new prime number $p > 2 \cdot N_{f,g}$.
(09) Apply the reduction φ_p to calculate the modular images $f_p(x), g_p(x) \in \mathbb{Z}_p[x]$.
(10) Calculate $e_p(x) = \gcd(f_p(x), g_p(x))$ in the Euclidean domain $\mathbb{Z}_p[x]$.
(11) If $D \le \deg e_p(x)$
(12) go to step (08);
(13) else
(14) choose a t such that the $\text{lc}(t \cdot e_p(x)) = w$;
(15) call Algorithm 1 to calculate the preimage $k(x)$ of $t \cdot e_p(x)$;
(16) calculate $\text{cont}(k(x))$ in the Euclidean domain $\mathbb{Z}$;
(17) set $d(x) = \text{pp}(k(x)) = k(x)/\text{cont}(k(x))$;
(18) if $d(x) \mid f(x)$ and $d(x) \mid g(x)$
(19) go to step (23);
(20) else
(21) set $D = \deg e_p(x)$;
(22) go to step (08).
(23) Output the result: $\gcd(f(x), g(x)) = r \cdot d(x)$.

ALGORITHM 2: Big prime modular gcd algorithm.

Input: non-zero polynomials $f(x), g(x) \in \mathbb{Z}[x]$.
Detect if $f(x)$ *and* $g(x)$ *are coprime.*
(01) Calculate $\text{cont}(f(x))$, $\text{cont}(g(x))$ in the Euclidean domain $\mathbb{Z}$.
(02) Calculate r in the Euclidean domain $\mathbb{Z}$ by (25).
(03) If $r \neq 1$
(04) output the result: $f(x)$ and $g(x)$ are not coprime and stop.
(05) Set $a_0 = \text{lc}(f(x))$ and $b_0 = \text{lc}(g(x))$.
(06) Calculate $w = \gcd(a_0, b_0)$ in the Euclidean domain $\mathbb{Z}$.
(07) Set $f(x) = \text{pp}(f(x))$ and $g(x) = \text{pp}(g(x))$.
(08) Compute the bound $A_{f,g}$ for polynomials $f(x), g(x)$ by (32).
(09) Find the maximal k for which $p_k\# \le A_{f,g}$.
(10) Set $i = 1$.
(11) While $i \neq k + 1$
(12) choose a new prime $p \nmid w$;
(13) apply the reduction φ_p to calculate the modular images $f_p(x), g_p(x) \in \mathbb{Z}_p[x]$;
(14) calculate $e_p = \gcd(f_p(x), g_p(x))$ in the Euclidean domain $\mathbb{Z}_p[x]$;
(15) if $\deg e_{q_i} = 0$
(16) output the result: $f(x)$ and $g(x)$ are coprime and stop.
(17) set $i = i + 1$.
(18) If $i < k + 1$
(19) go to step (12).
(20) else
(21) output the result: $f(x)$ and $g(x)$ are not coprime.

ALGORITHM 3: Coprime polynomials detection modular algorithm.

Input: non-zero polynomials $f(x), g(x) \in \mathbb{Z}[x]$.
Calculate their greatest common divisor $\gcd(f(x), g(x)) \in \mathbb{Z}[x]$.
(01) Calculate $\mathrm{cont}(f(x))$, $\mathrm{cont}(g(x))$ in the Euclidean domain $\mathbb{Z}$, choose their signs so that $\mathrm{sgn}\,\mathrm{cont}(f(x)) = \mathrm{sgn}\,\mathrm{lc}(f(x))$ and $\mathrm{sgn}\,\mathrm{cont}(g(x)) = \mathrm{sgn}\,\mathrm{lc}(g(x))$.
(02) Set $f(x) = \mathrm{pp}(f(x))$ and $g(x) = \mathrm{pp}(g(x))$.
(03) Calculate r in the Euclidean domain $\mathbb{Z}$ by (25).
(04) Set $a_0 = \mathrm{lc}(f(x))$ and $b_0 = \mathrm{lc}(g(x))$ (they are positive by our selection of signs for $\mathrm{cont}(f(x))$ and $\mathrm{cont}(g(x))$).
(05) Calculate the positive $w = \gcd(a_0, b_0)$ in the Euclidean domain $\mathbb{Z}$.
(06) Set $D = \min\{\deg f(x), \deg g(x)\} + 1$.
(07) Choose a prime number $q \nmid w$.
(08) Apply the reduction φ_q to calculate the modular images $f_q(x), g_q(x) \in \mathbb{Z}_q[x]$.
(09) Calculate $e_q(x) = \gcd(f_q(x), g_q(x))$ in the Euclidean domain $\mathbb{Z}_q[x]$.
(10) Set $s(q, f, g) = \deg e_q(x)$.
(11) Calculate $M_{q,f,g}$ by (40) using the value of $s(q, f, g)$.
(12) Choose a new prime number $p > 2 \cdot M_{q,f,g}$.
(13) Apply the reduction φ_p to calculate the modular images $f_p(x), g_p(x) \in \mathbb{Z}_p[x]$.
(14) Calculate $e_p(x) = \gcd(f_p(x), g_p(x))$ in the Euclidean domain $\mathbb{Z}_p[x]$.
(15) If $D \le \deg e_p(x)$
(16) go to step (12).
(17) else
(18) choose a t such that the $\mathrm{lc}(t \cdot e_p(x)) = w$;
(19) call Algorithm 1 to calculate the preimage $k(x)$ of $t \cdot e_p(x)$;
(20) calculate $\mathrm{cont}(k(x))$ in the Euclidean domain $\mathbb{Z}$;
(21) set $d(x) = \mathrm{pp}(k(x)) = k(x)/\mathrm{cont}(k(x))$;
(22) if $d(x) \mid f(x)$ and $d(x) \mid g(x)$
(23) go to step (27);
(24) else
(25) set $D = \deg e_p(x)$;
(26) go to step (12).
(27) Output the result: $\gcd(f(x), g(x)) = r \cdot d(x)$.

ALGORITHM 4: Big prime modular gcd algorithm with a preliminary estimate on divisor degree.

The following fact in a little different shape can be found in [4] or [5].

Corollary 14. *If the prime p does not divide at least one of the leading coefficients a_0, b_0 of polynomials, respectively, $f(x), g(x) \in \mathbb{Z}[x]$, then $\deg d(x) \le \deg e_p(x)$. If p also does not divide $R = \mathrm{res}(f(x)/d(x), g(x)/d(x))$, where $d(x) = \gcd(f(x), g(x))$, then*

$$\deg d(x) = \deg d_p(x) = \deg \gcd(f_p(x), g_p(x)) = \deg e_p(x). \tag{26}$$

Proof. Since $c_0 = \mathrm{lc}(d(x))$ divides $w = \gcd(a_0, b_0)$, then $\varphi_p(c_0) \ne 0$ by the choice of p. Thus, $\deg d(x) = \deg d_p(x) \le \deg \gcd(f_p(x), g_p(x))$.

Since $d_p(x) \ne 0$, we can consider the fractions $f_p(x)/d_p(x)$ and $g_p(x)/d_p(x)$ in $\mathbb{Z}_p[x]$. From unique factorizations of $f_p(x)$ and $g_p(x)$ in UFD $\mathbb{Z}_p[x]$, it is very easy to deduce that

$$e_p(x) \approx \gcd(f_p(x), g_p(x)) \approx d_p(x) \cdot \gcd\left(\frac{f_p(x)}{d_p(x)}, \frac{g_p(x)}{d_p(x)}\right). \tag{27}$$

In particular, $\deg d(x) = \deg d_p(x) \le \deg e_p(x)$. And the inequality $\deg d_p(x) \ne \deg e_p(x)$ may occur only if

$$\deg \gcd\left(\frac{f_p(x)}{d_p(x)}, \frac{g_p(x)}{d_p(x)}\right) > 0, \tag{28}$$

that is, when $f_p(x)/d_p(x)$ and $g_p(x)/d_p(x)$ are not coprime in $\mathbb{Z}_p[x]$ or, by Theorem 13, when $\mathrm{res}(f_p(x)/d_p(x), g_p(x)/d_p(x)) = 0$. The latter is the determinant of Sylvester matrix $S_{f_p/d_p, g_p/d_p}$. Consider the matrix rings homomorphism (matrix modular reduction)

$$\varphi_p : M_{m+n}(\mathbb{Z}) \longrightarrow M_{m+n}(\mathbb{Z}_p), \tag{29}$$

where $n = \deg f(x)$; $m = \deg g(x)$ (as mentioned earlier we use the same symbol φ_p for numeric, polynomial, and matrix reductions). Since, $\varphi_p(S_{f/d, g/d}) = S_{f_p/d_p, g_p/d_p}$ and since the determinant of a matrix is a sum of products of its elements, we get

$$R_p = \varphi_p(R) = \varphi_p\left(\mathrm{res}\left(\frac{f(x)}{d(x)}, \frac{g(x)}{d(x)}\right)\right) = \mathrm{res}\left(\frac{f_p(x)}{d_p(x)}, \frac{g_p(x)}{d_p(x)}\right). \tag{30}$$

Input: non-zero polynomials $f(x), g(x) \in \mathbb{Z}[x]$.

Calculate their greatest common divisor $\gcd(f(x), g(x)) \in \mathbb{Z}[x]$.

(01) Calculate $\text{cont}(f(x))$, $\text{cont}(g(x))$ in the Euclidean domain $\mathbb{Z}$, choose their signs so that $\text{sgn}\,\text{cont}(f(x)) = \text{sgn}\,\text{lc}(f(x))$ and $\text{sgn}\,\text{cont}(g(x)) = \text{sgn}\,\text{lc}(g(x))$.

(02) Set $f(x) = \text{pp}(f(x))$ and $g(x) = \text{pp}(g(x))$.

(03) Compute the bound $A_{f,g}$ for polynomials $f(x), g(x)$ by (32).

(04) Find the maximal k for which $p_k\# \le A_{f,g}$.

(05) Calculate r in the Euclidean domain $\mathbb{Z}$ by (25).

(06) Set $a_0 = \text{lc}(f(x))$ and $b_0 = \text{lc}(g(x))$ (they are positive by our selection of signs for $\text{cont}(f(x))$ and $\text{cont}(g(x))$).

(07) Calculate the positive $w = \gcd(a_0, b_0)$ in the Euclidean domain $\mathbb{Z}$.

(08) Set $s(f, g) = \min\{\deg f(x), \deg g(x)\}$.

(09) Set $i = 1$.

(10) While $i \ne k + 1$

(11)   choose a new prime $q_i \nmid w$;

(12)   apply the reduction φ_{q_i} to calculate the modular images $f_{q_i}(x), g_{q_i}(x) \in \mathbb{Z}_{q_i}[x]$;

(13)   calculate $e_{q_i} = \gcd(f_{q_i}(x), g_{q_i}(x))$ in the Euclidean domain $\mathbb{Z}_p[x]$;

(14)   if $\deg e_{q_i} \le s(f, g)$

(15)   set $s(f, g) = \deg e_{q_i}$;

(16)   if $\deg e_{q_i} = 0$

(17)   set $d(x) = 1$;

(18)   go to step (32);

(19)   set $i = i + 1$.

(20) Calculate $M_{f,g}$ by (43) using the value of $s(f, g)$.

(21) Choose a new prime number $p > 2 \cdot M_{f,g}$.

(22) Apply the reduction φ_p to calculate the modular images $f_p(x), g_p(x) \in \mathbb{Z}_p[x]$.

(23) Calculate $e_p(x) = \gcd(f_p(x), g_p(x))$ in the Euclidean domain $\mathbb{Z}_p[x]$.

(24) If $\deg e_p = s(f, g)$

(25)   choose a t such that the $\text{lc}(t \cdot e_p(x)) = w$;

(26)   call Algorithm 1 to calculate the preimage $k(x)$ of $t \cdot e_p(x)$;

(27)   calculate $\text{cont}(k(x))$ in the Euclidean domain $\mathbb{Z}$;

(28)   set $d(x) = \text{pp}(k(x)) = k(x)/\text{cont}(k(x))$;

(29)   go to step (32);

(30) else

(31)   go to step (21).

(32) Output the result: $\gcd(f(x), g(x)) = r \cdot d(x)$.

ALGORITHM 5: Big prime modular gcd algorithm with preliminary estimates on divisor degrees by multiple primes.

So R_p can be zero if and only if R is divisible by p. The polynomials $f(x)/d(x)$ and $g(x)/d(x)$ are coprime in $\mathbb{Z}[x]$ and their resultant is not zero by Theorem 13. And R cannot be a positive integer divisible by p since that contradicts the condition of this corollary. □

Corollary 14 shows that if for some p equality (22) does not hold for polynomials $f(x), g(x) \in \mathbb{Z}[x]$, then p either divides a_0 and b_0 or divides the resultant R. We do not know R, since we do not yet know $d(x)$ to calculate the resultant $R = \text{res}(f(x)/d(x), g(x)/d(x))$. But since the number of such primes p is just finite, we can arrive to the right p after trying the process for a few primes. We will turn to this again in Section 6.

5. The Big Prime Modular gcd Algorithm

Four steps of the previous section provide us with the following procedure. We keep all the notations from Section 4. Take the primitive polynomials $f(x), g(x) \in \mathbb{Z}[x]$. Without loss of generality, we may assume $a_0, b_0 > 0$. Take any $p > 2 \cdot N_{f,g}$. Then $p \nmid w = \gcd(a_0, b_0)$, since $a_0, b_0 \le N_{f,g}$. Calculate $e_p(x) = \gcd(f_p(x), g_p(x))$ in $\mathbb{Z}_p[x]$ by Euclidean algorithm. Then choose t so that (24) holds. Construct $k(x)$ applying Algorithm 1 to $t \cdot e_p(x)$. If the primitive part $d(x) = \text{pp}(k(x))$ divides both $f(x)$ and $g(x)$, then gcd for these primitive polynomials is found: $d(x) = \gcd(f(x), g(x))$. That follows from consideration about divisor degrees above: if $f(x), g(x)$ had a common divisor $h(x)$ of degree greater than $\deg d(x)$, then, since the degree of $h(x)$ is not altered by φ_p, we would get $\deg h_p(x) > \deg d_p(x) = \deg d(x) = \deg e_p(x)$, which contradicts the maximality of $\deg d_p(x)$ by Lemma 4.

This means that if for $p > 2 \cdot N_{f,g}$ we get $d(x) \nmid f(x)$ or $d(x) \nmid g(x)$, we have the case when p divides the resultant R. Then we just ignore the calculated polynomial, choose another $p > 2 \cdot N_{f,g}$, and redo the steps for it. Repeating these steps for finitely many times, we will eventually arrive to the correct $d(x)$ for the primitive polynomials $f(x), g(x)$.

The case of arbitrary nonzero polynomials can easily be reduced to this. By arguments mentioned earlier, we should calculate $d(x)$ for primitive polynomials $\mathrm{pp}(f(x))$ and $\mathrm{pp}(g(x))$ and then output the final answer as $r \cdot d(x)$, where r is defined by (25). The process we described is the traditional form of the Big prime modular gcd algorithm.

Remark 15. Since our approach in Section 4 evolved the maximality of degrees of the common divisors, we can shorten some of the steps of our algorithm. Let us store in a variable, say D, the minimal value for which we already know it is not $\deg \gcd(f(x), g(x))$. As an initial D, we may take, for example, $D = \min\{\deg f(x), \deg g(x)\}+1$. Each time we calculate $e_p(x) = \gcd(f_p(x), g_p(x))$, check if $\deg e_p(x)$ is equal to or larger than the current D. If yes, we already know that we have an "inappropriate" p. Then we no longer need to use Algorithm 1 to reconstruct $k(x)$ and to get $d(x) = \mathrm{pp}(k(x))$. We just skip these steps and proceed to the next p. Reconstruct $d(x)$ and check if $d(x) \mid f(x)$ and $d(x) \mid g(x)$ *only* when $\deg e_p(x) < D$. Then, if $d(x)$ does not divide $f(x)$ or $g(x)$, we have discovered a new bound D for $\deg(\gcd(f(x), g(x)))$. So set $D = \deg e_p(x)$ and proceed to the next p. If in next step we get $\deg e_p(x) \geq D$, we will again be aware that the steps of reconstruction of $d(x)$ need be skipped.

We constructed Algorithm 2.

Turning back to Remark 15, notice that for some prime numbers p we skip steps (14)–(18) of Algorithm 2 and directly jump to step (08). In fact, Remark 15 has mainly theoretical purpose to display how usage of UFD properties and comparison of divisor degrees may reduce some of the steps of the Big prime modular gcd algorithm. In practical examples, the set of primes we use contains few primes dividing $R = \mathrm{res}(f(x)/d(x), g(x)/d(x))$, so we may not frequently get examples where steps (14)–(18) are skipped.

Example 16. Let us apply Algorithm 2 to polynomials (1) mentioned in Knuth's example above. Since $\|f(x)\| = \sqrt{113}$ and $\|g(x)\| = \sqrt{570}$,

$$N_{f,g} = 2^{\min\{8,6\}} \cdot \gcd(1,3) \cdot \min\left\{\frac{\sqrt{113}}{1}, \frac{\sqrt{570}}{3}\right\} < 512. \tag{31}$$

And we can take the prime $p = 1031 > 2 \cdot N_{f,g}$. It is not hard to compute that $\gcd(f_{1031}(x), g_{1031}(x)) \approx 1$. So $f(x)$ and $g(x)$ are coprime. It is worth to compare $p = 1031$ with much smaller values $p = 67$ and $p = 37$ obtained below for the same polynomials in (1) in Example 27 using modified Algorithm 4.

In [1] we also apply Algorithm 2 to other polynomials with cases when the polynomials are not coprime.

6. Estimating the Prime Divisors of the Resultant

Although at the start of the Big prime modular gcd algorithm we cannot compute the resultant $R = \mathrm{res}(f(x)/d(x), g(x)/d(x))$ for the given $f(x), g(x) \in \mathbb{Z}[x]$ (we do not know $d(x)$), we can nevertheless estimate the value of R and the number of its prime divisors. Denote

$$\begin{aligned} A_{f,g} &= \sqrt{(n+1)^m (m+1)^n} \cdot N_f^m N_g^n \\ &= 2^{2nm-n-m}\sqrt{(n+1)^m (m+1)^n} \\ &\quad \cdot \|f(x)\|^m \|g(x)\|^n. \end{aligned} \tag{32}$$

Lemma 17. *For any polynomials $f(x), g(x) \in \mathbb{Z}[x]$ and for any of their common divisors $d(x)$, the following holds:*

$$\left|\mathrm{res}\left(\frac{f(x)}{d(x)}, \frac{g(x)}{d(x)}\right)\right| = \left|S_{f/d,g/d}\right| \leq A_{f,g}. \tag{33}$$

Proof. By Corollary 7, the coefficients of fractions $f(x)/d(x)$ and $g(x)/d(x)$ are bounded, respectively, by $N_f = 2^{n-1}\|f(x)\|$ and $N_g = 2^{m-1}\|g(x)\|$, where $n = \deg f(x)$; $m = \deg g(x)$. Since the numbers of summands in these fractions are at most $n+1$ and $m+1$, respectively, we get

$$\begin{aligned} \left\|\frac{f(x)}{d(x)}\right\| &\leq \sqrt{(n+1)N_f^2}, \\ \left\|\frac{g(x)}{d(x)}\right\| &\leq \sqrt{(m+1)N_g^2}. \end{aligned} \tag{34}$$

Applying Hadamard's maximal determinant bound [4] to the Sylvester matrix $S_{f/d,g/d}$, we get that

$$|R| = \left|S_{f/d,g/d}\right| \leq \left(\sqrt{(n+1)}N_f\right)^m \cdot \left(\sqrt{(m+1)}N_g\right)^n. \tag{35}$$

□

The bound of (32) is very rough. To see this, apply it to polynomials (1) of Knuth's example.

Example 18. For polynomials (1), we have $\|f(x)\| = \sqrt{113}$ and $\|g(x)\| = \sqrt{570}$. So we can estimate $N_f < 1408$, $N_g < 768$, and $N_{f,g} < 512$. Thus

$$\begin{aligned} |R| &\leq \sqrt{(8+1)^6 (6+1)^8} \cdot 1408^6 \cdot 570^8 = \omega \\ &= 1.6505374299582118582810249858265e+48, \end{aligned} \tag{36}$$

which is a too large number to comfortably operate with.

Remark 19. If in Algorithm 2 we use a prime

$$p > 2 \cdot A_{f,g}, \tag{37}$$

then we will get that $p \nmid R = \mathrm{res}(f(x)/d(x), g(x)/d(x))$ whatever the greatest common divisor $d(x)$ is. And, clearly, $p \nmid w$ holds for $w = \gcd(a_0, b_0)$. So, in this case, Algorithm 2 will output the correct $\mathrm{pp}(k(x))$ using just one p, and we will not have to take another $p \nmid w$ after step (18). However, Example 18 shows why it is *not* reasonable to choose p by rule (37) to have in Algorithm 2 one cycle only: it is easier to go via a few cycles for smaller p rather than to operate with a huge p, which is two times larger than the bound ω obtained in Example 18.

Nevertheless, the bound $A_{f,g}$ may be useful if we remember that the process in Algorithm 2 concerned not the value of $\text{res}(f(x)/d(x), g(x)/d(x))$ but the *number of its distinct prime divisors*. Let us denote by $p_k\#$ the product of the first k primes: $p_k\# = p_1 \cdot p_2 \cdots p_k$ (where $p_1 = 2$, $p_2 = 3$, etc.). They sometimes call $p_k\#$ the "*kth primorial.*" The following is essential.

Lemma 20. *The number of pairwise distinct prime divisors of a positive integer n is less than or equal to* $\max\{k \mid p_k\# \le n\}$.

From Lemmas 17 and 20, we easily get the following.

Corollary 21. *For any polynomials* $f(x), g(x) \in \mathbb{Z}[x]$ *and for any of their common divisors* $d(x)$, *the number of pairwise distinct prime divisors of* $\text{res}(f(x)/d(x), g(x)/d(x))$ *is at most* k, *where* k *is the largest number for which* $p_k\# \le A_{f,g}$.

Primorial (as a function on k) grows very rapidly. Say, for $k = 10$, it is more than six billions: $p_{10}\# = 6,469,693,230$. This observation allows using the bound $A_{f,g}$ in the following way: although the value of $A_{f,g}$ as a function on $n = \deg f(x)$ and $m = \deg g(x)$ and on the coefficients of $f(x)$ and $g(x)$ grows rapidly, the number of its distinct prime divisors may not be "very large" thanks to the fact that $p_k\#$ also grows rapidly. Consider this on polynomials and values from Example 18.

Example 22. It is easy to compute that

$$\begin{aligned} p_{30}\# &= 3.16100546404176077881452062915 44e+46 \\ &< \omega, \\ p_{31}\# &= 4.01447693933303618909444119902 6e+48 \\ &> \omega, \end{aligned} \tag{38}$$

where ω is the large number from Example 18. This means that the number of prime divisors of $R = \text{res}(f(x)/d(x), g(x)/d(x))$, whatever the divisor $d(x)$ is, is not greater than 30. And, whichever $30 + 1 = 31$ distinct primes we take, at least one of them will *not* be a divisor of R. That is, Algorithm 2 for the polynomials of Knuth's example will output the correct answer in not more than 31 cycles. We *cannot* find 31 primes $p \nmid w$ so that Algorithm 2 arrives to a wrong $d(x) = \text{pp}(k(x))$ on step (18) for all of them.

Remark 23. Let us stress that estimates on the number of prime divisors of the resultant and the analog of Algorithm 3 can be found elsewhere, for example, in [4]. So the only news we have is that here we use a slightly better value for N_f and N_g to get 2^{n+m} times smaller bound for $A_{f,g}$. Namely, in Corollary 7 we estimate $|c_i|$ not by $2^n\|f(x)\|$ but by $2^{n-1}\|f(x)\|$ (see (19) and Remark 8). This makes the bound $A_{f,g}$ in formula (32) 2^{n+m} times lower, since N_f and N_g appear m and n times, respectively.

7. An Algorithm to Check Coprime Polynomials

The first application of the bounds found in previous section is an algorithm checking if the given polynomials $f(x), g(x) \in \mathbb{Z}[x]$ are coprime. Present the polynomials as $f(x) = \text{cont}(f(x)) \cdot \text{pp}(f(x))$ and $g(x) = \text{cont}(g(x)) \cdot \text{pp}(g(x))$. If $r = \gcd(\text{cont}(f(x)), \text{cont}(g(x))) \not\approx 1$, then $f(x), g(x)$ are not coprime, and we do not have to check the primitive parts, at all.

If $r \approx 1$, then switch to the polynomials $f(x) = \text{pp}(f(x))$ and $g(x) = \text{pp}(g(x))$. By Corollary 21, the number of distinct prime divisors of $\text{res}(f(x)/d(x), g(x)/d(x))$ is less than or equal to k, where k is the largest number for which $p_k\# \le A_{f,g}$.

Consider any $k+1$ primes $p_1, \ldots, p_{k+1}$, each not dividing $w = \gcd(a_0, b_0)$, where $a_0 = \text{lc}(f(x))$ and $b_0 = \text{lc}(g(x))$. If $\gcd(f_{p_i}(x), g_{p_i}(x)) = 1$ for at least one p_i, then $f(x)$ and $g(x)$ are coprime because $0 = \deg\gcd(f_{p_i}(x), g_{p_i}(x)) \ge \deg\gcd(f(x), g(x))$ and $r \approx 1$.

And if $\gcd(f_{p_i}(x), g_{p_i}(x)) \ne 1$ for all $i = 1, \ldots, k+1$, then $f_{p_i}(x)$ and $g_{p_i}(x)$ are not coprime for at least one p_i, which is not dividing $\text{res}(f(x)/d(x), g(x)/d(x))$. This means that $f(x)$ and $g(x)$ are not coprime. We got Algorithm 3.

Two important advantages of this algorithm are that here we use much smaller primes p (we just require $p \nmid w$, not $p > 2 \cdot N_{f,g}$), and in Algorithm 3, unlike in Algorithm 2, we never need to find t, to compute the preimage $k(x)$ of $t \cdot \gcd(f_p(x), g_p(x))$ and the primitive part $\text{pp}(k(x))$.

Remark 24. As is mentioned by Knuth in [2], in a probabilistic sense, the polynomials are much more likely to be coprime than the integer numbers. So it is reasonable to first test by Algorithm 3 if the given polynomials $f(x), g(x)$ are coprime, and only after that apply Algorithm 2 to find their gcd in case if they are not coprime. See also Algorithm 5, where we combine both of these approaches with a better bound for prime p.

Example 25. Apply Algorithm 3 to polynomials (1) from Knuth's example. As we saw in Example 22, $k = 30$. For $p = 2$, we get $f_2(x) = x^8+x^6+x^4+x^3+1$, $g_2(x) = x^6+x^4+x+1$, which are not coprime, since $\gcd(f_2(x), g_2(x)) = x^2 + x + 1 \ne 1$. And for $p = 3$ we get $f_3(x) = x^8 + x^6 + 2x^2 + 2x + 1$, $g_3(x) = 2x^4+2x^2$, which are coprime. So $\gcd(f(x), g(x)) = 1$.

Example 26. If $f(x) = x^2+2x+1$ and $g(x) = x+1$. Then $N_f = 2\sqrt{6}$, $N_g = \sqrt{2}$ and $A_{f,g} < 39$. Since $2\cdot3\cdot5\cdot7 = 210 > 39$, we get that $k = 3$, and $\gcd(f(x), g(x)) \ne 1$ if $\gcd(f_p(x), g_p(x)) \ne 1$ for any *four* primes (not dividing w). It is easy to check that $\gcd(f_p(x), g_p(x)) \ne 1$ for $p = 2, 3, 5, 7$.

8. Other Modifications of Algorithms

The bounds mentioned in Section 6 can be applied to obtain modifications of Algorithm 2. Let us outline four ideas, of which only the last two will be written down as algorithms.

For the nonzero polynomials $f(x), g(x) \in \mathbb{Z}[x]$, let us again start by computing $r = \gcd(\text{cont}(f(x)), \text{cont}(g(x)))$

and switching to the primitive parts $f(x) = \mathrm{pp}(f(x))$ and $g(x) = \mathrm{pp}(g(x))$, assuming that their leading coefficients a_0 and b_0 are positive. Calculate N_f, N_g by Corollary 7, $N_{f,g}$ by Corollary 9, and $A_{f,g}$ by (32). Find the maximal k for which $p_k\# \le A_{f,g}$. Then take any $k+1$ primes $p_1, \ldots, p_{k+1}$, each greater than $2 \cdot N_{f,g}$. We do not know $d(x)$, but we are aware that the number of prime divisors of $R = \mathrm{res}(f(x)/d(x), g(x)/d(x))$ is less than or equal to k. So at least one of the primes $p_1, \ldots, p_{k+1}$ is not dividing R. To find it, compute the degrees of $e_{p_i}(x)$ for all $i = 1, \ldots, k+1$. Take any p_i, for which $\deg e_{p_i}(x)$ is the minimal (in case there is more than one p_i with this property, take one of them, preferably the smallest of all).

By our construction, $\deg e_{p_i}(x) = \deg \gcd(f(x), g(x))$ holds. So we can proceed to the next steps: choose t, such that $\mathrm{lc}(t \cdot e_{p_i}(x)) = w = \gcd(a_0, b_0)$; then find by Algorithm 1 the preimage $k(x)$ of $t \cdot e_{p_i}(x)$; then proceed to its primitive part $d(x) = \mathrm{pp}(k(x))$; and then output the final answer as $r \cdot d(x)$.

The advantage of this approach is that we do not have to go via steps (14)–(18) of Algorithm 2 for more than one prime p. Also, we do not have to take care of the variable D. But the disadvantage is that we have to compute $e_{p_i}(x)$ for large primes for $k+1$ times (whereas in Algorithm 2 the correct answer could be discovered after consideration of fewer primes). Clearly, the disadvantage is a serious obstacle, since repetitions for $k+1$ large primes consume more labour than steps (14)–(18) of Algorithm 2. So this is just a theoretical idea, not an approach for an effective algorithm.

The disadvantage can be reduced in the following way: in previous arguments, after we find $p_k\#$ and k, select the prime numbers $p_1, \ldots, p_{k+1}$ each satisfying the condition $p_i \nmid w$. This is a much weaker condition than the condition $p_i > 2 \cdot N_{f,g}$ used above, so we will surely get smaller primes. Take M to be the minimum of all degrees $\deg e_{p_i}(x)$ for all $i = 1, \ldots, k+1$. Since none of the primes $p_1, \ldots, p_{k+1}$ divides w, for any $i = 1, \ldots, k+1$, we have $\deg e_{p_i}(x) \ge \deg \gcd(f(x), g(x))$. On the other hand, since at least one of the primes $p_1, \ldots, p_{k+1}$ does not divide R, we know that, for that p_i, the degree of $e_{p_i}(x)$ is equal to $\deg \gcd(f(x), g(x))$. Combining these, we get that $\deg \gcd(f(x), g(x)) = M$. Since we know M, we can take a prime $p > 2 \cdot N_{f,g}$ and compute $e_p(x)$ and check its degree: if $\deg e_p(x) \ne M$, then we have a wrong p (we no longer need go to steps (14)–(18) of Algorithm 2 to discover that). Then choose a new value for p and repeat the step. And if $\deg e_p(x) = M$, then we have the right p. We calculate t, the preimage $k(x)$ of $t \cdot e_{p_i}(x)$, and then $d(x) = \mathrm{pp}(k(x))$, and output the answer $r \cdot d(x)$ (see Algorithm 5 for a better version of this idea).

The third modification, not depending on $A_{f,g}$, can be constructed by estimating $\deg \gcd(f(x), g(x)) = \deg d(x)$ by means of an auxiliary prime number q. By Landau-Mignotte Theorem 6, if $h(x) = c_0 x^k + \cdots + c_k$ is any divisor of the polynomials $f(x) = a_0 x^n + \cdots + a_n$ and $g(x) = b_0 x^m + \cdots + b_m$, then $|c_i| \le 2^k |c_0/a_0| \|f(x)\|$ and $|c_i| \le 2^k |c_0/b_0| \|g(x)\|$. Since $|c_0/a_0|$ is bounded by $|\gcd(a_0, b_0)/a_0|$ and $|c_0/b_0|$ is bounded by $|\gcd(a_0, b_0)/b_0|$, we get the following analog of (20):

$$|c_i| \le 2^k \cdot \gcd(a_0, b_0) \cdot \min\left\{\frac{\|f(x)\|}{|a_0|}, \frac{\|g(x)\|}{|b_0|}\right\}. \quad (39)$$

Now assume q is a prime not dividing w, and denote $s(q, f, g) = \deg \gcd(f_q(x), g_q(x))$. By Corollary 14, $\deg d(x) \le s(q, f, g)$. We get for the coefficients of $d(x)$ the following bound: $|c_i| \le M_{q,f,g}$, where

$$M_{q,f,g} = 2^{s(q,f,g)} \cdot \gcd(a_0, b_0) \cdot \min\left\{\frac{\|f(x)\|}{|a_0|}, \frac{\|g(x)\|}{|b_0|}\right\}. \quad (40)$$

$M_{q,f,g}$ is a better bound for the coefficients of $d(x)$ because $2^{s(q,f,g)}$ may be considerably less than $2^{\min\{n,m\}}$.

We can improve Algorithm 2, if we preliminarily find $s(q, f, g)$ by calculating $\gcd(f_q(x), g_q(x))$ for an "auxiliary" prime $q \nmid w$ and then choose the "main" prime p by the rule $p > 2 \cdot M_{q,f,g}$ (instead of $p > 2 \cdot N_{f,g}$). Observe that if $p > 2 \cdot M_{q,f,g}$ then also $q \nmid w = \gcd(a_0, b_0)$ because $\min\{\|f(x)\|/|a_0|, \|g(x)\|/|b_0|\} \ge 1$. Additionally, we can introduce the variable D to store the values $\deg \gcd(f_p(x), g_p(x))$ that we know are greater than $\deg d(x)$. We get Algorithm 4.

Example 27. Let us apply Algorithm 4 to polynomials (1) from Knuth's example. Since $w = 1$, take $q = 2$. We have already computed in Example 25 that $e_2(x) = \gcd(f_2(x), g_2(x)) = x^2 + x + 1$. Then $s(2, f, g) = \deg e_2(x) = 2$ and

$$M_{2,f,g} = 2^2 \cdot 1 \cdot \min\left\{\frac{\sqrt{113}}{1}, \frac{\sqrt{570}}{3}\right\} < 31.84. \quad (41)$$

Take $p = 67 > 2 \cdot M_{2,f,g}$. It is easy to calculate that $\gcd(f_{67}(x), g_{67}(x)) \approx 1$. Compare this with Example 16, where we had to use much larger prime $p = 1031$. Moreover, if we take as an auxiliary q, say $q = 3$, then $s(3, f, g) = \deg e_3(x) = 0$ and $M_{3,f,g} \le 15.92$. So we can take an even smaller prime $p = 37 > 2 \cdot M_{3,f,g}$.

The ideas of Algorithms 3 and 4 can be combined to work with more than one auxiliary prime q. Like we mentioned in Remark 24, Knuth in [2] recommends checking first if the polynomials $f(x), g(x) \in \mathbb{Z}[x]$ are coprime and proceeding to their gcd calculation only after we get that they are not coprime (this is motivated by probabilistic arguments). Compute $A_{f,g}$ by formula (32) and find k like we did in step (09) of Algorithm 3: k is the maximal number for which $p_k\# \le A_{f,g}$. Then choose any $k+1$ primes $q_1, \ldots, q_{k+1}$ not dividing w, and start computing the modular gcd's $e_{q_1}(x), e_{q_2}(x), \ldots$ ($k+1$ times). If at some step we find $\deg e_{q_i}(x) = 0$, then we are done: the polynomials $f(x), g(x)$ are coprime if $r \approx 1$, or their gcd is the nontrivial scalar $r \not\approx 1$. And if $\deg e_{q_i}(x) > 0$ for all q_i, then we know that

(1) these polynomials are not coprime,

(2) the positive degree of $\gcd(f(x), g(x))$ is the minimum

$$s(f, g) = \min\left\{\deg e_{q_1}(x), \ldots, \deg e_{q_{k+1}}(x)\right\} > 0. \quad (42)$$

This exact value of $s(f, g) = \deg \gcd(f(x), g(x))$ is a better result than the estimate $s(q, f, g)$ obtained earlier by just one q.

Like above, we can assume $f(x), g(x)$ to be primitive (if not, we can again denote $r = \gcd(\text{cont}(f(x)), \text{cont}(g(x)))$ and switch to the primitive parts $f(x) = \text{pp}(f(x))$ and $g(x) = \text{pp}(g(x))$). Applying the Landau-Mignotte Theorem 6 for the coefficients c_i of $\gcd(f(x), g(x))$, we get that $|c_i| \leq M_{f,g}$, where

$$M_{f,g} = 2^{s(f,g)} \cdot \gcd(a_0, b_0) \cdot \min\left\{\frac{\|f(x)\|}{|a_0|}, \frac{\|g(x)\|}{|b_0|}\right\}. \tag{43}$$

Now we can take $p > 2 \cdot M_{f,g}$ and by the Euclidean algorithm calculate $e_p(x)$ in $\mathbb{Z}_p[x]$. If $\deg e_p(x) > s(f, g)$, we drop this p and choose another prime $p > 2 \cdot M_{f,g}$. And if $\deg e_p(x) = s(f, g)$, then we proceed to the final steps: we choose t, then get the preimage $k(x)$ of $t \cdot e_p(x)$, then go to the primitive part $d(x) = \text{pp}(k(x))$, and output the final answer as $\gcd(f(x), g(x)) = r \cdot \text{pp}(k(x))$.

Remark 28. This approach has the following advantages: Firstly, the bound on primes p is better than formula (40) since here we have not $2^{s(q,f,g)}$ but $2^{s(f,g)}$. Secondly, we no longer need to calculate the number t, the preimage $k(x)$, and the primitive part $d(x)$ for *more than one* prime p. This is because, if the selected $p > 2 \cdot M_{f,g}$ is not appropriate, we already have an indicator of that: $\deg e_p(x) > s(f, g)$.

We built Algorithm 5.

Example 29. Let us apply Algorithm 5 again on polynomials of Knuth's example (1). As we saw in Example 22, $k = 30$. So we may have to consider at most 31 auxiliary primes q_i. But we in fact need just two of them, because $\deg \gcd(f_2(x), g_2(x)) = \deg(x^2 + x + 1) = 2$ and $\deg \gcd(f_3(x), g_3(x)) = \deg(1) = 0$ (see Example 25). So in Algorithm 5 we jump from step (16) to step (32) directly.

Competing Interests

The author declares that there is no conflict of interests regarding the publication of this paper.

Acknowledgments

The author was supported in part by joint grant 15RF-054 of RFBR and SCS MES RA (in frames of joint research projects SCS and RFBR) and by 15T-1A258 grant of SCS MES RA.

References

[1] V. H. Mikaelian, *Algorithmic Algebra, Commutative Rings and Fields*, Yerevan University Press, Yerevan, Armenia, Officially Approved by the Ministry of Higher Education and Science of RA As Textbook for Universities, 2015.

[2] D. Knuth, "The art of computer programming," in *Vol. 2. Seminumerical Algorithms*, Addison-Wesley Series in Computer Science and Information Processing, Addison-Wesley, 2nd edition, 1969.

[3] W. S. Brown, "On Euclid's algorithm and the computation of polynomial greatest common divisors," *Journal of the ACM*, vol. 18, pp. 478–504, 1971.

[4] J. von zur Gathen and J. Gerhard, *Modern Computer Algebra*, Cambridge University Press, Cambridge, UK, 3rd edition, 2013.

[5] J. H. Davenport, Y. Siret, and E. Tournier, *Computer Algebra. Systems and Algorithms for Algebraic Computation*, Academic Press, London, UK, 2nd edition, 1993.

[6] E. V. Pankratev, *Elements of Computer Algebra Study Guide*, Binom LZ, Moscow, Russia, 2007.

[7] S. Lang, *Algebra, Revised Third Edition*, vol. 211 of *Graduate Texts in Mathematics*, Springer, New York, NY, USA, 2002.

[8] P. B. Garrett, *Abstract Algebra*, Chapman & Hall/CRC, Boca Raton, Fla, USA, 2008.

[9] P. M. Cohn, *Introduction to Ring Theory*, Springer, London, UK, 2000.

[10] A. I. Kostrikin, *An Introduction to Algebra*, vol. 2 of *Main Structures*, FizMatLit, Moscow, Russia, 2004 (Russian).

[11] P. M. Cohn, *Basic Algebra. Groups, Rings and Fields*, Springer, London, UK, 2003.

9

New Branch of Intuitionistic Fuzzification in Algebras with Their Applications

Samaher Adnan Abdul-Ghani, Shuker Mahmood Khalil, Mayadah Abd Ulrazaq, and Abu Firas Muhammad Jawad Al-Musawi

Department of Mathematics, College of Science, Basrah University, Basrah 61004, Iraq

Correspondence should be addressed to Shuker Mahmood Khalil; shuker.alsalem@gmail.com

Academic Editor: Susana Montes

The intuitionistic fuzzification in ρ−algebras about the concepts of ideals and subalgebras given with several related characterizations is considered. Some new concepts like intuitionistic fuzzy ρ−ideal ($IF\rho i$), intuitionistic fuzzy ρ−subalgebra ($IF\rho s$), ρ−homomorphism, and intuitionistic fuzzy $\overline{\rho}$−ideal ($IF\overline{\rho}i$) are introduced and some of their descriptions are given in this work. Further, we show some applications on the family of all intuitionistic fuzzy ρ−subalgebras $IF\rho_S(\mathfrak{R})$ in ρ−algebra like the binary relations $\underset{\mu}{\approx}$, $\underset{\nu}{\approx}$ and Γ_r on $IF\rho_S(\mathfrak{R})$. Also, their equivalence classes are given and studied.

1. Introduction

The fuzzy set (FS) as suggested by Zadeh [1] in 1965 is a regulation to vagueness and encounter uncertainty. A FS maps each element of the universe of discourse to the interval $[0, 1]$. After the introduction of fuzzy sets theory by him, many mathematicians were conducted on the generalizations of the this concept and studied in the groups, algebras, and soft spaces (see [2–5]). By including a fuzzy set the degree of nonmembership, Atanassov [6] in 1986 suggested the intuitionistic fuzzy set (IFS), which seems more precise for provides opportunities and uncertainty quantification to accurately model a problem based on existing knowledge and monitoring. Also, this notion is discussed in different fields (see [7–11])

BCK−algebra, class of algebra of logic, was investigated by Imai and Iseki [12]. After that, the notion of d−algebras was investigated by Neggers and Kim [13]. In 2017, the concepts of ρ−algebra, $\overline{\rho}$−ideal, ρ−ideal, ρ−subalgebra, and permutation topological ρ−algebra were first proposed by Mahmood and Abud Alradha [14]. Next, they showed the notion of the soft ρ−algebra and soft edge ρ−algebra [15].

In the present work, the notions of intuitionistic fuzzy ρ−ideal ($IF\rho i$), intuitionistic fuzzy ρ−subalgebra ($IF\rho s$), ρ−homomorphism, and intuitionistic fuzzy $\overline{\rho}$−ideal ($IF\overline{\rho}i$) are introduced. Further, we show some applications on the family of all intuitionistic fuzzy ρ−subalgebras $IF\rho_S(\mathfrak{R})$ in ρ−algebra like the binary relations $\underset{\mu}{\approx}$, $\underset{\nu}{\approx}$ and Γ_r on $IF\rho_S(\mathfrak{R})$. Also, their equivalence classes are given and studied.

2. Preliminaries and Notations

We will recall basic definitions and results to obtain properties developed in this work.

Definition 1 (see [16]). An intuitionistic fuzzy set α (IFS, in short) over the universe $\mathfrak{R}$ is defined by $\alpha = \{\prec a, \mu_\alpha(a), \nu_\alpha(a) \succ \mid a \in \mathfrak{R}\}$, where $\mu_\alpha(a)$: $\mathfrak{R} \longrightarrow [0; 1]$, $\nu_\alpha(a)$: $\mathfrak{R} \longrightarrow [0; 1]$ with $0 \leq \mu_\alpha(a) + \nu_\alpha(a) \leq 1, \forall a \in \mathfrak{R}$. $\mu_\alpha(a)$ and $\nu_\alpha(a)$ are real numbers and their values represent the degree of membership and nonmembership of a to α, respectively.

Definition 2 (see [6]). The IF whole and empty sets of $\mathfrak{R}$ are defined by $\overline{1} = \{\prec a, (1, 0) \succ \mid a \in \mathfrak{R}\}$ and $\overline{0} = \{\prec a, (0, 1) \succ \mid a \in \mathfrak{R}\}$, respectively.

2.1. Basic Relations and Operations on Intuitionistic Fuzzy Sets [7]. Assume $\alpha = \{\prec a, (\mu_\alpha(a), \nu_\alpha(a)) \succ \mid a \in \mathfrak{R}\}$ and $\beta = \{\prec a, (\mu_\beta(a), \nu_\beta(a)) \succ \mid a \in \mathfrak{R}\}$ are two IF sets of $\mathfrak{R}$. We deduced the following relations:

(1) [inclusion] $\alpha \subseteq \beta$ iff $\mu_\alpha(a) \leq \mu_\beta(a)$ and $\nu_\alpha(a) \geq \nu_\beta(a)$, $\forall a \in \mathfrak{R}$,

(2) [equality] $\alpha = \beta$ iff $\alpha \subseteq \beta$ and $\beta \subseteq \alpha$,

(3) [intersection] $\alpha \tilde{\cap} \beta = \{(a, \min\{\mu_\alpha(a), \mu_\beta(a)\}, \max\{\nu_\alpha(a), \nu_\beta(a)\}) : a \in \mathfrak{R}\}$,

(4) [union] $\alpha \tilde{\cup} \beta = \{(a, \max\{\mu_\alpha(a), \mu_\beta(a)\}, \min\{\nu_\alpha(a), \nu_\beta(a)\}) : a \in \mathfrak{R}\}$,

(5) [complement] $\alpha^c = \{(a, \nu_\alpha(a), \mu_\alpha(a)), a \in \mathfrak{R}\}$.

Definition 3 (see [14]). We say $(\mathfrak{R}, \bullet, 0)$ is ρ–algebra if $(\bullet)$ is a binary operation on $\mathfrak{R}$ with a constant $0 \in \mathfrak{R}$ and such that

(1) $a \bullet a = 0$,

(2) $0 \bullet a = 0$,

(3) $a \bullet b = 0 = b \bullet a$ imply that $a = b$,

(4) For all $a \neq b \in \mathfrak{R} - \{0\}$ imply that $a \bullet b = b \bullet a \neq 0$.

Definition 4 (see [14]). Assume $(\mathfrak{R}, \bullet, 0)$ is a ρ–algebra and $\phi \neq K \subseteq \mathfrak{R}$. We say K is a ρ–subalgebra of $\mathfrak{R}$ if $a \bullet b \in K$, $\forall a, b \in K$.

Definition 5 (see [14]). Assume $(\mathfrak{R}, \bullet, 0)$ is ρ–algebra and $\phi \neq K \subseteq \mathfrak{R}$. We say K is ρ–ideal of $\mathfrak{R}$ if

(1) $a, b \in K$ imply $a \bullet b \in K$,

(2) $a \bullet b \in K$ and $b \in K$ imply $a \in K$, $\forall a, b \in \mathfrak{R}$.

Definition 6 (see [14]). Assume $(\mathfrak{R}, \bullet, 0)$ is a ρ–algebra and K subset of $\mathfrak{R}$. We say K is a $\overline{\rho}$–ideal of $\mathfrak{R}$ if

(1) $0 \in K$,

(2) $a \in K$ and $b \in \mathfrak{R} \longrightarrow a \bullet b \in K$, $\forall a, b \in \mathfrak{R}$.

Definition 7 (see [11]). Assume that $\alpha = \{\prec a, (\mu_\alpha(a), \nu_\alpha(a)) \succ \mid a \in \mathfrak{R}\}$ is an IFS in $\mathfrak{R}$ and $r \in [0,1]$. The set $W(\mu_\alpha, r) = \{a \in \mathfrak{R} \mid \mu_\alpha(a) \geq r\}$ (resp., $L(\nu_\alpha, r) = \{a \in \mathfrak{R} \mid \nu_\alpha(a) \leq r\}$) is said to be *$\mu$–level r–cut* (resp., *ν–level r–cut*) of α.

3. Intuitionistic Fuzzy ρ–Subalgebras in ρ–Algebras

In this section, we introduce some new concepts, such as $(IF\rho s)$, $(IF\rho i)$, $(IF\overline{\rho}i)$, and ρ–homomorphism which are introduced and discussed. Further, some binary relations $\underset{\mu}{\approx}$, $\underset{\nu}{\approx}$ and Γ_r on $IF\rho_S(\mathfrak{R})$ are given, and some basic properties are shown.

Definition 8. Assume $(\mathfrak{R}, \bullet, 0)$ is a ρ–algebra and $\alpha = \{\prec a, (\mu_\alpha(a), \nu_\alpha(a)) \succ \mid a \in \mathfrak{R}\}$ is IFS of $\mathfrak{R}$. We say α is an $(IF\rho s)$ of $\mathfrak{R}$ if $\mu_\alpha(a \bullet b) \geq \min\{\mu_\alpha(a), \mu_\alpha(b)\}$ and $\nu_\alpha(a \bullet b) \leq \max\{\nu_\alpha(a), \nu_\alpha(b)\}$, $\forall a, b \in \mathfrak{R}$.

Table 1

$*$	0	ω	∂	ℓ
0	0	0	0	0
ω	ω	0	∂	∂
∂	ω	∂	0	ω
ℓ	ℓ	∂	ω	0

Example 9. Let $\mathfrak{R} = \{0, \omega, \partial, \ell\}$ be ρ–algebra with Table 1.

Then, $\alpha = \{\prec a, (\mu_\alpha(a), \nu_\alpha(a)) \succ \mid a \in \mathfrak{R}\} = \{(0, 0.9, 0.1), (\omega, 0.4, 0.3), (\partial, 0.7, 0.3), (\ell, 0.4, 0.2)\}$ is an $(IF\rho s)$ of $\mathfrak{R}$.

Definition 10. Assume $(\mathfrak{R}, \bullet, 0)$ is ρ–algebra and $\alpha = \{\prec a, (\mu_\alpha(a), \nu_\alpha(a)) \succ \mid a \in \mathfrak{R}\}$ is IFS of $\mathfrak{R}$. We say α is $(IF\rho i)$ of $\mathfrak{R}$ if

(1) $\mu_\alpha(a \bullet b) \geq \min\{\mu_\alpha(a), \mu_\alpha(b)\}$ and $\nu_\alpha(a \bullet b) \leq \max\{\nu_\alpha(a), \nu_\alpha(b)\}$,

(2) $\mu_\alpha(a) \geq \min\{\mu_\alpha(a \bullet b), \mu_\alpha(b)\}$ and $\nu_\alpha(a) \leq \max\{\nu_\alpha(a \bullet b), \nu_\alpha(b)\}$, $\forall a, b \in \mathfrak{R}$.

Example 11. Let $(\mathfrak{R}, \bullet, 0)$ be ρ–algebra in Example 9 and let $\alpha = \{\prec a, (\mu_\alpha(a), \nu_\alpha(a)) \succ \mid a \in \mathfrak{R}\} = \{(0, 0.8, 0.2), (\omega, 0.3, 0.4), (\partial, 0.2, 0.7), (\ell, 0.4, 0.3)\}$ be IFS of $\mathfrak{R}$. Then, α is $(IF\rho i)$ of $\mathfrak{R}$.

Definition 12. Assume $(\mathfrak{R}, \bullet, 0)$ is ρ–algebra and $\alpha = \{\prec a, (\mu_\alpha(a), \nu_\alpha(a)) \succ \mid a \in \mathfrak{R}\}$ is IFS of $\mathfrak{R}$. We say α is $(IF\overline{\rho}i)$ of $\mathfrak{R}$ if

(1) $\mu_\alpha(0) \geq \mu_\alpha(a)$ and $\nu_\alpha(0) \leq \nu_\alpha(a)$,

(2) $\mu_\alpha(a \bullet b) \geq \min\{\mu_\alpha(a), \mu_\alpha(b)\}$ and $\nu_\alpha(a \bullet b) \leq \max\{\nu_\alpha(a), \nu_\alpha(b)\}$, $\forall a, b \in \mathfrak{R}$.

Example 13. Let $(\mathfrak{R}, \bullet, 0)$ be ρ–algebra in Example 9 and let $\alpha = \{\prec a, (\mu_\alpha(a), \nu_\alpha(a)) \succ \mid a \in \mathfrak{R}\} = \{(0, 0.9, 0.1), (\omega, 0.4, 0.3), (\partial, 0.7, 0.3), (\ell, 0.4, 0.2)\}$ be IFS of $\mathfrak{R}$. Then, α is $(IF\overline{\rho}i)$ of $\mathfrak{R}$.

Remark 14.

(1) If $\alpha = \{\prec a, (\mu_\alpha(a), \nu_\alpha(a)) \succ \mid a \in \mathfrak{R}\}$ is $(IF\rho i)$ of $\mathfrak{R}$, then α is $(IF\rho s)$.

(2) If $\alpha = \{\prec a, (\mu_\alpha(a), \nu_\alpha(a)) \succ \mid a \in \mathfrak{R}\}$ is $(IF\overline{\rho}i)$ of $\mathfrak{R}$, then α is $(IF\rho s)$.

(3) If $\alpha = \{\prec a, (\mu_\alpha(a), \nu_\alpha(a)) \succ \mid a \in \mathfrak{R}\}$ is $(IF\rho s)$ of $\mathfrak{R}$ and satisfies (2) in Definition 10, then α is $(IF\rho i)$.

(4) If $\alpha = \{\prec a, (\mu_\alpha(a), \nu_\alpha(a)) \succ \mid a \in \mathfrak{R}\}$ is $(IF\rho s)$ of $\mathfrak{R}$ and satisfies (1) in Definition 12, then α is $(IF\overline{\rho}i)$.

Lemma 15. *If $\alpha = \{\prec a, (\mu_\alpha(a), \nu_\alpha(a)) \succ \mid a \in \mathfrak{R}\}$ is $(IF\rho s)$ of $\mathfrak{R}$, then $\mu_\alpha(0) \geq \mu_\alpha(a)$ and $\nu_\alpha(0) \leq \nu_\alpha(a)$, $\forall a \in \mathfrak{R}$.*

Proof. Let $a \in \mathfrak{R}$. Then $\mu_\alpha(0) = \mu_\alpha(a \bullet a) \geq \min\{\mu_\alpha(a), \mu_\alpha(a)\} = \mu_\alpha(a)$ and $\nu_\alpha(0) = \nu_\alpha(a \bullet a) \leq \max\{\nu_\alpha(a), \nu_\alpha(a)\} = \nu_\alpha(a)$. □

Theorem 16. *If* $\{\alpha_i = \prec a, (\mu_{\alpha_i}(a), \nu_{\alpha_i}(a)) \succ \mid a \in \mathfrak{R}, i \in I\}$ *is any family of (IFρs) of* $\mathfrak{R}$, *then* $\widetilde{\bigcap}_{i\in I}\alpha_i$ *is (IF$\overline{\rho}$i) of* $\mathfrak{R}$, *where* $\widetilde{\bigcap}_{i\in I}\alpha_i = \{\prec a, (\min\{\mu_{\alpha_i}(a)\}, \max\{\nu_{\alpha_i}(a)\}) \succ \mid a \in \mathfrak{R}\}$.

Proof. Let $a, b \in \mathfrak{R}$. Thus we consider that $\min\{\mu_{\alpha_i}(a \bullet b)\} \geq \min\{\min\{\mu_{\alpha_i}(a), \mu_{\alpha_i}(b)\}\} = \min\{\min\{\mu_{\alpha_i}(a)\}, \min\{\mu_{\alpha_i}(b)\}\}$. Also $\max\{\nu_{\alpha_i}(a \bullet b)\} \leq \max\{\max\{\nu_{\alpha_i}(a), \nu_{\alpha_i}(b)\}\} = \max\{\max\{\nu_{\alpha_i}(a)\}, \max\{\nu_{\alpha_i}(b)\}\}$.

Thus $\widetilde{\bigcap}_{i\in I}\alpha_i = \{\prec a, (\min\{\mu_{\alpha_i}(a)\}, \max\{\nu_{\alpha_i}(a)\}) \succ \mid a \in \mathfrak{R}\}$ satisfies condition (2) in Definition 12. Also, let $a \in \mathfrak{R}$. Hence, we consider that $\min\{\mu_{\alpha_i}(0)\} = \min\{\mu_{\alpha_i}(a \bullet a)\} \geq \min\{\mu_{\alpha_i}(a), \mu_{\alpha_i}(a)\} = \min\{\mu_{\alpha_i}(a)\}$. Furthermore, $\max\{\nu_{\alpha_i}(0)\} = \max\{\nu_{\alpha_i}(a \bullet a)\} \leq \max\{\nu_{\alpha_i}(a), \nu_{\alpha_i}(a)\} = \max\{\nu_{\alpha_i}(a)\}$. Then (1) in Definition 12 is held and hence $\widetilde{\bigcap}_{i\in I}\alpha_i$ is ($I\overline{\rho}i$) of $\mathfrak{R}$. □

Theorem 17. *If* $\alpha = \{\prec a, (\mu_\alpha(a), \nu_\alpha(a)) \succ \mid a \in \mathfrak{R}\}$ *is (IFρi) of* $\mathfrak{R}$, *then* $K = \prec a, \mu_\alpha(a), 1 - \mu_\alpha(a) \succ$ *is (IFρi) of* $\mathfrak{R}$.

Proof. We need only to show that $1 - \mu_\alpha(a)$ satisfies the first and second condition in Definition 10. Assume $\forall a, b \in \mathfrak{R}$. Then $1 - \mu_\alpha(a \bullet b) \leq 1 - \min\{\mu_\alpha(a), \mu_\alpha(b)\} = \max\{1 - \mu_\alpha(a), 1 - \mu_\alpha(b)\}$. Furthermore, $1 - \mu_\alpha(a) \leq 1 - \min\{\mu_\alpha(a \bullet b), \mu_\alpha(b)\} = \max\{1 - \mu_\alpha(a \bullet b), 1 - \mu_\alpha(b)\}$. Hence K is (*IFρi*) of $\mathfrak{R}$. □

Theorem 18. *If* $\alpha = \{\prec a, (\mu_\alpha(a), \nu_\alpha(a)) \succ \mid a \in \mathfrak{R}\}$ *is (IFρs) of* $\mathfrak{R}$, *then the sets* $T_\mu = \{a \in \mathfrak{R} \mid \mu_\alpha(a) = \mu_\alpha(0)\}$ *and* $T_\nu = \{a \in \mathfrak{R} \mid \nu_\alpha(a) = \nu_\alpha(0)\}$ *are ρ–subalgebras of* $\mathfrak{R}$.

Proof. Let $a, b \in T_\mu$. Hence $\mu_\alpha(a) = \mu_\alpha(0) = \mu_\alpha(b)$, and $\mu_\alpha(a\bullet b) \geq \min\{\mu_\alpha(a), \mu_\alpha(b)\} = \mu_\alpha(0)$. By using Lemma 15, we consider that $\mu_\alpha(a\bullet b) = \mu_\alpha(0)$ or equivalently $a\bullet b \in T_\mu$. Now, let $a, b \in T_\nu$. This implies that $\nu_\alpha(a \bullet b) \leq \max\{\nu_\alpha(a), \nu_\alpha(b)\} = \nu_\alpha(0)$ and, by applying Lemma 15, we conclude that $\nu_\alpha(a\bullet b) = \nu_\alpha(0)$. Therefore $a \bullet b \in T_\nu$. □

Definition 19. Assume $\alpha = \{\prec a, (\mu_\alpha(a), \nu_\alpha(a)) \succ \mid a \in \mathfrak{R}\}$ is (*IFρs*) of $\mathfrak{R}$. We say α has finite image, if each image of μ_α and ν_α is with finite cardinality (i.e., $\mathrm{Im}(\mu_\alpha) = \{\mu_\alpha(a) \mid a \in \mathfrak{R}\}$ and $\mathrm{Im}(\nu_\alpha) = \{\nu_\alpha(a) \mid a \in \mathfrak{R}\}$ such that $|\mathrm{Im}(\mu_\alpha)| < \infty$ and $|\mathrm{Im}(\nu_\alpha)| < \infty$).

Definition 20. Assume that $\alpha = \{\prec a, (\mu_\alpha(a), \nu_\alpha(a)) \succ \mid a \in \mathfrak{R}\}$ is (*IFρs*) of $\mathfrak{R}$ and $r \in [0, 1]$. The set $W(\mu_\alpha, r) = \{a \in \mathfrak{R} \mid \mu_\alpha(a) \geq r\}$ (resp., $L(\nu_\alpha, r) = \{a \in \mathfrak{R} \mid \nu_\alpha(a) \leq r\}$) is said to be *μ–level r–cut* (resp., *ν–level r–cut*) of α.

Theorem 21. *If* $\alpha = \{\prec a, (\mu_\alpha(a), \nu_\alpha(a)) \succ \mid a \in \mathfrak{R}\}$ *is (IFρs) of* $\mathfrak{R}$, *then* $W(\mu_\alpha, r) = \{a \in \mathfrak{R} \mid \mu_\alpha(a) \geq r\}$ *and* $L(\nu_\alpha, r) = \{a \in \mathfrak{R} \mid \nu_\alpha(a) \leq r\}$) *of* α *are ρ–subalgebras of* $\mathfrak{R}$.

Proof. Let $a, b \in W(\mu_\alpha, r)$. Hence $\mu_\alpha(b) \geq r$ and $\mu_\alpha(b) \geq r$. This implies that $\mu_\alpha(a \bullet b) \geq \min\{\mu_\alpha(a), \mu_\alpha(b)\} \geq r$ so that $a \bullet b \in W(\mu_\alpha, r)$. Thus $W(\mu_\alpha, r)$ is ρ–subalgebra of $\mathfrak{R}$ Now let $a, b \in L(\nu_\alpha, r)$. Thus $\nu_\alpha(a \bullet b) \leq \max\{\nu_\alpha(a), \nu_\alpha(b)\} \leq r$ and $a \bullet b \in L(\nu_\alpha, r)$. Therefore $L(\nu_\alpha, r)$ is ρ–subalgebra of $\mathfrak{R}$. □

Theorem 22. *If* $\alpha = \{\prec a, (\mu_\alpha(a), \nu_\alpha(a)) \succ \mid a \in \mathfrak{R}\}$ *is IFS of ρ – algebra* $\mathfrak{R}$ *such that the sets* $W(\mu_\alpha, r)$ *and* $L(\nu_\alpha, r)$ *are ρ–subalgebras of* $\mathfrak{R}$, *then* α *is an (IFρs) of* $\mathfrak{R}$.

Proof. Suppose that there are two members t_1 and t_2 in $\mathfrak{R}$ with $\mu_\alpha(t_1 \bullet t_2) < \min\{\mu_\alpha(t_1), \mu_\alpha(t_2)\}$. Let $t = [\mu_\alpha(t_1 \bullet t_2) + \min\{\mu_\alpha(t_1), \mu_\alpha(t_2)\}]/2$. Hence $\mu_\alpha(t_1 \bullet t_2) < t < \min\{\mu_\alpha(t_1), \mu_\alpha(t_2)\}$ and so $t_1 \bullet t_2 \notin W(\mu_\alpha, t)$, but $t_1, t_2 \in W(\mu_\alpha, t)$. This is a contradiction, and therefore $\mu_\alpha(a \bullet b) \geq \min\{\mu_\alpha(a), \mu_\alpha(b)\}, \forall a, b \in \mathfrak{R}$. Now assume that $\nu_\alpha(t_1 \bullet t_2) > \min\{\nu_\alpha(t_1), \nu_\alpha(t_2)\}$ for some $t_1, t_2 \in \mathfrak{R}$. Taking $k = [\nu_\alpha(t_1 \bullet t_2) + \min\{\nu_\alpha(t_1), \nu_\alpha(t_2)\}]/2$, then we consider that $\nu_\alpha(t_1 \bullet t_2) > k > \max\{\nu_\alpha(t_1), \nu_\alpha(t_2)\}$. It follows that $t_1, t_2 \in L(\nu_\alpha, k)$ and $t_1 \bullet t_2 \notin L(\nu_\alpha, k)$. This is a contradiction. Therefore, we consider that $\nu_\alpha(a \bullet b) \leq \max\{\nu_\alpha(a), \nu_\alpha(b)\}, \forall a, b \in \mathfrak{R}$. Then A is (*IFρs*) of $\mathfrak{R}$. □

Theorem 23. *If H is ρ–subalgebra of* $\mathfrak{R}$, *then there exists (IFρs)* α *of* $\mathfrak{R}$, *where H satisfies both μ–level ρ–subalgebra and ν–level ρ–subalgebra of* α *in* $\mathfrak{R}$.

Proof. Assume H is ρ–subalgebra of $\mathfrak{R}$ and let μ_α and ν_α be fuzzy sets in $\mathfrak{R}$ defined by

$$\mu_\alpha(a) = \begin{cases} k, & if\ a \in H \\ 1, & Otherwise, \end{cases} \tag{1}$$

and

$$\nu_\alpha(a) = \begin{cases} m, & if\ \alpha \in H \\ 1, & Otherwise, \end{cases} \tag{2}$$

$\forall a \in \mathfrak{R}$, where $k, m \in (0, 1)$ are fixed real numbers with $k + m < 1$. Assume $a, b \in \mathfrak{R}$. Then $a \bullet b \in H$ whenever $a, b \in H$. This implies that $\mu_\alpha(a \bullet b) = \min\{\mu_\alpha(a), \mu_\alpha(b)\}$ and $\nu_\alpha(a \bullet b) \leq \max\{\nu_\alpha(a), \nu_\alpha(b)\}$. If at least one of a or b does not belong to H, then either $\mu_\alpha(a) = 0$ or $\mu_\alpha(b) = 0$ and hence either $\nu_\alpha(a) = 1$ or $\nu_\alpha(b) = 1$. It follows that $\mu_\alpha(a \bullet b) \geq 0 = \min\{\mu_\alpha(a), \mu_\alpha(b)\}$, $\nu_\alpha(a \bullet b) \leq 1 = \max\{\nu_\alpha(a), \nu_\alpha(b)\}$. Hence $\alpha = \{\prec a, (\mu_\alpha(a), \nu_\alpha(a)) \succ \mid a \in \mathfrak{R}\}$ is (*IFρs*) of $\mathfrak{R}$. Obviously, $W(\mu_\alpha, k) = H = L(\nu_\alpha, m)$. □

Definition 24. Assume $\Theta : \mathfrak{R} \longrightarrow Y$ is a mapping of ρ–algebras. We say Θ is ρ–homomorphism if $\Theta(a \bullet b) = \Theta(a) \bullet \Theta(b), \forall a, b \in \mathfrak{R}$. And $\Theta^{-1}(\beta) = \{\prec a, (\Theta^{-1}\mu_\beta(a), \Theta^{-1}\nu_\beta(a)) \succ \mid a \in \mathfrak{R}\}$ is IFS in ρ–algebra $\mathfrak{R}$ for any IFS $\beta = \{\prec c, (\mu_\beta(c), \nu_\beta(c)) \succ \mid c \in Y\}$ of ρ–algebra Y. Also, if $\alpha = \{\prec a, (\mu_\alpha(a), \nu_\alpha(a)) \succ \mid a \in \mathfrak{R}\}$ is IFS in ρ–algebra $\mathfrak{R}$, then $\Theta(\alpha)$ is IFS in Y and defined by
$\Theta(\alpha) = \{\prec c, (\Theta_{\sup}\mu_\alpha(y), \Theta_{\inf}\nu_\alpha(c)) \succ \mid c \in Y\}$, where

$$\Theta_{\sup}\mu_\alpha(c) = \begin{cases} \sup\{\mu_\alpha(a) \mid a \in \Theta^{-1}(c)\}, & if\ \Theta^{-1}(c) \neq 0, \\ 0, & Otherwise, \end{cases} \tag{3}$$

and

$$\Theta_{\inf}\nu_\alpha(c) = \begin{cases} \inf\left\{\nu_\alpha(a) \mid a \in \Theta^{-1}(c)\right\}, & if\ \Theta^{-1}(c) \neq 0, \\ 1, & Otherwise, \end{cases} \quad (4)$$

$\forall c \in Y$.

Theorem 25. *Let Θ be ρ–homomorphism of ρ–algebra $\mathfrak{R}$ into ρ–algebra Y and K be (IFρs) of Y. Then $\Theta^{-1}(K)$ is (IFρs) of $\mathfrak{R}$.*

Proof. Assuming $a, b \in \mathfrak{R}$, we have $\mu_{\Theta^{-1}(K)}(a \bullet b) = \mu_K(\Theta(a \bullet b)) = \mu_K(\Theta(a) \bullet \Theta(b)) \geq \min\{\mu_K(\Theta(a)), \mu_K(\Theta(b))\} = \min\{\mu_{\Theta^{-1}(K)}(a), \mu_{\Theta^{-1}(K)}(b)\}$ and $\nu_{\Theta^{-1}(K)}(a \bullet b) = \nu_K(\Theta(a \bullet b)) = \nu_K(\Theta(a) \bullet \Theta(b)) \leq \max\{\nu_K(\Theta(a)), \nu_K(\Theta(b))\} = \max\{\nu_{\Theta^{-1}(K)}(a), \nu_{\Theta^{-1}(K)}(b)\}$. Thus $\Theta^{-1}(K)$ is (IFρs) of $\mathfrak{R}$. □

Theorem 26. *Assume $\Theta : \mathfrak{R} \longrightarrow Y$ is ρ–homomorphism of ρ–algebra $\mathfrak{R}$ into ρ–algebra Y and $\alpha = \{\prec a, (\mu_\alpha(a), \nu_\alpha(a)) \succ \mid a \in \mathfrak{R}\}$ is (IFρs) of $\mathfrak{R}$. Then $\Theta(\alpha) = \prec b, (\Theta_{\sup}(\mu_\alpha), \Theta_{\inf}(\nu_\alpha)) \succ$ is (IFρs) of Y.*

Proof. Let $\alpha = \{\prec a, (\mu_\alpha(a), \nu_\alpha(a)) \succ \mid a \in \mathfrak{R}\}$ be (IFρs) of $\mathfrak{R}$ and let $t_1, t_2 \in Y$. Noticing that $\{a_1 \bullet a_2 \mid a_1 \in \Theta^{-1}(t_1)$ and $a_2 \in \Theta^{-1}(t_2)\} \subseteq \{a \in \mathfrak{R} \mid a \in \Theta^{-1}(t_1 \bullet t_2)\}$, we have $\Theta_{\sup}(\mu_\alpha)(t_1 \bullet t_2) = \sup\{\mu_\alpha(a) \mid a \in \Theta^{-1}(t_1 * t_2)\} \geq \sup\{\mu_\alpha(a_1 \bullet a_2) \mid a_1 \in \Theta^{-1}(t_1)$ and $a_2 \in \Theta^{-1}(t_2)\} \geq \sup\{\min\{\mu_\alpha(a_1), \mu_\alpha(a_2)\} \mid a_1 \in \Theta^{-1}(t_1)$ and $a_2 \in \Theta^{-1}(t_2)\} = \min\{\sup\{\mu_\alpha(a_1) \mid a_1 \in \Theta^{-1}(t_1)\}, \sup\{\mu_\alpha(a_2) \mid a_2 \in \Theta^{-1}(t_2)\}\} = \min\{\Theta_{\sup}(\mu_\alpha)(t_1), \Theta_{\sup}(\mu_\alpha)(t_2)\}$. Also, we consider that $\Theta_{\inf}(\nu_\alpha)(t_1 \bullet t_2) = \inf\{\nu_\alpha(a) \mid a \in \Theta^{-1}(t_1 \bullet t_2)\} \leq \inf\{\nu_\alpha(a_1 \bullet a_2) \mid a_1 \in \Theta^{-1}(t_1)$ and $a_2 \in \Theta^{-1}(t_2)\} \leq \inf\{\max\{\nu_\alpha(a_1), \nu_\alpha(a_2)\} \mid a_1 \in \Theta^{-1}(t_1)$ and $a_2 \in \Theta^{-1}(t_2)\} = \max\{\inf\{\nu_\alpha(a_1) \mid a_1 \in \Theta^{-1}(t_1)\}, \inf\{\nu_\alpha(a_2) \mid a_2 \in \Theta^{-1}(t_2)\}\} = \max\{\Theta_{\sup}(\nu_\alpha)(t_1), \Theta_{\sup}(\nu_\alpha)(t_2)\}$. Hence $\Theta(\alpha) = \prec b, (\Theta_{\sup}(\mu_\alpha), \Theta_{\inf}(\nu_\alpha)) \succ$ is (IFρs) of Y. □

Theorem 27. *Assume $\Theta : \mathfrak{R} \longrightarrow Y$ is ρ–homomorphism of ρ–algebra $\mathfrak{R}$ into ρ–algebra Y and $\alpha = \{\prec a, (\mu_\alpha(a), \nu_\alpha(a)) \succ \mid a \in \mathfrak{R}\}$ is (IFρi) of $\mathfrak{R}$. Then $\Theta(\alpha) = \prec b, (\Theta_{\sup}(\mu_\alpha), \Theta_{\inf}(\nu_\alpha)) \succ$ is (IFρi) of Y.*

Proof. Since $\alpha = \{\prec a, (\mu_\alpha(a), \nu_\alpha(a)) \succ \mid a \in \mathfrak{R}\}$ is (IFρi) of $\mathfrak{R}$, then by Theorem 26 and Remark 14 we have $\Theta(\alpha) = \prec b, (\Theta_{\sup}(\mu_\alpha), \Theta_{\inf}(\nu_\alpha)) \succ$ as (IFρi) of Y. Hence condition (1) in Definition 10 is held. Since Θ is surjective, then for any $t_1, t_2 \in Y$, $\exists a_1, a_2 \in \mathfrak{R}$ such that $a_1 \in \Theta^{-1}\Theta(a_1) = \Theta^{-1}(t_1)$ and $a_2 \in \Theta^{-1}\Theta(a_2) = \Theta^{-1}(t_2)$. Also, $a_1 \bullet a_2 \in \Theta^{-1}(t_1) \bullet \Theta^{-1}(t_1) = \Theta^{-1}(t_1 \bullet t_1)$. Further, noticing that $\mu_\alpha(a_1) \geq \min\{\mu_\alpha(a_1 \bullet a_2), \mu_\alpha(a_2)\}$ and $\nu_\alpha(a_1) \leq \max\{\nu_\alpha(a_1 \bullet a_2), \nu_\alpha(a_2)\}$, for any $t_1, t_2 \in Y$, we have $\Theta_{\sup}(\mu_\alpha)(t_1) = \sup\{\mu_\alpha(a) \mid a \in \Theta^{-1}(t_1)\} \geq \sup\{\min\{\mu_\alpha(a_1 \bullet a_2), \mu_\alpha(a_2)\} \mid a_1 \bullet a_2 \in \Theta^{-1}(t_1 \bullet t_2)$ and $a_2 \in \Theta^{-1}(t_2)\} = \min\{\sup\{\mu_\alpha(a_1 \bullet a_2) \mid a_1 \bullet a_2 \in \Theta^{-1}(t_1 \bullet t_2)\}, \sup\{\mu_\alpha(a_2) \mid a_2 \in \Theta^{-1}(t_2)\}\} = \min\{\Theta_{\sup}(\mu_\alpha)(t_1 \bullet t_2), \Theta_{\sup}(\mu_\alpha)(t_2)\}$. Also, $\Theta_{\sup}(\nu_\alpha)(t_1) = \sup\{\nu_\alpha(a) \mid a \in \Theta^{-1}(t_1)\} \leq \sup\{\max\{\nu_\alpha(a_1 \bullet a_2), \nu_\alpha(a_2)\} \mid a_1 \bullet a_2 \in \Theta^{-1}(t_1 \bullet t_2)$ and $a_2 \in \Theta^{-1}(t_2)\} = \max\{\sup\{\nu_\alpha(a_1 \bullet a_2) \mid a_1 \bullet a_2 \in \Theta^{-1}(t_1 \bullet t_2)\}, \sup\{\nu_\alpha(a_2) \mid a_2 \in \Theta^{-1}(t_2)\}\} = \max\{\Theta_{\sup}(\nu_\alpha)(t_1 \bullet t_2), \Theta_{\sup}(\nu_\alpha)(t_2)\}$. Thus we consider that $\Theta(\alpha) = \prec b, (\Theta_{\sup}(\mu_\alpha), \Theta_{\inf}(\nu_\alpha)) \succ$ is (IFρi) of Y. □

Theorem 28. *Assume $\Theta : \mathfrak{R} \longrightarrow Y$ is ρ–homomorphism of ρ–algebra $\mathfrak{R}$ into ρ–algebra Y and $\alpha = \{\prec a, (\mu_\alpha(a), \nu_\alpha(a)) \succ \mid a \in \mathfrak{R}\}$ is (IF$\overline{\rho}i$) of $\mathfrak{R}$. Then $\Theta(\alpha) = \prec b, (\Theta_{\sup}(\mu_\alpha), \Theta_{\inf}(\nu_\alpha)) \succ$ is (IF$\overline{\rho}i$) of Y.*

Proof. Since $\alpha = \{\prec a, (\mu_\alpha(a), \nu_\alpha(a)) \succ \mid a \in \mathfrak{R}\}$ is (IF$\overline{\rho}i$) of $\mathfrak{R}$. Then by Theorem 26 and Remark 14 we have $\Theta(\alpha) = \prec b, (\Theta_{\sup}(\mu_\alpha), \Theta_{\inf}(\nu_\alpha)) \succ$ as (IFρs) of Y. Hence condition (2) in Definition 12 is held. Assume that $0_\mathfrak{R}$ and 0_Y are constants of ρ – algebras $\mathfrak{R}$ and Y, respectively. Since $\alpha = \{\prec a, (\mu_\alpha(a), \nu_\alpha(a)) \succ \mid a \in \mathfrak{R}\}$ is (IF$\overline{\rho}i$) of $\mathfrak{R}$, hence $\mu_\alpha(0_\mathfrak{R}) \geq \mu_\alpha(a)$ and $\nu_\alpha(0_\mathfrak{R}) \leq \nu_\alpha(a)$, $\forall a \in \mathfrak{R}$. Since Θ is ρ–homomorphism of ρ–algebras, then $\Theta(0_\mathfrak{R}) = 0_Y$, where $0_\mathfrak{R}$ and 0_Y are constants for ρ–algebras $\mathfrak{R}$ and Y, respectively. Noticing that $0_\mathfrak{R} \in \Theta^{-1}(0_Y)$ and $\{a \mid a \in \Theta^{-1}(b)\} \subseteq \{a \mid a \in \mathfrak{R}\}$ for any $b \in Y$, then we have $\Theta_{\sup}(\mu_\alpha)(0_Y) = \sup\{\mu_\alpha(a) \mid a \in \Theta^{-1}(0_Y)\} = \mu_\alpha(0_\mathfrak{R}) \geq \sup\{\mu_\alpha(a) \mid a \in \mathfrak{R}\} \geq \sup\{\mu_\alpha(a) \mid a \in \Theta^{-1}(b)\} = \Theta_{\sup}(\mu_\alpha)(b)$. Also, $\Theta_{\sup}(\nu_\alpha)(0_Y) = \inf\{\nu_\alpha(a) \mid a \in \Theta^{-1}(0_Y)\} = \nu_\alpha(0_\mathfrak{R}) \leq \inf\{\nu_\alpha(a) \mid a \in \mathfrak{R}\} \leq \inf\{\nu_\alpha(a) \mid a \in \Theta^{-1}(b)\} = \Theta_{\sup}(\nu_\alpha)(b)$. Hence $\Theta(\alpha) = \prec b, (\Theta_{\sup}(\mu_\alpha), \Theta_{\inf}(\nu_\alpha)) \succ$ is (IF$\overline{\rho}i$) of Y. □

4. Some Applications on $IF\rho_S(\mathfrak{R})$

In this section, some applications on $IF\rho_S(\mathfrak{R})$ are shown like the binary relations $\underset{\mu}{\approx}$, $\underset{\nu}{\approx}$ and Γ_r on $IF\rho_S(\mathfrak{R})$. Also, in this section the equivalence classes for theses binary relations are given, and some of their basic properties are studied.

4.1. Equivalence Classes Modulo ($\underset{\mu}{\approx}/\underset{\nu}{\approx}$). Denote the collection of all ($IF\rho_S$) of $\mathfrak{R}$ by $IF\rho_S(\mathfrak{R})$ and let $r \in [0, 1]$. Define binary relations $\underset{\mu}{\approx}$ and $\underset{\nu}{\approx}$ on $IF\rho_S(\mathfrak{R})$ as follows.

$\alpha \underset{\mu}{\approx} \beta \Longleftrightarrow W(\mu_\alpha, r) = W(\mu_\beta, r)$ and $\alpha \underset{\nu}{\approx} \beta \Longleftrightarrow L(\nu_\alpha, r) = L(\nu_\beta, r)$, respectively, for $\alpha = \prec a, \mu_\alpha, \nu_\alpha \succ$ and $\beta = \prec a, \mu_\beta, \nu_\beta \succ$ in $IF\rho_S(\mathfrak{R})$. Moreover, it is clear that $\underset{\mu}{\approx}$ and $\underset{\nu}{\approx}$ are equivalence relations on $IF\rho_S(\mathfrak{R})$. If $\alpha = \prec a, \mu_\alpha, \nu_\alpha \succ \in IF\rho_S(\mathfrak{R})$, then we refer to the equivalence class of $\alpha = \prec a, \mu_\alpha, \nu_\alpha \succ$ modulo $\underset{\mu}{\approx}$ (resp., $\underset{\nu}{\approx}$) by $\langle\alpha\rangle_\mu$ (resp., $\langle\alpha\rangle_\nu$), and we refer to the family of all equivalence classes of α modulo $\underset{\mu}{\approx}$ (resp., $\underset{\nu}{\approx}$) by $IF\rho_S(\mathfrak{R})/\underset{\mu}{\approx}$ (resp., $IF\rho_S(X)/\underset{\nu}{\approx}$); i.e., $IF\rho_S(\mathfrak{R})/\underset{\mu}{\approx} = \{\langle\alpha\rangle_\mu \mid \alpha = \prec a, \mu_\alpha, \nu_\alpha \succ \in IF\rho_S(\mathfrak{R})\}$ (resp., $IF\rho_S(\mathfrak{R})/\underset{\nu}{\approx} = \{\langle\alpha\rangle_\mu \mid \alpha = \prec a, \mu_\alpha, \nu_\alpha \succ \in IF\rho_S(\mathfrak{R})\}$). Moreover, denote the collection of all ρ–ideals of $\mathfrak{R}$ by $\rho_I(\mathfrak{R})$ and let $r \in [0, 1]$. Let σ_r and η_r be maps from $IF\rho_S(\mathfrak{R})$ to $\rho_I(\mathfrak{R}) \cup \{\phi\}$ by $\sigma_r(\alpha) = W(\mu_\alpha, r)$ and $\eta_r(\alpha) = L(\nu_\alpha, r)$, respectively, $\forall \alpha = \prec a, \mu_\alpha, \nu_\alpha \succ \in IF\rho_S(\mathfrak{R})$. In other words, σ_r and η_r are well-defined.

Theorem 29. *Let σ_r and η_r be the maps from $IF\rho_S(\mathfrak{R})$ to $\rho_I(\mathfrak{R}) \cup \{\phi\}$. Then σ_r and η_r are surjective, for each $r \in (0, 1)$.*

Proof. Let $r \in (0,1)$. Then $\overline{0} =\prec a, \overline{0}, \overline{1} \succ$ is in $IF\rho_S(\mathfrak{R})$, where each one of $\overline{0}$ and $\overline{1}$ is (FS) in $\mathfrak{R}$ defined by $\overline{0}(a) = 0$ and $\overline{1}(a) = 1$, $\forall a \in \mathfrak{R}$. Furthermore, $\sigma_r(\overline{0}) = W(\overline{0}, r) = \phi = L(\overline{1}, r) = \eta_r(\overline{0})$. Let $\phi \neq H \in \rho_I(\mathfrak{R})$. $\forall a \in \mathfrak{R}$, let $\mu_H(a) = \left\{\begin{smallmatrix} 1, & if\ a\in H \\ 0, & if\ a\notin H \end{smallmatrix}\right.$, and $\nu_H(a) = 1 - \mu_H(a)$; thus $\sigma_r(\widehat{H}) = W(\mu_H, r) = H = L(\nu_H, r) = \eta_r(\widehat{H})$. Now, we want to prove that $\widehat{H} =\prec x, \mu_H, \nu_H \succ\in IF\rho_S(\mathfrak{R})$. Since $H \in \rho_I(\mathfrak{R})$, then by condition (1) in Definition 5 we have H as ρ−subalgebra of $\mathfrak{R}$ and this implies that $W(\mu_H, r)$ and $L(\nu_H, r)$ are ρ−subalgebras of $\mathfrak{R}$. By Theorem 22 we consider $\widehat{H} =\prec a, \mu_H, \nu_H \succ\in IF\rho_S(\mathfrak{R})$. Therefore, $\forall H \in \rho_I(\mathfrak{R})$ we consider $\sigma_r(\widehat{H}) = H$ and $\eta_r(\widehat{H}) = H$ for some $\widehat{H} \in IF\rho_S(\mathfrak{R})$. This completes the proof. □

Theorem 30. *Let $IF\rho_S(\mathfrak{R})/\underset{\mu}{\approx}$ and $IF\rho_S(\mathfrak{R})/\underset{\nu}{\approx}$ be quotient sets. Then they are equipotent to $\rho_I(\mathfrak{R}) \cup \{\phi\}$, $\forall r \in (0,1)$.*

Proof. Assume $r \in (0,1)$ and let σ_r'(resp. η_r') be a map from $IF\rho_S(\mathfrak{R})/\underset{\mu}{\approx}$ (resp., $IF\rho_S(\mathfrak{R})/\underset{\nu}{\approx}$) to $\rho_I(\mathfrak{R}) \cup \{\phi\}$ and they are defined by $\sigma_r'(\langle\alpha\rangle_\mu) = \sigma_r(\alpha)$ (resp. $\eta_r'(\langle\alpha\rangle_\nu) = \eta_r(\alpha)$), $\forall \alpha =\prec a, \mu_\alpha, \nu_\alpha \succ\in IF\rho_S(\mathfrak{R})$. Hence, $\alpha \underset{\mu}{\approx} \beta$ and $\alpha \underset{\nu}{\approx} \beta$, $\forall \alpha =\prec a, \mu_\alpha, \nu_\alpha \succ$ and $\beta =\prec a, \mu_\beta, \nu_\beta \succ$ in $IF\rho_S(\mathfrak{R})$, if $W(\mu_A, r) = W(\mu_B, r)$ and $L(\nu_A, r) = L(\nu_B, r)$. Then $\langle\alpha\rangle_\mu = \langle\beta\rangle_\mu$ and $\langle\alpha\rangle_\nu = \langle\beta\rangle_\nu$. This implies the maps σ_r' and η_r' are injective. Moreover, let $\phi \neq H \in \rho_I(\mathfrak{R})$ and $\forall a \in \mathfrak{R}$, let

$$\mu_H(a) = \begin{cases} 1, & if\ a \in H \\ 0, & if\ a \notin H, \end{cases} \tag{5}$$

$\nu_H(a) = 1 - \mu_H(a)$, and thus $\widehat{H} =\prec a, \mu_H, \nu_H \succ\in IF\rho_S(\mathfrak{R})$. We consider that $\sigma_r'(\langle\widehat{H}\rangle_\mu) = \sigma_r(\widehat{H}) = W(\mu_H, r) = H$, and $\eta_r'(\langle\widehat{H}\rangle_\nu) = \eta_r(\widehat{H}) = L(\nu_H, r) = H$. Finally, for $\overline{0} =\prec a, \overline{0}, \overline{1} \succ\in IF\rho_S(\mathfrak{R})$ we have $\sigma_r'(\langle\overline{0}\rangle_\mu) = \sigma_r(\overline{0}) = W(\overline{0}, r) = \phi$ and $\eta_r'(\langle\overline{0}\rangle_\nu) = \eta_r(\overline{0}) = L(\overline{1}, r) = \phi$. Therefore σ_r' and η_r' are surjective, and we are done. □

4.2. Equivalence Class Modulo Γ_r. Another relation Γ_r on $IF\rho_S(\mathfrak{R})$ is defined by $(\alpha, \beta) \in \Gamma_r \Longleftrightarrow W(\mu_\alpha, r) \cap L(\nu_\alpha, r) = W(\mu_\beta, r) \cap L(\nu_\beta, r)$, $\forall r \in [0,1]$ and, $\forall \alpha =\prec a, \mu_\alpha, \nu_\alpha \succ$, $\beta =\prec a, \mu_\beta, \nu_\beta \succ\in IF\rho_S(\mathfrak{R})$. Moreover, the relation Γ_r is also an equivalence relation on $IF\rho_S(\mathfrak{R})$. Let $\langle\alpha\rangle_{\Gamma_r}$ denote the equivalence class of $\alpha =\prec a, \mu_\alpha, \nu_\alpha \succ$ modulo Γ_r, $\forall \alpha =\prec a, \mu_\alpha, \nu_\alpha \succ\in IF\rho_S(\mathfrak{R})$.

Theorem 31. *For any $r \in (0,1)$, the map $\psi_r : IF\rho_S(\mathfrak{R}) \longrightarrow \rho_I(\mathfrak{R}) \cup \{\phi\}$ defined by $\psi_r(\mathfrak{R}) = \sigma_r(\mathfrak{R}) \cap \eta_r(\mathfrak{R})$, $\forall \alpha =\prec a, \mu_\alpha, \nu_\alpha \succ\in IF\rho_S(\mathfrak{R})$ is surjective.*

Proof. Let $r \in (0,1)$. For $\overline{0} =\prec a, \overline{0}, \overline{1} \succ\in IF\rho_S(\mathfrak{R})$, we get $\psi_r(\overline{0}) = \sigma_r(\overline{0}) \cap \eta_r(\overline{0}) = W(\overline{0}, r) \cap L(\overline{1}, r) = \phi$. For any $H \in IF\rho_S(\mathfrak{R})$, there exists $\widehat{H} =\prec a, \mu_H, \nu_H \succ\in IF\rho_S(\mathfrak{R})$, where

$$\mu_H(a) = \begin{cases} 1, & if\ a \in H \\ 0, & if\ a \notin H \end{cases} \tag{6}$$

and $\nu_H(a) = 1 - \mu_H(a)$ such that $\psi_r(\widehat{H}) = \sigma_r(\widehat{H}) \cap \eta_r(\widehat{H}) = W(\mu_H, r) \cap L(\nu_H, r) = H$. This completes the proof. □

Theorem 32. *For any $r \in (0,1)$, the quotient set $IF\rho_S(\mathfrak{R})/\Gamma_r$ is equipotent to $\rho_I(X) \cup \{\phi\}$.*

Proof. Assume $r \in (0,1)$ and $\psi_r' : IF\rho_S(\mathfrak{R})/\Gamma_r \longrightarrow \rho_I(\mathfrak{R}) \cup \{\phi\}$ is a map defined by $\psi_r'(\langle\alpha\rangle_{\Gamma_r}) = \psi_r(\alpha)$, $\forall \langle\alpha\rangle_{\Gamma_r} \in IF\rho_S(\mathfrak{R})/\Gamma_r$.

Suppose that $\psi_r'(\langle\alpha\rangle_{\Gamma_r}) = \psi_r'(\langle\beta\rangle_{\Gamma_r})$ for any $\langle\alpha\rangle_{\Gamma_r}, \langle\beta\rangle_{\Gamma_r} \in IF\rho_S(\mathfrak{R})/\Gamma_r$. We consider that $\sigma_r(\alpha) \cap \eta_r(\alpha) = \sigma_r(\beta) \cap \eta_r(\beta)$, i.e., $W(\mu_\alpha, r) \cap L(\nu_\alpha, r) = W(\mu_\beta, r) \cap L(\nu_\beta, r)$. Hence $(\alpha, \beta) \in \Gamma_r$, and so $\langle\alpha\rangle_{\Gamma_r} = \langle\beta\rangle_{\Gamma_r}$. Therefore ψ_r' is injective. Furthermore, for $\overline{0} =\prec a, \overline{0}, \overline{1} \succ\in IF\rho_S(\mathfrak{R})$ we get $\psi_r'(\langle\overline{0}\rangle_{\Gamma_r}) = \psi_r(\overline{0}) = \sigma_r(\overline{0}) \cap \eta_r(\overline{0}) = W(\overline{0}, r) \cap L(\overline{1}, r) = \phi$. Let $\widehat{H} =\prec a, \mu_H, \nu_H \succ\in IF\rho_S(\mathfrak{R})$, $\forall H \in IF\rho_S(\mathfrak{R})$, be the same $(IF\rho_S)$ of X that is defined in the proof of Theorem 22. Then we have $\psi_r'(\langle\widehat{H}\rangle_{\Gamma_r}) = \psi_r(\widehat{H}) = \sigma_r(\widehat{H}) \cap \eta_r(\widehat{H}) = W(\mu_H, r) \cap L(\nu_H, r) = H$. Hence ψ_r' is surjective. This completes the proof. □

5. Conclusion

In this work, we introduce the notions of $(IF\rho s)$, $(IF\rho i)$, $(IF\overline{\rho}i)$, and others; then we proved that for any ρ−subalgebra of X can be considered as both μ−level ρ−subalgebra and ν−level ρ−subalgebra of some $(IF\rho s)$ of $\mathfrak{R}$. At the same time, we proved that intersection of any family of $(IF\rho s)$ of X is $(IF\overline{\rho}i)$ of X. Also, we show that if IFS $\alpha = \{\prec a, (\mu_\alpha(a), \nu_\alpha(a)) \succ \mid a \in \mathfrak{R}\}$ of ρ − algebra X such that the sets $W(\mu_\alpha, r)$ and $L(\nu_\alpha, r)$ are ρ−subalgebras of $\mathfrak{R}$. Then α is $(IF\rho s)$ of $\mathfrak{R}$. Further, some interesting theorems about ρ−homomorphism are given. Finally, some binary relations $\underset{\mu}{\approx}$, $\underset{\nu}{\approx}$ and Γ_r on $IF\rho_S(\mathfrak{R})$ are obtained, and some of their basic properties are discussed. In future work, we will investigate IF in new types of algebras like BCL^+−algebras, BCL^+−subalgebras, BCL^+−ideals and others. Next, we will study their characteristics.

Conflicts of Interest

The authors declare no conflicts of interest.

References

[1] L. A. Zadeh, "Fuzzy sets," *Information and Control*, vol. 8, no. 3, pp. 338–353, 1965.

[2] M. Shabir and M. I. Ali, "Soft ideals and generalized fuzzy ideals in semigroups," *New Mathematics and Natural Computation*, vol. 5, no. 3, pp. 599–615, 2009.

[3] A. Aygünoğlu and H. Aygün, "Introduction to fuzzy soft groups," *Computers & Mathematics with Applications*, vol. 58, no. 6, pp. 1279–1286, 2009.

[4] S. Mahmood, "Dissimilarity fuzzy soft points and their applications," *Fuzzy Information and Engineering*, vol. 8, no. 3, pp. 281–294, 2016.

[5] P. K. Maji, R. Biswas, and A. R. Roy, "Fuzzy soft sets," *Journal of Fuzzy Mathematics*, vol. 9, no. 3, pp. 589–602, 2001.

[6] K. T. Atanassov, "Intuitionistic fuzzy sets," *Fuzzy Sets and Systems*, vol. 20, no. 1, pp. 87–96, 1986.

[7] P. A. Ejegwa, A. J. Akubo, and O. M. Joshua, "Intuitionistic fuzzy sets and its application in career determination via normalized euclidean distance method," *European Scientific Journal*, vol. 10, no. 12, pp. 349–365, 2014.

[8] S. M. Khalil, "On intuitionistic fuzzy soft b-closed sets in intuitionistic fuzzy soft topological spaces," *Annals of Fuzzy Mathematics and Informatics*, vol. 10, no. 2, pp. 221–233, 2015.

[9] P. K. Maji, R. Biswas, and A. R. Roy, "Intuitionistic fuzzy soft sets," *Journal of Fuzzy Mathematics*, vol. 9, no. 3, pp. 677–692, 2001.

[10] S. Mahmood and Z. Al-Batat, "Intuitionistic fuzzy soft LA-semigroups and intuitionistic fuzzy soft ideals," *International Journal of Applications of Fuzzy Sets and Artificial Intelligence*, vol. 6, pp. 119–132.

[11] Y. B. Jun, H. S. Kim, and D. S. Yoo, "Intuitionistic fuzzy d-algebras," *Scientiae Mathematicae Japonicae*, vol. 66, no. 1, pp. 117–125, 2007.

[12] Y. Imai and K. Iseki, "On axiom systems of propositional calculi, XIV," *Proceedings of the Japan Academy, Series A: Mathematical Sciences*, vol. 42, no. 1, pp. 19–22, 1966.

[13] J. Neggers and H. S. Kim, "On d-algebras," *Mathematica Slovaca*, vol. 49, no. 1, pp. 19–26, 1999.

[14] S. Mahmood and M. A. Alradha, "Characterizations of ρ-algebra and generation permutation topological ρ-algebra using permutation in symmetric group," *American Journal of Mathematics and Statistics*, vol. 7, no. 4, pp. 152–159, 2017.

[15] S. Mahmood and M. A. Alradha, "Soft Edge ρ-Algebras of the power sets," *International Journal of Applications of Fuzzy Sets and Artificial Intelligence*, vol. 7, pp. 231–243, 2017.

[16] H. Aktas and N. Cagman, "Soft sets and soft groups," *Information Sciences*, vol. 177, no. 13, pp. 2726–2735, 2007.

10

General Quadratic-Additive Type Functional Equation and Its Stability

Yang-Hi Lee[1] **and Soon-Mo Jung**[2]

[1]*Department of Mathematics Education, Gongju National University of Education, Gongju 32553, Republic of Korea*
[2]*Mathematics Section, College of Science and Technology, Hongik University, Sejong 30016, Republic of Korea*

Correspondence should be addressed to Soon-Mo Jung; smjung@hongik.ac.kr

Academic Editor: Vladimir V. Mityushev

We investigate the general functional equation of the form $f(ax+by+cz)-abf(x+y)-bcf(y+z)-acf(x+z)-a((a+1)/2-b-c)f(x)-b((b+1)/2-a-c)f(y)-c((c+1)/2-a-b)f(z)-(a(a-1)/2)f(-x)-(b(b-1)/2)f(-y)-(c(c-1)/2)f(-z)=0$, whose solutions are quadratic-additive mappings in connection with stability problems.

1. Introduction

In 1940, Ulam [1] posed an important problem concerning the stability of group homomorphisms. In the following year, Hyers [2] solved the problem for the case of Cauchy additive functional equation. After a period longer than two decades, Rassias [3] generalized Hyers' result and then Găvruta [4] extended Rassias' result by allowing unbounded control functions. The concept of stability introduced by Rassias and Găvruta is known today with the term "generalized Hyers-Ulam stability" of functional equations.

A solution to the functional equation

$$f(x+y)-f(x)-f(y)=0 \tag{1}$$

is called an additive mapping and a solution to the functional equation

$$f(x+y)+f(x-y)-2f(x)-2f(y)=0 \tag{2}$$

is called a quadratic mapping. If a mapping can be expressed by the sum of an additive mapping and a quadratic mapping, then we call the mapping a quadratic-additive mapping. Now, we consider the general quadratic-additive type functional equation

$$\begin{aligned}f(ax+by+cz)&-abf(x+y)-bcf(y+z)\\&-acf(x+z)-a\left(\frac{a+1}{2}-b-c\right)f(x)\\&-b\left(\frac{b+1}{2}-a-c\right)f(y)\\&-c\left(\frac{c+1}{2}-a-b\right)f(z)-\frac{a(a-1)}{2}f(-x)\\&-\frac{b(b-1)}{2}f(-y)-\frac{c(c-1)}{2}f(-z)=0\end{aligned} \tag{3}$$

with nonzero real constants a, b, and c. The mapping $f(x)=dx^2+ex$ is a solution to this functional equation, where d, e are real constants. For the case $a=b=c$, the stability of the functional equation (3) was investigated by some mathematicians (see [5] for $a=b=c=1/3$).

In this paper, we will prove that if a, b, and c are nonzero real constants, then every solution to the functional equation (3) is a quadratic-additive mapping and, conversely, we will also prove that every quadratic-additive mapping is a solution to (3) provided a, b, and c are rational constants. Moreover, we will prove the generalized Hyers-Ulam stability of the functional equation (3).

We remark here that (3) is a special form of the general linear equation, whose stability was investigated by Bahyrycz and Olko [6] via a different method from that we apply in this paper. In particular, the main result of this paper is a generalization of [6].

2. Preliminaries

Throughout this paper, let V and W be vector spaces over $\mathbb{R}$, let X be a real normed space, let Y be a real Banach space, and let a, b, and c be positive real constants.

For a given mapping $f : V \to W$, we use the following abbreviations:

$$f_o(x) := \frac{f(x) - f(-x)}{2},$$

$$f_e(x) := \frac{f(x) + f(-x)}{2},$$

$$Af(x, y) := f(x + y) - f(x) - f(y),$$

$$Qf(x, y) := f(x + y) + f(x - y) - 2f(x) - 2f(y),$$

$$\begin{aligned} D_{a,b}f(x, y) := {} & f(ax + by) - abf(x + y) - a\left(\frac{a+1}{2} - b\right)f(x) - b\left(\frac{b+1}{2} - a\right)f(y) \\ & - \frac{a(a-1)}{2}f(-x) - \frac{b(b-1)}{2}f(-y), \end{aligned}$$

$$\begin{aligned} C_{a,b}f(x, y) := {} & f(ax + by) + abf(x - y) - \frac{a(2b + a + 1)}{2}f(x) - \frac{b(b+1)}{2}f(y) \\ & - \frac{a(a-1)}{2}f(-x) - \frac{b(2a + b - 1)}{2}f(-y), \end{aligned}$$

$$\begin{aligned} D_{a,b,c}f(x, y, z) := {} & f(ax + by + cz) - abf(x + y) - bcf(y + z) - acf(x + z) \\ & - a\left(\frac{a+1}{2} - b - c\right)f(x) - b\left(\frac{b+1}{2} - a - c\right)f(y) - c\left(\frac{c+1}{2} - a - b\right)f(z) \\ & - \frac{a(a-1)}{2}f(-x) - \frac{b(b-1)}{2}f(-y) - \frac{c(c-1)}{2}f(-z) \end{aligned} \tag{4}$$

for all $x, y, z \in V$.

The following lemmas were proved in [7].

Lemma 1. *Let a, b be real numbers with $(ab+1)(a+b)ab(a^2+b^2+ab-1) \neq 0$. If a mapping $f : V \to W$ satisfies the functional equation $C_{a,b}f(x, y) = 0$ for all $x, y \in V$, then f_e is quadratic and f_o is additive.*

Lemma 2. *Let a, b be rational constants with $(ab + 1)(a + b)ab(a^2 + b^2 + ab - 1) \neq 0$. A mapping $f : V \to W$ satisfies $C_{a,b}f(x, y) = 0$ for all $x, y \in V$ if and only if f_e is quadratic and f_o is additive.*

Since the equality

$$C_{a,b}f(x, y) = D_{a,-b}f(x, -y) \tag{5}$$

holds for all $x, y \in V$, the next couple of lemmas are direct consequences of Lemmas 1 and 2.

Lemma 3. *Let a, b be real numbers with $(ab-1)(a-b)ab(a^2+b^2-ab-1) \neq 0$. If a mapping $f : V \to W$ satisfies the functional equation $D_{a,b}f(x, y) = 0$ for all $x, y \in V$, then f_e is quadratic and f_o is additive.*

Lemma 4. *Let a, b be rational constants with $(ab - 1)(a - b)ab(a^2 + b^2 - ab - 1) \neq 0$. A mapping $f : V \to W$ satisfies $D_{a,b}f(x, y) = 0$ for all $x, y \in V$ if and only if f_e is quadratic and f_o is additive.*

Lemma 5. *Let a, b be rational constants. If $f : V \to W$ is a quadratic-additive mapping, then f is a solution to the functional equation $D_{a,b}f(x, y) = 0$ for all $x, y \in V$.*

Proof. Let $f : V \to W$ be a quadratic-additive mapping. Then, f_e is a quadratic mapping and f_o is an additive mapping. Hence, we have $f_e(rx) = r^2 f_e(x)$, $f_o(rx) = rf_o(x)$, and $f_o(x + y) = f_o(x) + f_o(y)$, where r is an arbitrary rational number.

First, we will prove that the mapping f_e satisfies $D_{a,b}f_e(x, y) = 0$ for arbitrary rational numbers a and b. If $a = 0$ or $b = 0$, then the equality

$$D_{a,b}f_e(x, y) = 0 \tag{6}$$

holds for all $x, y \in V$. When $a = b$, the equality $D_{a,a}f_e(x, y) = 0$ follows from the equality

$$D_{a,a}f_e(x,y) = f_e(a(x+y)) - a^2 f_e(x+y) \quad (7)$$

for all $x, y \in V$. If a and b are nonzero rational constants with $a \neq b$, then there exists an integer k satisfying $(ak-bk)(akbk-1)((ak)^2+(bk)^2-akbk-1) \neq 0$. According to Lemma 4, the equality

$$D_{a,b}f_e(x,y) = D_{ak,bk}f_e\left(\frac{x}{k},\frac{y}{k}\right) \quad (8)$$

implies that f_e satisfies $D_{a,b}f_e(x,y) = 0$ for all $x, y \in V$. Altogether, the mapping f_e satisfies $D_{a,b}f_e(x,y) = 0$ for arbitrary rational numbers a and b.

On the other hand, the equality

$$D_{a,b}f_o(x,y) = Af_o(ax,by) - abAf_o(x,y) + f_o(ax) - af_o(x) + f_o(by) - bf_o(y) \quad (9)$$

implies that f_o satisfies $D_{a,b}f_o(x,y) = 0$ for arbitrary rational numbers a and b. Since the equality $D_{a,b}f(x,y) = D_{a,b}f_e(x,y) + D_{a,b}f_o(x,y)$ holds for all $x, y \in V$, we conclude that the mapping f satisfies $D_{a,b}f(x,y) = 0$ for all $x, y \in V$. □

Lemma 6. *If real numbers a, b, and c satisfy $(a-b)(b-c)(c-a) \neq 0$, then either $a^2+b^2-ab-1 \neq 0$, or $a^2+c^2-ac-1 \neq 0$, or $c^2+b^2-cb-1 \neq 0$.*

Proof. First, assume that $a = 0$ and a, b, and c are real constants satisfying $(a-b)(b-c)(c-a) \neq 0$, $a^2+b^2-ab-1 = 0$, $a^2+c^2-ac-1 = 0$, and $c^2+b^2-cb-1 = 0$. We then obtain the equalities $b^2 = 1$, $c^2 = 1$, and $cb = 1$, which contradict the fact $b \neq c$.

Second, assume that a, b, and c are nonzero real constants satisfying conditions $(a-b)(b-c)(c-a) \neq 0$, $a^2+b^2-ab-1 = 0$, $a^2+c^2-ac-1 = 0$, and $c^2+b^2-cb-1 = 0$. We then obtain the equalities $a+b = c$, $a+c = b$, and $a = 0$, which contradict the fact $a \neq 0$. □

We will now prove that f is a quadratic-additive mapping provided f is a solution to the functional equation $D_{a,b,c}f(x,y,z) = 0$ for all $x, y, z \in V$.

Remark 7. We remark that if $\sigma : \{a,b,c\} \to \{a,b,c\}$ and $\tau : \{x,y,z\} \to \{x,y,z\}$ are permutations, then we have

$$D_{a,b,c}f(x,y,z) = 0 \quad \forall x,y,z \in V$$
$$\text{iff } D_{\sigma(a),\sigma(b),\sigma(c)}f(\tau(x),\tau(y),\tau(z)) = 0 \quad \forall x,y,z \in V. \quad (10)$$

Theorem 8. *Let a, b, and c be nonzero real numbers. If a mapping $f : V \to W$ satisfies the functional equation $D_{a,b,c}f(x,y,z) = 0$ (with $f(0) = 0$, when $a^2+b^2+c^2-ab-bc-ac = 1$), then f is a quadratic-additive mapping.*

Proof. In view of Remark 7, it is enough to check the following three cases: (1), (2), and (3). Notice that $f(0) = D_{a,b,c}f(0,0,0)/(a^2+b^2+c^2-ab-bc-ac-1) = 0$, when $a^2+b^2+c^2-ab-bc-ac-1 \neq 0$.

(1) If a, b, and c are real constants with $(a-b)(a-c)(b-c) = 0$, then we can assume that $a = b$ without loss of generality by Remark 7. In view of Lemma 1, the equality $-acC_{1,1}f(x,y) = D_{a,a,c}f(y,-y,x) - D_{a,a,c}f(0,0,x)$ implies that f is a quadratic-additive mapping. In particular, we know that every solution to $D_{a,a,c}f(x,y,z) = 0$ is a quadratic-additive mapping.

(2) If a, b, and c are real constants with $(ab-1)(ac-1)(cb-1) = 0$, then we can assume that $ab = 1$ without loss of generality due to Remark 7. We can easily show the validity of the following equalities:

$$\begin{aligned} &f(ax) - \frac{a^2+a}{2}f(x) - \frac{a^2-a}{2}f(-x) \\ &\quad = D_{a,1/a,c}f(x,0,0), \\ &f\left(\frac{y}{a}\right) - \frac{1+a}{2a^2}f(y) - \frac{1-a}{2a^2}f(-y) \\ &\quad = D_{a,1/a,c}f(0,y,0), \\ &f\left(\frac{z}{c}\right) - \frac{1+c}{2c^2}f(c) - \frac{1-c}{2c^2}f(-z) = D_{a,1/a,c}f(0,0,z) \end{aligned} \quad (11)$$

for all $x, y, z \in V$. Since every solution to $D_{a,a,c}f(x,y,z) = 0$ is a quadratic-additive mapping, the equality

$$\begin{aligned} D_{1,1,1}f(x,y,z) &= D_{a,1/a,c}f\left(\frac{x}{a},ay,\frac{z}{c}\right) \\ &\quad - D_{a,1/a,c}f\left(\frac{x}{a},ay,0\right) \\ &\quad - D_{a,1/a,c}f\left(\frac{x}{a},0,\frac{z}{c}\right) \\ &\quad - D_{a,1/a,c}f\left(0,ay,\frac{z}{c}\right) \\ &\quad + D_{a,1/a,c}f\left(\frac{x}{a},0,0\right) \\ &\quad + D_{a,1/a,c}f\left(0,0,\frac{z}{c}\right) \\ &\quad + D_{a,1/a,c}f(0,ay,0) \end{aligned} \quad (12)$$

implies that f is a quadratic-additive mapping.

(3) By Lemma 6, we conclude that if a, b, and c are real constants with $(a-b)(b-c)(c-a)(ab-1)(ac-1)(cb-1) \neq 0$, then either $(a-b)(ab-1)(a^2+b^2-ab-1) \neq 0$, or $(a-c)(ac-1)(a^2+c^2-ac-1) \neq 0$, or $(b-c)(bc-1)(c^2+b^2-cb-1) \neq 0$. By Remark 7, we can assume that $(a-b)(ab-1)(a^2+b^2-ab-1) \neq 0$ without loss of generality. On account of Lemma 3, the

equality $D_{a,b}f(x, y) = D_{a,b,c}f(x, y, 0) = 0$ implies that f is a quadratic-additive mapping. □

Theorem 9. *Let a, b, and c be rational numbers. If* $f : V \to W$ *is a quadratic-additive mapping, then f satisfies the functional equation* $D_{a,b,c}f(x, y, z) = 0$.

Proof. Let $f : V \to W$ be a quadratic-additive mapping. Then we easily show that f_e is a quadratic mapping and f_o is an additive mapping. Hence, we have $f_e(rx) = r^2 f_e(x)$, $f_o(rx) = rf_o(x)$, and $f_o(x + y) = f_o(x) + f_o(y)$, where r is an arbitrary rational number. In view of Lemma 5, we see that the mappings f_e and f_o satisfy the equalities $D_{a,b}f_e(x, y) = 0$ and $D_{a,b}f_o(x, y) = 0$. Therefore, the equalities

$$\begin{aligned} D_{a,b,c}f_e(x, y, z) &= Qf_e\left(ax + \frac{cz}{2}, by + \frac{cz}{2}\right) \\ &\quad - Qf_e\left(ax + \frac{cz}{2}, \frac{cz}{2}\right) \\ &\quad - Qf_e\left(by + \frac{cz}{2}, \frac{cz}{2}\right) \\ &\quad - Qf_e(ax, by) + D_{a,b}f_e(x, y) \\ &\quad + D_{a,c}f_e(x, z) + D_{b,c}f_e(y, z) \\ &\quad - f_e(ax) + a^2 f_e(x) - f_e(by) \\ &\quad + b^2 f_e(y) - 4f_e\left(\frac{cz}{2}\right) \\ &\quad + c^2 f_e(z), \\ D_{a,b,c}f_o(x, y, z) &= Af_o(ax + by, cz) + Af_o(ax, by) \\ &\quad - abAf_o(x, y) - bcAf_o(y, z) \\ &\quad - acAf_o(x, z) + f_o(ax) \\ &\quad - af_o(x) + f_o(by) - bf_o(y) \\ &\quad + f_o(cz) - cf_o(z) \end{aligned} \tag{13}$$

imply that the equalities $D_{a,b,c}f_e(x, y, z) = 0$ and $D_{a,b,c}f_o(x, y, z) = 0$ hold for all $x, y, z \in V$. From the equality $D_{a,b,c}f(x, y, z) = D_{a,b,c}f_e(x, y, z) + D_{a,b,c}f_o(x, y, z)$ for all $x, y, z \in V$, we get the equality $D_{a,b,c}f(x, y, z) = 0$, as desired. □

The next theorem is a direct consequence of Theorems 8 and 9.

Theorem 10. *Let a, b, and c be nonzero rational constants. A mapping* $f : V \to W$ *satisfies* $D_{a,b,c}f(x, y, z) = 0$ *for all* $x, y, z \in V$ *if and only if f is a quadratic-additive mapping.*

3. Main Results

In the following theorems, Theorems 3.1–3.3 of [8] can be slightly modified for the case when $n = 3$ and $Df(x, y, z) = D_{a,b,c}f(x, y, z)$ without altering their proofs.

Theorem 11. *Given a real number k with* $|k| \neq 1$, *let* $\mu : V \setminus \{0\} \to [0, \infty)$ *be a function satisfying the condition*

$$\begin{aligned} &\sum_{i=0}^{\infty} \frac{\mu(k^i x)}{k^i} < \infty \quad \text{when } |k| > 1, \\ &\sum_{i=0}^{\infty} \frac{\mu(k^i x)}{k^{2i}} < \infty \quad \text{when } |k| < 1 \end{aligned} \tag{14}$$

for all $x \in V \setminus \{0\}$ *and let* $\varphi : (V \setminus \{0\})^3 \to [0, \infty)$ *be a function satisfying the condition*

$$\begin{aligned} &\sum_{i=0}^{\infty} \frac{\varphi(k^i x, k^i y, k^i z)}{|k|^i} < \infty \quad \text{when } |k| > 1, \\ &\sum_{i=0}^{\infty} \frac{\varphi(k^i x, k^i y, k^i z)}{k^{2i}} < \infty \quad \text{when } |k| < 1 \end{aligned} \tag{15}$$

for all $x, y, z \in V \setminus \{0\}$. *If a mapping* $f : V \to Y$ *satisfies* $f(0) = 0$ *and*

$$\left\| f(kx) - \frac{k^2 + k}{2} f(x) - \frac{k^2 - k}{2} f(-x) \right\| \leq \mu(x) \tag{16}$$

for all $x \in V \setminus \{0\}$ *and if f satisfies*

$$\left\| D_{a,b,c}f(x, y, z) \right\| \leq \varphi(x, y, z) \tag{17}$$

for all $x, y, z \in V \setminus \{0\}$, *then there exists a unique mapping* $F : V \to Y$ *such that*

$$D_{a,b,c}F(x, y, z) = 0 \tag{18}$$

holds for all $x, y, z \in V \setminus \{0\}$,

$$\begin{aligned} F_e(kx) &= k^2 F_e(x), \\ F_o(kx) &= kF_o(x) \end{aligned} \tag{19}$$

hold for all $x \in V$, *and*

$$\begin{aligned} &\| f(x) - F(x) \| \\ &\quad \leq \sum_{i=0}^{\infty} \frac{\left|k^{i+1} + 1\right| \mu(k^i x) + \left|k^{i+1} - 1\right| \mu(-k^i x)}{2k^{2i+2}} \end{aligned} \tag{20}$$

holds for all $x \in V \setminus \{0\}$.

Theorem 12. *Given a real constant k with $|k| \neq 1$, let $\mu : V \setminus \{0\} \to [0,\infty)$ be a function satisfying the condition*

$$\sum_{i=0}^{\infty} k^{2i}\mu\left(\frac{x}{k^i}\right) < \infty \quad when\ |k| > 1, \qquad \sum_{i=0}^{\infty} |k|^i \mu\left(\frac{x}{k^i}\right) < \infty \quad when\ |k| < 1 \tag{21}$$

for all $x \in V \setminus \{0\}$ and let $\varphi : (V\setminus\{0\})^3 \to [0,\infty)$ be a function satisfying the condition

$$\sum_{i=0}^{\infty} k^{2i}\varphi\left(\frac{x}{k^i},\frac{y}{k^i},\frac{z}{k^i}\right) < \infty \quad when\ |k| > 1, \qquad \sum_{i=0}^{\infty} |k|^i \varphi\left(\frac{x}{k^i},\frac{y}{k^i},\frac{z}{k^i}\right) < \infty \quad when\ |k| < 1 \tag{22}$$

for all $x, y, z \in V \setminus \{0\}$. If a mapping $f : V \to Y$ satisfies $f(0) = 0$ and (16) for all $x \in V \setminus \{0\}$ and if f satisfies (17) for all $x, y, z \in V \setminus \{0\}$, then there exists a unique mapping $F : V \to Y$ satisfying (18) for all $x, y, z \in V \setminus \{0\}$ and (19) for all $x \in V$ and satisfying

$$\|f(x) - F(x)\| \leq \sum_{i=0}^{\infty}\left(\frac{\left|k^{2i}+k^i\right|}{2}\mu\left(\frac{x}{k^{i+1}}\right) + \frac{\left|k^{2i}-k^i\right|}{2}\mu\left(\frac{-x}{k^{i+1}}\right)\right) \tag{23}$$

for all $x \in V \setminus \{0\}$.

Theorem 13. *Given a real constant k with $|k| \neq 1$, let $\mu : V \setminus \{0\} \to [0,\infty)$ be a function satisfying the conditions*

$$\sum_{i=0}^{\infty} \frac{\mu\left(k^i x\right)}{k^{2i}} < \infty, \quad \sum_{i=0}^{\infty} |k|^i \mu\left(\frac{x}{k^i}\right) < \infty \quad when\ |k| > 1, \qquad \sum_{i=0}^{\infty} \frac{\mu\left(k^i x\right)}{|k|^i} < \infty, \quad \sum_{i=0}^{\infty} k^{2i}\mu\left(\frac{x}{k^i}\right) < \infty \quad when\ |k| < 1 \tag{24}$$

for all $x \in V \setminus \{0\}$ and let $\varphi : (V\setminus\{0\})^3 \to [0,\infty)$ be a function satisfying the conditions

$$\sum_{i=0}^{\infty} \frac{\varphi\left(k^i x, k^i y, k^i z\right)}{k^{2i}} < \infty, \quad \sum_{i=0}^{\infty} |k|^i \varphi\left(\frac{x}{k^i},\frac{y}{k^i},\frac{z}{k^i}\right) < \infty \quad when\ |k| > 1,$$
$$\sum_{i=0}^{\infty} \frac{\varphi\left(k^i x, k^i y, k^i z\right)}{|k|^i} < \infty, \quad \sum_{i=0}^{\infty} k^{2i}\varphi\left(\frac{x}{k^i},\frac{y}{k^i},\frac{z}{k^i}\right) < \infty \quad when\ |k| < 1 \tag{25}$$

for all $x, y, z \in V \setminus \{0\}$. If a mapping $f : V \to Y$ satisfies $f(0) = 0$ and (16) for all $x \in V \setminus \{0\}$ and if f satisfies (17) for all $x, y, z \in V \setminus \{0\}$, then there exists a unique mapping $F : V \to Y$ satisfying (18) for all $x, y, z \in V \setminus \{0\}$ and the equalities in (19) for all $x \in V$, such that

$$\|f(x) - F(x)\| \leq \begin{cases} \sum_{i=0}^{\infty}\left[\frac{\mu\left(k^i x\right)+\mu\left(-k^i x\right)}{2k^{2i+2}} + \frac{|k|^i}{2}\left(\mu\left(\frac{x}{k^{i+1}}\right)+\mu\left(\frac{-x}{k^{i+1}}\right)\right)\right] & when\ |k| > 1, \\ \sum_{i=0}^{\infty}\left[\frac{k^{2i}}{2}\left(\mu\left(\frac{x}{k^{i+1}}\right)+\mu\left(\frac{-x}{k^{i+1}}\right)\right) + \frac{\mu\left(k^i x\right)+\mu\left(-k^i x\right)}{2|k|^{i+1}}\right] & when\ |k| < 1 \end{cases} \tag{26}$$

holds for all $x \in V \setminus \{0\}$.

It is to be noted that we can replace $V \setminus \{0\}$ with V in Theorems 11–13.

Lemma 14. *Let a, b, and c be real constants. If a mapping $f : V \to W$ satisfies $f(0) = 0$ and $D_{a,b,c}f(x,y,z) = 0$ for all $x, y, z \in V$, then*

(i) $f_e(ax) = a^2 f_e(x)$,

(ii) $f_e(bx) = b^2 f_e(x)$,

(iii) $f_e(cx) = c^2 f_e(x)$,

(iv) $f_e((a-b)x) = (a-b)^2 f_e(x)$,

(v) $f_e((a-c)x) = (a-c)^2 f_e(x)$,

(vi) $f_e((b-c)x) = (b-c)^2 f_e(x)$,

(vii) $f_o(ax) = af_o(x)$,

(viii) $f_o(bx) = bf_o(x)$,

(ix) $f_o(cx) = cf_o(x)$,

(x) $f_e((a-b)x) = (a-b)^2 f_e(x)$,

(xi) $f_e((a-c)x) = (a-c)^2 f_e(x)$,

(xii) $f_e((b-c)x) = (b-c)^2 f_e(x)$

for all $x \in V$.

Proof. Using the equalities

$$
\begin{aligned}
& f_e(ax) - a^2 f_e(x) = D_{a,b,c} f_e(x,0,0) = 0, \\
& f_o(ax) - af_o(x) = D_{a,b,c} f_o(x,0,0) = 0, \\
& f_e((a-b)x) - (a-b)^2 f_e(x) = D_{a,b,c} f_e(x,-x,0) \\
& \quad = 0, \\
& f_o((a-b)x) - (a-b) f_o(x) = D_{a,b,c} f_o(x,-x,0) \\
& \quad = 0
\end{aligned} \tag{27}
$$

for all $x \in V$ and by Remark 7, we can easily prove the assertions. □

Since every solution to the functional equation (3) is a quadratic-additive mapping according to Theorem 8 provided a, b, and c are nonzero real constants, we can prove the following set of theorems by using Theorems 11–13.

Theorem 15. *For nonzero real constants a, b, and c with* $a-b \notin \{-1,0,1\}$, *let* $\varphi : V^3 \to [0,\infty)$ *be a function satisfying the condition*

$$
\begin{aligned}
& \sum_{i=0}^{\infty} \frac{\varphi\left((a-b)^i x, (a-b)^i y, (a-b)^i z\right)}{|a-b|^i} < \infty \\
& \qquad \text{when } |a-b| > 1, \\
& \sum_{i=0}^{\infty} \frac{\varphi\left((a-b)^i x, (a-b)^i y, (a-b)^i z\right)}{(a-b)^{2i}} < \infty \\
& \qquad \text{when } |a-b| < 1
\end{aligned} \tag{28}
$$

for all $x, y, z \in V$. *If a mapping* $f : V \to Y$ *satisfies* $f(0) = 0$ *and*

$$\left\| D_{a,b,c} f(x,y,z) \right\| \le \varphi(x,y,z) \tag{29}$$

for all $x, y, z \in V$, *then there exists a unique quadratic-additive mapping* $F : V \to Y$ *such that (18) holds for all* $x, y, z \in V$ *and*

$$
\begin{aligned}
& \left\| f(x) - F(x) \right\| \\
& \quad \le \sum_{i=0}^{\infty} \frac{\left|(a-b)^{i+1} + 1\right| \varphi\left((a-b)^i x, -(a-b)^i x, 0\right)}{2(a-b)^{2i+2}} \\
& \quad + \sum_{i=0}^{\infty} \frac{\left|(a-b)^{i+1} - 1\right| \varphi\left(-(a-b)^i x, (a-b)^i x, 0\right)}{2(a-b)^{2i+2}}
\end{aligned} \tag{30}
$$

holds for all $x \in V$.

Proof. It follows from (29) that

$$
\begin{aligned}
& \left\| f((a-b)x) - \frac{(a-b)^2 + (a-b)}{2} f(x) \right. \\
& \quad \left. - \frac{(a-b)^2 - (a-b)}{2} f(-x) \right\| = \left\| D_{a,b,c} f(x,-x,0) \right\| \\
& \quad \le \varphi(x,-x,0)
\end{aligned} \tag{31}
$$

for all $x \in V$. If we put $\mu(x) = \varphi(x,-x,0)$ and $k = a-b$, then $\mu(x)$ satisfies condition (14) and inequality (16) for all $x \in V$. In view of Theorem 11, there exists a unique mapping $F : V \to Y$ satisfying equality (18) for all $x, y, z \in V$ and the equalities in (19) for all $x \in V$, such that inequality (30) holds for all $x \in V$.

In view of Lemma 14, the equalities in (19) follow from equality (18). Hence, there exists a unique mapping $F : V \to Y$ satisfying equality (18) for all $x, y, z \in V$ and inequality (30) for all $x \in V$. □

By the same way as in the proof of Theorem 15, we can prove the following couple of theorems. Hence, we omit their proofs.

Theorem 16. *Given nonzero real constants a, b, and c with* $a-b \notin \{-1,0,1\}$, *let* $\varphi : V^3 \to [0,\infty)$ *be a function satisfying the condition*

$$
\begin{aligned}
& \sum_{i=0}^{\infty} |a-b|^i \varphi\left(\frac{x}{(a-b)^i}, \frac{y}{(a-b)^i}, \frac{z}{(a-b)^i}\right) < \infty \\
& \qquad \text{when } |a-b| < 1, \\
& \sum_{i=0}^{\infty} (a-b)^{2i} \varphi\left(\frac{x}{(a-b)^i}, \frac{y}{(a-b)^i}, \frac{z}{(a-b)^i}\right) < \infty \\
& \qquad \text{when } |a-b| > 1
\end{aligned} \tag{32}
$$

for all $x, y, z \in V$. *If a mapping* $f : V \to Y$ *satisfies* $f(0) = 0$ *and inequality (29) for all* $x, y, z \in V$, *then there exists a unique quadratic-additive mapping* $F : V \to Y$ *satisfying (18) for all* $x, y, z \in V$ *and*

$$
\begin{aligned}
\left\| f(x) - F(x) \right\| \le \sum_{i=0}^{\infty} \Bigg(& \frac{\left|(a-b)^{2i} + (a-b)^i\right|}{2} \\
& \cdot \varphi\left(\frac{x}{(a-b)^{i+1}}, \frac{-x}{(a-b)^{i+1}}, 0\right)
\end{aligned}
$$

$$+\frac{\left|(a-b)^{2i}-(a-b)^{i}\right|}{2} \cdot \varphi\left(\frac{-x}{(a-b)^{i+1}},\frac{x}{(a-b)^{i+1}},0\right)\Bigg)\Bigg) \tag{33}$$

for all $x \in V$.

Theorem 17. *Given nonzero real constants a, b, and c satisfying* $a-b \notin \{-1,0,1\}$, *let* $\varphi : V^3 \to [0,\infty)$ *be a function satisfying the conditions*

$$\sum_{i=0}^{\infty}\frac{\varphi\left((a-b)^i x,(a-b)^i y,(a-b)^i z\right)}{(a-b)^{2i}}<\infty,$$

$$\sum_{i=0}^{\infty}|a-b|^i\varphi\left(\frac{x}{(a-b)^i},\frac{y}{(a-b)^i},\frac{z}{(a-b)^i}\right)<\infty \quad \text{when } |a-b|>1,$$

$$\sum_{i=0}^{\infty}\frac{\varphi\left((a-b)^i x,(a-b)^i y,(a-b)^i z\right)}{|a-b|^i}<\infty,$$

$$\sum_{i=0}^{\infty}|a-b|^{2i}\varphi\left(\frac{x}{(a-b)^i},\frac{y}{(a-b)^i},\frac{z}{(a-b)^i}\right)<\infty \quad \text{when } |a-b|<1 \tag{34}$$

for all $x,y,z \in V$. *If a mapping* $f : V \to Y$ *satisfies* $f(0)=0$ *and equality (29) for all* $x,y,z \in V$, *then there exists a unique quadratic-additive mapping* $F : V \to Y$ *satisfying equality (18) for all* $x,y,z \in V$ *and*

$$\|f(x)-F(x)\| \le \begin{cases} \displaystyle\sum_{i=0}^{\infty}\left[\frac{\varphi\left((a-b)^i x,-(a-b)^i x,0\right)+\varphi\left(-(a-b)^i x,(a-b)^i x,0\right)}{2(a-b)^{2i+2}}+\frac{|a-b|^i}{2}\left(\varphi\left(\frac{x}{(a-b)^{i+1}},\frac{-x}{(a-b)^{i+1}},0\right)+\varphi\left(\frac{-x}{(a-b)^{i+1}},\frac{x}{(a-b)^{i+1}},0\right)\right)\right] & \text{when } |a-b|>1, \\ \displaystyle\sum_{i=0}^{\infty}\left[\frac{\varphi\left((a-b)^i x,-(a-b)^i x,0\right)+\varphi\left(-(a-b)^i x,(a-b)^i x,0\right)}{2|a-b|^{i+1}}+\frac{(a-b)^{2i}}{2}\left(\varphi\left(\frac{x}{(a-b)^{i+1}},\frac{-x}{(a-b)^{i+1}},0\right)+\varphi\left(\frac{-x}{(a-b)^{i+1}},\frac{x}{(a-b)^{i+1}},0\right)\right)\right] & \text{when } |a-b|<1 \end{cases} \tag{35}$$

for all $x \in V$.

By the same way as in the proofs of Theorems 18–23, we can use Lemma 14 to prove Theorems 15–17.

Theorem 18. *For nonzero real constants a, b, and c with* $a \notin \{-1,0,1\}$, *let* $\varphi : V^3 \to [0,\infty)$ *be a function satisfying the condition*

$$\sum_{i=0}^{\infty}\frac{\varphi\left(a^i x,a^i y,a^i z\right)}{|a|^i}<\infty \quad \text{when } |a|>1,$$

$$\sum_{i=0}^{\infty}\frac{\varphi\left(a^i x,a^i y,a^i z\right)}{a^{2i}}<\infty \quad \text{when } |a|<1 \tag{36}$$

for all $x,y,z \in V$. *If a mapping* $f : V \to Y$ *satisfies* $f(0)=0$ *and (29) for all* $x,y,z \in V$, *then there exists a unique quadratic-additive mapping* $F : V \to Y$ *satisfying (18) for all* $x,y,z \in V$ *and*

$$\|f(x)-F(x)\| \le \sum_{i=0}^{\infty}\frac{\left|a^{i+1}+1\right|\varphi\left(a^i x,0,0\right)}{2a^{2i+2}} + \sum_{i=0}^{\infty}\frac{\left|a^{i+1}-1\right|\varphi\left(-a^i x,0,0\right)}{2a^{2i+2}} \tag{37}$$

for all $x \in V$.

Proof. If we put $\mu(x)=\varphi(x,0,0)$ and $k=a$, then it follows from (29) that

$$\left\|f(ax)-\frac{a^2+a}{2}f(x)-\frac{a^2-a}{2}f(-x)\right\| = \left\|D_{a,b,c}f(x,0,0)\right\| \le \varphi(x,0,0) \tag{38}$$

for all $x \in V$. For this case, conditions (14)–(17) in Theorem 11 are satisfied. Hence, there exists a unique mapping $F : V \to Y$ satisfying equality (18) for all $x,y,z \in V$ and the equalities in (19) for all $x \in V$ and satisfying inequality (37) for all $x \in V$. Since equality (18) implies equalities (19) in Lemma 14, there exists a unique mapping $F : V \to Y$ satisfying equality (18) for all $x,y,z \in V$ as well as inequality (37) for all $x \in V$. □

Theorem 19. *Given nonzero real constants a, b, and c with* $a \notin \{-1,0,1\}$, *assume that* $\varphi : V^3 \to [0,\infty)$ *is a function satisfying the condition*

$$\sum_{i=0}^{\infty}|a|^i\varphi\left(\frac{x}{a^i},\frac{y}{a^i},\frac{z}{a^i}\right)<\infty \quad \text{when } |a|<1,$$

$$\sum_{i=0}^{\infty}a^{2i}\varphi\left(\frac{x}{a^i},\frac{y}{a^i},\frac{z}{a^i}\right)<\infty \quad \text{when } |a|>1 \tag{39}$$

for all $x, y, z \in V$. If a mapping $f : V \to Y$ satisfies $f(0) = 0$ and equality (29) for all $x, y, z \in V$, then there exists a unique quadratic-additive mapping $F : V \to Y$ satisfying (18) for all $x, y, z \in V$ and

$$\|f(x) - F(x)\| \le \sum_{i=0}^{\infty}\left(\frac{\left|a^{2i}+a^{i}\right|}{2}\varphi\left(\frac{x}{a^{i+1}},0,0\right) + \frac{\left|a^{2i}-a^{i}\right|}{2}\varphi\left(\frac{-x}{a^{i+1}},0,0\right)\right) \tag{40}$$

for all $x \in V$.

Theorem 20. *Let a, b, and c be nonzero real constants with $a \notin \{-1,0,1\}$. Assume that $\varphi : V^3 \to [0,\infty)$ is a function satisfying the conditions*

$$\sum_{i=0}^{\infty}\frac{\varphi\left(a^{i}x,a^{i}y,a^{i}z\right)}{a^{2i}} < \infty, \qquad \sum_{i=0}^{\infty}|a|^{i}\varphi\left(\frac{x}{a^{i}},\frac{y}{a^{i}},\frac{z}{a^{i}}\right) < \infty \quad \text{when } |a| > 1,$$

$$\sum_{i=0}^{\infty}\frac{\varphi\left(a^{i}x,a^{i}y,a^{i}z\right)}{|a|^{i}} < \infty, \qquad \sum_{i=0}^{\infty}|a|^{2i}\varphi\left(\frac{x}{a^{i}},\frac{y}{a^{i}},\frac{z}{a^{i}}\right) < \infty \quad \text{when } |a| < 1 \tag{41}$$

for all $x, y, z \in V$. If a mapping $f : V \to Y$ satisfies $f(0) = 0$ and equality (29) for all $x, y, z \in V$, then there exists a unique quadratic-additive mapping $F : V \to Y$ satisfying (18) for all $x, y, z \in V$ and

$$\|f(x) - F(x)\| \le \begin{cases} \sum_{i=0}^{\infty}\left[\frac{\varphi\left(a^{i}x,0,0\right)+\varphi\left(-a^{i}x,0,0\right)}{2a^{2i+2}} + \frac{|a|^{i}}{2}\left(\varphi\left(\frac{x}{a^{i+1}},0,0\right)+\varphi\left(\frac{-x}{a^{i+1}},0,0\right)\right)\right] & \text{when } |a| > 1, \\ \sum_{i=0}^{\infty}\left[\frac{\varphi\left(a^{i}x,0,0\right)+\varphi\left(-a^{i}x,0,0\right)}{2a^{i+1}} + \frac{a^{2i}}{2}\left(\varphi\left(\frac{x}{a^{i+1}},0,0\right)+\varphi\left(\frac{-x}{a^{i+1}},0,0\right)\right)\right] & \text{when } |a| < 1 \end{cases} \tag{42}$$

for all $x \in V$.

Since every quadratic-additive mapping is a solution to the functional equation (3) and every solution to the functional equation (3) is a quadratic-additive mapping provided a, b, and c are nonzero rational numbers by Theorem 9, the following set of theorems are direct consequences of Theorems 15–20.

Theorem 21. *For nonzero rational constants a, b, and c, let $\varphi : V^3 \to [0,\infty)$ be a function satisfying condition (28) for all $x, y, z \in V$. If a mapping $f : V \to Y$ satisfies $f(0) = 0$ and (29) for all $x, y, z \in V$, then there exists a unique quadratic-additive mapping $F : V \to Y$ satisfying (30) for all $x \in V$.*

Theorem 22. *For nonzero rational constants a, b, and c, let $\varphi : V^3 \to [0,\infty)$ be a function satisfying condition (32) for all $x, y, z \in V$. If a mapping $f : V \to Y$ satisfies $f(0) = 0$ and inequality (29) for all $x, y, z \in V$, then there exists a unique quadratic-additive mapping $F : V \to Y$ satisfying (33) for all $x \in V$.*

Theorem 23. *Given nonzero rational constants a, b, and c, assume that $\varphi : V^3 \to [0,\infty)$ is a function satisfying the conditions in (34) for all $x, y, z \in V$. If a mapping $f : V \to Y$ satisfies $f(0) = 0$ and equality (29) for all $x, y, z \in V$, then there exists a unique quadratic-additive mapping $F : V \to Y$ satisfying (35) for all $x \in V$.*

Theorem 24. *Let a, b, and c be nonzero rational constants and let $\varphi : V^3 \to [0,\infty)$ be a function satisfying condition (36) for all $x, y, z \in V$. If a mapping $f : V \to Y$ satisfies $f(0) = 0$ and (29) for all $x, y, z \in V$, then there exists a unique quadratic-additive mapping $F : V \to Y$ satisfying (37) for all $x \in V$.*

Theorem 25. *For nonzero rational constants a, b, and c, let $\varphi : V^3 \to [0,\infty)$ be a function satisfying condition (39) for all $x, y, z \in V$. If a mapping $f : V \to Y$ satisfies $f(0) = 0$ and equality (29) for all $x, y, z \in V$, then there exists a unique quadratic-additive mapping $F : V \to Y$ satisfying (40) for all $x \in V$.*

Theorem 26. *Given nonzero rational constants a, b, and c, suppose $\varphi : V^3 \to [0,\infty)$ is a function satisfying the conditions in (41) for all $x, y, z \in V$. If a mapping $f : V \to Y$ satisfies $f(0) = 0$ and equality (29) for all $x, y, z \in V$, then there exists a unique quadratic-additive mapping $F : V \to Y$ satisfying (42) for all $x \in V$.*

We remember that X is a real normed space and Y is a real Banach space.

Corollary 27. *Assume that a, b, and c are nonzero real constants with $a - b \notin \{-1,0,1\}$, p is a real number with*

$p \notin \{1,2\}$, and θ is a positive real number. If a mapping $f : X \to Y$ satisfies $f(0) = 0$ and the inequality

$$\left\|D_{a,b,c} f(x,y,z)\right\| \le \theta\left(\|x\|^p + \|y\|^p + \|z\|^p\right) \quad (43)$$

for all $x, y, z \in X$, then there exists a unique quadratic-additive mapping $F : X \to Y$ such that $D_{a,b,c}F(x,y,z) = 0$ holds for all $x, y, z \in X$ and

$$\|f(x) - F(x)\| \le \begin{cases} \dfrac{2\theta\|x\|^p}{\left||a-b| - |a-b|^p\right|} & \text{when } p < 1, \\ \dfrac{2\theta\|x\|^p}{\left||a-b| - |a-b|^p\right|} + \dfrac{2\|x\|^p}{\left||a-b|^2 - |a-b|^p\right|} & \text{when } 1 < p < 2, \\ \dfrac{2\theta\|x\|^p}{\left||a-b|^2 - |a-b|^p\right|} & \text{when } p > 2 \end{cases} \quad (44)$$

holds for all $x \in X$.

Proof. If we put $\varphi(x,y,z) := \theta(\|x\|^p + \|y\|^p + \|z\|^p)$ for all $x, y, z \in X$, then φ satisfies either the first inequality in (28) when $p < 1$ and $|a-b| > 1$, or the second inequality in (28) when $p > 2$ and $|a-b| < 1$, or the first inequality in (32) when $p > 2$ and $|a-b| > 1$, or the second inequality in (32) when $p < 1$ and $|a-b| < 1$, or the first inequality in (34) when $1 < p < 2$ and $|a-b| > 1$, or the second inequality in (34) when $1 < p < 2$ and $|a-b| < 1$ for all $x, y, z \in X$. Therefore, by Theorems 15–17, we obtain the desired inequality (44). □

The following corollary follows from Theorems 18–20.

Corollary 28. *Assume that a, b, and c are nonzero real constants with $a \notin \{-1,0,1\}$, p is a real number with $p \notin \{1,2\}$, and θ is a positive real number. If a mapping $f : X \to Y$ satisfies $f(0) = 0$ and inequality (43) for all $x, y, z \in X$, then there exists a unique quadratic-additive mapping $F : X \to Y$ such that $D_{a,b,z}F(x,y,z) = 0$ holds for all $x, y, z \in X$ and*

$$\|f(x) - F(x)\| \le \begin{cases} \dfrac{\theta\|x\|^p}{\left||a| - |a|^p\right|} & \text{when } p < 1, \\ \dfrac{\theta\|x\|^p}{\left||a| - |a|^p\right|} + \dfrac{\|x\|^p}{\left||a|^2 - |a|^p\right|} & \text{when } 1 < p < 2, \\ \dfrac{\theta\|x\|^p}{\left||a|^2 - |a|^p\right|} & \text{when } p > 2 \end{cases} \quad (45)$$

holds for all $x \in X$.

Proof. If we put $\varphi(x,y,z) := \theta(\|x\|^p + \|y\|^p + \|z\|^p)$ for all $x, y, z \in X$, then φ satisfies either the first inequality in (36) when $p < 1$ and $|a| > 1$, or the second inequality in (36) when $p > 2$ and $|a| < 1$, or the first inequality in (39) when $p > 2$ and $|a| > 1$, or the second inequality in (39) when $p < 1$ and $|a| < 1$, or the first inequality in (41) when $1 < p < 2$ and $|a| > 1$, or the second inequality in (41) when $1 < p < 2$ and $|a| < 1$ for all $x, y \in X$. Therefore, by Theorems 18–20, we obtain the desired inequality (45). □

The following corollary follows from Theorems 21–23.

Corollary 29. *Let a, b, and c be nonzero rational constants with $a - b \notin \{-1,0,1\}$ and let θ, p be nonnegative real numbers with $p \notin \{1,2\}$. If a mapping $f : X \to Y$ satisfies $f(0) = 0$ and inequality (43) for all $x, y, z \in X$, then there exists a unique quadratic-additive mapping $F : X \to Y$ such that inequality (44) holds for all $x \in X$.*

The following corollary follows from Theorems 24–26.

Corollary 30. *Let a, b, and c be nonzero rational constants with $a \notin \{-1,0,1\}$ and let θ, p be nonnegative real numbers with $p \notin \{1,2\}$. If a mapping $f : X \to Y$ satisfies $f(0) = 0$ and inequality (43) for all $x, y, z \in X$, then there exists a unique quadratic-additive mapping $F : X \to Y$ such that inequality (45) holds for all $x \in X$.*

Competing Interests

The authors declare that there is no conflict of interests regarding the publication of this paper.

Authors' Contributions

All authors contributed equally to the writing of this paper. All authors read and approved the final paper.

Acknowledgments

Soon-Mo Jung was supported by Basic Science Research Program through the National Research Foundation of Korea (NRF) funded by the Ministry of Education (no. 2015R1D1A1A02061826).

References

[1] S. M. Ulam, *Problems in Modern Mathematics*, John Wiley & Sons, New York, NY, USA, 1964.

[2] D. H. Hyers, "On the stability of the linear functional equation," *Proceedings of the National Academy of Sciences of the United States of America*, vol. 27, pp. 222–224, 1941.

[3] T. M. Rassias, "On the stability of the linear mapping in Banach spaces," *Proceedings of the American Mathematical Society*, vol. 72, no. 2, pp. 297–300, 1978.

[4] P. Găvruta, "A generalization of the Hyers-Ulam-Rassias stability of approximately additive mappings," *Journal of Mathematical Analysis and Applications*, vol. 184, no. 3, pp. 431–436, 1994.

[5] Y.-H. Lee and S.-M. Jung, "Generalized Hyers-Ulam stability of a 3-dimensional quadratic-additive type functional equation," *International Journal of Mathematical Analysis*, vol. 9, no. 9–12, pp. 527–540, 2015.

[6] A. Bahyrycz and J. Olko, "On stability of the general linear equation," *Aequationes Mathematicae*, vol. 89, no. 6, pp. 1461–1474, 2015.

[7] Y.-H. Lee and S.-M. Jung, "Stability of some 2-dimensional functional equations," *International Journal of Mathematical Analysis*, vol. 10, no. 4, pp. 171–190, 2016.

[8] Y.-H. Lee and S.-M. Jung, "A general stability theorem for a class of functional equations including quadratic-additive functional equations," *Journal of Computational Analysis and Applications*, In press.

11

Introduction to Neutrosophic BCI/BCK-Algebras

A. A. A. Agboola[1] and B. Davvaz[2]

[1]*Department of Mathematics, Federal University of Agriculture, Abeokuta, Nigeria*
[2]*Department of Mathematics, Yazd University, Yazd, Iran*

Correspondence should be addressed to A. A. A. Agboola; aaaola2003@yahoo.com

Academic Editor: Sergejs Solovjovs

We introduce the concept of neutrosophic BCI/BCK-algebras. Elementary properties of neutrosophic BCI/BCK algebras are presented.

1. Introduction

Logic algebras are the algebraic foundation of reasoning mechanism in many fields such as computer sciences, information sciences, cybernetics, and artificial intelligence. In 1966, Imai and Iséki [1, 2] introduced the notions, called BCK-algebras and BCI-algebras. These notions are originated from two different ways: one of them is based on set theory; another is from classical and nonclassical propositional calculi. As is well known, there is a close relationship between the notions of the set difference in set theory and the implication functor in logical systems. Since then many researchers worked in this area and lots of literatures had been produced about the theory of BCK/BCI-algebra. On the theory of BCK/BCI-algebras, for example, see [2–6]. It is known that the class of BCK-algebras is a proper subclass of the class of BCI-algebras. MV-algebras were introduced by Chang in [7], in order to show that Lukasiewicz logic is complete with respect to evaluations of propositional variables in the real unit interval $[0, 1]$. It is well known that the class of MV-algebras is a proper subclass of the class of BCK- algebras.

By a BCI-algebra we mean an algebra $(X, *, 0)$ of type $(2, 0)$ satisfying the following axioms, for all $x, y, z \in X$:

(1) $((x * y) * (x * z)) * (z * y) = 0$,

(2) $(x * (x * y)) * y = 0$,

(3) $x * x = 0$,

(4) $x * y = 0$ and $y * x = 0$ imply $x = y$.

We can define a partial ordering $\leq$ by $x \leq y$ if and only if $x * y = 0$.

If a BCI-algebra X satisfies $0 * x = 0$ for all $x \in X$, then we say that X is a BCK-algebra. Any BCK-algebra X satisfies the following axioms for all $x, y, z \in X$:

(1) $(x * y) * z = (x * z) * y$,

(2) $((x * z) * (y * z)) * (x * y) = 0$,

(3) $x * 0 = x$,

(4) $x * y = 0 \Rightarrow (x * z) * (y * z) = 0, (z * y) * (z * x) = 0$.

Let $(X, *, 0)$ be a BCK-algebra. Consider the following:

(1) X is said to be commutative if for all $x, y \in X$ we have $x * (x * y) = y * (y * x)$;

(2) X is said to be implicative if for all $x, y \in X$, we have $x = x * (y * x)$.

In 1995, Smarandache introduced the concept of neutrosophic logic as an extension of fuzzy logic; see [8–10]. In 2006, Kandasamy and Smarandache introduced the concept of neutrosophic algebraic structures; see [11, 12]. Since then, several researchers have studied the concepts and a great deal of literature has been produced. Agboola et al. in [13–17] continued the study of some types of neutrosophic algebraic structures.

Let X be a nonempty set. A set $X(I) = \langle X, I \rangle$ generated by X and I is called a neutrosophic set. The elements of $X(I)$ are of the form (x, yI), where x and y are elements of X.

In the present paper, we introduce the concept of neutrosophic BCI/BCK-algebras. Elementary properties of neutrosophic BCI/BCK-algebras are presented.

2. Main Results

Definition 1. Let $(X,*,0)$ be any BCI/BCK-algebra and let $X(I) = \langle X, I\rangle$ be a set generated by X and I. The triple $(X(I),*,(0,0))$ is called a neutrosophic BCI/BCK-algebra. If (a,bI) and (c,dI) are any two elements of $X(I)$ with $a,b,c,d \in X$, we define

$$(a,bI)*(c,dI) = (a*c,(a*d\wedge b*c\wedge b*d)I). \qquad (1)$$

An element $x\in X$ is represented by $(x,0)\in X(I)$ and $(0,0)$ represents the constant element in $X(I)$. For all $(x,0),(y,0)\in X$, we define

$$(x,0)*(y,0) = (x*y,0) = (x\wedge\neg y,0), \qquad (2)$$

where $\neg y$ is the negation of y in X.

Example 2. Let $(X(I),+)$ be any commutative neutrosophic group. For all $(a,bI),(c,dI)\in X(I)$ define

$$(a,bI)*(c,dI) = (a,bI)-(c,dI) = (a-c,(b-d)I). \qquad (3)$$

Then $(X(I),*,(0,0))$ is a neutrosophic BCI-algebra.

Example 3. Let $X(I)$ be a neutrosophic set and let $A(I)$ and $B(I)$ be any two nonempty subsets of $X(I)$. Define

$$A(I)*B(I) = A(I)-B(I) = A(I)\cap B'(I). \qquad (4)$$

Then $(X(I),*,\emptyset)$ is a neutrosophic BCK-algebra.

Theorem 4. *Every neutrosophic BCK-algebra $(X(I),*,(0,0))$ is a neutrosophic BCI-algebra.*

Proof. It is straightforward. □

Theorem 5. *Every neutrosophic BCK-algebra $(X(I),*,(0,0))$ is a BCK-algebra and not the converse.*

Proof. Suppose that $(X(I),*,(0,0))$ is a neutrosophic BCK-algebra. Let $x=(a,bI)$, $y=(c,dI)$, and $z=(e,fI)$ be arbitrary elements of $X(I)$. Then we have the following.

(1) We have

$$\begin{aligned}&((x*y)*(x*z))*(z*y)\\&=(((a,bI)*(c,dI))*((a,bI)*(e,fI)))\\&\quad*((e,fI)*(c,dI))\\&\equiv[(r,sI)*(p,qI)]*(u,vI),\end{aligned} \qquad (5)$$

where

$$\begin{aligned}(r,sI)&=(a*c,(a*d\wedge b*c\wedge b*d)I)\\&=(a\wedge\neg c,(a\wedge\neg d\wedge b\wedge\neg c)I),\\(p,qI)&=(a*e,(a*f\wedge b*e\wedge b*f)I)\\&=(a\wedge\neg e,(a\wedge\neg f\wedge b\wedge\neg e)I),\\(u,vI)&=(e*c,(e*d\wedge f*c\wedge f*d)I)\\&=(e\wedge\neg c,(e\wedge\neg d\wedge f\wedge\neg c)I).\end{aligned} \qquad (6)$$

Hence,

$$\begin{aligned}(r,sI)*(p,qI)&=(r*p,(r*q\wedge s*p\wedge s*q)I)\\&=(r\wedge\neg p,(r\wedge\neg q\wedge s\wedge\neg p)I)\\&\equiv(m,kI),\\(m,kI)*(u,vI)&=(m*u,(m*v\wedge k*u\wedge k*v)I)\\&=(m\wedge\neg u,(m\wedge\neg v\wedge k\wedge\neg u)I)\\&\equiv(g,hI).\end{aligned} \qquad (7)$$

Now, we obtain

$$\begin{aligned}g&=m\wedge\neg u=r\wedge\neg p\wedge\neg u\\&=(a\wedge\neg c\wedge e)(\neg e\vee c)=0.\end{aligned} \qquad (8)$$

Also, we have

$$\begin{aligned}h&=m\wedge\neg v\wedge k\wedge\neg u\\&=r\wedge\neg p\wedge\neg q\wedge s\wedge\neg v\wedge\neg u\\&=a\wedge\neg c\wedge\neg d\wedge b\wedge\neg p\wedge\neg q\wedge\neg v\wedge\neg u\\&=a\wedge\neg c\wedge\neg d\wedge b\wedge(\neg a\vee e)\wedge(\neg e\vee c)\wedge\neg q\wedge\neg v\\&=a\wedge\neg c\wedge\neg d\wedge b\wedge e\wedge(\neg e\vee c)\wedge\neg q\wedge\neg v\\&=0.\end{aligned} \qquad (9)$$

This shows that $(g,hI)=(0,0)$ and, consequently, $((x*y)*(x*z))*(z*y)=0$.

(2) We have

$$\begin{aligned}&(x*(x*y))*y\\&=((a,bI)*((a,bI)*(c,dI)))*(c,dI)\\&=((a,bI)*(a*c,(a*d\wedge b*c\wedge b*d)I))*(c,dI)\\&=((a,bI)*(r,sI))*(c,dI),\end{aligned} \qquad (10)$$

where

$$\begin{aligned}(r,sI)&=(a,bI)*(c,dI)\\&=(a*c,(a*d\wedge b*c\wedge b*d)I)\\&=(a\wedge\neg c,(a\wedge\neg d\wedge b\wedge\neg c)I).\end{aligned} \qquad (11)$$

Then,

$$\begin{aligned}(a,bI)*(r,sI) &= (a*r,(a*s\wedge b*r\wedge b*s)I)\\ &= (a\wedge\neg r,(a\wedge\neg s\wedge b\wedge\neg r)I)\\ &\equiv (u,vI).\end{aligned} \tag{12}$$

Therefore, we obtain

$$\begin{aligned}(u,vI)*(c,dI) &= (u*c,(u*d\wedge v*c\wedge v*d)I)\\ &= (u\wedge\neg c,u\wedge\neg d\wedge v\wedge\neg c)\\ &\equiv (p,qI),\end{aligned} \tag{13}$$

where

$$\begin{aligned}p &= u\wedge\neg c = a\wedge\neg r\wedge\neg c\\ &= a\wedge(\neg a\vee c)\wedge\neg c = a\wedge c\wedge\neg c = 0,\\ q &= u\wedge\neg d\wedge v\wedge\neg c = a\wedge\neg r\wedge\neg d\wedge v\wedge\neg c\\ &= a\wedge(\neg a\vee c)\wedge\neg d\wedge v\wedge\neg c\\ &= a\wedge c\wedge\neg d\wedge v\wedge\neg c = 0.\end{aligned} \tag{14}$$

Since $(p,qI)=(0,0)$, it follows that $(x*(x*y))*y=0$.

(3) We have

$$\begin{aligned}x*x &= (a,bI)*(a,bI)\\ &= (a*a,(a*b\wedge b*a\wedge b*b)I)\\ &= (a\wedge\neg a,(a\wedge\neg b\wedge b\wedge\neg a\wedge b\wedge\neg b)I)\\ &= (0,0).\end{aligned} \tag{15}$$

(4) Suppose that $x*y=0$ and $y*x=0$. Then $(a,bI)*(c,dI)=(0,0)$ and $(c,dI)*(a,bI)=(0,0)$ from which we obtain $(a*c,(a*d\wedge b*c\wedge b*d)I)=(0,0)$ and $(c*a,(c*b\wedge d*a\wedge d*b)I)=(0,0)$. These imply that $(a\wedge\neg c,(a\wedge\neg d\wedge b\neg c)I)=(0,0)$ and $(c\wedge\neg a,(c\wedge\neg b\wedge d\neg a)I)=(0,0)$ and therefore, $a\wedge\neg c=0, a\wedge\neg d\wedge b\neg c=0$, $c\wedge\neg a=0$, and $c\wedge\neg b\wedge d\neg a=0$ from which we obtain $a=c$ and $b=d$. Hence, $(a,bI)=(c,dI)$; that is, $x=y$.

(5) We have

$$\begin{aligned}0*x &= (0,0)*(a,bI) = (0*a,(0*b\wedge 0*a)I)\\ &= (0,(0\wedge 0)I) = (0,0).\end{aligned} \tag{16}$$

Items (1)–(5) show that $(X(I),*,(0,0))$ is a BCK-algebra. □

Lemma 6. *Let $(X(I),*,(0,0))$ be a neutrosophic BCK-algebra. Then $(a,bI)*(0,0)=(a,bI)$, if and only if $a=b$.*

Proof. Suppose that $(a,bI)*(0,0)=(a,bI)$. Then $(a*0,(a*0\wedge b*0)I)=(a,bI)$ which implies that $(a,(a\wedge b)I)=(a,bI)$ from which we obtain $a=b$. The converse is obvious. □

Lemma 7. *Let $(X(I),*,(0,0))$ be a neutrosophic BCI-algebra. Then for all $(a,bI),(c,dI)\in X(I)$,*

(1) $(0,0)*((a,bI)*(c,dI))=((0,0)*(a,bI))*((0,0)*(c,dI))$;

(2) $(0,0)*((0,0)*((a,bI)*(c,dI)))=(0,0)*((a,bI)*(c,dI))$.

Theorem 8. *Let $(X(I),*,(0,0))$ be a neutrosophic BCK-algebra. Then for all $(a,bI),(c,dI),(e,fI)\in X(I)$,*

(1) $(a,bI)*(c,dI)=(0,0)$ *implies that* $((a,bI)*(e,fI))*((c,dI)*(e,fI))=(0,0)$ *and* $((e,fI)*(c,dI))*((e,fI)*(a,bI))=(0,0)$;

(2) $((a,bI)*(c,dI))*(e,fI)=((a,bI)*(e,fI))*(c,dI)$;

(3) $((a,bI)*(e,fI))*((c,dI)*(e,fI))*((a,bI)*(c,dI))=(0,0)$.

Proof. (1) Suppose that $(a,bI)*(c,dI)=(0,0)$. Then $(a*c,(a*d\wedge b*c\wedge b*d)I)=(0,0)$ from which we obtain

$$a\wedge\neg c=0,\qquad a\wedge\neg d\wedge b\wedge\neg c=0. \tag{17}$$

Now,

$$\begin{aligned}(x,yI) &= (a,bI)*(e,fI)\\ &= (a\wedge\neg e,(a\wedge\neg f\wedge b\wedge\neg e)I),\\ (p,qI) &= (c,dI)*(e,fI)\\ &= (c\wedge\neg e,(c\wedge\neg f\wedge d\wedge\neg e)I).\end{aligned} \tag{18}$$

Hence,

$$\begin{aligned}(x,yI)*(p,qI) &= (x\wedge\neg p,(x\wedge\neg q\wedge y\wedge\neg p)I)\\ &\equiv (u,vI),\end{aligned} \tag{19}$$

where

$$\begin{aligned}u &= x\wedge\neg p = a\wedge\neg e\wedge(\neg c\vee e)\\ &= a\wedge\neg e\wedge\neg c = 0,\\ v &= x\wedge\neg q\wedge y\wedge\neg p\\ &= a\wedge\neg e\wedge\neg f\wedge b\wedge\neg e\wedge(\neg c\vee e)\wedge\neg q\\ &= a\wedge\neg e\wedge\neg f\wedge b\wedge\neg e\wedge\neg c\wedge(\neg c\vee f\vee\neg d\vee e)\\ &= (a\wedge\neg c\wedge\neg e\wedge\neg f\wedge b)\vee(a\wedge\neg d\wedge b\wedge\neg c\wedge\neg e\wedge\neg f)\\ &= 0\vee 0 = 0.\end{aligned} \tag{20}$$

This shows that $(u,vI)=(0,0)$ and so $((a,bI)*(e,fI))*((c,dI)*(e,fI))=(0,0)$. Similar computations show that $((e,fI)*(c,dI))*((e,fI)*(a,bI))=(0,0)$.

(2) Put

$$\text{LHS} = ((a,bI)*(c,dI))*(e,fI) = (x,yI)*(e,fI), \tag{21}$$

where

$$(x, yI) = (a, bI) * (c, dI) = (a \wedge \neg c, (a \wedge \neg d \wedge b \wedge \neg c) I). \quad (22)$$

Therefore,

$$(x, yI) * (e, fI) = (x \wedge \neg e, (x \wedge \neg f \wedge y \wedge \neg e) I) \equiv (u, vI). \quad (23)$$

Now, we have

$$\begin{aligned} u &= x \wedge \neg e = a \wedge \neg c \wedge \neg e, \\ v &= x \wedge \neg f \wedge y \wedge \neg e = x \wedge \neg f \wedge y \wedge \neg e \\ &= a \wedge \neg c \wedge \neg f \wedge \neg e \wedge \neg d \wedge b. \end{aligned} \quad (24)$$

Thus,

$$\text{LHS} = (a \wedge \neg c \wedge \neg e, (a \wedge \neg c \wedge \neg f \wedge \neg e \wedge \neg d \wedge b) I). \quad (25)$$

Similarly, it can be shown that

$$\begin{aligned} \text{RHS} &= ((a, bI) * (e, fI)) * (c, dI) \\ &= (a \wedge \neg c \wedge \neg e, (a \wedge \neg c \wedge \neg f \wedge \neg e \wedge \neg d \wedge b) I). \end{aligned} \quad (26)$$

(3) Put

$$\begin{aligned} \text{LHS} &= ((a, bI) * (e, fI)) \\ &\quad * ((c, dI) * (e, fI)) * ((a, bI) * (c, dI)) \\ &\equiv ((x, yI) * (p, qI)) * (u, vI), \end{aligned} \quad (27)$$

where

$$\begin{aligned} (x, yI) &= (a, bI) * (e, fI) = (a \wedge \neg c, (a \wedge \neg f \wedge b \wedge \neg e) I), \\ (p, qI) &= (c, dI) * (e, fI) = (c \wedge \neg e, (c \wedge \neg f \wedge d \wedge \neg e) I), \\ (u, vI) &= (a, bI) * (c, dI) = (a \wedge \neg c, (a \wedge \neg d \wedge b \wedge \neg c) I). \end{aligned} \quad (28)$$

Thus, we have

$$(x, yI) * (p, qI) = (x \wedge \neg p, (x \wedge \neg q \wedge y \wedge \neg p) I) \equiv (g, hI). \quad (29)$$

Now,

$$(g, hI) * (u, vI) = (g \wedge \neg u, (g \wedge \neg v \wedge h \wedge \neg u) I) \equiv (m, kI), \quad (30)$$

where

$$\begin{aligned} m &= g \wedge \neg u \\ &= x \wedge \neg p \wedge (\neg a \wedge c) \\ &= a \wedge \neg e \wedge (\neg c \vee e) \wedge (\neg a \vee c) \\ &= a \wedge \neg e \wedge \neg c \wedge (\neg a \vee c) \\ &= 0 \vee 0 = 0, \\ k &= g \wedge \neg v \wedge h \wedge \neg u \\ &= x \wedge \neg p \wedge \neg v \wedge \neg q \wedge y \wedge (\neg a \vee c) \\ &= a \wedge \neg e \wedge (\neg c \vee e) \wedge (\neg a \vee c) \wedge y \wedge \neg v \wedge \neg q \\ &= a \wedge \neg e \wedge \neg c \wedge (\neg a \vee c) \wedge y \wedge \neg v \wedge \neg q \\ &= (0 \vee 0) \wedge y \wedge \neg v \wedge \neg q = 0. \end{aligned} \quad (31)$$

Since $(m, kI) = (0, 0)$, it follows that LHS = $(0, 0)$. Hence this completes the proof. □

Theorem 9. *Let $(X(I), *, (0, 0))$ be a neutrosophic BCI/BCK-algebra. Then*

(1) *$X(I)$ is not commutative even if X is commutative;*

(2) *$X(I)$ is not implicative even if X is implicative.*

Proof. (1) Suppose that X is commutative. Let $(a, bI), (c, dI) \in X(I)$. Then

$$\begin{aligned} &(a, bI) * ((a, bI) * (c, dI)) \\ &\quad = (a, bI) * (a * c, (a * d \wedge b * c \wedge b * d) I) \\ &\quad = (a * (a * c), (a * (a * d \wedge b * c \wedge b * d) \wedge b * (a * c) \\ &\qquad \wedge b * (a * d \wedge b * c \wedge b * d)) I) \\ &\quad \equiv (u, vI), \end{aligned} \quad (32)$$

where

$$\begin{aligned} u &= a * (a * c) = c * (c * a), \\ v &= a * (a * d \wedge b * c \wedge b * d) \wedge b * (a * c) \\ &\quad \wedge b * (a * d \wedge b * c \wedge b * d) \\ &= a * (a * d) \wedge a * (b * c) \wedge a * (b * d) \\ &\quad \wedge b * (a * c) \wedge b * (a * d) \\ &\quad \wedge b * (b * c) \wedge b * (b * d) \\ &= d * (d * a) \wedge a * (b * c) \wedge a * (b * d) \\ &\quad \wedge b * (a * c) \wedge b * (a * d) \\ &\quad \wedge c * (c * b) \wedge d * (d * b). \end{aligned} \quad (33)$$

Also,

$$\begin{aligned}
&(c, dI) * ((c, dI) * (a, bI)) \\
&\quad = (c, dI) * (c * a, (c * b \wedge d * a \wedge d * b) I) \\
&\quad = (c * (c * a), (c * (c * b \wedge d * a \wedge d * b) \wedge d * (c * a) \\
&\qquad\qquad \wedge d * (c * b \wedge d * a \wedge d * b)) I) \\
&\quad \equiv (p, qI),
\end{aligned} \tag{34}$$

where

$$\begin{aligned}
p &= c * (c * a) = u, \\
q &= c * (c * b \wedge d * a \wedge d * b) \wedge d * (c * a) \\
&\quad \wedge d * (c * b \wedge d * a \wedge d * b) \\
&= c * (c * b) \wedge c * (d * a) \wedge c * (d * b) \wedge d * (c * a) \\
&\quad \wedge d * (c * b) \wedge d * (d * a) \wedge d * (d * b) \\
&\neq v.
\end{aligned} \tag{35}$$

This shows that $(a, bI) * ((a, bI) * (c, dI)) \neq (c, dI) * ((c, dI) * (a, bI))$ and therefore $X(I)$ is not commutative.

(2) Suppose that X is implicative. Let $(a, bI), (c, dI) \in X(I)$. Then

$$\begin{aligned}
&(a, bI) * ((c, dI) * (a, bI)) \\
&\quad = (a, bI) * (c * a, (c * b \wedge d * a \wedge d * b) I) \\
&\quad = (a * (c * a), (a * (c * b \wedge d * a \wedge d * b) \wedge b * (c * a) \\
&\qquad\qquad \wedge b * (c * b \wedge d * a \wedge d * b)) I) \\
&\quad \equiv (u, vI),
\end{aligned} \tag{36}$$

where

$$\begin{aligned}
u &= a * (c * a) = a, \\
v &= a * (c * b \wedge d * a \wedge d * b) \wedge b * (c * a) \\
&\quad \wedge b * (c * b \wedge d * a \wedge d * b) \\
&= a * (c * b) \wedge a * (d * a) \wedge a * (d * b) \wedge b * (c * a) \\
&\quad \wedge b * (c * b) \wedge b * (d * a) \wedge b * (d * b) \\
&\neq b.
\end{aligned} \tag{37}$$

Hence, $(a, bI) \neq (a, bI) * ((c, dI) * (a, bI))$ and so $X(I)$ is not implicative. □

Definition 10. Let $(X(I), *, (0, 0))$ be a neutrosophic BCI/BCK-algebra. A nonempty subset $A(I)$ is called a neutrosophic subalgebra of $X(I)$ if the following conditions hold:

(1) $(0, 0) \in A(I)$;

(2) $(a, bI) * (c, dI) \in A(I)$ for all $(a, bI), (c, dI) \in A(I)$;

(3) $A(I)$ contains a proper subset which is a BCI/BCK-algebra.

If $A(I)$ does not contain a proper subset which is a BCI/BCK-algebra, then $A(I)$ is called a pseudo neutrosophic subalgebra of $X(I)$.

Example 11. Any neutrosophic subgroup of the commutative neutrosophic group $(X(I), +)$ of Example 2 is a neutrosophic BCI-subalgebra.

Theorem 12. *Let $(X(I), *, (0, 0))$ be a neutrosophic BCK-algebra and for $a \neq 0$ let $A_{(a,aI)}(I)$ be a subset of $X(I)$ defined by*

$$A_{(a,aI)}(I) = \{(x, yI) \in X(I) : (x, yI) * (a, aI) = (0, 0)\}. \tag{38}$$

Then,

(1) *$A_{(a,aI)}(I)$ is a neutrosophic subalgebra of $X(I)$;*

(2) *$A_{(a,aI)}(I) \subseteq A_{(0,0)}(I)$.*

Proof. (1) Obviously, $(0, 0) \in A_{(a,aI)}(I)$ and $A_{(a,aI)}(I)$ contains a proper subset which is a BCK-algebra. Let $(x, yI), (p, qI) \in A_{(a,aI)}(I)$. Then $(x, yI) * (a, aI) = (0, 0)$ and $(p, qI) * (a, aI) = (0, 0)$ from which we obtain $x*a = 0, x*a \wedge y*a = 0, p*a = 0$, and $p * a \wedge q * a = 0$. Since $a \neq 0$, we have $x = y = p = q = a$. Now,

$$\begin{aligned}
&((x, yI) * (p, qI)) * (a, aI) \\
&\quad = ((x * p), (x * q \wedge y * p \wedge y * q) I) * (a, aI) \\
&\quad = ((x * p) * a, (((x * p) * a) \wedge (x * q \wedge y * p \\
&\qquad\qquad \wedge y * q) * a) I) \\
&\quad = ((a * a) * a, ((a * a) * a) I) \\
&\quad = (0 * a, (0 * a) I) \\
&\quad = (0, 0).
\end{aligned} \tag{39}$$

This shows that $(x, yI) * (p, qI) \in A_{(a,aI)}(I)$ and the required result follows.

(2) Follows. □

Definition 13. Let $(X(I), *, (0, 0))$ and $(X'(I), \circ, (0', 0'))$ be two neutrosophic BCI/BCK-algebras. A mapping $\phi : X(I) \to X'(I)$ is called a neutrosophic homomorphism if the following conditions hold:

(1) $\phi((a, bI) * (c, dI)) = \phi((a, bI)) \circ \phi((c, dI)), \forall (a, bI), (c, dI) \in X(I)$;

(2) $\phi((0, I)) = (0, I)$.

In addition,

(3) if ϕ is injective, then ϕ is called a neutrosophic monomorphism;

(4) if ϕ is surjective, then ϕ is called a neutrosophic epimorphism;

(5) if ϕ is a bijection, then ϕ is called a neutrosophic isomorphism. A bijective neutrosophic homomorphism from $X(I)$ onto $X(I)$ is called a neutrosophic automorphism.

Definition 14. Let $\phi : X(I) \to Y(I)$ be a neutrosophic homomorphism of neutrosophic BCK/BCI-algebras. Consider the following:

(1) $\operatorname{Ker}\phi = \{(a, bI) \in X(I) : \phi((a, bI)) = (0, 0)\}$;

(2) $\operatorname{Im}\phi = \{\phi((a, bI)) \in Y(I) : (a, bI) \in X(I)\}$.

Example 15. Let $(X(I), *, (0, 0))$ be a neutrosophic BCI/BCK-algebra and let $\phi : X(I) \to X(I)$ be a mapping defined by

$$\phi((a, bI)) = (a, bI) \quad \forall (a, bI) \in X(I). \tag{40}$$

Then ϕ is a neutrosophic isomorphism.

Lemma 16. *Let $\phi : X(I) \to X'(I)$ be a neutrosophic homomorphism from a neutrosophic BCI/BCK-algebra $X(I)$ into a neutrosophic BCI/BCK-algebra $X'(I)$. Then $\phi((0, 0)) = (0', 0')$.*

Proof. It is straightforward. □

Theorem 17. *Let $\phi : X(I) \to Y(I)$ be a neutrosophic homomorphism of neutrosophic BCK/BCI-algebras. Then ϕ is a neutrosophic monomorphism if and only if $\operatorname{Ker}\phi = \{(0, 0)\}$.*

Proof. The proof is the same as the classical case and so is omitted. □

Theorem 18. *Let $X(I)$, $Y(I)$, and $Z(I)$ be neutrosophic BCI/BCK-algebras. Let $\phi : X(I) \to Y(I)$ be a neutrosophic epimorphism and let $\psi : X(I) \to Z(I)$ be a neutrosophic homomorphism. If $\operatorname{Ker}\phi \subseteq \operatorname{Ker}\psi$, then there exists a unique neutrosophic homomorphism $\nu : Y(I) \to Z(I)$ such that $\nu\phi = \psi$. The following also hold:*

(1) $\operatorname{Ker}\nu = \phi(\operatorname{Ker}\psi)$;

(2) $\operatorname{Im}\nu = \operatorname{Im}\psi$;

(3) *ν is a neutrosophic monomorphism if and only if $\operatorname{Ker}\phi = \operatorname{Ker}\psi$;*

(4) *ν is a neutrosophic epimorphism if and only if ψ is a neutrosophic epimorphism.*

Proof. The proof is similar to the classical case and so is omitted. □

Theorem 19. *Let $X(I)$, $Y(I)$, $Z(I)$ be neutrosophic BCI/BCK-algebras. Let $\phi : X(I) \to Z(I)$ be a neutrosophic homomorphism and let $\psi : Y(I) \to Z(I)$ be a neutrosophic monomorphism such that $\operatorname{Im}\phi \subseteq \operatorname{Im}\psi$. Then there exists a unique neutrosophic homomorphism $\mu : X(I) \to Y(I)$ such that $\phi = \psi\mu$. Also,*

(1) $\operatorname{Ker}\mu = \operatorname{Ker}\phi$;

(2) $\operatorname{Im}\mu = \psi^{-1}(\operatorname{Im}\phi)$;

(3) *μ is a neutrosophic monomorphism if and only if ϕ is a neutrosophic monomorphism;*

(4) *μ is a neutrosophic epimorphism if and only if $\operatorname{Im}\psi = \operatorname{Im}\phi$.*

Proof. The proof is similar to the classical case and so is omitted. □

Conflict of Interests

The authors declare that there is no conflict of interests regarding the publication of this paper.

References

[1] Y. Imai and K. Iséki, "On axiom systems of propositional calculi, XIV," *Proceedings of the Japan Academy*, vol. 42, pp. 19–22, 1966.

[2] K. Iséki, "An algebra related with a propositional calculus," *Proceedings of the Japan Academy*, vol. 42, pp. 26–29, 1966.

[3] Y. S. Huang, *BCI-Algebra*, Science Press, Beijing, China, 2006.

[4] K. Iséki, "On BCI-algebras," *Kobe University. Mathematics Seminar Notes*, vol. 8, no. 1, pp. 125–130, 1980.

[5] K. Iséki and S. Tanaka, "An introduction to the theory of BCK-algebras," *Mathematica Japonica*, vol. 23, no. 1, pp. 1–26, 1978.

[6] J. M. Jie and Y. B. Jun, *BCK-Algebras*, Kyung Moon Sa Co., Seoul, Republic of Korea, 1994.

[7] C. C. Chang, "Algebraic analysis of many valued logics," *Transactions of the American Mathematical Society*, vol. 88, no. 2, pp. 467–490, 1958.

[8] F. Smarandache, *A Unifying Field in Logics: Neutrosophic Logic, Neutrosophy, Neutrosophic Set, Neutrosophic Probability*, American Research Press, Rehoboth, NM, USA, 2003.

[9] F. Smarandache, *Introduction to Neutrosophic Statistics*, Sitech and Education Publishing, Craiova, Romania, 2014.

[10] F. Smarandache, *Neutrosophy/Neutrosophic Probability, Set, and Logic*, American Research Press, Rehoboth, Mass, USA, 1998.

[11] W. B. V. Kandasamy and F. Smarandache, *Some Neutrosophic Algebraic Structures and Neutrosophic N-Algebraic Structures*, Hexis, Phoenix, Ariz, USA, 2006.

[12] W. B. Vasantha Kandasamy and F. Smarandache, *Neutrosophic Rings*, Hexis, Phoenix, Ariz, USA, 2006.

[13] A. A. A. Agboola, A. D. Akinola, and O. Y. Oyebola, "Neutrosophic rings I," *International Journal of Mathematical Combinatorics*, vol. 4, pp. 1–14, 2011.

[14] A. A. A. Agboola, E. O. Adeleke, and S. A. Akinleye, "Neutrosophic rings II," *International Journal of Mathematical Combinatorics*, vol. 2, pp. 1–8, 2012.

[15] A. A. A. Agboola, A. O. Akwu, and Y. T. Oyebo, "Neutrosophic groups and neutrosopic subgroups," *International Journal of Mathematical Combinatorics*, vol. 3, pp. 1–9, 2012.

[16] A. A. Agboola and B. Davvaz, "Introduction to neutrosophic hypergroups," *ROMAI Journal*, vol. 9, no. 2, pp. 1–10, 2013.

[17] A. A. A. Agboola and B. Davvaz, "On neutrosophic canonical hypergroups and neutrosophic hyperrings," *Neutrosophic Sets and Systems*, vol. 2, pp. 34–41, 2014.

12

Annular Bounds for the Zeros of a Polynomial

Le Gao[1] and N. K. Govil[2]

[1]Department of Mechanical Engineering, Auburn University, Auburn, AL 36849, USA
[2]Department of Mathematics and Statistics, Auburn University, Auburn, AL 36849, USA

Correspondence should be addressed to N. K. Govil; govilnk@auburn.edu

Academic Editor: Irena Lasiecka

The problem of obtaining the smallest possible region containing all the zeros of a polynomial has been attracting more and more attention recently, and in this paper, we obtain several results providing the annular regions that contain all the zeros of a complex polynomial. Using MATLAB, we construct specific examples of polynomials and show that for these polynomials our results give sharper regions than those obtainable from some of the known results.

1. Introduction

The Fundamental Theorem of Algebra states that every nonzero, single-variable, polynomial

$$p(z) = a_0 + a_1 z + a_2 z^2 + a_3 z^3 + \cdots + a_n z^n, \qquad (1)$$

of degree n with complex coefficients, has exactly n complex zeros. The zeros may however be coincident. Although the above theorem tells about the number of zeros, it does not mention anything about the location of these zeros. The problem of obtaining the smallest possible region containing all the zeros of a polynomial has been attracting more and more attention recently, since the results are very useful in engineering applications as well as in many areas of applied mathematics such as control theory, cryptography, mathematical biology, and combinatorics [1–4].

For any second-degree polynomial equation, $ax^2 + bx + c = 0, a \neq 0$, the roots can be found by using the familiar quadratic formula, $x = (-b \pm \sqrt{b^2 - 4ac})/2a$, and for polynomials of third and fourth degree, there are analogous formulas to find the zeros. However, for polynomials of degree five or higher with arbitrary coefficients, the Abel-Ruffini theorem states that there is no algebraic solution. Several methods, for example, Aberth-Ehrlich method [5, 6], have been proposed for the simultaneous determination of zeros of algebraic polynomials and there are studies [7, 8] to accelerate convergence and increase computational efficiency of these methods. Approximations to the zeros of a polynomial can be drawn by these methods, and these methods can become more efficient when an annulus containing all the zeros of the polynomial is provided. So this paper is focused on finding new theorems that can provide smaller annuli containing all the zeros of a polynomial.

2. Preliminaries

The first contributor to this subject was probably Gauss, who proved the following.

Theorem 1 (Gauss). *A polynomial*

$$p(z) = a_0 + a_1 z + a_2 z^2 + a_3 z^3 + \cdots + a_n z^n, \qquad (2)$$

with all a_k real, has no zeros outside the circle $|z| = R$, where

$$R = \max_{1 \le k \le n} \left(n\sqrt{2}\,|a_k|\right)^{1/k}. \qquad (3)$$

Later in 1850, Gauss [9] proved that, for any real or complex a_k, the radius R may be taken as the positive root of the equation

$$z^n - 2^{1/2}\left(|a_1|\, z^{n-1} + \cdots + |a_n|\right) = 0. \qquad (4)$$

Theorem 1 of Gauss was improved in 1829 by Cauchy [10] who derived more exact bounds for the moduli of the zeros of a polynomial than those given by Gauss, by proving the following.

Theorem 2 (Cauchy). *If*

$$p(z) = a_0 + a_1 z + a_2 z^2 + a_3 z^3 + \cdots + a_n z^n \quad (5)$$

is a complex polynomial of degree n, then all the zeros lie in the disc

$$\{z : |z| < \eta\} \subset \{z : |z| < 1 + A\}, \quad (6)$$

where $A = \max_{1\le k\le n-1} |a_k|$ *and* η *is the unique positive root of the real-coefficient equation*

$$z^n - |a_{n-1}| z^{n-1} - |a_{n-2}| z^{n-2} - \cdots - |a_1| z - |a_0| = 0. \quad (7)$$

By applying Theorem 2 to the polynomial $p(z) = z^n p(1/z)$, *we easily get Theorem 3.*

Theorem 3. *All the zeros of the polynomial*

$$p(z) = a_0 + a_1 z + a_2 z^2 + a_3 z^3 + \cdots + a_n z^n, \quad a_n \neq 0, \quad (8)$$

lie in the annulus $r_1 \le |z| \le r_2$, *where* r_1 *is the unique positive root of the equation*

$$|a_n| z^n + |a_{n-1}| z^{n-1} + |a_{n-2}| z^{n-2} + \cdots + |a_1| z - |a_0| = 0, \quad (9)$$

and r_2 *is the unique positive root of the equation*

$$|a_0| + |a_1| z + \cdots + |a_{n-1}| z^{n-1} - |a_n| z^n = 0. \quad (10)$$

Diaz-Barrero [11] gave the following results, providing circular regions containing all the zeros of a polynomial in terms of the binomial coefficients and Fibonacci's numbers. Note that the binomial coefficients are defined by

$$C(n,k) = \frac{n!}{k!(n-k)!}, \quad 0! = 1, \quad (11)$$

and Fibonacci's numbers are defined by

$$\begin{aligned} F_0 &= 0, \\ F_1 &= 1, \\ F_j &= F_{j-1} + F_{j-2}, \quad \text{for } j \ge 2. \end{aligned} \quad (12)$$

Theorem 4. *Let* $p(z) = \sum_{j=0}^{n} a_j z^j$ $(a_j \neq 0,\ 0 \le j \le n)$ *be a nonconstant complex polynomial. Then all its zeros lie in the disc* $C = \{z : r_1 \le |z| \le r_2\}$, *where*

$$\begin{aligned} r_1 &= \frac{3}{2} \min_{1\le j\le n} \left\{ \frac{2^n F_j C_j^n}{F_{4n}} \left| \frac{a_0}{a_j} \right| \right\}^{1/k}, \\ r_2 &= \frac{2}{3} \max_{1\le j\le n} \left\{ \frac{F_{4n}}{2^n F_j C_j^n} \left| \frac{a_{n-j}}{a_n} \right| \right\}^{1/k}. \end{aligned} \quad (13)$$

Kim [12] provides another annulus containing all the zeros of a polynomial with the binomial coefficients.

Theorem 5. *Let* $p(z) = \sum_{k=0}^{n} a_k z^k$ $(a_k \neq 0,\ 0 \le k \le n)$ *be a nonconstant polynomial with complex coefficients. Then all the zeros of* $p(z)$ *lie in the annulus* $A = \{z : r_1 \le |z| \le r_2\}$, *where*

$$\begin{aligned} r_1 &= \min_{1\le k\le n} \left\{ \frac{C(n,k)}{2^n - 1} \left| \frac{a_0}{a_k} \right| \right\}^{1/k}, \\ r_2 &= \max_{1\le k\le n} \left\{ \frac{2^n - 1}{C(n,k)} \left| \frac{a_{n-k}}{a_n} \right| \right\}^{1/k}. \end{aligned} \quad (14)$$

There are in fact many results in this direction, and for some more results, see, for example, [13, 14].

Recently, Dalal and Govil [15] unified the above theorems and proved the following theorem.

Theorem 6. *Let* $A_k > 0$ *for* $1 \le k \le n$ *such that* $\sum_{k=1}^{n} A_k = 1$. *If* $p(z) = a_0 + a_1 z + a_2 z^2 + a_3 z^3 + \cdots + a_n z^n$ *is a nonconstant complex polynomial of degree n, with* $a_k \neq 0$ *for* $1 \le k \le n$, *then all the zeros of* $p(z)$ *lie in the annulus* $C = \{z : r_1 \le |z| \le r_2\}$, *where*

$$\begin{aligned} r_1 &= \min_{1\le k\le n} \left\{ A_k \left| \frac{a_0}{a_k} \right| \right\}^{1/k}, \\ r_2 &= \max_{1\le k\le n} \left\{ \frac{1}{A_k} \left| \frac{a_{n-k}}{a_n} \right| \right\}^{1/k}. \end{aligned} \quad (15)$$

Using the above theorems, Dalal and Govil [15] also gave the following.

Theorem 7. *Let* $p(z) = a_0 + a_1 z + a_2 z^2 + a_3 z^3 + \cdots + a_n z^n$ *be a nonconstant complex polynomial of degree n, with* $a_k \neq 0$ *for* $1 \le k \le n$. *Then all the zeros of* $p(z)$ *lie in the annulus* $C = \{z : r_1 \le |z| \le r_2\}$, *where*

$$\begin{aligned} r_1 &= \min_{1\le k\le n} \left\{ \frac{C_{k-1} C_{n-k}}{C_n} \left| \frac{a_0}{a_k} \right| \right\}^{1/k}, \\ r_2 &= \max_{1\le k\le n} \left\{ \frac{C_n}{C_{k-1} C_{n-k}} \left| \frac{a_{n-k}}{a_n} \right| \right\}^{1/k}, \end{aligned} \quad (16)$$

where

$$C_k = \frac{C(2k,k)}{k+1} \quad (17)$$

is the kth Catalan number in which $C(2k,k)$ *are the binomial coefficients.*

Theorem 6 of Dalal and Govil [15] can generate infinitely many results, including Theorems 4 and 5, giving annulus containing all the zeros of a polynomial, and over the years, mathematicians have shown the usefulness of their results by comparing their bounds with the existing bounds in the literature by giving some examples and thus showing that their bounds are better in some special cases. In this connection, Dalal and Govil in [16] have shown that no matter what result you obtain as a corollary to Theorem 6, one can always generate polynomials for which the corollary so obtained gives better bound than the existing ones, implying that every result obtained by Theorem 6 can be useful. Since the results obtained as corollaries of Theorem 6 cannot in general be compared, more recently Dalal and Govil [17] have given results that help to compare the bounds for a subclass of polynomials. For this, they provide a class of polynomials with some conditions on degree or absolute range of coefficients of the polynomial, and for this class of polynomials, the bound obtained by one corollary is always better than the bound obtained from the other.

3. Main Results and Their Proofs

In this section, we obtain some new results which provide annuli containing all the zeros of a polynomial. These results have been obtained by making use of some identities and Theorem 6. Then, in Section 4, we use MATLAB to obtain examples of polynomials for which our results give bounds that are sharper than those obtainable from Theorems 5 and 7. In fact, it is not difficult to construct polynomials for which our theorems also give better bounds than those obtainable from Theorem 4.

Theorem 8. *Let $p(z) = a_0 + a_1 z + a_2 z^2 + a_3 z^3 + \cdots + a_n z^n$ be a nonconstant complex polynomial of degree n, with $a_k \neq 0$ for $1 \leq k \leq n$. Then all the zeros of $p(z)$ lie in the annulus $C = \{z : r_1 \leq |z| \leq r_2\}$, where*

$$r_1 = \min_{1\leq k\leq n} \left\{ \frac{kC(n,k)^2}{n(n-k+1)C_n} \left| \frac{a_0}{a_k} \right| \right\}^{1/k}, \quad r_2 = \max_{1\leq k\leq n} \left\{ \frac{n(n-k+1)C_n}{kC(n,k)^2} \left| \frac{a_{n-k}}{a_n} \right| \right\}^{1/k}. \tag{18}$$

Theorem 9. *Let $p(z) = a_0 + a_1 z + a_2 z^2 + a_3 z^3 + \cdots + a_n z^n$ be a nonconstant complex polynomial of degree n, with $a_k \neq 0$ for $1 \leq k \leq n$. Then all the zeros of $p(z)$ lie in the annulus $C = \{z : r_1 \leq |z| \leq r_2\}$, where*

$$r_1 = \min_{1\leq k\leq n} \left\{ \frac{C(n,2k)C_k 2^{n-2k}}{C_{n+1} - 2^n} \left| \frac{a_0}{a_k} \right| \right\}^{1/k}, \quad r_2 = \max_{1\leq k\leq n} \left\{ \frac{C_{n+1} - 2^n}{C(n,2k)C_k 2^{n-2k}} \left| \frac{a_{n-k}}{a_n} \right| \right\}^{1/k}. \tag{19}$$

Theorem 10. *Let $p(z) = a_0 + a_1 z + a_2 z^2 + a_3 z^3 + \cdots + a_n z^n$ be a nonconstant complex polynomial of degree n, with $a_k \neq 0$ for $1 \leq k \leq n$. Then all the zeros of $p(z)$ lie in the annulus $C = \{z : r_1 \leq |z| \leq r_2\}$, where*

$$r_1 = \min_{1\leq k\leq n} \left\{ \frac{F_{2k}}{F_{2n+1} - 1} \left| \frac{a_0}{a_k} \right| \right\}^{1/k}, \quad r_2 = \max_{1\leq k\leq n} \left\{ \frac{F_{2n+1} - 1}{F_{2k}} \left| \frac{a_{n-k}}{a_n} \right| \right\}^{1/k}. \tag{20}$$

Theorem 11. *Let $p(z) = a_0 + a_1 z + a_2 z^2 + a_3 z^3 + \cdots + a_n z^n$ be a nonconstant complex polynomial of degree n, with $a_k \neq 0$ for $1 \leq k \leq n$. Then all the zeros of $p(z)$ lie in the annulus $C = \{z : r_1 \leq |z| \leq r_2\}$, where*

$$r_1 = \min_{1\leq k\leq n} \left\{ \frac{F_k^{\,2}}{F_n F_{n+1}} \left| \frac{a_0}{a_k} \right| \right\}^{1/k}, \quad r_2 = \max_{1\leq k\leq n} \left\{ \frac{F_n F_{n+1}}{F_k^{\,2}} \left| \frac{a_{n-k}}{a_n} \right| \right\}^{1/k}. \tag{21}$$

Theorem 12. *Let $p(z) = a_0 + a_1 z + a_2 z^2 + a_3 z^3 + \cdots + a_n z^n$ be a nonconstant complex polynomial of degree n, with $a_k \neq 0$ for $1 \leq k \leq n$. Then all the zeros of $p(z)$ lie in the annulus $C = \{z : r_1 \leq |z| \leq r_2\}$, where*

$$r_1 = \min_{1\leq k\leq n} \left\{ \frac{k^2 C(n,k)}{n(n+1)2^{n-2}} \left| \frac{a_0}{a_k} \right| \right\}^{1/k}, \quad r_2 = \max_{1\leq k\leq n} \left\{ \frac{n(n+1)2^{n-2}}{k^2 C(n,k)} \left| \frac{a_{n-k}}{a_n} \right| \right\}^{1/k}. \tag{22}$$

Theorem 13. *Let $p(z) = a_0 + a_1 z + a_2 z^2 + a_3 z^3 + \cdots + a_n z^n$ be a nonconstant complex polynomial of degree n, with $a_k \neq 0$ for $1 \leq k \leq n$. Then all the zeros of $p(z)$ lie in the annulus $C = \{z : r_1 \leq |z| \leq r_2\}$, where*

$$r_1 = \min_{1\leq k\leq n} \left\{ \frac{kC(n,k)^2}{(n/2)C(2n,n)} \left| \frac{a_0}{a_k} \right| \right\}^{1/k}, \quad r_2 = \max_{1\leq k\leq n} \left\{ \frac{(n/2)C(2n,n)}{kC(n,k)^2} \left| \frac{a_{n-k}}{a_n} \right| \right\}^{1/k}. \tag{23}$$

Theorem 14. *Let $p(z) = a_0 + a_1 z + a_2 z^2 + a_3 z^3 + \cdots + a_n z^n$ be a nonconstant complex polynomial of degree n, with $a_k \neq 0$ for $1 \leq k \leq n$. Then all the zeros of $p(z)$ lie in the annulus $C = \{z : r_1 \leq |z| \leq r_2\}$, where*

$$r_1 = \min_{1\leq k\leq n} \left\{ \frac{k^2 C(n,k)^2}{n^2 C(2n-2,n-1)} \left| \frac{a_0}{a_k} \right| \right\}^{1/k}, \quad r_2 = \max_{1\leq k\leq n} \left\{ \frac{n^2 C(2n-2,n-1)}{k^2 C(n,k)^2} \left| \frac{a_{n-k}}{a_n} \right| \right\}^{1/k}. \tag{24}$$

3.1. Lemmas. For the proofs of the above theorems, we will need the following lemmas.

Lemma 15. *If*

$$C_n = \frac{C(2n,n)}{n+1} \tag{25}$$

is the nth Catalan number in which $C(2k,k)$ and $C(n,k)$ are binomial coefficients, then for $n \geq 1$,

$$\frac{1}{n}\sum_{k=1}^{n} \frac{kC(n,k)^2}{n-k+1} = C_n. \tag{26}$$

Proof of Lemma 15. Note that

$$\begin{aligned}\frac{1}{n}\sum_{k=1}^{n}\frac{kC(n,k)^2}{n-k+1} &= \frac{1}{n}\sum_{k=1}^{n}C(n,k)\frac{k}{n-k+1}C(n,k)\\ &= \frac{1}{n}\sum_{k=1}^{n}C(n,k)\frac{k}{n-k+1}\frac{n!}{k!(n-k)!}\\ &= \frac{1}{n}\sum_{k=1}^{n}C(n,k)\frac{n!}{(k-1)!(n-k+1)!}\\ &= \frac{1}{n}\sum_{k=1}^{n}C(n,k)C(n,k-1) = C_n.\end{aligned} \quad (27)$$

□

Lemma 16. *If*

$$C_n = \frac{C(2n,n)}{n+1} \quad (28)$$

is the nth Catalan number in which $C(2k,k)$ *and* $C(n,k)$ *are binomial coefficients, then for* $n \geq 1$,

$$C_{n+1} = \sum_{k=0}^{n}C(n,2k)C_k 2^{n-2k}. \quad (29)$$

Proof of Lemma 16. Note that, by [18, p. 292], we have

$$C_{n+1} = \sum_{j=0}^{n}C(n,j)M_j, \quad (30)$$

where M_j is the jth Motzkin number defined by

$$\begin{aligned}&M_0 = M_1 = M_{-1} = 1;\\ &M_{j+1} = \frac{2j+3}{j+3}M_j + \frac{3j}{j+3}M_{j-1}, \quad j \geq 1.\end{aligned} \quad (31)$$

Also, by [18, p. 292], we have

$$M_j = \sum_{k=0}^{j}C(j,2k)C_k. \quad (32)$$

Therefore,

$$\begin{aligned}C_{n+1} &= \sum_{j=0}^{n}C(n,j)\sum_{k=0}^{j}C(j,2k)C_k\\ &= \sum_{k=0}^{n}C(n,2k)C_k 2^{n-2k}.\end{aligned} \quad (33)$$

□

Lemma 17. *If* F_n *is the nth Fibonacci number, then for* $n \geq 1$,

$$\sum_{k=1}^{n}F_{2k} = F_{2n+1} - 1. \quad (34)$$

Proof of Lemma 17. Since $F_0 = 0$, $F_1 = 1$, and $F_j = F_{j-1} + F_{j-2}$ for $j \geq 2$, hence

$$\begin{aligned}\sum_{k=1}^{n}F_{2k} &= F_2 + F_4 + F_6 + \cdots + F_{2n-2} + F_{2n}\\ &= F_1 + F_2 + F_4 + F_6 + \cdots + F_{2n-2} + F_{2n} - F_1\\ &= F_3 + F_4 + F_6 + \cdots + F_{2n-2} + F_{2n} - F_1\\ &\vdots\\ &= F_{2n-1} + F_{2n} - F_1\\ &= F_{2n+1} - F_1.\end{aligned} \quad (35)$$

□

Lemma 18. *If* F_n *is the nth Fibonacci number, then for* $n \geq 1$,

$$\sum_{k=1}^{n}F_k{}^2 = F_n F_{n+1}. \quad (36)$$

Proof of Lemma 18. One has

$$\begin{aligned}F_nF_{n+1} &= F_n(F_n + F_{n-1})\\ &= F_n^2 + F_nF_{n-1}\\ &= F_n^2 + F_{n-1}(F_{n-1} + F_{n-2})\\ &= F_n^2 + F_{n-1}^2 + F_{n-1}F_{n-2}\\ &\vdots\\ &= F_n^2 + F_{n-1}^2 + \cdots + F_1^2\\ &= \sum_{k=1}^{n}F_k^2.\end{aligned} \quad (37)$$

□

Lemma 19. *If*

$$C_n = \frac{C(2n,n)}{n+1} \quad (38)$$

is the nth Catalan number in which $C(2k,k)$ *and* $C(n,k)$ *are binomial coefficients, then for* $n \geq 1$,

$$\sum_{k=0}^{n}k^2C(n,k) = n(1+n)2^{n-2}. \quad (39)$$

Proof of Lemma 19. Note that

$$\begin{aligned}\sum_{k=0}^{n}k^2C(n,k) = \sum_{k=1}^{n}k^2C(n,k) &= \sum_{k=1}^{n}knC(n-1,k-1)\\ &\text{since } kC(n,k) = nC(n-1,k-1)\\ &= n\sum_{k=0}^{n-1}(k+1)C(n-1,k)\end{aligned}$$

$$= n\sum_{k=0}^{n-1} kC(n-1,k) + n\sum_{k=0}^{n-1} C(n-1,k)$$

$$= n\sum_{k=1}^{n-1} kC(n-1,k) + n\sum_{k=0}^{n-1} C(n-1,k)$$

$$= n(n-1)2^{n-2} + n2^{n-1},$$

$$\text{since } \sum_{k=1}^{n-1} kC(n-1,k) = (n-1)2^{n-2}$$

$$= n(1+n)2^{n-2}. \tag{40}$$

□

Lemma 20. *If*

$$C_n = \frac{C(2n,n)}{n+1} \tag{41}$$

is the nth Catalan number in which $C(2k,k)$ and $C(n,k)$ are binomial coefficients, then for $n \geq 1$,

$$\sum_{k=0}^{n} kC(n,k)^2 = \frac{n}{2}C(2n,n). \tag{42}$$

Proof of Lemma 20. One has

$$\begin{aligned}\sum_{k=0}^{n} kC(n,k)^2 &= \sum_{k=0}^{n} nC(n-1,k-1)C(n,k)\\ &= \sum_{k=0}^{n} nC(n-1,k-1)C(n,n-k)\\ &= nC(2n-1,n-1)\\ &= n\frac{(2n-1)!}{(n-1)!n!}\\ &= \frac{n}{2}C(2n,n).\end{aligned} \tag{43}$$

□

Lemma 21. *If*

$$C_n = \frac{C(2n,n)}{n+1} \tag{44}$$

is the nth Catalan number in which $C(2k,k)$ and $C(n,k)$ are binomial coefficients, then for $n \geq 1$,

$$\sum_{k=0}^{n} k^2C(n,k)^2 = n^2C(2n-2,n-1). \tag{45}$$

Proof of Lemma 21. One has

$$\begin{aligned}\sum_{k=0}^{n} k^2C(n,k)^2 &= \sum_{k=0}^{n} n^2C(n-1,k-1)^2\\ &= n^2\sum_{k=0}^{n} C(n-1,k-1)C(n-1,n-k)\\ &= n^2C(2n-2,n-1).\end{aligned} \tag{46}$$

□

3.2. Proofs of the Theorems

Proof of Theorem 8. By Lemma 15, we have that

$$\frac{1}{n}\sum_{k=1}^{n}\frac{kC(n,k)^2}{n-k+1} = C_n. \tag{47}$$

If we take

$$A_k = \frac{1}{n}\frac{kC(n,k)^2}{(n-k+1)C_n}, \tag{48}$$

then $A_k > 0$ and $\sum_{k=1}^{n} A_k = 1$. Hence applying Theorem 6 for this set of values of A_k, we get our desired result. □

Proof of Theorem 9. From Lemma 16, we have that

$$C_{n+1} = \sum_{k=0}^{n} C(n,2k)C_k2^{n-2k}. \tag{49}$$

So $C_{n+1} - 2^n = \sum_{k=1}^{n} C(n,2k)C_k2^{n-2k}$. If we take

$$A_k = \frac{C(n,2k)C_k2^{n-2k}}{C_{n+1} - 2^n}, \tag{50}$$

then $A_k > 0$ and $\sum_{k=1}^{n} A_k = 1$. Hence by applying Theorem 6 for this set of values of A_k, Theorem 9 can be proved. □

Proof of Theorem 10. From Lemma 17, we have that

$$\sum_{k=1}^{n} F_{2k} = F_{2n+1} - 1. \tag{51}$$

If we take

$$A_k = \frac{F_{2k}}{F_{2n+1} - 1}, \tag{52}$$

then $A_k > 0$ and $\sum_{k=1}^{n} A_k = 1$. Hence by applying Theorem 6 for this set of values of A_k, Theorem 10 can be proved. □

Proof of Theorem 11. From Lemma 18, we have that

$$\sum_{k=1}^{n} F_k^2 = F_nF_{n+1}. \tag{53}$$

If we take

$$A_k = \frac{F_k^2}{F_nF_{n+1}}, \tag{54}$$

then $A_k > 0$ and $\sum_{k=1}^{n} A_k = 1$. Hence by applying Theorem 6 for this set of values of A_k, Theorem 11 can be proved. □

Proof of Theorem 12. From Lemma 19, we have that

$$\sum_{k=0}^{n} k^2C(n,k) = n(1+n)2^{n-2}. \tag{55}$$

If we take

$$A_k = \frac{k^2C(n,k)}{n(1+n)2^{n-2}}, \tag{56}$$

then $A_k > 0$ and $\sum_{k=1}^{n} A_k = 1$. Hence by applying Theorem 6 for this set of values of A_k, Theorem 12 can be proved. □

TABLE 1

Theorems	r_1	r_2	Area of the annulus
5	0.4641	1.6984	8.382
7	0.6542	1.205	3.2159
8	0.5192	1.5183	6.395
10	0.5833	1.20	3.4549
11	0.7757	1.0164	1.3553
12	0.6403	1.2313	3.4748
14	0.4886	1.6134	7.428

Proof of Theorem 13. From Lemma 20, we have that

$$\sum_{k=0}^{n} kC(n,k)^2 = \frac{n}{2}C(2n,n). \quad (57)$$

If we take

$$A_k = \frac{kC(n,k)^2}{(n/2)C(2n,n)}, \quad (58)$$

then $A_k > 0$ and $\sum_{k=1}^{n} A_k = 1$. Hence by applying Theorem 6 for this set of values of A_k, Theorem 13 can be proved. □

Proof of Theorem 14. From Lemma 21, we have that

$$\sum_{k=0}^{n} k^2C(n,k)^2 = n^2C(2n-2,n-1). \quad (59)$$

If we take

$$A_k = \frac{k^2C(n,k)^2}{n^2C(2n-2,n-1)}, \quad (60)$$

then $A_k > 0$ and $\sum_{k=1}^{n} A_k = 1$. Hence by applying Theorem 6 for this set of values of A_k, Theorem 14 can be proved. □

4. Computational Results and Analysis

In this section, we present two examples of polynomials in order to compare our theorems with some of the above stated known theorems and show that for these polynomials our theorems give better bounds than those obtainable by these known theorems.

Example 1. Consider the polynomial $p(z) = z^3+0.1z^2+0.1z+0.7$.

Table 1 suggests that our Theorem 11 gives significantly better bounds than those obtained from any result, including Theorems 5 and 7. As can be seen, the area of the annulus obtained by Theorem 11 is about 16.17% of the area of the annulus obtained by Theorem 5 and about 42.14% of the area obtained by Theorem 7. The inner radius obtained from Theorem 11 is almost 167% of the inner radius obtained from Theorem 5, and similarly the outer radius obtained from Theorem 11 is almost 59.84% of the outer radius obtained from Theorem 5.

TABLE 2

Theorems	r_1	r_2	Area of the annulus
5	5.56×10^{-4}	1.158	4.2125
7	0.0011	1.5608	7.6529
8	8.21×10^{-5}	1.1036	3.8265
12	7.18×10^{-5}	1.3191	5.4664
13	1.37×10^{-4}	1.1626	4.2464
14	4.93×10^{-5}	1.1264	3.9859

Example 2. Consider the polynomial

$$p(z) = z^5 + 0.006z^4 + 0.29z^3 + 0.29z^2 + 0.29z + 0.001. \quad (61)$$

As shown in Table 2, the best upper bound of the annular region containing all the zeros of the polynomial $p(z)$ comes from Theorem 8, which also gives the smallest area of the annulus containing all the zeros. The area of the annulus obtained by Theorem 8 is about 90.84% of the area of the annulus obtained by Theorem 5 and about 50% of the area of the annulus obtained by Theorem 7.

Conflicts of Interest

The authors declare that there are no conflicts of interest regarding the publication of this paper.

References

[1] H. Anai and K. Horimoto, *Algebraic Biology 2005, Proceedings of the First Conference on Algebraic Biology, Tokyo, Japan*, 2005.

[2] C. Bissel, *Control Engineering*, CRC Press, 2nd edition, 2009.

[3] N. K. Govil and E. R. Nwaeze, "On geometry of the zeros of a polynomial," in *Computation, Cryptography, and Network Security*, N. J. Daras and M. Th. Rassias, Eds., pp. 253–287, Springer, Cham, Switzerland, 2015.

[4] L. Pachter and B. Sturmdfels, *Algebraic Statistics for Computational Biology*, Cambridge University Press, 2005.

[5] O. Aberth, "Iteration methods for finding all zeros of a polynomial simultaneously," *Mathematics of Computation*, vol. 27, pp. 339–344, 1973.

[6] L. W. Ehrlich, "A modified Newton method for polynomials," *Communications of the ACM*, vol. 10, no. 2, pp. 107-108, 1967.

[7] G. V. Milovanović and M. S. Petković, "On computational efficiency of the iterative methods for the simultaneous approximation of polynomial zeros," *ACM Transactions on Mathematical Software*, vol. 12, no. 4, pp. 295–306, 1986.

[8] M. S. Petković, "A highly efficient zero-solver of very fast convergence," *Applied Mathematics and Computation*, vol. 205, no. 1, pp. 298–302, 2008.

[9] K. F. Gauss, *Beiträge zur Theorie der algebraischen Gleichungen*, vol. 4, Abhandlungen der Königlichen Gesellschaft der Wissenschaften zu Göttingen, 1850, Ges. Werke, vol. 3, pp. 73–102, 1850.

[10] A. L. Cauchy, *Excercises de Mathématique*, IV Année de Bure Freres, Paris, France, 1829.

[11] J. L. Diaz-Barrero, "An annulus for the zeros of polynomials," *Journal of Mathematical Analysis and Applications*, vol. 273, no. 2, pp. 349–352, 2002.

[12] S.-H. Kim, "On the moduli of the zeros of a polynomial," *The American Mathematical Monthly*, vol. 112, no. 10, pp. 924-925, 2005.

[13] M. Bidkham and E. Shashahani, "An annulus for the zeros of polynomials," *Applied Mathematics Letters*, vol. 24, no. 2, pp. 122–125, 2011.

[14] N. A. Rather and S. G. Mattoo, "On annulus containing all the zeros of a polynomial," *Applied Mathematics E-Notes*, vol. 13, pp. 155–159, 2013.

[15] A. Dalal and N. K. Govil, "On region containing all the zeros of a polynomial," *Applied Mathematics and Computation*, vol. 219, no. 17, pp. 9609–9614, 2013.

[16] A. Dalal and N. K. Govil, "Annulus containing all the zeros of a polynomial," *Applied Mathematics and Computation*, vol. 249, pp. 429–435, 2014.

[17] A. Dalal and N. K. Govil, "On comparison of annuli containing all the zeros of a polynomial," *Applicable Analysis and Discrete Mathematics*, vol. 11, no. 1, pp. 232–241, 2017.

[18] R. Donaghey and L. W. Shapiro, "Motzkin numbers," *Journal of Combinatorial Theory, Series A*, vol. 23, no. 3, pp. 291–301, 1977.

13

The Regular Part of a Semigroup of Full Transformations with Restricted Range: Maximal Inverse Subsemigroups and Maximal Regular Subsemigroups of Its Ideals

Worachead Sommanee

Department of Mathematics and Statistics, Faculty of Science and Technology, Chiang Mai Rajabhat University, Chiang Mai 50300, Thailand

Correspondence should be addressed to Worachead Sommanee; worachead_som@cmru.ac.th

Academic Editor: Pentti Haukkanen

Let $T(X)$ be the full transformation semigroup on a set X. For a fixed nonempty subset Y of a set X, let $T(X,Y)$ be the semigroup consisting of all full transformations from X into Y. In a paper published in 2008, Sanwong and Sommanee proved that the set $F(X,Y) = \{\alpha \in T(X,Y) : X\alpha = Y\alpha\}$ is the largest regular subsemigroup of $T(X,Y)$. In this paper, we describe the maximal inverse subsemigroups of $F(X,Y)$ and completely determine all the maximal regular subsemigroups of its ideals.

1. Introduction

Let $T(X)$ be the set of all full transformations from a nonempty set X into itself. It is well-known that $T(X)$ is a regular semigroup under composition of functions; see [1, p. 63]. Moreover, every semigroup can be embedded in $T(X)$ for some nonempty set X; see [1, p. 7]. In [2, 3], Schein posed the problem to determine all maximal inverse subsemigroups of $T(X)$; this problem is still unsolved. In 1976, Nichols [4] characterized one class of maximal inverse subsemigroup of $T(X)$. Later in 1978, Reilly [5] has generalized Nichols construction and obtained a much wider class of maximal inverse subsemigroups of $T(X)$. In 1999, Yang [6] described all of the maximal inverse subsemigroups of the finite symmetric inverse semigroup. Later in 2001, Yang [7] obtained the maximal subsemigroups of the finite singular transformation semigroups. In 2002, You [8] determined all the maximal regular subsemigroups of all ideals of the finite full transformation semigroup. In 2004, H. B. Yang and X. L. Yang [9] completely described the maximal subsemigroups of ideals of the finite full transformation semigroup. In 2014, Zhao et al. [10] showed that any maximal regular subsemigroup of ideals of the finite full transformation semigroup is idempotent generated. After that in 2015, East et al. [11] classified the maximal subsemigroups of $T(X)$ when X is an infinite set containing certain subgroups of the symmetric group on X.

Let Y be a fixed nonempty subset of X and let $T(X,Y)$ be the subsemigroup of $T(X)$ of all elements with ranges contained in Y. In 1975, Symons [12] introduced and studied the semigroup $T(X,Y)$. He described all the automorphisms of $T(X,Y)$ and also determined the isomorphism theorem for two semigroups of type $T(X,Y)$. In 2005, Nenthein et al. [13] characterized the regular elements of $T(X,Y)$. In general, $T(X,Y)$ is not a regular semigroup. In 2008, Sanwong and Sommanee [14] defined

$$F(X,Y) = \{\alpha \in T(X,Y) : X\alpha = Y\alpha\} \tag{1}$$

and showed that $F(X,Y)$ is the largest regular subsemigroup of $T(X,Y)$. Obviously, $F(X,Y) = T(X)$ when $X = Y$. Hence, we may regard $F(X,Y)$ as a generalization of $T(X)$. The authors also characterized Green's relations on $T(X,Y)$ and gave one class of maximal inverse subsemigroups of $T(X,Y)$. Later in 2009, Sanwong et al. [15] described all maximal and minimal congruences on $T(X,Y)$. In 2011, Mendes-Gonçalves and Sullivan [16] obtained all the ideals of $T(X,Y)$. In the same year, Sanwong [17] described Green's relations and ideals and all maximal regular subsemigroups of $F(X,Y)$. The author also proved that every regular semigroup S can be embedded in $F(S^1, S)$. After that, in 2013, Sommanee and Sanwong [18] computed the rank of $F(X,Y)$ when X is a finite set. They also obtained the rank and the idempotent rank of its ideals when X is finite. In 2014, Fernandes and Sanwong [19] calculated the rank of $T(X,Y)$ when X is finite. In 2016, L. Sun and J. Sun [20] characterised the natural partial order on $T(X,Y)$. In the same year, Tinpun and Koppitz [21] determined the relative rank of $T(X,Y)$ modulo the semigroup $S(X,Y)$ of all extensions of the bijections on Y when X is finite.

Let X be a chain and $\mathrm{OT}(X)$ the full order-preserving transformation semigroup on X. In 2011, Dimitrova and Koppitz [22] characterized the maximal regular subsemigroups of ideals of $\mathrm{OT}(X)$ when X is a finite chain. In 2015, Sommanee and Sanwong [23] investigated the regularity and Green's relations of the order-preserving transformation semigroup

$$\begin{aligned} OF(X,Y) &= F(X,Y) \cap \mathrm{OT}(X) \\ &= \{\alpha \in F(X,Y) : \alpha \text{ is order-preserving}\}. \end{aligned} \tag{2}$$

They also proved that $OF(X,Y)$ is idempotent generated when Y is finite.

Let V be a vector space and let $T(V)$ denote the linear transformation semigroup from V into V. For a fixed subspace W of V, let $T(V,W)$ be the semigroup consisting of all linear transformations from V into W. Recently in 2017, Sommanee and Sangkhanan [24] determined all the maximal regular subsemigroups of the semigroup

$$Q = \{\alpha \in T(V,W) : V\alpha = W\alpha\}, \tag{3}$$

when W is a finite dimensional subspace of V over a finite field.

In this paper, we describe the maximal inverse subsemigroups of $F(X,Y)$ and completely determine all the maximal regular subsemigroups of its ideals.

2. Preliminaries and Notations

For all undefined notions, the reader is referred to [1].

For any set A, $|A|$ means the cardinality of the set A. If A is a subset of a semigroup S, then $\langle A\rangle$ denotes the subsemigroup of S generated by A. An element e of a semigroup S is called *idempotent* if $e^2 = e$. If U is a subset of a semigroup S, then $E(U)$ denotes the set of all idempotents in the set U. An element a of a semigroup S is called *regular* if there exists $x \in S$ such that $a = axa$. The semigroup S is called regular if all its elements are regular. If a is an element of a semigroup S, we say that a' is an *inverse* of a if $a = aa'a$ and $a' = a'aa'$. We denote the set of inverses of an element a by $V(a)$. A semigroup S is called an inverse semigroup if every element a in S has a unique inverse a^{-1} in S; it is equivalent to Sbeing regular and idempotent elements commute. A proper *(regular, inverse)* subsemigroup M of a semigroup S is a *maximal (regular, inverse)* subsemigroup of S if whenever $M \subseteq T \subseteq S$ for some a (regular, inverse) subsemigroup T of S, then $M = T$ or $T = S$.

The Green's relations $\mathscr{L}, \mathscr{R}, \mathscr{H}, \mathscr{D}$, and $\mathscr{J}$ on a semigroup S are defined as follows. For $a, b \in S$,

(1) $a\mathscr{L}b$ if and only if $S^1a = S^1b$;

(2) $a\mathscr{R}b$ if and only if $aS^1 = bS^1$;

(3) $a\mathscr{J}b$ if and only if $S^1aS^1 = S^1bS^1$;

(4) $\mathscr{H} = \mathscr{L} \cap \mathscr{R}$ and $\mathscr{D} = \mathscr{L} \circ \mathscr{R}$.

For each $a \in S$, we denote $\mathscr{L}$-class, $\mathscr{R}$-class, $\mathscr{H}$-class, $\mathscr{D}$-class, and $\mathscr{J}$-class containing a by L_a, R_a, H_a, D_a, and J_a, respectively.

Lemma 1 (see [5, Lemma 1]). *Let M be a maximal inverse subsemigroup of a semigroup S. If M contains a minimum idempotent e, then $H_e \subseteq M$.*

Lemma 2. *Let S be any inverse semigroup. If e is a minimum idempotent in S, then $ea = ae$ for all $a \in S$.*

Proof. Assume that e is a minimum idempotent in S. Then $e \le f$ for all $f \in E(S)$; that is, $e = ef = fe$. Let $a \in S$. Then $aea^{-1}, a^{-1}ea \in E(S)$ and $ea = e(aea^{-1})a = (aea^{-1})ea = a(ea^{-1}ea) = ae$. □

Throughout the paper we assume that Y is a nonempty subset of a set X. Let $T(X,Y) = \{\alpha \in T(X) : X\alpha \subseteq Y\}$, where $X\alpha$ denotes the image of α. Define

$$F(X,Y) = \{\alpha \in T(X,Y) : X\alpha = Y\alpha\}. \tag{4}$$

It is known that $F(X,Y)$ is the largest regular subsemigroup of $T(X,Y)$.

If $\alpha \in T(X)$ and $x \in X$, then the image of x under α is written as $x\alpha$. The set of all inverse images of x under α is denoted by $x\alpha^{-1}$.

Lemma 3 (see [17, Lemma 1]). *Let $\alpha \in T(X,Y)$. Then $\alpha \in F(X,Y)$ if and only if $a\alpha^{-1} \cap Y \neq \emptyset$ for all $a \in X\alpha$.*

In [17], the author gave a complete description of Green's relations on $F(X,Y)$ as the following lemma.

Lemma 4. *Let $\alpha, \beta \in F(X,Y)$. Then*

(1) *$\alpha\mathscr{L}\beta$ if and only if $X\alpha = X\beta$;*

(2) *$\alpha\mathscr{R}\beta$ if and only if $\pi_\alpha = \pi_\beta$, where $\pi_\alpha = \{x\alpha^{-1} : x \in X\alpha\}$;*

(3) *$\alpha\mathscr{D}\beta$ if and only if $|X\alpha| = |X\beta|$;*

(4) *$\mathscr{D} = \mathscr{J}$.*

In addition, $\alpha\mathscr{H}\beta$ if and only if $X\alpha = X\beta$ and $\pi_\alpha = \pi_\beta$.

Lemma 5. *$F(X,Y)$ is an inverse semigroup if and only if $|Y| = 1$.*

Proof. If $|Y| = 1$, then $|F(X,Y)| = 1$ and so it is inverse. Conversely, assume that $F(X,Y)$ is an inverse semigroup. Let a and b be elements in Y. Then there exist constant maps X_a and X_b with ranges $\{a\}$ and $\{b\}$, respectively. Thus they are idempotents in $F(X,Y)$ such that $X_a = X_bX_a = X_aX_b = X_b$; it follows that $a = b$. Hence, $|Y| = 1$. □

Proposition 6. *M is a maximal inverse subsemigroup of $T(X,Y)$ if and only if M is a maximal inverse subsemigroup of $F(X,Y)$.*

Proof. One direction is clear. Indeed, each inverse subsemigroup of $T(X,Y)$ is contained in $F(X,Y)$, the largest regular subsemigroup of $T(X,Y)$. Conversely, assume that M is a maximal inverse subsemigroup of $F(X,Y)$. Then M is an inverse subsemigroup of $T(X,Y)$. Let V be an inverse subsemigroup of $T(X,Y)$ such that $M \subseteq V \subsetneq T(X,Y)$. Since $F(X,Y)$ is the largest regular subsemigroup of $T(X,Y)$, it follows that $M \subseteq V \subseteq F(X,Y)$. We get that $M = V$ or $V = F(X,Y)$ by the maximality of M in $F(X,Y)$. If $F(X,Y) = V$, then $F(X,Y)$ is an inverse semigroup and so $|Y| = 1$ by Lemma 5; this implies that $T(X,Y) = F(X,Y)$. Therefore, M is a maximal inverse subsemigroup of $T(X,Y)$. □

For each $a \in Y$, let

$$F_a = \{\alpha \in F(X,Y) : a\alpha = a,\ \alpha \text{ is injective on } X \setminus a\alpha^{-1}\}. \quad (5)$$

The authors [14] showed that F_a is a maximal inverse subsemigroup of $T(X,Y)$. Thus by Proposition 6, we have that F_a is also a maximal inverse subsemigroup of $F(X,Y)$.

3. Maximal Inverse Subsemigroups

In this section, we write $G(A)$ for the permutation group on a set A.

Let e be any idempotent in $T(X)$. In 1978, Reilly defined I_e to be the set of those elements $\alpha \in T(X)$ which satisfy the following three conditions:

(1) The restriction of α to Xe is a permutation of Xe; that is, $\alpha|_{Xe} \in G(Xe)$.

(2) For $x \in X$, if $x\alpha \notin Xe$, then $|(x\alpha)\alpha^{-1}| = 1$.

(3) $\alpha e = e\alpha$.

Lemma 7 (see [5, Theorem 2]). *Let e be an idempotent in $T(X)$ and let I_e be defined as above. Then I_e is a maximal inverse subsemigroup of $T(X)$ with minimum idempotent e.*

Notice that if e is a constant map with range $\{a\} \subseteq X$, then $I_e = I_a$, where

$$I_a = \{\alpha \in T(X) : a\alpha = a,\ \alpha \text{ is injective on } X \setminus a\alpha^{-1}\}; \quad (6)$$

see [4, 5] for more details.

Now, let e be any idempotent in $F(X,Y)$ and define

$$F_e = I_e \cap F(X,Y). \quad (7)$$

Since $e \in I_e \cap F(X,Y) = F_e$, we obtain $F_e \neq \emptyset$ and so F_e is a subsemigroup of $F(X,Y)$. We note that if $X = Y$, then $F(X,Y) = T(X)$ and $F_e = I_e \cap F(X,Y) = I_e \cap T(X) = I_e$.

Here, we aim to prove that F_e is the maximal inverse subsemigroups of $F(X,Y)$ with minimum idempotent e.

To prove our main result we need the following five lemmas.

Lemma 8. *Let e be any idempotent in $F(X,Y)$ and $\alpha \in F(X,Y)$. Then $\alpha \in F_e$ if and only if*

(1) $\alpha|_{Xe} \in G(Xe)$,

(2) *for $y \in Y$, if $y\alpha \notin Xe$, then $|(y\alpha)\alpha^{-1}| = 1$,*

(3) $\alpha e = e\alpha$.

Proof. One direction is clear. Conversely, assume that conditions (1), (2), and (3) hold. We prove that $\alpha \in I_e$. Suppose that $x \in X$ and $x\alpha \notin Xe$. Since $\alpha \in F(X,Y)$, it follows that $x\alpha = y\alpha$ for some $y \in Y$. By condition (2), we get that $|(y\alpha)\alpha^{-1}| = 1$. Thus $|(x\alpha)\alpha^{-1}| = 1$ and so $\alpha \in I_e$. □

Lemma 9. *Let e be any idempotent in $F(X,Y)$ and $\alpha \in F(X,Y)$. Then condition (2) in Lemma 8 is equivalent to the statement*

$$(X \setminus Y)\alpha \subseteq Xe, \quad \alpha \text{ is injective on } Y \setminus (Xe)\alpha^{-1}. \quad (8)$$

Proof. Assume that, for $y \in Y$, if $y\alpha \notin Xe$, then $|(y\alpha)\alpha^{-1}| = 1$. We prove $(X\setminus Y)\alpha \subseteq Xe$. Let $x\alpha \in (X\setminus Y)\alpha$ for some $x \in X\setminus Y$. Since $(X \setminus Y)\alpha \subseteq X\alpha = Y\alpha$, there exists $y \in Y$ such that $x\alpha = y\alpha$. Thus $x, y \in (y\alpha)\alpha^{-1}$. If $y\alpha \notin Xe$, then $|(y\alpha)\alpha^{-1}| = 1$ by the assumption. Whence $x = y$, this is a contradiction. Hence $y\alpha \in Xe$; that is, $x\alpha \in Xe$. Next, we prove α is injective on $Y \setminus (Xe)\alpha^{-1}$. Let $w, z \in Y \setminus (Xe)\alpha^{-1}$ be such that $w\alpha = z\alpha$. Then $w \in Y$ and $w\alpha \notin Xe$. Since $w, z \in (w\alpha)\alpha^{-1}$, we obtain $w = z$ by our assumption.

Conversely, suppose that $(X \setminus Y)\alpha \subseteq Xe$ and α is injective on $Y \setminus (Xe)\alpha^{-1}$. Let $y \in Y$ and $y\alpha \notin Xe$. To show that $|(y\alpha)\alpha^{-1}| = 1$, we let $u, v \in (y\alpha)\alpha^{-1}$. Then $u\alpha = v\alpha = y\alpha \notin Xe$; this implies that $u, v \notin (Xe)\alpha^{-1}$. Since $(X \setminus Y)\alpha \subseteq Xe$, we obtain $u, v \in Y$. Thus $u, v \in Y \setminus (Xe)\alpha^{-1}$ and so $u = v$ since α is injective on $Y \setminus (Xe)\alpha^{-1}$. □

Lemma 10. *Let α be any element in $T(X)$ and let e be an idempotent in $T(X)$. If there is $\alpha' \in V(\alpha)$ such that $e = e\alpha'\alpha = \alpha\alpha' e$, $e\alpha = \alpha e$ and $e\alpha' = \alpha' e$, then $\alpha|_{Xe} \in G(Xe)$.*

Proof. Assume that the conditions hold. Since

$$Xe = X(e\alpha'\alpha) = X(e\alpha')\alpha = X(\alpha' e)\alpha \subseteq (Xe)\alpha = X(e\alpha) = X(\alpha e) \subseteq Xe, \quad (9)$$

it follows that $(Xe)\alpha = Xe$; that is, α maps Xe onto Xe. We prove that $\alpha|_{Xe}$ is injective. Let $x, y \in Xe$ be such that $x\alpha = y\alpha$. Then

$$x = xe = x(\alpha\alpha' e) = (x\alpha)\alpha' e = (y\alpha)\alpha' e = y(\alpha\alpha' e) = ye = y. \quad (10)$$

Hence $\alpha|_{Xe} : Xe \to Xe$ is bijective. □

Lemma 11. *Let e be an idempotent in $F(X,Y)$. Then F_e is an inverse subsemigroup of $F(X,Y)$ with minimum idempotent e.*

Proof. Let α be any element in F_e. Then $\alpha \in I_e$. Since I_e is an inverse semigroup, there exists $\beta \in I_e$ such that $\alpha = \alpha\beta\alpha$

and $\beta = \beta\alpha\beta$. We prove $\beta \in F(X,Y)$. Since $\beta = \beta\alpha\beta$, we obtain $X\beta = X\beta\alpha\beta \subseteq X\alpha\beta \subseteq Y\beta \subseteq X\beta$; that is, $X\beta = Y\beta$. To show $X\beta \subseteq Y$, we let $x\beta \in X\beta$ for some $x \in X$. Since $\alpha \in F_e$, $(X \setminus Y)\alpha \subseteq Xe$ by Lemma 9. If $x\beta \in X \setminus Y$, then $(x\beta)\alpha \in (X \setminus Y)\alpha \subseteq Xe$. It follows that $x\beta = (x\beta\alpha)\beta \in (Xe)\beta = Xe \subseteq Y$; this is a contradiction. Hence, $x\beta \in Y$ and so $X\beta \subseteq Y$. Thus, $\beta \in I_e \cap F(X,Y) = F_e$. Whence F_e is a regular subsemigroup of I_e, it follows from Lemma 7 that F_e is an inverse subsemigroup of $F(X,Y)$ and e is the minimum idempotent in F_e. □

Lemma 12. *Let e be an idempotent in $F(X,Y)$. If M is any inverse subsemigroup of $F(X,Y)$ such that $F_e \subseteq M$, then $E(M) \subseteq F_e$.*

Proof. Assume that M is an inverse subsemigroup of $F(X,Y)$ containing F_e. Let f be any idempotent in M. Since e, f are idempotents in M, we get $ef = fe$. We prove that $Xe \subseteq Xf$. Suppose that $Xe \nsubseteq Xf$. Then $Xef \subseteq Xe \cap Xf \subsetneq Xe$, so there exists $z \in Xe \setminus Xef$. We see that $zf \in Xef$ and $zf \neq z$. Now, we have $z, zef \in Xe$ such that $z \neq zef$. Let $\sigma \in G(Xe)$ be such that $z\sigma = zef$ and $(zef)\sigma = z$. We define $\alpha \in T(X)$ by

$$x\alpha = (xe)\sigma \quad \forall x \in X. \tag{11}$$

We see that $X\alpha = (Xe)\sigma = (Ye)\sigma = Y\alpha$ and $X\alpha = (Xe)\sigma = Xe \subseteq Y$; we obtain $\alpha \in F(X,Y)$. We prove $\alpha|_{Xe} \in G(Xe)$. Let $x, y \in X$ be such that $(xe)\alpha = (ye)\alpha$. Then $((xe)e)\sigma = ((ye)e)\sigma$; that is, $(xe)\sigma = (ye)\sigma$. Thus $xe = ye$ since σ is injective on Xe. Since $(Xe)\alpha = ((Xe)e)\sigma = (Xe)\sigma = Xe$, α is surjective on Xe. Hence, $\alpha|_{Xe} \in G(Xe)$. Since $y\alpha = (ye)\sigma \in Xe$ for all $y \in Y$, whence condition (2) in Lemma 8 holds. To prove $\alpha e = e\alpha$. Let $x \in X$. Since $(xe)\sigma \in Xe$ and e is an idempotent, we get that $x\alpha e = ((xe)\sigma)e = (xe)\sigma = ((xe)e)\sigma = xe\alpha$. So, $\alpha \in F_e$. By Lemma 11, we have F_e as an inverse subsemigroup of $F(X,Y)$. Then there exists $\alpha^{-1} \in V(\alpha) \cap F_e$ and $\sigma^{-1} = \alpha^{-1}|_{Xe} \in G(Xe)$. Since $zef = zfe \in Xe$, we obtain $(zef)\alpha^{-1} = (zef)\sigma^{-1} = z$. It is clear that $\alpha^{-1}f\alpha$ is an idempotent in M. We see that

$$\begin{aligned}(zef)\,f\left(\alpha^{-1}f\alpha\right) &= (zef)\left(\alpha^{-1}f\alpha\right) = (zf)\,\alpha \\ &= \left[(zf)\,e\right]\sigma = (zef)\,\sigma = z, \\ (zef)\left(\alpha^{-1}f\alpha\right)f &= (zf)\,\alpha f = \left[(zfe)\,\sigma\right]f \\ &= \left[(zef)\,\sigma\right]f = zf.\end{aligned} \tag{12}$$

Since $z \neq zf$, this implies that $f(\alpha^{-1}f\alpha) \neq (\alpha^{-1}f\alpha)f$ which is a contradiction. Therefore, $Xe \subseteq Xf$. Hence $ef = e$ and so $e = ef = fe$. It is easy to see that $f|_{Xe} = id_{Xe}$, the identity map on Xe, which implies that $f|_{Xe} \in G(Xe)$. Thus, conditions (1) and (3) in Lemma 8 are satisfied for f.

In order to verify condition (2) in Lemma 8 for f to be an element of F_e, let $y \in Y$ be such that $yf \notin Xe$. We show that $|(yf)f^{-1}| = 1$. Let $u, v \in (yf)f^{-1}$. Then $uf = vf = yf \notin Xe$. Suppose that $u \neq v$. Assume that $v \in Y$. We define $d \in T(X)$ by

$$xd = \begin{cases} x & \text{if } x \in Y \setminus \{u\}, \\ xe & \text{if } x \in \{u\} \cup (X \setminus Y). \end{cases} \tag{13}$$

Obviously, $Xd \subseteq Y$ and it is easy to verify that $(xe)d = xe = x(de)$ for all $x \in X$. Hence, $(Xe)d = Xe$ and $ed = de$. It is easy to show that $d|_{Xe}$ is injective, whence $d|_{Xe} \in G(Xe)$. We prove $Xd \subseteq Yd$. Let $xd \in Xd$ for some $x \in X$. If $x \in Y \setminus \{u\}$, then $xd \in Yd$. But if $x \in \{u\} \cup (X \setminus Y)$, then $xd = xe = (xe)d \in Yd$. Thus, $Xd \subseteq Yd$ and so $d \in F(X,Y)$. We prove d satisfies the condition in Lemma 9. By the definition of d, we have $(X \setminus Y)d \subseteq Xe$. Let $a, b \in Y \setminus (Xe)d^{-1}$. Then $a, b \in Y$ and $ad, bd \notin Xe$, which implies that $a, b \in Y \setminus \{u\}$. So, if $ad = bd$, then $a = b$. Hence, d is injective on $Y \setminus (Xe)d^{-1}$ and $d \in F_e$. Next, we prove d is an idempotent. For $x \in Y \setminus \{u\}$, $(xd)d = xd$. If $x \in \{u\} \cup (X \setminus Y)$, then $(xd)d = (xe)d = xe = xd$. Thus, $d^2 = d$. Since d and f are idempotents in M, $df = fd$. However,

$$\begin{aligned} udf &= (ue)\,f = ue \in Xe, \\ ufd &= (vf)\,d = vdf = vf \notin Xe, \end{aligned} \tag{14}$$

a contradiction. Hence, $v \in X \setminus Y$. Define $g \in T(X)$ by

$$xg = \begin{cases} x & \text{if } x \in Y, \\ xe & \text{if } x \in X \setminus Y. \end{cases} \tag{15}$$

Similarly, by using the same argument as in the proof above for d, we can show that g is an idempotent in F_e. Since $f, g \in E(M)$, we obtain $fg = gf$. We see that

$$\begin{aligned} vfg &= vf \notin Xe \\ \text{but } vgf &= (ve)\,f = ve \in Xe, \end{aligned} \tag{16}$$

and this is a contradiction. Thus, $u = v$ and so condition (2) in Lemma 8 is also satisfied. Whence $f \in F_e$, therefore, $E(M) \subseteq F_e$ as required. □

Now, we are ready to prove our main result (Theorem 13).

Theorem 13. *Let e be an idempotent in $F(X,Y)$. Then F_e is a maximal inverse subsemigroup of $F(X,Y)$ with minimum idempotent e.*

Proof. From Lemma 11, F_e is an inverse subsemigroup of $F(X,Y)$ with minimum idempotent e. To prove the maximality, let M be any inverse semigroup of $F(X,Y)$ such that $F_e \subseteq M \subsetneq F(X,Y)$. Then by Lemma 12, we have $E(M) \subseteq F_e$, which implies that e is the minimum idempotent in M. We prove $M = F_e$ by letting $\alpha \in M$. Since $\alpha, \alpha^{-1} \in M$, we obtain $e\alpha = \alpha e$ and $e\alpha^{-1} = \alpha^{-1}e$ by Lemma 2. And since $\alpha\alpha^{-1}$ and $\alpha^{-1}\alpha$ are idempotents in M, $e = \alpha\alpha^{-1}e$ and $e = e\alpha^{-1}\alpha$. Then by Lemma 10, we get that $\alpha|_{Xe} \in G(Xe)$. Thus, conditions (1) and (3) in Lemma 8 are satisfied for α to be an element of F_e. Next, we prove that α satisfies condition (2) in Lemma 8.

Let $y \in Y$ be such that $y\alpha \notin Xe$. Let $u, v \in (y\alpha)\alpha^{-1}$. Then $u\alpha = v\alpha = y\alpha \notin Xe$; this implies that $u\alpha\alpha^{-1} = v\alpha\alpha^{-1} = y\alpha\alpha^{-1} \notin Xe$. Now, we have $\alpha\alpha^{-1} \in E(F_e)$, $y \in Y$ and $y(\alpha\alpha^{-1}) \notin Xe$. Since $u, v \in [y(\alpha\alpha^{-1})](\alpha\alpha^{-1})^{-1}$, we obtain that $u = v$ by using condition (2) in Lemma 8 for $\alpha\alpha^{-1} \in F_e$. Hence, $\alpha \in F_e$ and so $M \subseteq F_e$. Therefore, $M = F_e$ and F_e is a maximal inverse subsemigroup of $F(X, Y)$. □

Corollary 14. *Let e be an idempotent in $F(X, Y)$. Then F_e is a maximal inverse subsemigroup of $T(X, Y)$.*

Proof. It follows directly from Theorem 13 and Proposition 6. □

Remark 15. $F_e \neq F_f$ for all distinct idempotents $e, f \in F(X, Y)$. In fact, if there exist idempotents e and f in $F(X, Y)$ such that $F_e = F_f$, then $f \in F_e$ and $e \in F_f$ such that e and f are minimum idempotents of F_e and F_f, respectively. Thus, we get $e \le f$ and $f \le e$; it follows that $e = f$.

Corollary 16. *Let e be an idempotent in $F(X, Y)$ such that $Xe = Y$. Then $F_e = H_e$ is a maximal inverse subsemigroup of $F(X, Y)$.*

Proof. Since F_e is a maximal inverse subsemigroup of $F(X, Y)$ with the minimum idempotent e, we obtain that $H_e \subseteq F_e$ by Lemma 1. To show that $F_e \subseteq H_e$, let $\alpha \in F_e$. Then $\alpha|_{Xe} \in G(Xe)$ and $\alpha e = e\alpha$. Since $(Xe)\alpha = Xe$, it follows that $X\alpha = Y\alpha = (Xe)\alpha = Xe$. We prove that $\pi_\alpha = \pi_e$. Let $x, y \in X$ such that $x\alpha = y\alpha$. Then $(xe)\alpha = (x\alpha)e = (y\alpha)e = (ye)\alpha$; this implies that $xe = ye$ since $\alpha|_{Xe}$ is injective. Hence $\pi_\alpha \subseteq \pi_e$. On the other hand, let $xe = ye$. Then $(x\alpha)e = (xe)\alpha = (ye)\alpha = (y\alpha)e$. Since $x\alpha, y\alpha \in Y = Xe$ and e is an idempotent, we obtain $x\alpha = y\alpha$. Hence $\pi_e \subseteq \pi_\alpha$ and $\pi_\alpha = \pi_e$. Then by Lemma 4, we get that $\alpha \in H_e$. Thus, $F_e \subseteq H_e$. Therefore, $H_e = F_e$ is the maximal inverse subsemigroup of $F(X, Y)$. □

Notice that if e is a constant map with range $\{a\} \subseteq Y$, then $I_e = I_a$ and

$$F_e = F(X, Y) \cap I_a = \{\alpha \in F(X, Y) : a\alpha = a,\ \alpha \text{ is injective on } X \setminus a\alpha^{-1}\}; \tag{17}$$

that is, $F_e = F_a$. Thus, we get the following corollary which appeared in [14, Theorem 4.3].

Corollary 17. *If $e \in F(X, Y)$ is a constant map with range $\{a\} \subseteq Y$, then $F_e = F_a$ is a maximal inverse subsemigroup of $F(X, Y)$ and $T(X, Y)$.*

Next, we count the number of elements in F_e. Let e be any idempotent in $F(X, Y)$. We know that $H_e \cong G(Xe)$; see [17, Section 4]. Thus, if $Xe = Y$, then $F_e = H_e \cong G(Xe) = G(Y)$ by Corollary 16. Hence, $|F_e| = |G(Y)|$ when $Xe = Y$. In what follows, we assume that $Xe \subsetneq Y$.

Recall that the number of ways that r objects can be chosen from n distinct objects written as $\binom{n}{r}$ is given by

$$\binom{n}{r} = \frac{n!}{(n-r)!r!}. \tag{18}$$

Let A and B be (possibly empty) sets and let $\mathrm{PI}(A, B)$ be the set of all partial injective maps from A into B. We note that $\mathrm{PI}(A, B)$ contains the empty map. Moreover, if $A = \emptyset$ or $B = \emptyset$, then $|\mathrm{PI}(A, B)| = 1$. And if A and B are finite sets such that $|A| = m$ and $|B| = n$, then we shall write $\mathrm{PI}(m, n)$ instead of $\mathrm{PI}(A, B)$. By a consideration of the cardinality of a domain of partial injective maps, we can verify that

$$|\mathrm{PI}(m, n)| = \sum_{j=0}^{\min\{m,n\}} \binom{m}{j}\binom{n}{j} j! \quad \text{for } m, n \ge 1. \tag{19}$$

Since $Xe \subsetneq Y$, there exists $x \in Y \setminus Xe$ such that $xe \in Xe$. We see that $x \in ((xe)e^{-1} \setminus \{xe\}) \cap Y$. For each $y \in Xe$, we define

$$\begin{aligned} A(y) &= (ye^{-1} \setminus \{y\}) \cap Y, \\ B(y) &= ye^{-1} \cap (X \setminus Y). \end{aligned} \tag{20}$$

Then $ye^{-1} = \{y\} \cup A(y) \cup B(y)$ is a disjoint union such that $ye^{-1} \cap Xe = \{y\}$ and $A(y) \cap Xe = \emptyset = B(y) \cap Xe$. Moreover, $A(y) \neq \emptyset$ for some $y \in Xe$ since $Xe \subsetneq Y$. Let

$$\begin{aligned} \mathscr{A} &= \{y_i \in Xe : A(y_i) \neq \emptyset,\ i \in I\}, \\ \mathscr{B} &= \{y_j \in Xe : B(y_j) \neq \emptyset,\ j \in J\}. \end{aligned} \tag{21}$$

Then $y_i e = y_i$ and $y_i e^{-1} \cap Xe = \{y_i\}$ for all $i \in I$. It is easy to see that

$$X \setminus Xe = \left(\bigcup_{i\in I} A(y_i)\right) \cup \left(\bigcup_{j\in J} B(y_j)\right) \text{ is adisjoint union.} \tag{22}$$

Let α be any element in F_e. Then $\alpha e = e\alpha$ and $\alpha|_{Xe} = \sigma$ for some $\sigma \in G(Xe)$. For each $x \in X \setminus Xe$, $x \in A(y_i) \cup B(y_j)$ for some $y_i \in \mathscr{A}$ and $y_j \in \mathscr{B}$. If $x \in A(y_i)$, then $xe = y_i \in Xe$. Since $\alpha e = e\alpha$, we obtain $(x\alpha)e = (xe)\alpha = y_i\alpha = y_i\sigma$; that is, $x\alpha \in (y_i\sigma)e^{-1} = \{y_i\sigma\} \cup A(y_i\sigma) \cup B(y_i\sigma)$. Since $x\alpha \in Y$ and $B(y_i\sigma) \subseteq X \setminus Y$, we obtain that $x\alpha \in \{y_i\sigma\} \cup A(y_i\sigma)$. If $x \in B(y_j) \subseteq X \setminus Y$, then $x\alpha \in (X \setminus Y)\alpha \subseteq Xe$ by Lemma 9. Since e is an idempotent, we get $x\alpha = (x\alpha)e = (xe)\alpha = y_j\alpha = y_j\sigma$. By condition (2) in Lemma 8, we see that α is injective on all elements of $A(y_i)$ which are not mapped into $\{y_i\sigma\}$ for all $i \in I$. Thus for each $i \in I$, $\alpha|_{A(y_i)}$ corresponds with an element θ_i of $\mathrm{PI}(A(y_i), A(y_i\sigma))$; that is, for $u \in A(y_i)$,

$$u\alpha = \begin{cases} u\theta_i & \text{if } u \in \mathrm{dom}(\theta_i), \\ y_i\sigma & \text{if } u \in A(y_i) \setminus \mathrm{dom}(\theta_i). \end{cases} \tag{23}$$

We have shown the following. Given $\alpha \in F_e$. Then α is a union of a permutation $\sigma \in G(Xe)$, a union of partial injections $\theta_i : A(y_i) \to A(y_i\sigma)$ for each $y_i \in \mathscr{A}$, and a union of functions from $B(y_j)$ into $\{y_j\sigma\}$ for each $y_j \in \mathscr{B}$.

Theorem 18. *Let e be any idempotent in $F(X, Y)$ with $Xe \subsetneq Y$. Let $G(Xe) = \{\sigma_k : k \in \Lambda\}$ and $\mathscr{A} = \{y_i \in Xe : A(y_i) \neq \emptyset,\ i \in I\}$. Then there is a one-to-one correspondence between F_e and $\bigcup_{k\in\Lambda}[\{\sigma_k\} \times \prod_{i\in I} PI(A(y_i), A(y_i\sigma_k))]$.*

Proof. For each $\alpha \in F_e$, we have $\alpha|_{Xe} = \sigma_k$ for some $k \in \Lambda$ and partial injections $\theta_i : A(y_i) \to A(y_i\sigma_k)$ for all $y_i \in \mathcal{A}$. We define

$$\Psi : F_e \longrightarrow \bigcup_{k\in\Lambda} \left[\{\sigma_k\} \times \prod_{i\in I} \mathrm{PI}\left(A\left(y_i\right), A\left(y_i\sigma_k\right)\right) \right], \quad (24)$$

by $\alpha\Psi = (\sigma_k, (\theta_i)_{i\in I})$ for all $\alpha \in F_e$. We verify that Ψ is injective. Let $\alpha, \beta \in F_e$ be such that $\alpha\Psi = \beta\Psi$. Then $(\sigma_k, (\theta_i)_{i\in I}) = (\sigma_l, (\tau_i)_{i\in I})$ for some $\sigma_k, \sigma_l \in G(Xe), (\theta_i)_{i\in I} \in \prod_{i\in I}\mathrm{PI}(A(y_i), A(y_i\sigma_k))$ and $(\tau_i)_{i\in I} \in \prod_{i\in I}\mathrm{PI}(A(y_i), A(y_i\sigma_l))$. Thus, $\sigma_k = \sigma_l$ and $\theta_i = \tau_i$ for all $i \in I$; it follows that $x\alpha = x\beta$ for all $x \in Xe \cup (\bigcup_{i\in I} A(y_i)) = Y$. For $x \in X \setminus Y$, $x \in B(y_j)$ for some $y_j \in \mathcal{B} = \{y_j \in Xe : B(y_j) \neq \emptyset,\ j \in J\}$. We get that $x\alpha = y_j\sigma_k = y_j\sigma_l = x\beta$. Whence $\alpha = \beta$, next, we prove Ψ is surjective. Let $\sigma_k \in G(Xe)$ and $(\theta_i)_{i\in I} \in \prod_{i\in I}\mathrm{PI}(A(y_i), A(y_i\sigma_k))$. We define $\alpha \in T(X,Y)$ which is determined by σ_k and θ_i for all $i \in I$ as follows. For $x \in X$,

$$x\alpha = \begin{cases} x\sigma_k & \text{if } x \in Xe, \\ x\theta_i & \text{if } x \in \mathrm{dom}\left(\theta_i\right) \neq \emptyset, \\ y_i\sigma_k & \text{if } x \in A\left(y_i\right),\ \left(x \notin \mathrm{dom}\left(\theta_i\right) \text{ or } \theta_i \text{ is the empty map}\right), \\ y_j\sigma_k & \text{if } x \in B\left(y_j\right). \end{cases} \quad (25)$$

We see that, for each $x \in X \setminus Y$, $x \in B(y_j)$ for some $y_j \in \mathcal{B}$. Then $y_j \in Xe \subseteq Y$ and $x\alpha = y_j\sigma_k = y_j\alpha$. So, $X\alpha \subseteq Y\alpha$, that is $\alpha \in F(X,Y)$. By the construction of α, conditions (1) and (2) in Lemma 8 are satisfied for α to be an element of F_e. To prove $\alpha e = e\alpha$, we let $x \in X$. If $x \in Xe$, then $(x\alpha)e = (x\sigma_k)e = x\sigma_k = x\alpha = (xe)\alpha$, since $x, x\sigma_k \in Xe$ and e is an idempotent. If $x \in \mathrm{dom}(\theta_i) \neq \emptyset$, then $x \in A(y_i)$; it follows that $xe = y_i$ and $x\theta_i \in A(y_i)\theta_i \subseteq A(y_i\sigma_k)$. Thus, $(x\theta_i)e = y_i\sigma_k$ and $(x\alpha)e = (x\theta_i)e = y_i\sigma_k = y_i\alpha = (xe)\alpha$ since $y_i \in Xe$. If $x \in A(y_i)$ and ($x \notin \mathrm{dom}(\theta_i)$ or θ_i is the empty map), then $xe = y_i$ and $(x\alpha)e = (y_i\sigma_k)e = y_i\sigma_k = y_i\alpha = (xe)\alpha$. And if $x \in B(y_j)$, then $xe = y_j$ and $(x\alpha)e = (y_j\sigma_k)e = y_j\sigma_k = y_j\alpha = (xe)\alpha$, since $y_j \in Xe$. Hence, $\alpha e = e\alpha$. Whence $\alpha \in F_e$ and $\alpha\Psi = (\sigma_k, (\theta_i)_{i\in I})$, therefore, Ψ is a bijection. □

The following corollary is a straightforward consequence of Theorem 18.

Corollary 19. *Let e be an idempotent in $F(X,Y)$ such that $Xe \subsetneq Y$ and $|Xe| = p \in \mathbb{Z}^+$. Let $\mathcal{A} = \{y_i \in Xe : A(y_i) \neq \emptyset,\ 1 \le i \le t \le p\}$. For each $y_i \in \mathcal{A}$, we let $|A(y_i)| = m_i \ge 1$ and $|A(y_i\sigma_k)| = n_{i,k} \ge 0$ $(1 \le i \le t, 1 \le k \le p!)$. Then*

$$\left|F_e\right| = \sum_{k=1}^{p!} \left(\prod_{i=1}^{t} \left|PI\left(m_i, n_{i,k}\right)\right| \right). \quad (26)$$

It is known that if $X = Y$, then $F(X,Y) = T(X)$ and $F_e = I_e$. Then by Corollary 19, we have the following corollary.

Corollary 20. *Let e be an idempotent in $T(X)$ such that $Xe \subsetneq X$ and $|Xe| = p \in \mathbb{Z}^+$. Let $\mathcal{A} = \{y_i \in Xe : A(y_i) \neq \emptyset,\ 1 \le i \le t \le p\}$. For each $y_i \in \mathcal{A}$, we let $|A(y_i)| = m_i \ge 1$ and $|A(y_i\sigma_k)| = n_{i,k} \ge 0$ $(1 \le i \le t, 1 \le k \le p!)$. Then*

$$\left|I_e\right| = \sum_{k=1}^{p!} \left(\prod_{i=1}^{t} \left|PI\left(m_i, n_{i,k}\right)\right| \right). \quad (27)$$

If e is a constant map with range $\{a\} \subseteq Y$, then we get the following corollary which appeared in [14, Theorem 4.5].

Corollary 21. *Let Y be a finite subset of X with $|Y| = r \ge 2$ and e a constant map in $F(X,Y)$ with range $\{a\} \subseteq Y$; then $|F_a| = \sum_{j=0}^{r-1} j! \binom{r-1}{j}^2$.*

Proof. We note that $F_e = F_a$ by Corollary 17. Since $Xe = \{a\}$, we obtain $|Xe| = 1$ and there is a unique $A(a) = Y \setminus \{a\}$; that is, $|A(a)| = r - 1$. By Corollary 19, we have

$$\begin{aligned} \left|F_a\right| = \left|F_e\right| &= \sum_{k=1}^{1!} \left(\prod_{i=1}^{1} \left|\mathrm{PI}\left(A\left(a\right), A\left(a\right)\right)\right| \right) \\ &= \left|\mathrm{PI}\left(A\left(a\right), A\left(a\right)\right)\right| = \left|\mathrm{PI}\left(r-1, r-1\right)\right|, \end{aligned} \quad (28)$$

where

$$\begin{aligned} \left|\mathrm{PI}\left(r-1, r-1\right)\right| &= \sum_{j=0}^{\min\{r-1,r-1\}} \binom{r-1}{j}\binom{r-1}{j} j! \\ &= \sum_{j=0}^{r-1} \binom{r-1}{j}^2 j!. \end{aligned} \quad (29)$$

Therefore, $|F_a| = \sum_{j=0}^{r-1} j! \binom{r-1}{j}^2$. □

4. Maximal Regular of Ideals

Throughout this section, X is a finite set with n elements and Y a nonempty subset of X with $r \ge 2$ elements. For convenience, we write $F_{n,r}$ for $F(X,Y)$.

For each $\alpha \in F_{n,r}$ with $|X\alpha| = k \in \mathbb{Z}^+$, it follows from Lemma 3 that we can write

$$\alpha = \begin{pmatrix} A_1 & A_2 & \cdots & A_k \\ a_1 & a_2 & \cdots & a_k \end{pmatrix}, \quad (30)$$

where $a_1, a_2, \ldots, a_k \in Y$ and $\{A_1, A_2, \ldots, A_k\}$ is a partition of X such that $A_i \cap Y \neq \emptyset$ for all $1 \le i \le k$.

For $1 \leq k \leq r$, we define $J(F;k) = \{\alpha \in F_{n,r} : |X\alpha| = k\}$. Then $J(F;k)$ is a $\mathcal{J}$-class of the semigroup $F_{n,r}$. Let $Q(F;k) = J(F;1) \cup J(F;2) \cup \cdots \cup J(F;k) = \{\alpha \in F_{n,r} : |X\alpha| \leq k\}$, where $1 \leq k \leq r$. Then $Q(F;k)$ is an ideal of $F_{n,r}$ and it is a regular subsemigroup of $F_{n,r}$; see [17, Lemma 7].

Recall that the principal factor of $F_{n,r}$ is the Rees quotient

$$P_k = Q(F;k) / Q(F;k-1), \quad (31)$$

where $2 \leq k \leq r$. It is usually convenient to think of it as $J(F;k) \cup \{0\}$, and the product of two elements of P_k is taken to be zero if it falls in $Q(F;k-1)$. Since P_k is finite and it is not a zero semigroup, we obtain that P_k is a completely 0-simple semigroup; see [18, p. 233] for more details.

We need the following lemmas for proof of our main result.

Lemma 22 (see [1, p. 98]). *Let S be a completely 0-simple semigroup. If $a, b \in S$ and $ab \neq 0$, then $ab \in R_a \cap L_b$.*

Lemma 23 (see [1, Proposition 2.3.7]). *Let a, b be elements in $\mathcal{D}$-class D. Then $ab \in R_a \cap L_b$ if and only if $L_a \cap R_b$ contains an idempotent.*

Lemma 24. *Let a be any element in a semigroup S and T a regular subsemigroup of S.*

(1) *If $L_a \cap T \neq \emptyset$, then $L_a \cap T$ contains an idempotent.*

(2) *If $R_a \cap T \neq \emptyset$, then $R_a \cap T$ contains an idempotent.*

Proof. (1) Assume that $L_a \cap T \neq \emptyset$. Then there exists $b \in L_a \cap T$. So, $b\mathcal{L}a$ and $b = bb'b$ for some $b' \in T$. We get that $b'b$ is an idempotent such that $b'b \in T$ and $b'b\mathcal{L}b\mathcal{L}a$. Therefore, $b'b$ is an idempotent in $L_a \cap T$.

(2) If there is $b \in R_a \cap T$, then we can write $b = bb'b$ for some $b' \in T$ and show that bb' is an idempotent in $R_a \cap T$. □

Lemma 25 (see [18, Lemma 3.5]). *For $1 \leq k \leq r-1$, $Q(F;k) = \langle J(F;k) \rangle$.*

Lemma 26 (see [18, Lemma 4.2]). *If $\alpha \in J(F;k)$, then α can be written as a finite product of idempotents in $J(F;k)$.*

The following fact can be obtained from Lemmas 25 and 26 immediately.

Lemma 27. *For $1 \leq k \leq r-1$, $Q(F;k) = \langle E(J(F;k)) \rangle$.*

Lemma 28. *Let $1 \leq k \leq r-1$ and $\alpha \in J(F;k)$. If $\beta \in L_\alpha$, then $\beta = \gamma\alpha$ for some $\gamma \in J(F;k)$ such that $L_\gamma \neq L_\alpha$.*

Proof. Suppose that $\beta \in L_\alpha$. Thus $X\alpha = X\beta$; we write

$$\alpha = \begin{pmatrix} A_1 & A_2 & \cdots & A_k \\ a_1 & a_2 & \cdots & a_k \end{pmatrix}, \quad \beta = \begin{pmatrix} B_1 & B_2 & \cdots & B_k \\ a_1 & a_2 & \cdots & a_k \end{pmatrix}, \quad (32)$$

where $a_1, a_2, \ldots, a_k \in Y$ and $A_i \cap Y \neq \emptyset \neq B_i \cap Y$ for all $1 \leq i \leq k$. Since $X\alpha \neq Y$, there is $y \in Y \setminus X\alpha$ such that $y \in A_j$ for some $j \in \{1, 2, \ldots, k\}$. We choose $d_{a_i} \in A_i \cap Y$ for all $i \in \{1, 2, \ldots, j-1, j+1, \ldots, k\}$ and define

$$\gamma = \begin{pmatrix} B_1 & \cdots & B_{j-1} & B_j & B_{j+1} & \cdots & B_k \\ d_{a_1} & \cdots & d_{a_{j-1}} & y & d_{a_{j+1}} & \cdots & d_{a_k} \end{pmatrix}. \quad (33)$$

Then $\gamma \in J(F;k)$ such that $X\gamma \neq X\alpha$ and $\beta = \gamma\alpha$. □

Lemma 29. *Let $2 \leq k \leq r-1$ and $\alpha \in J(F;k)$. If $\beta \in R_\alpha$, then $\beta = \alpha\gamma$ for some $\gamma \in J(F;k)$ such that $R_\gamma \neq R_\alpha$.*

Proof. Assume that $\beta \in R_\alpha$. Thus $\pi_\alpha = \pi_\beta$; we write

$$\alpha = \begin{pmatrix} A_1 & A_2 & \cdots & A_k \\ a_1 & a_2 & \cdots & a_k \end{pmatrix}, \quad \beta = \begin{pmatrix} A_1 & A_2 & \cdots & A_k \\ b_1 & b_2 & \cdots & b_k \end{pmatrix}, \quad (34)$$

where $a_i, b_i \in Y$ and $A_i \cap Y \neq \emptyset$ for all $1 \leq i \leq k$. Define

$$\gamma = \begin{pmatrix} a_1 & a_2 & \cdots & a_{k-1} & \{a_k\} \cup (X \setminus X\alpha) \\ b_1 & b_2 & \cdots & b_{k-1} & b_k \end{pmatrix}. \quad (35)$$

It is clear that $\gamma \in J(F;k)$. If $\gamma \notin R_\alpha$, then $R_\gamma \neq R_\alpha$ and $\beta = \alpha\gamma$. If such $\gamma \in R_\alpha$, we define

$$\gamma' = \begin{pmatrix} \{a_1, y\} & a_2 & \cdots & a_{k-1} & \{a_k\} \cup ((X \setminus X\alpha) \setminus \{y\}) \\ b_1 & b_2 & \cdots & b_{k-1} & b_k \end{pmatrix}, \quad (36)$$

where $y \in Y \setminus X\alpha$. So, $\gamma' \in J(F;k)$ such that $R_{\gamma'} \neq R_\gamma = R_\alpha$ and $\beta = \alpha\gamma'$. □

For the case $k = 1$, the ideal $Q(F;1) = J(F;1)$ consisting of all constant maps X_a with range $\{a\} \subseteq Y$. It is easy to verify that $J(F;1) \setminus \{X_a\}$ is a maximal regular subsemigroup of $Q(F;1)$ for all $a \in Y$.

Lemma 30. *For $2 \leq k \leq r-1$, $M = Q(F;k-1) \cup (J(F;k) \setminus L_\alpha)$ is a maximal regular subsemigroup of $Q(F;k)$ for all $\alpha \in J(F;k)$.*

Proof. Let α be any element in $J(F;k)$. Let $\beta, \gamma \in J(F;k) \setminus L_\alpha$. If $\beta\gamma \in Q(F;k-1)$, then $\beta\gamma \in M$. If $\beta\gamma \notin Q(F;k-1)$, then $\beta\gamma \neq 0$ in $P_k = Q(F;k)/Q(F;k-1)$; that is, $\beta\gamma \in R_\beta \cap L_\gamma$ by Lemma 22. Thus, $\beta\gamma \in L_\gamma \neq L_\alpha$ and so $\beta\gamma \in M$. Hence, M is a subsemigroup of $Q(F;k)$. We prove M is regular. Let $\lambda \in M$. If $\lambda \in Q(F;k-1)$, then λ is a regular element in $Q(F;k-1)$. If $\lambda \in J(F;k) \setminus L_\alpha$, we write

$$\lambda = \begin{pmatrix} A_1 & A_2 & \cdots & A_k \\ a_1 & a_2 & \cdots & a_k \end{pmatrix}, \quad (37)$$

where $a_1, a_2, \ldots, a_k \in Y$ and $A_i \cap Y \neq \emptyset$ for all $1 \leq i \leq k$. Since $X\alpha \neq Y$, there exists $y \in Y \setminus X\alpha$ such that $y \in A_j$

for some $j \in \{1,2,\dots,k\}$. We choose $d_{a_i} \in A_i \cap Y$ for all $i \in \{1,2,\dots,j-1,j+1,\dots,k\}$ and define

$$\mu = \begin{pmatrix} a_1 & \cdots & a_{j-1} & a_j & a_{j+1} & \cdots & a_{k-1} & \{a_k\} \cup (X \setminus X\lambda) \\ d_{a_1} & \cdots & d_{a_{j-1}} & y & d_{a_{j+1}} & \cdots & d_{a_{k-1}} & d_{a_k} \end{pmatrix}. \quad (38)$$

Then $\mu \in J(F;k)$, $X\mu \neq X\alpha$, and $\lambda = \lambda\mu\lambda$. Hence, M is a regular semigroup. Next, we prove M is maximal. Let S be a regular subsemigroup of $Q(F;k)$ such that $M \subsetneq S \subseteq Q(F;k)$. Then there exists $\xi \in S \setminus M$; that is, $\xi \in L_\alpha \cap S$. Let τ be any element in $L_\alpha = L_\xi$. Then there is $\gamma \in J(F;k)$ such that $\gamma \notin L_\xi$ and $\tau = \gamma\xi$ by Lemma 28; that is, $\gamma \in M \subseteq S$. Thus, $\tau = \gamma\xi \in S$. Hence, $L_\alpha \subseteq S$ and therefore $S = Q(F;k)$. □

Lemma 31. *For $2 \le k \le r-1$, $M = Q(F;k-1) \cup (J(F;k) \setminus R_\alpha)$ is a maximal regular subsemigroup of $Q(F;k)$ for all $\alpha \in J(F;k)$.*

Proof. Let α be any element in $J(F;k)$. Let $\beta,\gamma \in J(F;k)\setminus R_\alpha$. If $\beta\gamma \in Q(F;k-1)$, then $\beta\gamma \in M$. If $\beta\gamma \notin Q(F;k-1)$, then $\beta\gamma \neq 0$ in P_k and so $\beta\gamma \in R_\beta \cap L_\gamma$ by Lemma 22. Thus, $\beta\gamma \in R_\beta \neq R_\alpha$. So, $\beta\gamma \in M$. Hence, M is a subsemigroup of $Q(F;k)$. We prove M is regular. Let $\lambda \in M$. If $\lambda \in Q(F;k-1)$, then λ is a regular element in $Q(F;k-1)$. If $\lambda \in J(F;k) \setminus R_\alpha$, we write

$$\lambda = \begin{pmatrix} A_1 & A_2 & \cdots & A_k \\ a_1 & a_2 & \cdots & a_k \end{pmatrix}, \quad (39)$$

where $a_1, a_2, \dots, a_k \in Y$ and $A_i \cap Y \neq \emptyset$ for all $1 \le i \le k$. We choose $d_{a_i} \in A_i \cap Y$ for all $1 \le i \le k$ and define

$$\lambda' = \begin{pmatrix} a_1 & a_2 & \cdots & a_{k-1} & \{a_k\} \cup (X \setminus X\lambda) \\ d_{a_1} & d_{a_2} & \cdots & d_{a_{k-1}} & d_{a_k} \end{pmatrix}. \quad (40)$$

Then $\lambda' \in J(F;k)$ and $\lambda = \lambda\lambda'\lambda$. If $\lambda' \notin R_\alpha$, then λ' is a regular element in M. If $\lambda' \in R_\alpha$, we define

$$\lambda'' = \begin{pmatrix} \{a_1, y\} & a_2 & \cdots & a_{k-1} & \{a_k\} \cup ((X \setminus X\alpha) \setminus \{y\}) \\ d_{a_1} & d_{a_2} & \cdots & d_{a_{k-1}} & d_{a_k} \end{pmatrix}, \quad (41)$$

where $y \in Y \setminus X\lambda$ since $k < r$. We obtain $\lambda'' \notin R_{\lambda'} = R_\alpha$ and $\lambda = \lambda\lambda''\lambda$. Hence, M is a regular semigroup. For a maximality of M, we let S be a regular subsemigroup of $Q(F;k)$ such that $M \subsetneq S \subseteq Q(F;k)$. Then there exists $\xi \in S \setminus M$; that is, $\xi \in R_\alpha \cap S$. Let τ be any element in $R_\alpha = R_\xi$. Then there is $\gamma \in J(F;k)$ such that $\gamma \notin R_\xi = R_\alpha$ and $\tau = \xi\gamma$ by Lemma 29; that is, $\gamma \in M \subseteq S$. Thus, $\tau \in S$ and so $R_\alpha \subseteq S$. Therefore, $S = Q(F;k)$. □

Theorem 32. *Each maximal regular subsemigroup of $Q(F;k)$, $2 \le k \le r-1$ must be one of the following forms:*

(1) $Q(F;k-1) \cup (J(F;k) \setminus L_\alpha)$;

(2) $Q(F;k-1) \cup (J(F;k) \setminus R_\beta)$,

where $\alpha, \beta \in J(F;k)$, L_α is the $\mathscr{L}$-class containing α in $J(F;k)$, and R_β is the $\mathscr{R}$-class containing β in $J(F;k)$.

Proof. By Lemmas 30 and 31, both (1) and (2) are maximal regular subsemigroups of $Q(F;k)$.

On the other hand, let M be an arbitrary maximal regular subsemigroup of $Q(F;k)$. It is easy to see that $M \cup Q(F;k-1)$ is a regular subsemigroup of $Q(F;k)$ such that $M \subseteq M \cup Q(F;k-1) \subseteq Q(F;k)$. Then $M = M \cup Q(F;k-1)$ or $M \cup Q(F;k-1) = Q(F;k)$ by the maximality of M. If $M \cup Q(F;k-1) = Q(F;k)$, then $J(F;k) = M \cap J(F;k) \subseteq M$. We obtain $M \subseteq Q(F;k) = \langle J(F;k) \rangle \subseteq M$ by Lemma 25; this implies that $M = Q(F;k)$, a contradiction. Whence $M = M \cup Q(F;k-1)$ and so $Q(F;k-1) \subseteq M$, assume that $L_\alpha \cap M \neq \emptyset$ and $R_\alpha \cap M \neq \emptyset$ for all $\alpha \in J(F;k)$. We first prove that $E(J(F;k)) \subseteq M$. Let e be an idempotent in $J(F;k)$. Thus, $L_e \cap M \neq \emptyset \neq R_e \cap M$. By Lemma 24, $L_e \cap M$ and $R_e \cap M$ contain idempotents, say f and g respectively. So, $f\mathscr{L}e$ and $g\mathscr{R}e$. Since $L_f \cap R_g = L_e \cap R_e$ contains the idempotent e, $fg \in R_f \cap L_g \subseteq J(F;k)$ by Lemma 23. Since $fg \in M$ and M is regular, $fg = (fg)h(fg)$ for some $h \in M$. We see that $f(ghf)g = fg \in J(F;k)$, which implies that $ghf \in J(F;k)$; that is, $(gh)f \neq 0$ in P_k. Then by Lemma 22, we get $(gh)f \in R_{gh} \cap L_f$. Also, we have $g(hf) \in R_g \cap L_{hf}$. Whence $ghf \in L_f \cap R_g = L_e \cap R_e = H_e$, the group $\mathscr{H}$-class contains e. So, there exists a positive integer t such that $e = (ghf)^t \in M$ since $g,h,f \in M$. Therefore, $E(J(F;k)) \subseteq M$. It follows from Lemma 27 that $Q(F;k) = \langle E(J(F;k)) \rangle \subseteq M \subseteq Q(F;k)$; hence $M = Q(F;k)$, a contradiction. Therefore, $L_\alpha \cap M = \emptyset$ for some $\alpha \in J(F;k)$ or $R_\beta \cap M = \emptyset$ for some $\beta \in J(F;k)$. If $M \cap L_\alpha = \emptyset$ for some $\alpha \in J(F;k)$, then

$$M \subseteq Q(F;k-1) \cup (J(F;k) \setminus L_\alpha) \subsetneq Q(F;k). \quad (42)$$

Thus, $M = Q(F;k-1) \cup (J(F;k) \setminus L_\alpha)$ by the maximality of M. If $M \cap R_\beta = \emptyset$ for some $\beta \in J(F;k)$, then

$$M \subseteq Q(F;k-1) \cup (J(F;k) \setminus R_\beta) \subsetneq Q(F;k). \quad (43)$$

Thus, $M = Q(F;k-1) \cup (J(F;k) \setminus R_\beta)$ by the maximality of M. □

Notice that the number of $\mathscr{L}$-classes in $J(F;k)$ equals $\binom{r}{k}$ and the number of $\mathscr{R}$-classes in $J(F;k)$ equals $S(r,k)k^{n-r}$ where $S(r,k)$ is the Stirling number of the second kind; see [18, Section 2] for details. Therefore, the number of maximal regular subsemigroups of $Q(F;k)$ is equal to $\binom{r}{k} + S(r,k)k^{n-r}$ when $2 \le k \le r-1$.

We shall normally write T_n instead of $T(X)$. For $1 \le k \le n$, define

$$J_k = \{\alpha \in T_n : |X\alpha| = k\}, \quad K(n,k) = \{\alpha \in T_n : |X\alpha| \le k\}. \quad (44)$$

It is well-known that $K(n,k) = J_1 \cup J_2 \cup \cdots \cup J_k$ and $K(n,k)$ is an ideal of T_n for all $1 \le k \le n$. If $n = r$, then $F_{n,r} = F_{n,n} = T_n$, $J(F;k) = J_k$, and $Q(F;k) = K(n,k)$. Therefore, we establish the following corollary which first appeared in [8, Theorem 2].

Corollary 33. *Each maximal regular subsemigroup of $K(n,k)$, $2 \le k \le n-1$ must be one of the following forms:*

(1) $K(n,k-1) \cup (J_k \setminus L_\alpha)$;

(2) $K(n,k-1) \cup (J_k \setminus R_\beta)$,

where $\alpha, \beta \in J_k$, L_α *is the* $\mathscr{L}$*-class containing* α *in* J_k *of* T_n, *and* R_β *is the* $\mathscr{R}$*-class containing* β *in* J_k *of* T_n.

We note that if $n = r$, then we immediately obtain that the number of maximal regular subsemigroups of $K(n,k)$ is equal to $\binom{n}{k} + S(n,k)$ when $2 \leq k \leq n-1$.

Conflicts of Interest

The author declares that they have no conflicts of interest.

Acknowledgments

Financial support from the Coordinating Center for Thai Government Science and Technology Scholarship Students (CSTS), National Science and Technology Development Agency (NSTDA), is acknowledged.

References

[1] J. M. Howie, *Fundamentals of Semigroup Theory*, Oxford University Press, New York, NY, USA, 1995.

[2] B. M. Schein, "A symmetric semigroup of transformations is covered by its inverse subsemigroups," *Acta Mathematica Hungarica*, vol. 22, pp. 163–171, 1971/72.

[3] B. M. Schein, "Research problems," *Semigroup Forum*, vol. 1, pp. 91-92, 1970.

[4] J. W. Nichols, "A class of maximal inverse subsemigroups of T_X," *Semigroup Forum*, vol. 13, no. 2, pp. 187-188, 1976.

[5] N. R. Reilly, "Maximal inverse subsemigroups of T_X," *Semigroup Forum*, vol. 15, no. 4, pp. 319–326, 1978.

[6] X. Yang, "A classification of maximal inverse subsemigroups of the finite symmetric inverse semigroups," *Communications in Algebra*, vol. 27, no. 8, pp. 4089–4096, 1999.

[7] X. Yang, "Maximal subsemigroups of the finite singular transformation semigroup," *Communications in Algebra*, vol. 29, no. 3, pp. 1175–1182, 2001.

[8] T. You, "Maximal regular subsemigroups of certain semigroups of transformations," *Semigroup Forum*, vol. 64, no. 3, pp. 391–396, 2002.

[9] H. B. Yang and X. L. Yang, "Maximal subsemigroups of finite transformation semigroups *K(n, r)*," *Acta Mathematica Sinica*, vol. 20, no. 3, pp. 475–482, 2004.

[10] P. Zhao, H. Hu, and T. You, "A note on maximal regular subsemigroups of the finite transformation semigroups $T_{(n,r)}$," *Semigroup Forum*, vol. 88, no. 2, pp. 324–332, 2014.

[11] J. East, J. D. Mitchell, and Y. Péresse, "Maximal subsemigroups of the semigroup of all mappings on an infinite set," *Transactions of the American Mathematical Society*, vol. 367, no. 3, pp. 1911–1944, 2015.

[12] J. S. Symons, "Some results concerning a transformation semigroup," *Journal of the Australian Mathematical Society: Pure Mathematics and Statistics. Series A*, vol. 19, no. 4, pp. 413–425, 1975.

[13] S. Nenthein, P. Youngkhong, and Y. Kemprasit, "Regular elements of some transformation semigroups," *Pure Mathematics and Applications*, vol. 16, no. 3, pp. 307–314 (2006), 2005.

[14] J. Sanwong and W. Sommanee, "Regularity and Green's relations on a semigroup of transformations with restricted range," *International Journal of Mathematics and Mathematical Sciences*, vol. 2008, Article ID 794013, 11 pages, 2008.

[15] J. Sanwong, B. Singha, and R. P. Sullivan, "Maximal and minimal congruences on some semigroups," *Acta Mathematica Sinica*, vol. 25, no. 3, pp. 455–466, 2009.

[16] S. Mendes-Gonçalves and R. P. Sullivan, "The ideal structure of semigroups of transformations with restricted range," *Bulletin of the Australian Mathematical Society*, vol. 83, no. 2, pp. 289–300, 2011.

[17] J. Sanwong, "The regular part of a semigroup of transformations with restricted range," *Semigroup Forum*, vol. 83, no. 1, pp. 134–146, 2011.

[18] W. Sommanee and J. Sanwong, "Rank and idempotent rank of finite full transformation semigroups with restricted range," *Semigroup Forum*, vol. 87, no. 1, pp. 230–242, 2013.

[19] V. H. Fernandes and J. Sanwong, "On the ranks of semigroups of transformations on a finite set with restricted range," *Algebra Colloquium*, vol. 21, no. 3, pp. 497–510, 2014.

[20] L. Sun and J. Sun, "A natural partial order on certain semigroups of transformations with restricted range," *Semigroup Forum*, vol. 92, no. 1, pp. 135–141, 2016.

[21] K. Tinpun and J. Koppitz, "Relative rank of the finite full transformation semigroup with restricted range," *Acta Mathematica Universitatis Comenianae*, vol. 85, no. 2, pp. 347–356, 2016.

[22] I. Dimitrova and J. Koppitz, "On the maximal regular subsemigroups of ideals of order-preserving or order-reversing transformations," *Semigroup Forum*, vol. 82, no. 1, pp. 172–180, 2011.

[23] W. Sommanee and J. Sanwong, "Order-preserving transformations with restricted range: regularity, Green's relations, and ideals," *Algebra Universalis*, vol. 74, no. 3-4, pp. 277–291, 2015.

[24] W. Sommanee and K. Sangkhanan, "The regular part of a semigroup of linear transformations with restricted range," *Journal of the Australian Mathematical Society*, vol. 103, no. 3, pp. 402–419, 2017.

14

Explicit Formulas for Meixner Polynomials

Dmitry V. Kruchinin[1,2] **and Yuriy V. Shablya**[1]

[1] *Tomsk State University of Control Systems and Radioelectronics, 40 Lenina Avenue, Tomsk 634050, Russia*
[2] *National Research Tomsk Polytechnic University, 30 Lenin Avenue, Tomsk 634050, Russia*

Correspondence should be addressed to Dmitry V. Kruchinin; kruchinindm@gmail.com

Academic Editor: Hari M. Srivastava

Using notions of composita and composition of generating functions, we show an easy way to obtain explicit formulas for some current polynomials. Particularly, we consider the Meixner polynomials of the first and second kinds.

1. Introduction

There are many authors who have studied polynomials and their properties (see [1–10]). The polynomials are applied in many areas of mathematics, for instance, continued fractions, operator theory, analytic functions, interpolation, approximation theory, numerical analysis, electrostatics, statistical quantum mechanics, special functions, number theory, combinatorics, stochastic processes, sorting, and data compression.

The research area of obtaining explicit formulas for polynomials has received much attention from Srivastava [11, 12], Cenkci [13], Boyadzhiev [14], and Kruchinin [15–17].

The main purpose of this paper is to obtain explicit formulas for the Meixner polynomials of the first and second kinds.

In this paper we use a method based on a notion of composita, which was presented in [18].

Definition 1. Suppose $F(t) = \sum_{n>0} f(n)t^n$ is the generating function, in which there is no free term $f(0) = 0$. From this generating function we can write the following condition:

$$[F(t)]^k = \sum_{n>0} F(n,k)\, t^n. \quad (1)$$

The expression $F(n,k)$ is *composita* [19] and it is denoted by $F^{\Delta}(n,k)$. Below we show some required rules and operations with compositae.

Theorem 2. *Suppose $F(t) = \sum_{n>0} f(n)t^n$ is the generating function, and $F^{\Delta}(n,k)$ is composita of $F(t)$, and α is constant. For the generating function $A(t) = \alpha F(t)$ composita is equal to*

$$A^{\Delta}(n,k) = \alpha^k F^{\Delta}(n,k). \quad (2)$$

Theorem 3. *Suppose $F(t) = \sum_{n>0} f(n)t^n$ is the generating function, and $F^{\Delta}(n,k)$ is composita of $F(t)$, and α is constant. For the generating function $A(t) = F(\alpha t)$ composita is equal to*

$$A^{\Delta}(n,k) = \alpha^n F^{\Delta}(n,k). \quad (3)$$

Theorem 4. *Suppose $F(t) = \sum_{n>0} f(n)t^n$, $R(t) = \sum_{n>0} r(n)t^n$ are generating functions, and $F^{\Delta}(n,k)$, $R^{\Delta}(n,k)$ are their compositae. Then for the composition of generating functions $A(t) = R(F(t))$ composita is equal to*

$$A^{\Delta}(n,k) = \sum_{m=k}^{n} F^{\Delta}(n,m)\, G^{\Delta}(m,k). \quad (4)$$

Theorem 5. *Suppose $F(t) = \sum_{n>0} f(n)t^n$, $R(t) = \sum_{n\geqslant 0} r(n)t^n$ are generating functions, and $F^{\Delta}(n,k)$ is composita of $F(t)$. Then for the composition of generating functions $A(t) = R(F(t))$ coefficients of generating functions $A(t) = \sum_{n\geq 0} a(n)t^n$ are*

$$a(n) = \sum_{k=1}^{n} F^{\Delta}(n,k)\, r(k),$$
$$a(0) = r(0). \quad (5)$$

Theorem 6. *Suppose* $F(t) = \sum_{n>0} f(n)t^n$, $R(t) = \sum_{n\geq 0} r(n)t^n$ *are generating functions. Then for the product of generating functions* $A(t) = F(t)R(t)$ *coefficients of generating functions* $A(t) = \sum_{n\geq 0} a(n)t^n$ *are*

$$a(n) = \sum_{k=0}^{n} f(k)\, r(n-k). \quad (6)$$

In this paper we consider an application of this mathematical tool for the Bessel polynomials and the Meixner polynomials of the first and second kinds.

2. Bessel Polynomials

Krall and Frink [20] considered a new class of polynomials. Since the polynomials connected with the Bessel function, they called them the Bessel polynomials. The explicit formula for the Bessel polynomials is

$$y_n(x) = \sum_{k=0}^{n} \frac{(n+k)!}{(n-k)!k!} \left(\frac{x}{2}\right)^k. \quad (7)$$

Then Carlitz [21] defined a class of polynomials associated with the Bessel polynomials by

$$p_n(x) = x^n y_{n-1}\left(\frac{1}{x}\right). \quad (8)$$

The $p_n(x)$ is defined by the explicit formula [22]

$$p_n(x) = \sum_{k=1}^{n} \frac{(2n-k-1)!}{(k-1)!\,(n-k)!} \left(\frac{1}{2}\right)^{n-k} x^k \quad (9)$$

and by the following generating function:

$$\sum_{n=0}^{\infty} \frac{p_n(x)}{n!} t^n = e^{x(1-\sqrt{1-2t})}. \quad (10)$$

Using the notion of composita, we can obtain an explicit formula (9) from the generating function (10).

For the generating function $C(t) = (1 - \sqrt{1-4t})/2$ composita is given as follows (see [19]):

$$C^{\Delta}(n,k) = \frac{k}{n}\binom{2n-k-1}{n-1}. \quad (11)$$

We represent $e^{x(1-\sqrt{1-2t})}$ as the composition of generating functions $A(B(t))$, where

$$A(t) = x\left(1 - \sqrt{1-2t}\right) = 2xC\left(\frac{t}{2}\right),$$
$$B(t) = e^t = \sum_{n=0}^{\infty} \frac{1}{n!} t^n. \quad (12)$$

Then, using rules (2) and (3), we obtain composita of $A(t)$:

$$A^{\Delta}(n,k) = (2x)^k \left(\frac{1}{2}\right)^n C^{\Delta}(n,k) = 2^{k-n} x^k \frac{k}{n}\binom{2n-k-1}{n-1}. \quad (13)$$

Coefficients of the generating function $B(t) = \sum_{n=0}^{\infty} b(n)t^n$ are equal to

$$b(n) = \frac{1}{n!}. \quad (14)$$

Then, using (5), we obtain the explicit formula

$$p_n(x) = n! \sum_{k=1}^{n} A^{\Delta}(n,k)\, b(k) = \sum_{k=1}^{n} \frac{(2n-k-1)!}{(k-1)!\,(n-k)!} 2^{k-n} x^k, \quad (15)$$

which coincides with the explicit formula (9).

3. Meixner Polynomials of the First Kind

The Meixner polynomials of the first kind are defined by the following recurrence relation [23, 24]:

$$m_{n+1}(x;\beta,c) = \frac{(c-1)^n}{c^n} \cdot \left(\left(x - b_n\right) m_n(x;\beta,c) - \lambda_n m_{n-1}(x;\beta,c)\right), \quad (16)$$

where

$$m_0(x;\beta,c) = 1,$$
$$m_1(x;\beta,c) = x - b_0,$$
$$b_n = \frac{(1+c)\,n + \beta c}{1-c}, \quad (17)$$
$$\lambda_n = \frac{cn\,(n+\beta-1)}{(1-c)^2}.$$

Using (16), we obtain the first few Meixner polynomials of the first kind:

$$m_0(x;\beta,c) = 1;$$
$$m_1(x;\beta,c) = \frac{x(c-1) + \beta c}{c};$$
$$m_2(x;\beta,c) = \frac{x^2(c-1)^2 + x\left((2\beta+1)c^2 - 2\beta c - 1\right) + (\beta+1)\beta c^2}{c^2}. \quad (18)$$

The Meixner polynomials of the first kind are defined by the following generating function [22]:

$$\sum_{n=0}^{\infty} \frac{m_n(x;\beta,c)}{n!} t^n = \left(1 - \frac{t}{c}\right)^x (1-t)^{-x-\beta}. \quad (19)$$

Using the notion of composita, we can obtain an explicit formula $m_n(x;\beta,c)$ from the generating function (19).

First, we represent the generating function (19) as a product of generating functions $A(t)\ B(t)$, where the functions $A(t)$ and $B(t)$ are expanded by binomial theorem:

$$A(t) = \left(1 - \frac{t}{c}\right)^x = \sum_{n=0}^{\infty} \binom{x}{n} (-1)^n c^{-n} t^n,$$
$$B(t) = (1-t)^{-x-\beta} = \sum_{n=0}^{\infty} \binom{-x-\beta}{n} (-1)^n t^n. \tag{20}$$

Coefficients of the generating functions $A(t) = \sum_{n=0}^{\infty} a(n)t^n$ and $B(t) = \sum_{n=0}^{\infty} b(n)t^n$ are, respectively, given as follows:

$$a(n) = \binom{x}{n} (-1)^n c^{-n},$$
$$b(n) = \binom{-x-\beta}{n} (-1)^n. \tag{21}$$

Then, using (6), we obtain a new explicit formula for the Meixner polynomials of the first kind:

$$m_n(x;\beta,c) = n! \sum_{k=0}^{n} a(k)\, b(n-k)$$
$$= (-1)^n n! \sum_{k=0}^{n} \binom{x}{k} \binom{-x-\beta}{n-k} c^{-k}. \tag{22}$$

4. Meixner Polynomials of the Second Kind

The Meixner polynomials of the second kind are defined by the following recurrence relation [23, 24]:

$$M_{n+1}(x;\delta,\eta) = (x - b_n) M_n(x;\delta,\eta) - \lambda_n M_{n-1}(x;\delta,\eta), \tag{23}$$

where

$$M_0(x;\delta,\eta) = 1,$$
$$M_1(x;\delta,\eta) = x - b_0,$$
$$b_n = (2n+\eta)\delta, \tag{24}$$
$$\lambda_n = (\delta^2+1)\, n\, (n+\eta-1).$$

Using (23), we get the first few Meixner polynomials of the second kind:

$$M_0(x;\delta,\eta) = 1;$$
$$M_1(x;\delta,\eta) = x - \delta\eta; \tag{25}$$
$$M_2(x;\delta,\eta) = x^2 - x2\delta(1+\eta) + \eta\left((\eta+1)\delta^2 - 1\right).$$

The Meixner polynomials of the second kind are defined by the following generating function [22]:

$$\sum_{n=0}^{\infty} \frac{M_n(x;\delta,\eta)}{n!} t^n = \left((1+\delta t)^2 + t^2\right)^{-\eta/2} \cdot \exp\left(x \tan^{-1}\left(\frac{t}{1+\delta t}\right)\right). \tag{26}$$

Using the notion of composita, we can obtain an explicit formula $M_n(x;\delta,\eta)$ from the generating function (26).

We represent the generating function (26) as a product of generating functions $A(t)\ B(t)$, where

$$A(t) = \left((1+\delta t)^2 + t^2\right)^{-\eta/2}$$
$$= \left(1 + 2\delta t + (\delta^2+1) t^2\right)^{-\eta/2}, \tag{27}$$
$$B(t) = \exp\left(x \tan^{-1}\left(\frac{t}{1+\delta t}\right)\right).$$

Next we represent $A(t)$ as a composition of generating functions $A_1(A_2(t))$ and we expand $A_1(t)$ by binomial theorem:

$$A_1(t) = (1+t)^{-\eta/2} = \sum_{n=0}^{\infty} \binom{-\frac{\eta}{2}}{n} t^n,$$
$$A_2(t) = 2\delta t + (\delta^2+1) t^2. \tag{28}$$

Coefficients of generating function $A_1(t) = \sum_{n=0}^{\infty} a_1(n)t^n$ are

$$a_1(n) = \binom{-\frac{\eta}{2}}{n}. \tag{29}$$

The composita for the generating function $ax + bx^2$ is given as follows (see [15]):

$$a^{2k-n} b^{n-k} \binom{k}{n-k}. \tag{30}$$

Then composita of the generating function $A_2(t)$ equals

$$A_2^{\Delta}(n,k) = (2\delta)^{2k-n} (\delta^2+1)^{n-k} \binom{k}{n-k}. \tag{31}$$

Using (5), we obtain coefficients of the generating function $A(t) = \sum_{n=0}^{\infty} a(n)t^n$:

$$a(0) = a_1(0),$$
$$a(n) = \sum_{k=1}^{n} A_2^{\Delta}(n,k)\, a_1(k). \tag{32}$$

Therefore, we get the following expression:

$$a(0) = 1,$$
$$a(n) = \sum_{k=1}^{n} (2\delta)^{2k-n} (\delta^2+1)^{n-k} \binom{k}{n-k} \binom{-\frac{\eta}{2}}{k}. \tag{33}$$

Next we represent $B(t)$ as a composition of generating functions $B_1(B_2(t))$, where

$$B_1(t) = e^t = \sum_{n=0}^{\infty} \frac{1}{n!} t^n,$$
$$B_2(t) = x \tan^{-1}\left(\frac{t}{1+\delta t}\right). \tag{34}$$

We also represent $B_2(t)$ as a composition of generating functions $xC_1(C_2(t))$, where

$$C_1(t) = \tan^{-1}(t),$$
$$C_2(t) = \frac{t}{1+\delta t}. \quad (35)$$

The composita for the generating function $\tan^{-1}(t)$ is given as follows (see [19]):

$$\left((-1)^{(3n+k)/2} + (-1)^{(n-k)/2}\right) \frac{k!}{2^{k+1}} \sum_{j=k}^{n} \frac{2^j}{j!} \binom{n-1}{j-1} \left[{j \atop k} \right], \quad (36)$$

where $\left[{j \atop k} \right]$ is the Stirling number of the first kind.

Then composita of generating function $C_1(t)$ equals

$$C_1^{\Delta}(n,k) = \left((-1)^{(3n+k)/2} + (-1)^{(n-k)/2}\right) \frac{k!}{2^{k+1}} \cdot \sum_{j=k}^{n} \frac{2^j}{j!} \binom{n-1}{j-1} \left[{j \atop k} \right]. \quad (37)$$

The composita for the generating function $at/(1-bt)$ is given as follows (see [19]):

$$\binom{n-1}{k-1} a^k b^{n-k}. \quad (38)$$

Therefore, composita of generating function $C_2(t)$ is equal to

$$C_2^{\Delta}(n,k) = \binom{n-1}{k-1} (-\delta)^{n-k}. \quad (39)$$

Using rules (4) and (2), we obtain composita of generating function $B_2(t)$:

$$B_2^{\Delta}(n,k) = x^k \sum_{m=k}^{n} C_2^{\Delta}(n,m)\, C_1^{\Delta}(m,k). \quad (40)$$

Coefficients of generating function $B_1(t) = \sum_{n=0}^{\infty} b_1(n) t^n$ are defined by

$$b_1(n) = \frac{1}{n!}. \quad (41)$$

Then, using rule (5), we obtain coefficients of generating function $B(t) = \sum_{n=0}^{\infty} b(n) t^n$:

$$b(0) = b_1(0),$$
$$b(n) = \sum_{k=1}^{n} B_2^{\Delta}(n,k)\, b_1(k). \quad (42)$$

After some transformations, we obtain the following expression:

$$b(0) = 1,$$
$$b(n) = \sum_{k=1}^{n} \frac{x^k}{n} \sum_{m=k}^{n} \binom{n}{m} \delta^{n-m} \left(\frac{1+(-1)^{m+k}}{(-1)^{(m+k)/2-n}} \right) \cdot \sum_{j=k}^{m} \frac{2^{j-k-1}}{(j-1)!} \binom{m}{j} \left[{j \atop k} \right]. \quad (43)$$

Therefore, using (6), we obtain a new explicit formula for the Meixner polynomials of the second kind:

$$M_n(x;\delta,\eta) = n! \sum_{k=0}^{n} a(k)\, b(n-k). \quad (44)$$

Conflict of Interests

The authors declare that there is no conflict of interests regarding the publication of this paper.

Acknowledgment

This work was partially supported by the Ministry of Education and Science of Russia, Government Order no. 3657 (TUSUR).

References

[1] R. P. Boas Jr. and R. C. Buck, *Polynomial Expansions of Analytic Functions*, Springer, 1964.

[2] H. M. Srivastava and H. L. Manocha, *A Treatise on Generating Functions*, Halsted Press, Ellis Horwood Limited, Chichester, UK; John Wiley & Sons, New York, NY, USA, 1984.

[3] H. M. Srivastava and J. Choi, *Zeta and Q-Zeta Functions and Associated Series and Integrals*, Elsevier Science Publishers, Amsterdam, The Netherlands, 2012.

[4] H. M. Srivastava, "Some generalizations and basic (or q-) extensions of the Bernoulli, Euler and Genocchi polynomials," *Applied Mathematics and Information Sciences*, vol. 5, pp. 390–444, 2011.

[5] Y. Simsek and M. Acikgoz, "A new generating function of (q-) Bernstein-type polynomials and their interpolation function," *Abstract and Applied Analysis*, vol. 2010, Article ID 769095, 12 pages, 2010.

[6] H. Ozden, Y. Simsek, and H. M. Srivastava, "A unified presentation of the generating functions of the generalized Bernoulli, Euler and Genocchi polynomials," *Computers & Mathematics with Applications.*, vol. 60, no. 10, pp. 2779–2787, 2010.

[7] R. Dere and Y. Simsek, "Applications of umbral algebra to some special polynomials," *Advanced Studies in Contemporary Mathematics*, vol. 22, no. 3, pp. 433–438, 2012.

[8] Y. Simsek, "Complete sum of products of (h,q)-extension of Euler polynomials and numbers," *Journal of Difference Equations and Applications*, vol. 16, no. 11, pp. 1331–1348, 2010.

[9] Y. He and S. Araci, "Sums of products of APOstol-Bernoulli and APOstol-Euler polynomials," *Advances in Difference Equations*, vol. 2014, article 155, 2014.

[10] S. Araci, "Novel identities involving Genocchi numbers and polynomials arising from applications of umbral calculus," *Applied Mathematics and Computation*, vol. 233, pp. 599–607, 2014.

[11] H. M. Srivastava and G.-D. Liu, "Explicit formulas for the Norlund polynomials $B_n^{(x)}$ and $b_n^{(x)}$," *Computers & Mathematics with Applications*, vol. 51, pp. 1377–1384, 2006.

[12] H. M. Srivastava and P. G. Todorov, "An explicit formula for the generalized Bernoulli polynomials," *Journal of Mathematical Analysis and Applications*, vol. 130, no. 2, pp. 509–513, 1988.

[13] M. Cenkci, "An explicit formula for generalized potential polynomials and its applications," *Discrete Mathematics*, vol. 309, no. 6, pp. 1498–1510, 2009.

[14] K. N. Boyadzhiev, "Derivative polynomials for tanh, tan, sech and sec in explicit form," *The Fibonacci Quarterly*, vol. 45, no. 4, pp. 291–303 (2008), 2007.

[15] D. V. Kruchinin and V. V. Kruchinin, "Application of a composition of generating functions for obtaining explicit formulas of polynomials," *Journal of Mathematical Analysis and Applications*, vol. 404, no. 1, pp. 161–171, 2013.

[16] D. V. Kruchinin and V. V. Kruchinin, "Explicit formulas for some generalized polynomials," *Applied Mathematics & Information Sciences*, vol. 7, no. 5, pp. 2083–2088, 2013.

[17] D. V. Kruchinin, "Explicit formula for generalized Mott polynomials," *Advanced Studies in Contemporary Mathematics*, vol. 24, no. 3, pp. 327–322, 2014.

[18] D. V. Kruchinin and V. V. Kruchinin, "A method for obtaining expressions for polynomials based on a composition of generating functions," in *Proceedings of the AIP Conference on Numerical Analysis and Applied Mathematics (ICNAAM '12)*, vol. 1479, pp. 383–386, 2012.

[19] V. V. Kruchinin and D. V. Kruchinin, "Composita and its properties," *Journal of Analysis and Number Theory*, vol. 2, no. 2, pp. 37–44, 2014.

[20] H. L. Krall and O. Frink, "A new class of orthogonal polynomials: the Bessel polynomials," *Transactions of the American Mathematical Society*, vol. 65, pp. 100–115, 1949.

[21] L. Carlitz, "A note on the Bessel polynomials," *Duke Mathematical Journal*, vol. 24, pp. 151–162, 1957.

[22] S. Roman, *The Umbral Calculus*, Academic Press, 1984.

[23] G. Hetyei, "Meixner polynomials of the second kind and quantum algebras representing su(1,1)," *Proceedings of the Royal Society A: Mathematical, Physical and Engineering Sciences*, vol. 466, no. 2117, pp. 1409–1428, 2010.

[24] T. S. Chihara, *An Introduction to Orthogonal Polynomials*, Gordon and Breach Science, 1978.

Characterizations of Regular Ordered Semirings by Ordered Quasi-Ideals

Pakorn Palakawong na Ayutthaya[1,2] and Bundit Pibaljommee[1,2]

[1]*Department of Mathematics, Faculty of Science, Khon Kaen University, Khon Kaen 40002, Thailand*
[2]*Centre of Excellence in Mathematics CHE, Si Ayutthaya Road, Bangkok 10400, Thailand*

Correspondence should be addressed to Bundit Pibaljommee; banpib@kku.ac.th

Academic Editor: Howard E. Bell

We introduce the notion of an ordered quasi-ideal of an ordered semiring and show that ordered quasi-ideals and ordered bi-ideals coincide in regular ordered semirings. Then we give characterizations of regular ordered semirings, regular ordered duo-semirings, and left (right) regular ordered semirings by their ordered quasi-ideals.

1. Introduction

The concept of a quasi-ideal was defined first by Steinfeld for semigroups and for rings [1–3] as a generalization of a right ideal and a left ideal. Then Iséki [4] introduced the notion of a quasi-ideal in a semiring without zero and investigated some of its properties. In 1994, Dönges [5] studied quasi-ideals of a semiring with zero, investigated connections between left (right) ideals, bi-ideals, and quasi-ideals and characterized regular semirings using their quasi-ideals. Later, Shabir et al. [6] have studied some properties of quasi-ideals, using quasi-ideals to characterize regular and intraregular semirings and regular duo-semirings. As a generalization of quasi-ideals of semirings the quasi-ideals of Γ-semirings were investigated by many authors; see, for example, [7–9].

In 2011, the notion of an ordered semiring was introduced by Gan and Jiang [10] as a semiring with a partially ordered relation on the semiring such that the relation is compatible to the operations of the semiring. In the paper, the concept of a left (right) ordered ideal, a minimal ordered ideal, and a maximal ordered ideal was defined. Then Mandal [11] studied fuzzy ideals in an ordered semiring with the least element zero and gave a characterization of regular ordered semirings by their fuzzy ideals.

In this paper, we introduce the notion of an ordered quasi-ideal of an ordered semiring and show that ordered quasi-ideals and ordered bi-ideals coincide in regular ordered semirings. Then characterizations of regular ordered semirings, regular ordered duo-semirings, and left (right) regular ordered semirings by their ordered quasi-ideals have been investigated.

2. Preliminaries

An ordered semiring is a system $(S, +, \cdot, \leq)$ consisting of a nonempty set S such that $(S, +, \cdot)$ is a semiring, $(S, \leq)$ is a partially ordered set, and for any $a, b, x \in S$ the following conditions are satisfied:

(i) if $a \leq b$ then $a + x \leq b + x$ and $x + a \leq x + b$;

(ii) if $a \leq b$ then $ax \leq bx$ and $xa \leq xb$.

An ordered semiring S is said to be *additively commutative* if $a + b = b + a$ for all $a, b \in S$. An element $0 \in S$ is said to be an *absorbing zero* if $0a = 0 = a0$ and $a + 0 = a = 0 + a$ for all $a \in S$. In this paper we assume that S is an additively commutative ordered semiring with an absorbing zero 0.

For any subsets A, B of S and $a \in S$, we denote

$$(A] = \{x \in S \mid x \leq a \text{ for some } a \in A\},$$

$$AB = \{ab \in S \mid a \in A, b \in B\},$$

$$\Sigma A = \left\{ \sum_{i\in I} a_i \in S \mid a_i \in A \text{ and } I \text{ is a finite subset of } \mathbb{N} \right\},$$

$$\Sigma AB = \left\{ \sum_{i\in I} a_i b_i \in S \mid a_i \in A, b_i \in B \text{ and } I \text{ is a finite subset of } \mathbb{N} \right\},$$

$$\mathbb{N}a = \Sigma\{a\}. \tag{1}$$

Now, we mention some properties of finite sums on an ordered semiring.

Remark 1. For any subsets A, B of S, the following statements hold:

(i) $\Sigma(A] \subseteq (\Sigma A]$;

(ii) $\Sigma(\Sigma A) = \Sigma A$;

(iii) $A(\Sigma B) \subseteq \Sigma AB$ and $(\Sigma A)B \subseteq \Sigma AB$;

(iv) $\Sigma(A\Sigma B) \subseteq \Sigma AB$ and $\Sigma(\Sigma A)B \subseteq \Sigma AB$;

(v) $\Sigma(A + B) = \Sigma A + \Sigma B$.

We note that, for any $A \subseteq S$, $\Sigma A = A$ if and only if $A + A \subseteq A$ ($(A, +)$ is a subsemigroup of $(S, +)$).

Now, we give the basic properties of the operator $(\,]$ which are not difficult to verify.

Lemma 2. *Let A, B, C be subsets of an ordered semiring S. Then the following statements hold:*

(i) $A \subseteq (A]$ *and* $((A]] = (A]$;

(ii) *If* $A \subseteq B$ *then* $(A] \subseteq (B]$;

(iii) $A(B] \subseteq (A](B] \subseteq (AB]$ *and* $(A]B \subseteq (A](B] \subseteq (AB]$;

(iv) $A + (B] \subseteq (A] + (B] \subseteq (A + B]$ *and* $(A] + B \subseteq (A] + (B] \subseteq (A + B]$;

(v) $A(B + C] \subseteq (AB + AC]$ *and* $(A + B]C \subseteq (AC + BC]$;

(vi) $(A \cup B] = (A] \cup (B]$;

(vii) $(A \cap B] \subseteq (A] \cap (B]$.

In (vii) of the above lemma, we have $(A \cap B] = (A] \cap (B]$ when $(A] = A$ and $(B] = B$.

Lemma 3. *Let S be an ordered semiring and $\emptyset \neq A \subseteq S$. If $A \subseteq (\Sigma A^2 + \Sigma ASA]$ then $\Sigma A^2 \subseteq (\Sigma ASA]$.*

Proof. Assume that $A \subseteq (\Sigma A^2 + \Sigma ASA]$. Then

$$\begin{aligned} \Sigma A^2 &\subseteq \Sigma\left(\Sigma A^2 + \Sigma ASA\right] A \\ &\subseteq \Sigma\left(\left(\Sigma A^2\right) A + (\Sigma ASA) A\right] \subseteq \Sigma\left(\Sigma A^3 + \Sigma ASA\right] \\ &= \left(\Sigma\left(\Sigma A^3\right) + \Sigma(\Sigma ASA)\right] = \left(\Sigma A^3 + \Sigma ASA\right] \\ &= (\Sigma AAA + \Sigma ASA] \subseteq (\Sigma ASA + \Sigma ASA] \\ &= (\Sigma ASA]. \end{aligned} \tag{2}$$

□

Definition 4 (see [10]). Let S be an ordered semiring and $\emptyset \neq A \subseteq S$. Then A is said to be a *left ordered ideal (right ordered ideal)* if the following conditions are satisfied.

(1) A is a left ideal (right ideal) of S.

(2) If $x \leq a$ for some $a \in A$ then $x \in A$ (i.e., $A = (A]$).

We call A an *ordered ideal* if it is both left ordered ideal and right ordered ideal of S.

Example 5 (see [10]). Let $[0, 1]$ be the unit interval of real numbers. Define binary operations $\oplus$ and $\odot$ on $[0, 1]$ by letting $a, b \in [0, 1]$,

$$\begin{aligned} a \oplus b &= \max\{a, b\}, \\ a \odot b &= \max\{a + b - 1, 0\}, \end{aligned} \tag{3}$$

and an ordered relation $\leq$ is the natural order on real numbers. It is easy to show that $L = ([0, 1], \oplus, \odot, \leq)$ is an ordered semiring. Let $I = [0, 1/2]$. Then we can prove that I is an ordered ideal of L.

Lemma 6. *Let A be a nonempty subset of an ordered semiring S. Then*

(i) $(\Sigma SA]$ *is a left ordered ideal of* S;

(ii) $(\Sigma AS]$ *is a right ordered ideal of* S;

(iii) $(\Sigma SAS]$ *is an ordered ideal of* S.

Proof. (i) Let $x, y \in (\Sigma SA]$. Then $x \leq x'$ and $y \leq y'$ for some $x', y' \in \Sigma SA$. It is clear that $x + y \leq x' + y' \in \Sigma SA$, and so $x + y \in (\Sigma SA]$. By Remark 1 and Lemma 2, we obtain $S(\Sigma SA] \subseteq (S\Sigma SA] \subseteq (\Sigma SSA] \subseteq (\Sigma SA]$. We have $((\Sigma SA]] = (\Sigma SA]$. Hence, $(\Sigma SA]$ is a left ordered ideal of S.

(ii) and (iii) can be proved similar to (i). □

Corollary 7. *Let S be an ordered semiring. Then, for any $a \in S$,*

(i) $(Sa]$ *is a left ordered ideal of* S;

(ii) $(aS]$ *is a right ordered ideal of* S;

(iii) $(\Sigma SaS]$ *is an ordered ideal of* S.

Let A be a nonempty subset of an ordered semiring S. We denote $L(A), R(A)$ and $I(A)$ as the smallest left ordered ideal, right ordered ideal, and ordered ideal of S containing

A, respectively. In particular, we can show that if A is a left ideal (right ideal, ideal) of S then $(A]$ is the smallest left ordered ideal (resp., right ordered ideal and ordered ideal) of S containing A.

Lemma 8. *Let A be a nonempty subset of an ordered semiring S. Then*

(i) $L(A) = (\Sigma A + \Sigma SA]$;

(ii) $R(A) = (\Sigma A + \Sigma AS]$;

(iii) $I(A) = (\Sigma A + \Sigma SA + \Sigma AS + \Sigma SAS]$.

Proof. (i) Since S has an absorbing zero, we have, for every $a \in A$, $a = a + 0 \in \Sigma A + \Sigma SA \subseteq (\Sigma A + \Sigma SA]$. Hence, $A \subseteq (\Sigma A + \Sigma SA]$. Let $x, y \in (\Sigma A + \Sigma SA]$. Then $x \leq x'$ and $y \leq y'$ for some $x', y' \in \Sigma A + \Sigma SA$. Thus $x' = a_1 + b_1$ and $y' = a_2 + b_2$ for some $a_1, a_2 \in \Sigma A$ and $b_1, b_2 \in \Sigma SA$. It is easy to show that $a_1 + a_2 \in \Sigma A$ and $b_1 + b_2 \in \Sigma SA$. It follows that $x + y \leq x' + y' \in \Sigma A + \Sigma SA$, and so $x + y \in (\Sigma A + \Sigma SA]$. By Remark 1 and Lemma 2, we obtain

$$\begin{aligned} S(\Sigma A + \Sigma SA] &\subseteq (S(\Sigma A + \Sigma SA)] \subseteq (S\Sigma A + S\Sigma SA] \\ &\subseteq (\Sigma SA + \Sigma SSA] \subseteq (\Sigma SA + \Sigma SA] \\ &= (\Sigma SA] \subseteq (\Sigma A + \Sigma SA]. \end{aligned} \tag{4}$$

Since $((\Sigma A + \Sigma SA]] = (\Sigma A + \Sigma SA]$, L is a left ordered ideal of S. Let K be any left ordered ideal of S containing A. It turns out $\Sigma A \subseteq K$ and $\Sigma SA \subseteq K$, so $\Sigma A + \Sigma SA \subseteq K$. It follows that $(\Sigma A + \Sigma SA] \subseteq (K] = K$. Therefore, $(\Sigma A + \Sigma SA]$ is the smallest left ordered ideal of S containing A.

(ii) and (iii) can be proved similar to (i). □

As a special case of Lemma 8, if $A = \{a\}$ then we have the following corollary.

Corollary 9. *Let S be an ordered semiring. Then, for any $a \in S$,*

(i) $L(a) = (\mathbb{N}a + Sa]$;

(ii) $R(a) = (\mathbb{N}a + aS]$;

(iii) $I(a) = (\mathbb{N}a + Sa + aS + \Sigma SaS]$.

An element e of an ordered semiring S is said to be an *identity* if $ea = a = ae$ for all $a \in S$. If S has an identity, then we denote 1 as the identity of S.

It is not difficult to show that if S has an identity, then $L(A) = (\Sigma SA]$, $R(A) = (\Sigma AS]$ and $I(A) = (\Sigma SAS]$ for any $A \subseteq S$. In particular case, we have $L(a) = (Sa]$, $R(a) = (aS]$ and $I(a) = (\Sigma SaS]$ for any $a \in S$.

3. Ordered Quasi-Ideals in Ordered Semirings

Here, we present a notion of an ordered quasi-ideal of an ordered semiring. Then, in ordered semiring with an identity, we show that every ordered quasi-ideal can be expressed as an intersection of an ordered left ideal and an ordered right ideal.

Definition 10. Let $(S, +, \cdot, \leq)$ be an ordered semiring and let $(Q, +)$ be a subsemigroup of $(S, +)$. Then Q is said to be an *ordered quasi-ideal* of S if the following conditions are satisfied:

(1) $(\Sigma SQ] \cap (\Sigma QS] \subseteq Q$;

(2) if $x \leq q$ for some $q \in Q$ then $x \in Q$ (i.e., $Q = (Q]$).

It is clear that every left ordered ideal (right ordered ideal and ordered ideal) of an ordered semiring S is an ordered quasi-ideal of S. Moreover, each ordered quasi-ideal of S is a subsemiring of S; indeed, $QQ \subseteq (QQ] \subseteq (SQ] \cap (QS] \subseteq (\Sigma SQ] \cap (\Sigma QS] \subseteq Q$.

Example 11. Let $S = \{a, b, c, d\}$. Define binary operations $+$ and $\cdot$ on S by the following equations:

+	a	b	c	d
a	a	b	c	d
b	b	b	b	b
c	c	b	c	d
d	d	b	d	d

·	a	b	c	d
a	a	a	a	a
b	a	b	b	b
c	a	c	c	c
d	a	b	b	b

(5)

Then $(S, +, \cdot)$ is an additively commutative semiring with an absorbing zero a. Define a binary relation $\leq$ on S by

$$\leq := \{(a, a), (b, b), (c, c), (d, d), (b, d)\}. \tag{6}$$

We give the covering relation "$\prec$" and the figure of S:

$$\prec := \{(b, d)\}. \tag{7}$$

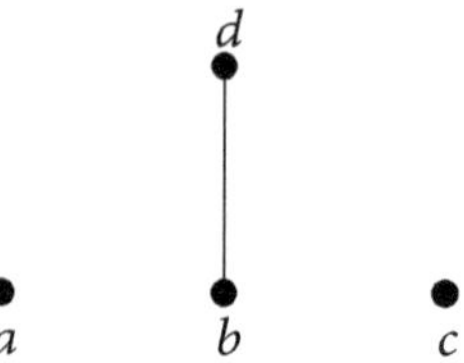

Now, $(S, +, \cdot, \leq)$ is an ordered semiring. Let $Q = \{a, b\}$. We have $(\Sigma SQ] \cap (\Sigma QS] = \{a, b, c\} \cap \{a, b\} = Q$ and $(Q] = Q$. Hence, Q is an ordered quasi-ideal of S but is not a left ordered ideal of S, since $SQ = \{a, b, c\} \nsubseteq Q$.

Lemma 12. *Let S be an ordered semiring and let $\{Q_i \mid i \in I\}$ be a family of ordered quasi-ideals of S. Then $\bigcap_{i \in I} Q_i$ is an ordered quasi-ideal of S.*

Let A be a nonempty subset of an ordered semiring S. We denote $Q(A)$ the smallest ordered quasi-ideal of S containing A.

Theorem 13. *Let S be an ordered semiring and let A be a nonempty subset of S. Then $Q(A) = (\Sigma A + ((\Sigma SA] \cap (\Sigma AS])]$.*

Proof. Let $Q = (\Sigma A + ((\Sigma SA] \cap (\Sigma AS]))]$. Since S has an absorbing zero, we have $a = a+0 \in \Sigma A+((\Sigma SA]\cap(\Sigma AS]) \subseteq Q$ for every $a \in A$. Hence, $A \subseteq Q$. Let $x, y \in Q$. Then $x \leq x'$ and $y \leq y'$ for some $x', y' \in \Sigma A + ((\Sigma SA] \cap (\Sigma AS])$. Thus $x' = a_1 + b_1$ and $y' = a_2 + b_2$ for some $a_1, a_2 \in \Sigma A$ and $b_1, b_2 \in (\Sigma SA] \cap (\Sigma AS]$. Clearly, $a_1 + a_2 \in \Sigma A$ and $b_1 + b_2 \in (\Sigma SA] \cap (\Sigma AS]$. It follows that $x + y \leq x' + y' \in \Sigma A + ((\Sigma SA] \cap (\Sigma AS])$, and so $x + y \in Q$. By Remark 1 and Lemma 2, we obtain

$$\begin{aligned}
(\Sigma SQ] \cap (\Sigma QS] &\subseteq (\Sigma SQ] \\
&= (\Sigma S(\Sigma A + ((\Sigma SA] \cap (\Sigma AS]))]] \\
&\subseteq (\Sigma S(\Sigma A + (\Sigma SA]]] \\
&\subseteq (\Sigma (S\Sigma A + S(\Sigma SA]]] \\
&\subseteq (\Sigma (\Sigma SA + (\Sigma SSA]]] \\
&\subseteq (\Sigma (\Sigma SA + \Sigma SSA]] \\
&\subseteq (\Sigma (\Sigma SA + \Sigma SA]] \subseteq (\Sigma (\Sigma SA]] \\
&\subseteq ((\Sigma SA]] = (\Sigma SA].
\end{aligned} \tag{8}$$

Similarly, we can show that $(\Sigma SQ] \cap (\Sigma QS] \subseteq (\Sigma AS]$. Thus $(\Sigma SQ]\cap(\Sigma QS] \subseteq (\Sigma SA]\cap(\Sigma AS] \subseteq \Sigma A+((\Sigma SA]\cap(\Sigma AS]) \subseteq Q$. Since $(Q] = Q$, we obtain that Q is an ordered quasi-ideal of S containing A. Let K be any ordered quasi-ideal of S containing A. It follows that $(\Sigma SA]\cap(\Sigma AS] \subseteq (\Sigma SK]\cap(\Sigma KS] \subseteq K$. So $\Sigma A+((\Sigma SA]\cap(\Sigma AS]) \subseteq K$. Hence, $Q = (\Sigma A+((\Sigma SA]\cap(\Sigma AS])] \subseteq (K] = K$. Therefore, Q is the smallest ordered quasi-ideal of S containing A. □

As a special case of Theorem 13, if $A = \{a\}$ then we have the following corollary.

Corollary 14. *Let S be an ordered semiring. Then $Q(a) = (\mathbb{N}a + ((Sa] \cap (aS])]$ for any $a \in S$.*

If S has an identity, then it is easy to check that $Q(A) = (\Sigma SA] \cap (\Sigma AS]$ for any $A \subseteq S$. In particular case, we have $Q(a) = (Sa] \cap (aS]$ for any $a \in S$.

Let $\mathbb{Q}(S)$ be the set of all ordered quasi-ideals of an ordered semiring S. Using Lemma 12, we define the operations $\wedge$ and $\vee$ on $\mathbb{Q}(S)$ by letting $P_1, P_2 \in \mathbb{Q}(S)$,

$$\begin{aligned}
P_1 \wedge P_2 &= P_1 \cap P_2, \\
P_1 \vee P_2 &= Q(P_1 \cup P_2).
\end{aligned} \tag{9}$$

Then we obtain the following theorem.

Theorem 15. *Let S be an ordered semiring. Then $(\mathbb{Q}(S), \wedge, \vee)$ is a complete lattice.*

Theorem 16. *The intersection of a left ordered ideal L and a right ordered ideal R of an ordered semiring S is an ordered quasi-ideal of S.*

Proof. It is easy to show that $L \cap R$ is a subsemigroup of $(S, +)$. By Remark 1 and Lemma 2, we obtain

$$\begin{aligned}
(\Sigma S(L \cap R)] \cap (\Sigma (L \cap R) S] &\subseteq (\Sigma S(L \cap R)] \\
&= (\Sigma (SL \cap SR)] \subseteq (\Sigma SL] \subseteq L, \\
(\Sigma S(L \cap R)] \cap (\Sigma (L \cap R) S] &\subseteq (\Sigma (L \cap R) S] \\
&= (\Sigma (LS \cap RS)] \subseteq (\Sigma RS] \subseteq R.
\end{aligned} \tag{10}$$

Hence, $(\Sigma S(L \cap R)] \cap (\Sigma(L \cap R)S] \subseteq L \cap R$. Let $s \in S$ such that $s \leq x$ for some $x \in L \cap R$. Then $s \in (L \cap R] \subseteq (L] \cap (R] = L \cap R$. □

The converse of Theorem 16 is not true as Example 2.1 page 8 in [2] given by A. H. Clifford.

Corollary 17. *Let S be an ordered semiring. Then the following statements hold.*

(i) $(\Sigma SA] \cap (\Sigma AS]$ *is an ordered quasi-ideal of S, for any $A \subseteq S$.*

(ii) $(Sa] \cap (aS]$ *is an ordered quasi-ideal of S, for any $a \in S$.*

Proof. (i) By Lemma 6, we have $(\Sigma SA]$ and $(\Sigma AS]$ a left and a right ordered ideal of S, respectively. Then by Theorem 16, we have that $(\Sigma SA] \cap (\Sigma AS]$ is an ordered quasi-ideal of S.

(ii) It is a particular case of (i). □

Now, we will show that the converse of Theorem 16 is true if S contains an identity as the following theorem.

Theorem 18. *Let S be an ordered semiring with identity. Then every ordered quasi-ideal Q of S can be written in the form $Q = R \cap L$ for some right ordered ideal R and left ordered ideal L of S.*

Proof. Assume that S has an identity. Let Q be an ordered quasi-ideal of S. Then $R(Q) = (\Sigma QS]$ and $L(Q) = (\Sigma SQ]$. We obtain $Q \subseteq R(Q) \cap L(Q)$ and $R(Q) \cap L(Q) = (\Sigma QS] \cap (\Sigma SQ] \subseteq Q$. Hence, $Q = R(Q) \cap L(Q)$. □

4. Regular Ordered Semirings

In this section, we show that in regular ordered semirings the converse of Theorem 16 is true and ordered quasi-ideals coincide with ordered bi-ideals. Then we give characterizations of regular ordered semirings, regular ordered duo-semirings, and left regular and right regular ordered semirings by their ordered quasi-ideals.

Definition 19 (see [11]). An element a of an ordered semiring S is said to be *regular* if $a \leq axa$ for some $x \in S$. An ordered semiring S is said to be *regular* if every element $a \in S$ is regular.

The following lemma is characterizations of regular ordered semiring which directly follows Definition 19.

Lemma 20. *Let S be an ordered semiring. Then the following statements are equivalent:*

(i) *S is regular;*

(ii) $A \subseteq (\Sigma ASA]$ *for each* $A \subseteq S$;

(iii) $a \in (aSa]$ *for any* $a \in S$.

Now, we will show that the converse of Theorem 16 is true in regular ordered semirings.

Theorem 21. *Every ordered quasi-ideal of a regular ordered semiring S can be written in the form* $Q = R \cap L$ *for some right ordered ideal R and left ordered ideal L of S.*

Proof. Let Q be an ordered quasi-ideal of S. By Lemma 8, we have $R(Q) = (\Sigma Q + \Sigma QS]$ and $L(Q) = (\Sigma Q + \Sigma SQ]$. Now, $Q \subseteq R(Q) \cap L(Q)$. Let $q \in Q$. Since S is regular, there exists $x \in S$ such that $q \leq qxq \in \Sigma QS$. So $Q \subseteq (\Sigma QS]$. Since $Q + Q \subseteq Q$, $\Sigma Q = Q$. It follows that

$$(\Sigma QS] \subseteq (\Sigma Q + \Sigma QS] = (Q + \Sigma QS] \subseteq ((\Sigma QS] + \Sigma QS] \subseteq (\Sigma QS]. \quad (11)$$

This implies that $R(Q) = (\Sigma QS]$. Similarly, we can show that $L(Q) = (\Sigma SQ]$. Hence, $R(Q) \cap L(Q) = (\Sigma QS] \cap (\Sigma SQ] \subseteq Q$. Therefore, $Q = R(Q) \cap L(Q)$. □

Definition 22. Let $(S, +, \cdot, \leq)$ be an ordered semiring. A subsemigroup $(B, +)$ of $(S, +)$ is said to be an *ordered bi-ideal* of S if the following conditions hold:

(1) $BSB \subseteq B$;

(2) if $x \leq b$ for some $b \in B$, then $x \in B$ (i.e., $B = (B]$).

We note that condition (1) of Definition 22 is equivalent to $\Sigma BSB \subseteq B$.

Theorem 23. *Every ordered quasi-ideal of an ordered semiring S is an ordered bi-ideal of S.*

Proof. Let Q be an ordered quasi-ideal of S. Then $\Sigma QSQ \subseteq \Sigma QS \subseteq (\Sigma QS]$ and $QSQ \subseteq \Sigma SQ \subseteq (\Sigma SQ]$. So, $\Sigma QSQ \subseteq (\Sigma SQ] \cap (\Sigma QS] \subseteq Q$. Hence, Q is an ordered bi-ideal of S. □

The converse of Theorem 23 is not generally true as the following example.

Example 24. Let $S = \{a, b, c, d, e\}$. Define binary operations + and · by the following equations:

+	a	b	c	d	e
a	a	b	c	d	e
b	b	b	d	d	d
c	c	d	d	d	d
d	d	d	d	d	d
e	e	d	d	d	e

·	a	b	c	d	e
a	a	a	a	a	a
b	a	a	a	a	a
c	a	a	b	b	b
d	a	a	b	b	b
e	a	a	b	b	b

(12)

Then $(S, +, \cdot)$ is an additively commutative semiring with an absorbing zero a. Define a binary relation $\leq$ on S by

$$\leq := \{(a,a),(b,b),(c,c),(d,d),(e,e),(a,b),(a,c),(a,e),(a,d),(b,d),(c,d),(e,d)\}. \quad (13)$$

We give the covering relation "$\prec$" and the figure of S:

$$\prec := \{(a,b),(a,c),(a,e),(b,d),(c,d),(e,d)\}. \quad (14)$$

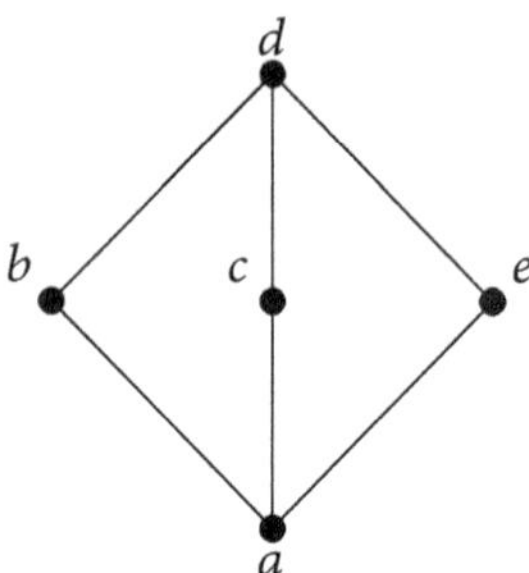

Then $(S, +, \cdot, \leq)$ is an ordered semiring but not regular, since $d \nleq dxd$ for any $x \in S$. Let $B = \{a, e\}$. It is easy to show that B is an ordered bi-ideal but not an ordered quasi-ideal of S, since $(\Sigma SB] \cap (\Sigma BS] = \{a, b\} \nsubseteq B$.

Now, we show that in regular ordered semirings, ordered bi-ideals and ordered quasi-ideals coincide as the following theorem.

Theorem 25. *Let S be a regular ordered semiring. Then ordered bi-ideals and ordered quasi-ideals coincide in S.*

Proof. By Theorem 23, we have that every ordered quasi-ideal of S is an ordered bi-ideal of S. Now, we show that every ordered bi-ideal of S is an ordered quasi-ideal of S. Let B be an ordered bi-ideal of S. Let $a \in (\Sigma SB] \cap (\Sigma BS]$. By Lemma 20, Remark 1, and Lemma 2, we obtain $a \in (aSa] \subseteq ((\Sigma BS]S(\Sigma SB]] \subseteq ((\Sigma BSS](\Sigma SB]] \subseteq ((\Sigma BS)(\Sigma SB)] \subseteq (\Sigma(BS(\Sigma SB))] \subseteq (\Sigma(\Sigma BSSB)] \subseteq (\Sigma BSB] \subseteq B$. Hence, B is an ordered quasi-ideal of S. □

Theorem 26. *Let S be an ordered semiring. Then the following statements are equivalent:*

(i) *S is regular;*

(ii) $(\Sigma RL] = R \cap L$ *for every right ordered ideal R and left ordered ideal L of S;*

(iii) $B = (\Sigma BSB]$ *for each ordered bi-ideal B of S;*

(iv) $Q = (\Sigma QSQ]$ *for each ordered quasi-ideal Q of S.*

Proof. (i) ⇒ (ii): assume that S is regular and let R and L be a right ordered ideal and a left ordered ideal of S, respectively. So, $(\Sigma RL] \subseteq (\Sigma R] = R$ and $(\Sigma RL] \subseteq (\Sigma L] = L$. Hence, $(\Sigma RL] \subseteq R \cap L$. Let $a \in R \cap L$. Since S is regular, $a \leq axa$ for some $x \in S$. Since $a \in R$, $xa \in L$. It follows that $a \leq a(xa) \in RL$. This means $a \in (RL] \subseteq (\Sigma RL]$. Therefore, $(\Sigma RL] = R \cap L$.

(ii) ⇒ (iii): assume that (ii) holds. Let B be an ordered bi-ideal of S. It is clear that $(\Sigma BSB] \subseteq B$. By assumption, $B \subseteq R(B) \cap L(B) = (\Sigma R(B)L(B)]$. By Lemma 8, Remark 1, and Lemmas 2 and 3, we have

$$\begin{aligned} B &\subseteq (\Sigma B + \Sigma BS] \cap (\Sigma B + \Sigma SB] \\ &= (\Sigma((\Sigma B + \Sigma BS](\Sigma B + \Sigma SB])] \\ &\subseteq (\Sigma((\Sigma B + \Sigma BS)(\Sigma B + \Sigma SB))] \\ &\subseteq (\Sigma(\Sigma B(\Sigma B + \Sigma SB) + \Sigma BS(\Sigma B + \Sigma SB))] \\ &\subseteq \left(\Sigma\left(\Sigma B^2 + \Sigma BSB + \Sigma BSB + \Sigma BSSB\right)\right] \\ &\subseteq \left(\Sigma\left(\Sigma B^2 + \Sigma BSB\right)\right] = \left(\Sigma\left(\Sigma B^2\right) + \Sigma(\Sigma BSB)\right] \\ &= \left(\Sigma B^2 + \Sigma BSB\right] \subseteq ((\Sigma BSB] + \Sigma BSB] \subseteq (\Sigma BSB]. \end{aligned} \tag{15}$$

(iii) ⇒ (iv): it follows from Theorem 23.

(iv) ⇒ (i): let $a \in S$. Then $Q(a) = (Q(a)SQ(a)]$. By Corollary 14, Remark 1, and Lemma 2, we have

$$\begin{aligned} a &\in (\mathbb{N}a + ((Sa] \cap (aS])] \\ &= (\Sigma((\mathbb{N}a + ((Sa] \cap (aS])]S(\mathbb{N}a + ((Sa] \cap (aS])])] \\ &\subseteq (\Sigma((\mathbb{N}a + (aS]]S(\mathbb{N}a + (Sa]])] \\ &\subseteq (\Sigma((\mathbb{N}a + aS]S(\mathbb{N}a + Sa])] \\ &\subseteq (\Sigma(((\mathbb{N}a + aS)S](\mathbb{N}a + Sa])] \\ &\subseteq (\Sigma((aS](\mathbb{N}a + Sa])] \subseteq (\Sigma(aS(\mathbb{N}a + Sa))] \\ &\subseteq (\Sigma aSa] = (aSa]. \end{aligned} \tag{16}$$

By Lemma 20, S is regular. □

Theorem 27. *Let S be a regular ordered semiring. Then the following statements hold:*

(i) *every ordered quasi-ideal Q of S can be written in the form $Q = R \cap L = (RL]$ for some right ordered ideal R and left ordered ideal L of S;*

(ii) $(Q^2] = (Q^3]$ *for each ordered quasi-ideal Q of S.*

Proof. (i) It is obvious by Theorems 21 and 26.

(ii) Let Q be an ordered quasi-ideal of S. Clearly, $((QQ)Q] \subseteq (QQ]$. Let $x \in (QQ]$. Then $x \leq q_1q_2$ for some $q_1, q_2 \in Q$. Since S is regular, there exists $s \in S$ such that $x \leq q_1q_2 \leq (q_1q_2)s(q_1q_2) \in QQSQQ$. Hence, $x \in (Q(QSQ)Q] \subseteq (QQQ]$. Therefore, $(Q^2] = (Q^3]$. □

Theorem 28. *Let S be an ordered semiring. Then S is regular if and only if $B \cap I \cap L \subseteq (BIL]$ for every ordered bi-ideal B, every ordered ideal I, and every left ordered ideal L of S.*

Proof. Let B, I, and L be an ordered bi-ideal, an ordered ideal, and a left ordered ideal of S, respectively. Let $a \in B \cap I \cap L$. Since S is regular, $a \leq axa \leq axaxaxa \in BIL$. Hence, $B \cap I \cap L \subseteq (BIL]$.

Conversely, assume that $B \cap I \cap L \subseteq (BIL]$ for every ordered bi-ideal B, every ordered ideal I, and every left ordered ideal L of S. Then we obtain $R \cap L = R \cap S \cap L \subseteq (RSL] \subseteq (RL] \subseteq (\Sigma RL]$ for every right ordered ideal R and left ordered ideal L of S. On the other hand, we have $(\Sigma RL] \subseteq R \cap L$. Hence, $(\Sigma RL] = R \cap L$. By Theorem 26, S is regular. □

Definition 29. An ordered semiring S is said to be an *ordered duo-semiring* if every one-sided (right or left) ordered ideal of S is an ordered ideal of S.

We note that every multiplicatively commutative ordered semiring is an ordered duo-semiring, but the converse is not generally true. Now, we give an example of a multiplicatively noncommutative ordered semiring which is an ordered duo-semiring.

Example 30. Let $S = \{a, b, c, d, e\}$. Define binary operations $+$ and $\cdot$ by the following equations:

+	a	b	c	d	e
a	a	b	c	d	e
b	b	c	c	c	c
c	c	c	c	c	c
d	d	c	c	c	c
e	e	c	c	c	c

,

·	a	b	c	d	e
a	a	a	a	a	a
b	a	e	c	e	c
c	a	c	c	c	c
d	a	c	c	e	c
e	a	c	c	c	c

. (17)

Then $(S, +, \cdot)$ is an additively commutative semiring with an absorbing zero a. Define a binary relation $\leq$ on S by

$$\leq := \{(a,a), (b,b), (c,c), (d,d), (e,e), (e,c)\}. \tag{18}$$

We give the covering relation "$\prec$" and the figure of S:

$$\prec := \{(e,c)\}. \tag{19}$$

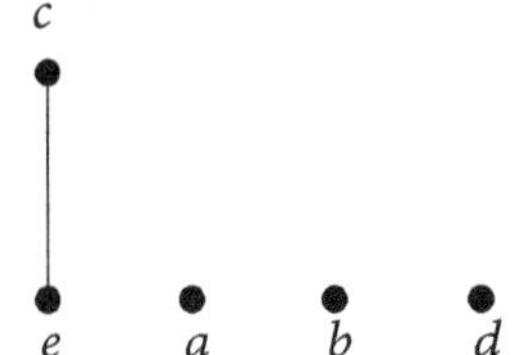

Then $(S, +, \cdot, \leq)$ is an ordered semiring which is not multiplicatively commutative, since $bd \neq db$. We have all one-sided ordered ideals of S which are as follows:

$$\{a\}, \{a,c\}, \{a,c,e\}, \{a,b,c,e\}, \{a,c,d,e\}, S. \tag{20}$$

It is not difficult to check that all of them are ordered ideals of S. This shows that S is an ordered duo-semiring.

Lemma 31. *Let S be an ordered semiring. Then the following conditions are equivalent:*

(i) *S is an ordered duo-semiring;*

(ii) *$R(A) = L(A)$ for each $A \subseteq S$;*

(iii) *$R(a) = L(a)$ for each $a \in S$.*

Proof. (i) $\Rightarrow$ (ii) and (ii) $\Rightarrow$ (iii) are obvious.

(iii) $\Rightarrow$ (i): let L be a left ordered ideal of S and let $x \in L, s \in S$. By assumption, we have $xs \in R(x)S \subseteq R(x) = L(x) \subseteq L(L) = L$. It follows that L is a right ordered ideal of S. Similarly, we have that every right ordered ideal of S is a left ordered ideal of S. Hence, S is an ordered duo-semiring. □

Theorem 32. *Let S be an ordered duo-semiring. Then S is regular if and only if $(\Sigma Q_1Q_2] = Q_1 \cap Q_2$ for each two ordered quasi-ideals Q_1 and Q_2 of S.*

Proof. Assume that S is a regular ordered semiring. Let Q_1 and Q_2 be ordered quasi-ideals of S. By Theorem 21, Q_1 and Q_2 can be written in the forms

$$\begin{aligned} Q_1 &= R_1 \cap L_1, \\ Q_2 &= R_2 \cap L_2 \end{aligned} \tag{21}$$

for some R_1, R_2 and L_1, L_2 which are right ordered ideals and left ordered ideals of S, respectively. Since S is an ordered duo-semiring, R_1, R_2, L_1, and L_2 are ordered ideals of S. It follows that Q_1 and Q_2 are ordered ideals of S. By Theorem 26, we have $(\Sigma Q_1Q_2] = Q_1 \cap Q_2$.

Conversely, assume that $(\Sigma Q_1Q_2] = Q_1 \cap Q_2$ for each two ordered quasi-ideals Q_1 and Q_2 of S. Let $A \subseteq S$. By assumption, $A \subseteq Q(A) \cap Q(A) = (\Sigma Q(A)Q(A)]$. By Theorem 13, Remark 1, and Lemmas 2 and 3, we have

$$\begin{aligned} A &\subseteq (\Sigma((\Sigma A + ((\Sigma SA] \cap (\Sigma AS])] \\ &\quad \cdot (\Sigma A + ((\Sigma SA] \cap (\Sigma AS])])] \subseteq (\Sigma((\Sigma A + (\Sigma AS]] \\ &\quad \cdot (\Sigma A + (\Sigma SA]])] \subseteq (\Sigma((\Sigma A + \Sigma AS](\Sigma A + \Sigma SA])] \\ &\subseteq (\Sigma((\Sigma A + \Sigma AS)(\Sigma A + \Sigma SA))] \\ &\subseteq (\Sigma(\Sigma A(\Sigma A + \Sigma SA) + \Sigma AS(\Sigma A + \Sigma SA))] \\ &\subseteq \left(\Sigma\left(\Sigma A^2 + \Sigma ASA + \Sigma ASA + \Sigma ASSA\right)\right] \\ &\subseteq \left(\Sigma\left(\Sigma A^2 + \Sigma ASA\right)\right] = \left(\Sigma\left(\Sigma A^2\right) + \Sigma(\Sigma ASA)\right] \\ &= \left(\Sigma A^2 + \Sigma ASA\right] \subseteq ((\Sigma ASA] + \Sigma ASA] \subseteq (\Sigma ASA]. \end{aligned} \tag{22}$$

By Lemma 20, S is a regular ordered semiring. □

Theorem 33. *Let S be an ordered duo-semiring. Then the following conditions are equivalent:*

(i) *S is regular;*

(ii) *$(\Sigma L_1L_2] = L_1 \cap L_2$ and $(\Sigma R_1R_2] = R_1 \cap R_2$ for each two left ordered ideals L_1, L_2 and right ordered ideals R_1, R_2 of S;*

(iii) *$(\Sigma RL] = R \cap L = (\Sigma LR]$, for each right ordered ideal R and left ordered ideal L of S.*

Proof. It is obvious by Theorem 26. □

Definition 34. Let S be an ordered semiring. Then an element $a \in S$ is said to be *left regular (right regular)* if $a \leq xa^2$ ($a \leq a^2x$) for some $x \in S$. An ordered semiring S is said to be *left regular (right regular)* if every element $a \in S$ is left regular (right regular).

Example 35. Let $S = \{a, b, c, d, e, f\}$. Define binary operations $+$ and $\cdot$ on S by the following equations:

(23)

$+$	a	b	c	d	e	f
a	a	b	c	d	e	f
b	b	b	b	b	b	b
c	c	c	c	c	c	c
d	d	d	d	d	d	d
e	e	e	e	e	e	e
f	f	f	f	f	f	f

$\cdot$	a	b	c	d	e	f
a	a	a	a	a	a	a
b	a	b	b	b	b	b
c	a	b	b	b	b	b
d	a	b	b	d	b	d
e	a	e	e	e	e	e
f	a	e	e	f	e	f

Then $(S, +, \cdot)$ is a semiring with an absorbing zero a. Define a binary relation $\leq$ on S by

$$\begin{aligned} \leq := \{&(a,a),(b,b),(c,c),(d,d),(e,e),(f,f),(b,c), \\ &(b,d),(b,e),(b,f),(c,d),(c,e),(c,f),(d,f),(e,f)\}. \end{aligned} \tag{24}$$

We give the covering relation "$\prec$" and the figure of S:

$$\prec := \{(b,c),(c,d),(c,e),(d,f),(e,f)\}. \tag{25}$$

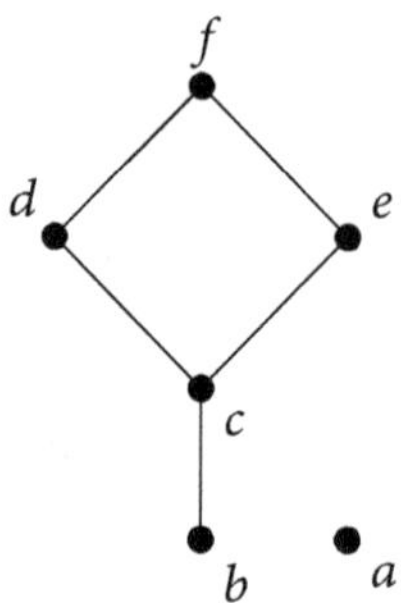

Now, $(S, +, \cdot, \leq)$ is an ordered semiring. Clearly, a, b, d, e, and f are left regular. We consider $c \leq ec^2 = eb = e$. This implies that S is left regular. Since there does not exist $x \in S$ such that $c \leq cxc$, S is not regular.

Example 36. Consider the ordered semiring $S = (\mathbb{N} \cup \{0\}, \max, \min, \leq)$ where $\leq$ is the natural order relation on numbers. Since $n \leq \min\{n, n, n\}$ for any $n \in \mathbb{N}$, we get S a regular, left regular, and right regular ordered semiring.

The following lemmas can be easily proved using Definition 34.

Lemma 37. *Let S be an ordered semiring. Then the following statements are equivalent:*

(i) *S is left regular;*

(ii) *$A \subseteq (\Sigma SA^2]$ for each $A \subseteq S$;*

(iii) *$a \in (Sa^2]$ for each $a \in S$.*

Lemma 38. *Let S be an ordered semiring. Then the following statements are equivalent:*

(i) *S is right regular;*

(ii) *$A \subseteq (\Sigma A^2 S]$ for each $A \subseteq S$;*

(iii) *$a \in (a^2 S]$ for each $a \in S$.*

Definition 39. Let T be a nonempty subset of an ordered semiring S. Then T is said to be *semiprime* if for any $a \in S$, if $a^2 \in T$, then $a \in T$.

We note that a nonempty subset T of S is semiprime if and only if, for any $\emptyset \neq A \subseteq S$, $A^2 \subseteq T$ implies $A \subseteq T$. Because, if T is semiprime, $\emptyset \neq A \subseteq T$, and $a \in A$ then $a^2 \in T$ and so $a \in T$; that is, $A \subseteq T$.

Theorem 40. *Let S be an ordered semiring. Then S is left regular and right regular if and only if every ordered quasi-ideal of S is semiprime.*

Proof. Let Q be an ordered quasi-ideal of S. Let A be a nonempty subset of S such that $A^2 \subseteq Q$. Since S is left regular and by Lemma 37, we have $A \subseteq (\Sigma SA^2]$. Since S is right regular and by Lemma 38, we have $A \subseteq (\Sigma A^2 S]$. Hence, $A \subseteq (\Sigma SA^2] \cap (\Sigma A^2 S] \subseteq (\Sigma SQ] \cap (\Sigma QS] \subseteq Q$. Therefore, Q is semiprime.

Conversely, assume that every ordered quasi-ideal of S is semiprime. Let $A \subseteq S$. By Theorem 13, we have $Q(A^2) = (\Sigma A^2 + ((\Sigma SA^2] \cap (\Sigma A^2 S])]$. Since $A^2 \subseteq Q(A^2)$ is semiprime, $A \subseteq Q(A^2) = (\Sigma A^2 + ((\Sigma SA^2] \cap (\Sigma A^2 S])]$. Then we obtain

$$\begin{aligned} A &\subseteq \left(\Sigma A^2 + \left(\left(\Sigma SA^2\right] \cap \left(\Sigma A^2 S\right]\right)\right] \\ &\subseteq \left(\Sigma A^2 + \left(\Sigma SA^2\right]\right] \subseteq \left(\Sigma A^2 + \Sigma SA^2\right], \\ A &\subseteq \left(\Sigma A^2 + \left(\left(\Sigma SA^2\right] \cap \left(\Sigma A^2 S\right]\right)\right] \\ &\subseteq \left(\Sigma A^2 + \left(\Sigma A^2 S\right]\right] \subseteq \left(\Sigma A^2 + \Sigma A^2 S\right]. \end{aligned} \tag{26}$$

In case $A \subseteq (\Sigma A^2 + \Sigma SA^2]$, we get

$$\begin{aligned} \Sigma A^2 &\subseteq \Sigma A\left(\Sigma A^2 + \Sigma SA^2\right] \subseteq \Sigma\left(A\Sigma A^2 + A\Sigma SA^2\right] \\ &\subseteq \Sigma\left(\Sigma A^3 + \Sigma ASA^2\right] \subseteq \Sigma\left(\Sigma A^3 + \Sigma SA^2\right] \\ &\subseteq \left(\Sigma\left(\Sigma A^3\right) + \Sigma\left(\Sigma SA^2\right)\right] \subseteq \left(\Sigma A^3 + \Sigma SA^2\right] \\ &\subseteq \left(\Sigma SA^2 + \Sigma SA^2\right] = \left(\Sigma SA^2\right] \end{aligned} \tag{27}$$

and so $A \subseteq (\Sigma A^2 + \Sigma SA^2] \subseteq ((\Sigma SA^2] + \Sigma SA^2] \subseteq (\Sigma SA^2]$. By Lemma 37, S is left regular. Similarly, in case $A \subseteq (\Sigma A^2 + \Sigma A^2 S]$, we get $A \subseteq (A^2 S]$. By Lemma 38, S is right regular. □

Conflict of Interests

The authors declare that there is no conflict of interests regarding the publication of this paper.

Acknowledgment

This work has been supported by the Centre of Excellence in Mathematics, the Commission on Higher Education, Thailand.

References

[1] O. Steinfeld, "On ideal-quotients and prime ideals," *Acta Mathematica Academiae Scientiarum Hungaricae*, vol. 4, pp. 289–298, 1953.

[2] O. Steinfeld, *Quasi-Ideals in Rings and Semigroups*, vol. 10 of *Disquisitiones Mathematicae Hungaricae*, Akadémiai Kiadó, Budapest, Hungary, 1978.

[3] O. Steinfeld, "Über die Quasiideale von Halbgruppen," *Publicationes Mathematicae Debrecen*, vol. 4, pp. 262–275, 1956.

[4] K. Iséki, "Quasi-ideals in semirings without zero," *Proceedings of the Japan Academy*, vol. 34, pp. 79–81, 1958.

[5] C. Dönges, "On Quasi-ideals of semirings," *International Journal of Mathematics and Mathematical Sciences*, vol. 17, no. 1, pp. 47–58, 1994.

[6] M. Shabir, A. Ali, and S. Batool, "A note on quasi-ideals in semirings," *Southeast Asian Bulletin of Mathematics*, vol. 27, no. 5, pp. 923–928, 2004.

[7] R. Chinram, "A note on quasi-ideals in Γ-semirings," *International Mathematical Forum*, vol. 3, no. 25–28, pp. 1253–1259, 2008.

[8] R. Chinram, "A note on quasi-ideals in regular Γ-semirings," *International Journal of Contemporary Mathematical Sciences*, vol. 3, no. 35, pp. 1725–1731, 2008.

[9] R. D. Jagatap and Y. S. Pawar, "Quasi-ideals and minimal quasi-ideals in Γ-semirings," *Novi Sad Journal of Mathematics*, vol. 39, no. 2, pp. 79–87, 2009.

[10] A. P. Gan and Y. L. Jiang, "On ordered ideals in ordered semirings," *Journal of Mathematical Research & Exposition*, vol. 31, no. 6, pp. 989–996, 2011.

[11] D. Mandal, "Fuzzy ideals and fuzzy interior ideals in ordered semirings," *Fuzzy Information and Engineering*, vol. 6, no. 1, pp. 101–114, 2014.

16

On Generalized Semiderivations of Prime Near Rings

Abdelkarim Boua,[1] A. Raji,[2] Asma Ali,[3] and Farhat Ali[3]

[1]*Department of Mathematics, Faculty of Sciences of Agadir, Ibn Zohr University, P.O. Box 8106, Agadir, Morocco*
[2]*Département de Mathématiques, Faculté des Sciences et Techniques, Université Moulay Ismaïl, Groupe d'Algèbre et Applications, BP 509, Boutalamine, Errachidia, Morocco*
[3]*Department of Mathematics, Aligarh Muslim University, Aligarh 202002, India*

Correspondence should be addressed to Asma Ali; asma_ali2@rediffmail.com

Academic Editor: Kaiming Zhao

Let N be a near ring. An additive mapping $F : N \to N$ is said to be a generalized semiderivation on N if there exists a semiderivation $d : N \to N$ associated with a function $g : N \to N$ such that $F(xy) = F(x)y + g(x)d(y) = d(x)g(y) + xF(y)$ and $F(g(x)) = g(F(x))$ for all $x, y \in N$. In this paper we prove that prime near rings satisfying identities involving semiderivations are commutative rings, thereby extending some known results on derivations, semiderivations, and generalized derivations. We also prove that there exist no nontrivial generalized semiderivations which act as a homomorphism or as an antihomomorphism on a 3-prime near ring N.

1. Introduction

Throughout the paper, N denotes a zero-symmetric left near ring with multiplicative centre Z; and for any pair of elements $x, y \in N$, $[x, y]$ denotes the commutator $xy - yx$ while the symbol (x, y) denotes the additive commutator $x + y - x - y$. An element x of N is said to be distributive if $(y + z)x = yx + zx$, for all $y, z \in N$. A near ring N is called zero-symmetric if $0x = 0$, for all $x \in N$ (recall that left distributivity yields that $x0 = 0$). The near ring N is said to be 3-prime if $xNy = \{0\}$ for $x, y \in N$ implies that $x = 0$ or $y = 0$. A near ring N is called 2-torsion free if $(N, +)$ has no element of order 2. An additive mapping $f : N \to N$ is said to be a right (resp., left) generalized derivation with associated derivation D if $f(xy) = f(x)y + xD(y)$ (resp., $f(xy) = D(x)y + xf(y)$), for all $x, y \in N$, and f is said to be a generalized derivation with associated derivation D on N if it is both a right generalized derivation and a left generalized derivation on N with associated derivation D. Motivated by a definition given by Bergen [1] for rings, we define an additive mapping $d : N \to N$ to be a semiderivation on a near ring N if there exists a function $g : N \to N$ such that (i) $d(xy) = d(x)g(y) + xd(y) = d(x)y + g(x)d(y)$ and (ii) $d(g(x)) = g(d(x))$, for all $x, y \in N$. In case g is the identity map on N, d is of course just a derivation on N, so the notion of semiderivation generalizes that of derivation. But the generalization is not trivial; for example, take $N = N_1 \oplus N_2$, where N_1 is a zero-symmetric near ring and N_2 is a ring. Then the map $d : N \to N$ defined by $d((x, y)) = (0, y)$ is a semiderivation associated with function $g : N \to N$ such that $g(x, y) = (x, 0)$. However d is not a derivation on N. An additive mapping $F : N \to N$ is said to be a generalized semiderivation of N if there exists a semiderivation $d : N \to N$ associated with a map $g : N \to N$ such that (i) $F(xy) = F(x)y + g(x)d(y) = d(x)g(y) + xF(y)$ and (ii) $F(g(x)) = g(F(x))$ for all $x, y \in N$. All semiderivations are generalized semiderivations. Moreover, if g is the identity map on N, then all generalized semiderivations are merely generalized derivations; again the notion of generalized semiderivation generalizes that of generalized derivation. Moreover, the generalization is not trivial as the following example shows.

Example 1. Let S be a 2-torsion free left near ring and let

$$N = \left\{ \begin{pmatrix} 0 & x & y \\ 0 & 0 & 0 \\ 0 & 0 & z \end{pmatrix} \mid x, y, z \in S \right\}. \qquad (1)$$

Define maps $F, d, g : N \to N$ by

$$F\begin{pmatrix} 0 & x & y \\ 0 & 0 & 0 \\ 0 & 0 & z \end{pmatrix} = \begin{pmatrix} 0 & xy & 0 \\ 0 & 0 & 0 \\ 0 & 0 & 0 \end{pmatrix};$$

$$d\begin{pmatrix} 0 & x & y \\ 0 & 0 & 0 \\ 0 & 0 & z \end{pmatrix} = \begin{pmatrix} 0 & 0 & y \\ 0 & 0 & 0 \\ 0 & 0 & z \end{pmatrix}, \qquad (2)$$

$$g\begin{pmatrix} 0 & x & y \\ 0 & 0 & 0 \\ 0 & 0 & z \end{pmatrix} = \begin{pmatrix} 0 & x & 0 \\ 0 & 0 & 0 \\ 0 & 0 & 0 \end{pmatrix}.$$

It can be verified that N is a left near ring and F is a generalized semiderivation with associated semiderivation d and a map g associated with d. However F is not a generalized derivation on N.

2. Preliminary Results

We begin with the following Lemmas which are extensively used to prove our main theorems. Unless it is stated otherwise, it will be assumed that N is a zero-symmetric 3-prime near ring.

Lemma 2 (see [2, Lemma 1.2]). *Let N be a 3-prime near ring.*

(i) *If $z \in Z \setminus \{0\}$ and $xz \in Z$, then $x \in Z$.*

(ii) *If $x \in Z \setminus \{0\}$ then x is not a zero divisor.*

Lemma 3 (see [2, Lemma 1.5]). *If N is a 3-prime near ring and Z contains a nonzero left semigroup ideal, then N is a commutative ring.*

Lemma 4 (see [3, Theorem 2.1]). *Let N be a 2-torsion free 3-prime near ring with a nonzero semiderivation d associated with a map g. If $d(N) \subseteq Z$, then N is a commutative ring.*

Lemma 5. *Let N be a 3-prime near ring admitting a generalized semiderivation F associated with a semiderivation d. If g is the map associated with d such that $g(xy) = g(x)g(y)$ for all $x, y \in N$, then N satisfies the following partial distributive laws:*

(i) *$(F(x)y + g(x)d(y))z = F(x)yz + g(x)d(y)z$ for all $x, y, z \in N$.*

(ii) *$(d(x)y + g(x)d(y))z = d(x)yz + g(x)d(y)z$ for all $x, y, z \in N$.*

Proof. (i) Let $x, y, z \in N$, and by defining F we have

$$\begin{aligned} F(xyz) &= F(xy)z + g(xy)d(z) \\ &= (F(x)y + g(x)d(y))z + g(x)g(y)d(z). \end{aligned} \qquad (3)$$

On the other hand,

$$\begin{aligned} F(xyz) &= F(x)yz + g(x)d(yz) \\ &= F(x)yz + g(x)(d(y)z + g(y)d(z)) \\ &= F(x)yz + g(x)d(y)z + g(x)g(y)d(z). \end{aligned} \qquad (4)$$

Combining both expressions of $F(xyz)$, we obtain

$$(F(x)y + g(x)d(y))z = F(x)yz + g(x)d(y)z, \quad \forall x, y, z \in N. \qquad (5)$$

(ii) With a simple calculation of $d((xy)z) = d(x(yz))$, we obtain the required result. □

Lemma 6. *Let N be a 2-torsion free zero-symmetric 3-prime near ring. If d is a nonzero semiderivation of N associated with a map g which is onto, then $d^2 \neq 0$.*

Proof. Suppose $d^2(N) = 0$. Then for $x, y \in N$, we may write

$$\begin{aligned} 0 &= d^2(xy) \\ &= d(d(xy)) \\ &= d(d(x)y + g(x)d(y)), \quad \forall x, y \in N \\ &= d^2(x)y + d(x)d(y) + d(g(x))d(y) \\ &\quad + g(x)d^2(y) \\ &= d(x)d(y) + d(g(x))d(y). \end{aligned} \qquad (6)$$

Note that $g(d(x)) = d(g(x))$ and g is onto; we get

$$2d(x)d(y) = 0, \quad \forall x, y \in N. \qquad (7)$$

Since N is 2-torsion free, we get

$$d(x)d(y) = 0, \quad \forall x, y \in N. \qquad (8)$$

Replacing y by ry in the above relation, we get

$$\begin{aligned} d(x)d(ry) &= 0, \quad \forall x, y, r \in N. \\ d(x)(d(r)y + g(r)d(y)) &= 0, \quad \forall x, y, r \in N. \\ d(x)d(r)y + d(x)g(r)d(y) &= 0, \quad \forall x, y, r \in N. \end{aligned} \qquad (9)$$

This implies that

$$\begin{aligned} d(x)g(r)d(y) &= 0, \quad \forall x, y, r \in N. \\ d(x)rd(y) &= 0, \quad \forall x, y, r \in N. \\ d(N)Nd(N) &= \{0\}. \end{aligned} \qquad (10)$$

Thus we obtain that $d = 0$, a contradiction. □

The following lemma extends results of Herstein [4, Theorem 2] and Bell and Mason [2, Theorem 3].

Lemma 7. *Let N be a 2-torsion free 3-prime near ring admitting a nonzero semiderivation d and a map g associated with d such that g is onto and $g(xy) = g(x)g(y)$ for all $x, y \in N$. If $[d(x), d(y)] = 0$ for all $x, y \in N$, then N is a commutative ring.*

Proof. Suppose that

$$d(x)d(y) = d(y)d(x), \quad \forall x, y \in N. \tag{11}$$

Replacing y by yz in (11) and using Lemma 5(ii), we obtain

$$\begin{aligned} &d(x)d(y)z + d(x)g(y)d(z) \\ &= d(y)zd(x) + g(y)d(z)d(x), \quad \forall x, y, z \in N. \end{aligned} \tag{12}$$

Substituting $d(y)$ for y in (12) and using (11), we find that

$$d(x)d^2(y)z = d^2(y)zd(x), \quad \forall x, y, z \in N. \tag{13}$$

Taking zt instead of z in (13) after using (13), we arrive at

$$d^2(y)z[d(x), t] = 0, \quad \forall x, y, t, z \in N \tag{14}$$

which can be rewritten as

$$d^2(y)N[d(x), t] = \{0\}, \quad \forall x, y, t \in N. \tag{15}$$

In the light of the 3-primeness of N, (15) implies that

$$\begin{aligned} &d^2 = 0 \\ &\text{or } d(N) \subseteq Z. \end{aligned} \tag{16}$$

But $d^2 = 0$ contradicts Lemma 6, so $d(N)$ is contained in Z and N is a commutative ring by Lemma 4. □

3. The Condition $[F(N),F(N)] = \{0\}$

The theorems that we prove in this section are motivated by the results proved in [2, Theorem 2], [5, Theorem 2.1], [6, Theorem 2.1 and 4.1], and [3, Theorem 2.1].

Theorem 8. *Let N be a 2-torsion free 3-prime near ring with a generalized semiderivation F associated with a nonzero semiderivation d and onto map g associated with d such that $g(xy) = g(x)g(y)$ for all $x, y \in N$. If $F(N) \subseteq Z$, then N is a commutative ring.*

Proof. Assume that

$$F(x) \in Z, \quad \forall x \in N. \tag{17}$$

Taking into account Lemma 5(i), we have for all $x, n \in N$, $z \in Z$,

$$\begin{aligned} F(xz)n &= F(x)zn + g(x)d(z)n, \\ nF(xz) &= nF(x)z + ng(x)d(z) \\ &= F(x)zn + ng(x)d(z). \end{aligned} \tag{18}$$

Since $F(xz)n = nF(xz)$, we have $g(x)d(z)n = ng(x)d(z)$ for all $x, n \in N$, $z \in Z \setminus \{0\}$. Thus $g(x)d(z) \in Z$ for all $x \in N$. Let $d(Z) \neq \{0\}$. Choosing z such that $d(z) \neq 0$ and noting that $d(z) \in Z$, we have $g(x) \in Z$. Since g is onto, we have $N \subseteq Z$. Hence N is a commutative ring by Lemma 3. On the other hand if $d(z) = 0$, then for all $x, y \in N$

$$\begin{aligned} &0 = d(F(xy)), \\ &0 = d(F(x)y + g(x)d(y)), \\ &0 = F(x)d(y) + g(x)d^2(y) + d(g(x))g(d(y)), \\ &\qquad \forall x, y \in Z. \end{aligned} \tag{19}$$

Hence $F(xd(y)) = -d(g(x))g(d(y)) \in Z$ for all $x, y \in N$. Since g is onto, we have $d(x)d(y) \in Z$ for all $x, y \in N$. This implies that

$$d(x)(d(x)d(y) - d(y)d(x)) = 0, \quad \forall x, y \in N. \tag{20}$$

Left multiplying by $d(y)$, we arrive at

$$\begin{aligned} &d(y)d(x)N(d(x)d(y) - d(y)d(x)) = \{0\}, \\ &\qquad \forall x, y \in N. \end{aligned} \tag{21}$$

Since N is a 3-prime near ring, we get

$$[d(x), d(y)] = 0, \quad \forall x, y \in N. \tag{22}$$

We conclude that N is a commutative ring by Lemma 7. □

Corollary 9 (see [6, Theorem 3.2]). *Let N be a 2-torsion free 3-prime near ring. If N admits a nonzero generalized derivation F such that $F(N) \subseteq Z$, then N is a commutative ring.*

Theorem 10. *Let N be a 3-prime near ring admitting a generalized semiderivation F associated with a nonzero semiderivation d and onto map g associated with d such that $g(xy) = g(x)g(y)$ for all $x, y \in N$. If $[F(N), F(N)] = 0$, then $(N, +)$ is abelian.*

Proof. Assume that

$$F(x)F(y) = F(y)F(x), \quad \forall x, y \in N. \tag{23}$$

Then

$$\begin{aligned} &F(x + x)F(y + z) = F(y + z)F(x + x), \\ &\qquad \forall x, y, z \in N. \end{aligned} \tag{24}$$

By (23), the last equation yields that

$$\begin{aligned} &F(y)F(x + x) + F(z)F(x + x) \\ &= F(x)F(y + z) + F(x)F(y + z), \\ &\qquad \forall x, y, z \in N. \end{aligned} \tag{25}$$

Hence,

$$F(y)F(x)+F(y)F(x)+F(z)F(x)+F(z)F(x) = F(x)F(y)+F(x)F(z)+F(x)F(y) + F(x)F(z), \quad (26)$$

that is $F(y)F(x)+F(z)F(x) = F(x)F(z)+F(x)F(y), \quad \forall x,y,z \in N,$

which implies that

$$F(y+z-y-z)F(x)=0, \quad \forall x,y,z \in N. \quad (27)$$

Putting xr instead of x in (27), we get

$$F(y,z)g(x)d(r)=0, \quad \forall x,y,z,r \in N, \quad (28)$$

which can be rewritten as

$$F(y,z)Nd(r)=\{0\}, \quad \forall y,z,r \in N. \quad (29)$$

Since N is a 3-prime near ring and $d \neq 0$, we get

$$F(y,z)=0, \quad \forall y,z \in N. \quad (30)$$

Replacing y and z by ry and rz, respectively, in (30), we obtain

$$\begin{aligned} 0 &= F(r(y,z)) \\ &= rF(y,z)+d(r)g(y,z) \quad (31) \\ &= d(r)g(y,z), \quad \forall y,z,r \in N. \end{aligned}$$

Taking rt instead of r in the last equation and using Lemma 5(ii), we get

$$d(r)tg(y,z)=0, \quad \forall y,z,r,t \in N. \quad (32)$$

Thus,

$$d(r)Ng(y,z)=\{0\}, \quad \forall y,z,r \in N. \quad (33)$$

Again using the fact that N is 3-prime and $d \neq 0$, we find that $g(y)+g(z)=g(z)+g(y)$ for all $y,z \in N$. Since g is onto, $(N,+)$ is abelian. □

Theorem 11. *Let N be a 2-torsion free 3-prime near ring admitting a nonzero generalized semiderivation F associated with a nonzero semiderivation d and onto map g associated with d such that $g(xy)=g(x)g(y)$ for all $x,y \in N$. If $[F(N),F(N)]=0$, then N is a commutative ring.*

Proof. By the hypothesis

$$F(x)F(y)=F(y)F(x), \quad \forall x,y \in N. \quad (34)$$

Replace y by $F(z)y$ in the above relation, and we get

$$F(x)F(F(z)y)=F(F(z)y)F(x), \quad \forall x,y,z \in N. \quad (35)$$

This implies that

$$F(x)(d(F(z))g(y)+F(z)F(y)) = (d(F(z))g(y)+F(z)F(y))F(x), \quad \forall x,y,z \in N. \quad (36)$$

Using Lemma 5(i), we find that

$$F(x)d(F(z))g(y)=d(F(z))g(y)F(x), \quad \forall x,y,z \in N. \quad (37)$$

Taking yw instead of y in (37) and using (37)

$$d(F(z))g(y)F(x)g(w) = d(F(z))g(y)g(w)F(x), \quad \forall x,y,z,w \in N. \quad (38)$$

Since g is onto, we get

$$d(F(z))yF(x)w=d(F(z))ywF(x), \quad \forall x,y,z,w \in N. \quad (39)$$

This implies that

$$d(F(z))N(F(x)w-wF(x))=\{0\}, \quad \forall z,w \in N. \quad (40)$$

Since N is a 3-prime near ring, we have

$$d(F(N))=\{0\} \text{ or } F(N)\subseteq Z. \quad (41)$$

If $F(N)$ is contained in Z, then N is a commutative ring by Theorem 8. On the other hand, we see that if $d(F(N))=0$, then

$$d(F(xy))=d(d(x)g(y)+xF(y))=0, \quad \forall x,y \in N. \quad (42)$$

Thus

$$d^2(x)g(y)+d(x)d(g(y))+d(x)F(y) + xd(F(y))=0, \quad \forall x,y \in N. \quad (43)$$

This implies that

$$d^2(x)g(y)+d(x)d(g(y))+d(x)F(y)=0, \quad \forall x,y \in N. \quad (44)$$

Replacing y by yz and using the fact that g is onto, we get

$$d^2(x)yz + d(x)d(yz) + d(x)F(yz) = 0, \quad \forall x, y, z \in N.$$

$$d^2(x)yz + d(x)(yd(z) + d(y)g(z)) + d(x)(F(y)z + g(y)d(z)) = 0, \quad \forall x, y, z \in N,$$

$$d^2(x)yz + d(x)yd(z) + d(x)d(y)g(z) + d(x)F(y)z + d(x)g(y)d(z) = 0, \quad \forall x, y, z \in N, \tag{45}$$

$$d^2(x)yz + d(x)yd(z) + d(x)d(y)z + d(x)F(y)z + d(x)yd(z) = 0, \quad \forall x, y, z \in N,$$

$$\{d^2(x)y + d(x)d(y) + d(x)F(y)\}z + 2d(x)yd(z) = 0, \quad \forall x, y, z \in N.$$

Since N is 2-torsion free, using (44) we get

$$d(x)yd(z) = 0, \quad \forall x, y, z \in N, \qquad d(N)Nd(N) = \{0\}. \tag{46}$$

Thus we obtain that $d = 0$, a contradiction which completes the proof. □

Corollary 12 ([6, Theorem 4.1]). *Let N be a 2-torsion free prime near ring. If N admits a generalized derivation F associated with a nonzero derivation d such that $[F(x), F(y)] = 0$ for all $x, y \in N$, then N is a commutative ring.*

Theorem 13. *Let N be a 2-torsion free 3-prime near ring. If F is a generalized semiderivation of N associated with a nonzero semiderivation d and an automorphism g associated with d, then the following assertions are equivalent:*

(i) *$F([x, y]) = [F(x), y]$ for all $x, y \in N$.*

(ii) *$F([x, y]) = -[F(x), y]$ for all $x, y \in N$.*

(iii) *N is a commutative ring.*

Proof. It is obvious that (iii) implies both (i) and (ii).

Now we prove that (i) ⇒ (iii). By hypothesis

$$F([x, y]) = [F(x), y], \quad \forall x, y \in N. \tag{47}$$

Taking xy instead of y in (47) and noting that $[x, xy] = x[x, y]$, we get

$$xF([x, y]) + d(x)g([x, y]) = F(x)xy - xyF(x), \quad \forall x, y \in N. \tag{48}$$

Using (47) and noting that $F(x)x = xF(x)$ by (47), then the last equation yields

$$d(x)g([x, y]) = 0, \quad \forall x, y \in N. \tag{49}$$

Since g is an automorphism, we get

$$d(x)g(x)g(y) = d(x)g(y)g(x), \quad \forall x, y \in N. \tag{50}$$

Replacing y by yt in (50) and using (50), we arrive at

$$d(x)y[x, t] = 0, \quad \forall x, y, t \in N. \tag{51}$$

This implies that

$$d(x)N[x, t] = \{0\}, \quad \forall x, t \in N. \tag{52}$$

3-primeness of N yields that either $N \subseteq Z$ or $d(N) = \{0\}$. In both the cases N is a commutative ring by Lemmas 3 and 4, respectively.

Using the similar techniques as above we can show that (ii) ⇒ (iii). □

Corollary 14 (see [7, Theorem 2.6]). *Let N be a 3-prime near ring. If N admits a generalized derivation F associated with a nonzero derivation d such that $F([x, y]) = [F(x), y]$ for all $x, y \in N$, then N is a commutative ring.*

Theorem 15. *Let N be a 2-torsion free 3-prime near ring. If F is a generalized semiderivation of N with associated semiderivation d and an automorphism g associated with d, then the following assertions are equivalent:*

(i) *$F([x, y]) = [x, F(y)]$ for all $x, y \in N$.*

(ii) *$F([x, y]) = -[x, F(y)]$ for all $x, y \in N$.*

(iii) *N is a commutative ring.*

Proof. Obviously, (iii) implies both (i) and (ii).

Now we prove that (i) ⇒ (iii). By hypothesis

$$F([x, y]) = [x, F(y)], \quad \forall x, y \in N. \tag{53}$$

Replacing x by yx in (53), we arrive at

$$yF([x, y]) + d(y)g([x, y]) = yxF(y) - F(y)yx, \quad \forall x, y \in N. \tag{54}$$

Using (53) and noting that $yF(y) = F(y)y$ by (53), we find that

$$d(y)g([x, y]) = 0, \quad \forall x, y \in N. \tag{55}$$

Arguing in the similar manner as in Theorem 13, we get the result.

Similarly we can prove that (ii) ⇒ (iii). □

Corollary 16 (see [7, Theorem 2.7]). *Let N be 3-prime near ring. If N admits a generalized derivation F associated with a nonzero derivation d such that $F([x, y]) = [x, F(y)]$ for all $x, y \in N$, then N is a commutative ring.*

The following example shows that the conditions on the hypothesis of the above theorems are not superfluous.

Example 17. Let S be a 2-torsion free left near ring and let

$$N = \left\{ \begin{pmatrix} 0 & 0 & x \\ 0 & 0 & y \\ 0 & 0 & 0 \end{pmatrix} \mid x, y \in S \right\}. \quad (56)$$

Define $F, d, g : N \to N$ by

$$F\begin{pmatrix} 0 & 0 & x \\ 0 & 0 & y \\ 0 & 0 & 0 \end{pmatrix} = \begin{pmatrix} 0 & 0 & 0 \\ 0 & 0 & y \\ 0 & 0 & 0 \end{pmatrix};$$
$$d\begin{pmatrix} 0 & 0 & x \\ 0 & 0 & y \\ 0 & 0 & 0 \end{pmatrix} = \begin{pmatrix} 0 & 0 & x \\ 0 & 0 & 0 \\ 0 & 0 & 0 \end{pmatrix}, \quad (57)$$
$$g\begin{pmatrix} 0 & 0 & x \\ 0 & 0 & y \\ 0 & 0 & 0 \end{pmatrix} = \begin{pmatrix} 0 & 0 & y \\ 0 & 0 & x \\ 0 & 0 & 0 \end{pmatrix}.$$

It can be checked that N is a left near ring and F is a generalized semiderivation of N associated with a semiderivation d and onto map g associated with d satisfying

(i) $[d(A), d(B)] = 0$,

(ii) $F(N) \subseteq Z$,

(iii) $F([A, B]) = \pm[F(A), B]$,

(iv) $F([A, B]) = \pm[A, F(B)]$,

(v) $[F(A), F(B)] = 0$

for all $A, B \in N$. However, N is not a commutative ring.

4. Generalized Semiderivations Acting as a Homomorphism or as an Antihomomorphism

In [8], Bell and Kappe proved that if R is a semiprime ring and d is a derivation on R which is either an endomorphism or an antiendomorphism on R, then $d = 0$. Of course, derivations which are not endomorphisms or antiendomorphisms on R may behave as such on certain subsets of R; for example, any derivation d behaves as the zero endomorphism on the subring C consisting of all constants (i.e., the elements x for which $d(x) = 0$). In fact in a semiprime ring R, d may behave as an endomorphism on a proper ideal of R. However as noted in [8], the behaviour of d is somewhat restricted in the case of a prime ring. Recently the authors in [9] considered (θ, ϕ)-derivation d acting as a homomorphism or an antihomomorphism on a nonzero Lie ideal of a prime ring and concluded that $d = 0$. In this section we establish similar results in the setting of a 3-prime near ring admitting a generalized semiderivation.

Theorem 18. *Let N be a 3-prime near ring. Suppose that F is a generalized semiderivation of N associated with a semiderivation d and onto map g associated with d such that $g(xy) = g(x)g(y)$ for all $x, y \in N$. If F acts as a homomorphism on N, then either F is identity map or $F = 0$.*

Proof. By the hypothesis

$$F(xy) = d(x)g(y) + xF(y) = F(x)F(y), \quad \forall x, y \in N. \quad (58)$$

Replacing y by yz in the above relation, we get

$$F(xyz) = d(x)g(yz) + xF(yz), \quad \forall x, y, z \in N.$$
$$F(xy)F(z) = d(x)g(yz) + xF(yz), \quad \forall x, y, z \in N. \quad (59)$$

This implies that

$$(d(x)g(y) + xF(y))F(z) = d(x)g(yz) + x(d(y)g(z) + yF(z)), \quad \forall x, y, z \in N. \quad (60)$$

Using Lemma 5(ii), we obtain

$$d(x)g(y)F(z) + xF(y)F(z) = d(x)g(yz) + xd(y)g(z) + xyF(z), \quad \forall x, y, z \in N.$$
$$d(x)g(y)F(z) + xF(yz) = d(x)g(yz) + xd(y)g(z) + xyF(z), \quad \forall x, y, z \in N. \quad (61)$$
$$d(x)g(y)F(z) + xd(y)g(z) + xyF(z) = d(x)g(yz) + xd(y)g(z) + xyF(z), \quad \forall x, y, z \in N.$$

This implies that

$$d(x)g(y)F(z) = d(x)g(y)g(z), \quad \forall x, y, z \in N, \quad (62)$$

that is $d(x)y(F(z) - z) = 0, \quad \forall x, y, z \in N.$

Thus

$$d(x)N(F(z) - z) = \{0\}, \quad \forall x, y, z \in N. \quad (63)$$

Therefore, $d(N) = \{0\}$ or $F(z) = z$ for all $z \in N$.

In the later case F is an identity map. On the other hand suppose that $d(N) = \{0\}$. Then $F(xy) = F(x)y = F(x)F(y)$; that is, $F(x)(y - F(y)) = 0$ for all $x, y \in N$. Replacing y by zy, $z \in N$, and noting that $F(zy) = zF(y)$, we have $F(x)N(y - F(y)) = \{0\}$ for all $x, y \in N$. Therefore, $F(N) = \{0\}$ or F is an identity map. □

Theorem 19. *Let N be a 2-torsion free 3-prime near ring. Suppose that F is a generalized semiderivation of N associated with a semiderivation d and onto map g such that $g(xy) = g(x)g(y)$ for all $x, y \in N$. If F acts as antihomomorphism on N, then $F = 0$ or F is the identity map on N and N is a commutative ring.*

Proof. By the hypothesis

$$F(xy) = d(x)g(y) + xF(y) = F(y)F(x), \quad \forall x, y \in N. \tag{64}$$

Thus

$$F(y)F(x) = d(x)g(y) + xF(y), \quad \forall x, y \in N. \tag{65}$$

Replacing y by xy in the above relation, we obtain

$$\begin{aligned} &F(xy)F(x) = d(x)g(xy) + xF(xy), \quad \forall x, y \in N, \\ &\text{that is } (d(x)g(y) + xF(y))F(x) \\ &\quad = d(x)g(xy) + xF(y), \quad \forall x, y \in N. \end{aligned} \tag{66}$$

By Lemma 5(ii), we have

$$\begin{aligned} &d(x)g(y)F(x) + xF(y)F(x) \\ &\quad = d(x)g(xy) + xF(y), \quad \forall x, y \in N. \end{aligned} \tag{67}$$

This implies that

$$d(x)g(y)F(x) = d(x)g(x)g(y), \quad \forall x, y \in N. \tag{68}$$

Replacing y by yr in the above relation, we get

$$d(x)g(y)g(r)F(x) = d(x)g(x)g(y)g(r), \quad \forall x, y, r \in N. \tag{69}$$

Using (68) in the above relation, we get

$$d(x)g(y)g(r)F(x) = d(x)g(y)F(x)g(r), \quad \forall x, y, r \in N. \tag{70}$$

Since g is onto, we have

$$d(x)N[F(x), r] = \{0\}, \quad \forall x, r \in N. \tag{71}$$

Therefore, either $d(N) = \{0\}$ or $F(N) \subseteq Z$. Hence in either case F acts as a homomorphism by Lemma 4 and Theorem 8 which completes the proof. □

Conflict of Interests

The authors declare that there is no conflict of interests regarding the publication of this paper.

References

[1] J. Bergen, "Derivations in prime rings," *Canadian Mathematical Bulletin*, vol. 26, no. 3, pp. 267–270, 1983.

[2] H. E. Bell and G. Mason, "On derivations in near-rings," *North-Holland Mathematics Studies*, vol. 137, pp. 31–35, 1987.

[3] A. Boua and L. Oukhtite, "Semiderivations satisfying certain algebraic identities on prime near-rings," *Asian-European Journal of Mathematics*, vol. 6, no. 3, Article ID 1350043, 8 pages, 2013.

[4] I. N. Herstein, "A note on derivations," *Canadian Mathematical Bulletin*, vol. 21, pp. 369–370, 1978.

[5] H. E. Bell, "On derivations in near-rings, II," in *Nearrings, Nearfields and K-Loops*, vol. 426 of *Mathematics and Its Applications*, pp. 191–197, Springer, Dordrecht, The Netherlands, 1997.

[6] H. E. Bell, "On prime near-rings with generalized derivation," *International Journal of Mathematics and Mathematical Sciences*, vol. 2008, Article ID 490316, 5 pages, 2008.

[7] A. Boua and L. Oukhtite, "Some conditions under which near-rings are rings," *Southeast Asian Bulletin of Mathematics*, vol. 37, no. 3, pp. 325–331, 2013.

[8] H. E. Bell and L.-C. Kappe, "Rings in which derivations satisfy certain algebraic conditions," *Acta Mathematica Hungarica*, vol. 53, no. 3-4, pp. 339–345, 1989.

[9] A. Asma, N. Rehman, and A. Shakir, "On Lie ideals with derivations as homomorphisms and anti-homomorphisms," *Acta Mathematica Hungarica*, vol. 101, no. 1-2, pp. 79–82, 2003.

17

Characterization and Enumeration of Good Punctured Polynomials over Finite Fields

Somphong Jitman,[1] Aunyarut Bunyawat,[2] Supanut Meesawat,[2] Arithat Thanakulitthirat,[2] and Napat Thumwanit[2]

[1]Department of Mathematics, Faculty of Science, Silpakorn University, Nakhon Pathom 73000, Thailand
[2]Department of Mathematics, Mahidol Wittayanusorn School, Nakhon Pathom 73170, Thailand

Correspondence should be addressed to Somphong Jitman; sjitman@gmail.com

Academic Editor: Shyam L. Kalla

A family of good punctured polynomials is introduced. The complete characterization and enumeration of such polynomials are given over the binary field $\mathbb{F}_2$. Over a nonbinary finite field $\mathbb{F}_q$, the set of good punctured polynomials of degree less than or equal to 2 are completely determined. For $n \geq 3$, constructive lower bounds of the number of good punctured polynomials of degree n over $\mathbb{F}_q$ are given.

1. Introduction

From the fundamental theorem of algebra, every polynomial over the rational numbers $\mathbb{Q}$ (or over the real numbers $\mathbb{R}$) has a root in $\mathbb{C}$. However, it is not guaranteed that a polynomial has a root in $\mathbb{Q}$ or in $\mathbb{R}$. Therefore, for a given polynomial over $\mathbb{Q}$ (resp., $\mathbb{R}$), it is of natural interest to determine whether it has a root in $\mathbb{Q}$ (resp., $\mathbb{R}$). In general, determining whether a given polynomial has a root in a nonalgebraically closed field is an interesting problem and has been extensively studied (see, e.g., [1–4]).

In this paper, we introduce punctured forms of a polynomial $f(x)$ over a field $\mathbb{F}$ (see the definition below) and focus on determining whether the punctured parts of $f(x)$ have a root in $\mathbb{F}$. Due to the rich algebraic structures and various applications of polynomials over finite fields (see [5–9] and references therein), their properties such as factorization, root finding, and irreducibility have extensively been studied (see [10–13]). In this paper, we mainly focus on punctured polynomials over a finite field which is not an algebraically closed field. The readers may refer to [5] for more details on finite fields and polynomials over finite fields.

Let q be a prime power and let $\mathbb{F}_q$ denote the finite field of q elements. Denote by $\mathbb{F}_q^*$ the set of nonzero elements in $\mathbb{F}_q$. Let

$$\mathbb{F}_q[x] = \left\{ \sum_{i=0}^{m} a_i x^i \mid a_i \in \mathbb{F}_q \ \forall i = 0, 1, 2, \ldots, m, \ m \in \mathbb{N} \cup \{0\} \right\} \tag{1}$$

be the set of all polynomials with indeterminate x over $\mathbb{F}_q$. Let

$$\begin{aligned} \mathbb{F}_q[x]_n &= \left\{ f(x) \in \mathbb{F}_q[x] \mid \deg(f(x)) \leq n \right\}, \\ \widehat{\mathbb{F}}_q[x]_n &= \left\{ f(x) \in \mathbb{F}_q[x] \mid \deg(f(x)) = n \right\} \end{aligned} \tag{2}$$

be the set of all polynomials of degree less than or equal to n over $\mathbb{F}_q$ and the set of all polynomials of degree n over $\mathbb{F}_q$, respectively.

Given a polynomial $f(x) = \sum_{i=0}^{m} a_i x^i \in \mathbb{F}_q[x]$ of degree m, for each $j \in \{0,1,2,\ldots,m\}$, the jth *punctured polynomial of* $f(x)$ is defined to be

$$f_{(j)}(x) = \sum_{i=0}^{j-1} a_i x^i + \sum_{i=j+1}^{m} a_i x^{i-1}. \tag{3}$$

For convenience, by abuse of notation, the degree of zero polynomial is defined to be 0. Hence, we can write $f_{(0)}(x) = 0$ for all constant polynomials $f(x) = a \in \mathbb{F}_q$.

A polynomial $f(x) \in \mathbb{F}_q[x]$ of degree m is said to be *good punctured* if $f_{(j)}(x)$ has a root in $\mathbb{F}_q$ for all $j \in \{0,1,2,\ldots,m\}$. Otherwise, $f(x)$ is said to be *bad punctured*. The constant polynomials $f(x) = a \in \mathbb{F}_q$ are always good punctured and referred to as *trivial good punctured polynomials*. A good punctured polynomial is called *nontrivial* if it is not a trivial good punctured polynomial.

Example 1. Let $f(x) = x^2 + 2x + 2$ be a polynomial in $\mathbb{F}_3[x]$. Then $f_{(2)}(x) = 2x + 2$, $f_{(1)}(x) = x + 2$, and $f_{(0)}(x) = x + 2$. It is not difficult to see that $f(x)$ is good punctured.

Given a positive integer n and a prime power q, let $T_{(q,n)}$ and $\widehat{T}_{(q,n)}$ denote the set of good punctured polynomials of degree less than or equal to n over $\mathbb{F}_q$ and the set of all good punctured polynomials of degree n over $\mathbb{F}_q$, respectively. Precisely,

$$\begin{aligned} T_{(q,n)} &= \{f(x) \in \mathbb{F}_q[x]_n \mid f_{(j)}(x) \text{ has a root in } \mathbb{F}_q\ \forall 0 \\ &\quad \le j \le \deg(f(x))\}, \\ \widehat{T}_{(q,n)} &= \{f(x) \\ &\quad \in \widehat{\mathbb{F}}_q[x]_n \mid f_{(j)}(x) \text{ has a root in } \mathbb{F}_q\ \forall 0 \le j \le n\}. \end{aligned} \tag{4}$$

By convention, since $\deg(0) = 0$, we have $T_{(q,0)} = \widehat{T}_{(q,0)} = \mathbb{F}_q$ and $|T_{(q,0)}| = |\widehat{T}_{(q,0)}| = q$.

Remark 2. From the definitions of $T_{(q,n)}$ and $\widehat{T}_{(q,n)}$, we have the following facts:

(1) $\widehat{T}_{(q,n)}$ is a subset of $T_{(q,n)}$.

(2) $T_{(q,n)} = T_{(q,n-1)} \cup \widehat{T}_{(q,n)}$ is a disjoint union for all $n \ge 2$.

(3) $|T_{(q,n)}| = |T_{(q,n-1)}| + |\widehat{T}_{(q,n)}|$ for all $n \ge 2$.

Example 3. Over the finite fields $\mathbb{F}_2$, we have

$$\begin{aligned} T_{(2,2)} &= \{0, 1, x^2, x^2 + x, x^2 + x + 1\}, \\ \widehat{T}_{(2,2)} &= \{x^2, x^2 + x, x^2 + x + 1\}. \end{aligned} \tag{5}$$

Hence, $|T_{(2,2)}| = 5$ and $|\widehat{T}_{(2,2)}| = 3$, respectively.

In this paper, we focus on the characterization and enumeration of the good punctured polynomials of degree n over $\mathbb{F}_q$. The complete characterization and enumeration of good punctured polynomials over the binary field $\mathbb{F}_2$ are given in Section 2. In Section 3, good punctured polynomials of degree n over $\mathbb{F}_q$, where $q > 2$, are studied. The good punctured polynomials of degree less than or equal to 2 over fields $\mathbb{F}_q$ are completely determined. Lower bounds of the size of the set of good punctured polynomials of degree greater than 2 are provided as well. Conclusion and some discussions about future researches on punctured polynomials are provided in Section 4.

2. Good Punctured Polynomials over the Binary Field

In this section, we focus on good punctured polynomials over the finite field $\mathbb{F}_2$. The characterization and enumeration of such polynomials are completely determined.

First, we determine the set $T_{(2,n)}$ of good punctured polynomials of degree less than or equal to n over the binary field $\mathbb{F}_2$. It is not difficult to see that $T_{(2,1)} = \mathbb{F}_2$. For $n \ge 2$, the set $T_{(2,n)}$ is given as follows.

Theorem 4. *Let $n \ge 2$ be a positive integer. Then*

$$T_{(2,n)} = A \cup B \cup C, \tag{6}$$

where $A = \{\sum_{i=0}^{2k} x^i \mid 0 \le k \le n/2\}$, $B = \{xf(x) \mid f(x) \in \mathbb{F}_2[x]_{n-1}$ and $f(1) = 0\}$, and $C = \{x^2 f(x) \mid f(x) \in \mathbb{F}_2[x]_{n-2}\}$.

Proof. First, we prove that $A \cup B \cup C \subseteq T_{(2,n)}$. Let $g(x) \in A \cup B \cup C$. We distinguish the proof into three cases.

Case 1 ($g(x) \in A$). Then $g(x) = \sum_{i=0}^{2k} x^i$ for some $0 \le k \le n/2$. It follows that 1 is a root of $g_{(j)}(x) = \sum_{i=0}^{2k-1} x^i$ for all $0 \le j \le 2k$. Hence, $g(x) \in T_{(2,n)}$.

Case 2 ($g(x) \in B$). Then $g(x) = xf(x)$ for some $f(x) \in \mathbb{F}_2[x]_{n-1}$. We have $g_{(0)}(1) = f(1) = 0$ and $g_{(i)}(0) = g(0) = 0$ for all $0 < i \le \deg(g(x))$. Hence, $g_{(i)}(x)$ has a root in $\mathbb{F}_2$ for all $0 \le i \le \deg(g(x))$. Therefore, $g(x) \in T_{(2,n)}$.

Case 3 ($g(x) \in C$). Then $g(x) = x^2 h(x)$ for some $h(x) \in \mathbb{F}_2[x]_{n-2}$. It follows that 0 is a root of $g_{(0)}(x) = g_{(1)}(x) = xh(x)$ and $g_{(i)}(x) = x^2 h_{(i-2)}(x)$ for all $2 \le i \le \deg(g(x))$. Therefore, $f_{(i)}(x)$ has a root in $\mathbb{F}_2$ for all $0 \le i \le \deg(g(x))$. As desired, $g(x) \in T_{(2,n)}$.

On the other hand, let $g(x) \in T_{(2,n)}$. Write $g(x) = \sum_{i=0}^{\deg(g(x))} g_i x^i$ and consider the following two cases.

Case 1 ($g_0 = 0$)

Case 1.1 ($g_1 = 0$). Then $g(x) \in C$.

Case 1.2 ($g_1 = 1$). Then $\deg(g(x)) \ge 1$. Since $g_{(0)}(x) = \sum_{i=0}^{\deg(g(x))-1} g_{i+1} x^i$, we have $g_{(0)}(0) = g_1 = 1$. It follows that $0 = g_{(0)}(1) = \sum_{i=0}^{\deg(g(x))-1} g_{i+1} = g(1)$. Hence, $g(x) \in B$.

Case 2 ($g_0 = 1$). Since $g_{(0)}(0) = g_0 = 1$, we have $0 = g_{(0)}(1) = \sum_{i=1}^{\deg(g(x))} g_i$. Suppose that there exists $1 \le j < \deg(g(x))$ such that $g_j = 0$. Since $g_{(j)}(0) = g_0 = 1$ and $g_{(\deg(g(x)))}(0) = g_0 = 1$, we have

$$0 = g_{(j)}(1) = \sum_{i=0}^{j-1} g_i + \sum_{i=j+1}^{\deg(g(x))} g_i, \qquad 0 = g_{(\deg(g(x)))}(1) = \sum_{i=0}^{\deg(g(x))-1} g_i. \tag{7}$$

It follows that $0 = g_j = g_{\deg(g(x))} = 1$, a contradiction. Hence, $g_i = 1$ for all $0 \le i \le \deg(g(x))$. Since $0 = g_{(0)}(1) = \sum_{i=1}^{\deg(g(x))} g_i$, the degree of $g(x)$ must be even. We conclude that $g(x) \in A$.

From the two cases, we have $g(x) \in A \cup B \cup C$, and, hence, $T_{(2,n)} \subseteq A \cup B \cup C$.

Therefore, $T_{(2,n)} = A \cup B \cup C$ as desired. □

Corollary 5. *If n is a positive integer, then*

$$|T_{(2,n)}| = \begin{cases} 2 & \text{if } n = 1, \\ 3 \cdot 2^{n-2} + \left\lfloor \frac{n}{2} \right\rfloor + 1 & \text{if } n \ge 2. \end{cases} \tag{8}$$

Proof. By direct calculation, we have $T_{(2,1)} = \{0, 1\}$ and $|T_{(2,1)}| = 2$.

Next, assume that $n \ge 2$. By Theorem 4, we have

$$|T_{(2,n)}| = |A \cup B \cup C|, \tag{9}$$

where $A = \{\sum_{i=0}^{2k} x^i \mid 0 \le k \le n/2\}$, $B = \{xf(x) \mid f(x) \in \mathbb{F}_2[x]_{n-1}$ and $f(1) = 0\}$, and $C = \{x^2 f(x) \mid f(x) \in \mathbb{F}_2[x]_{n-2}\}$.

Since A and $B \cup C$ are disjoint, by the inclusion-exclusion principle, we have

$$|T_{(2,n)}| = |A \cup B \cup C| = |A| + |B| + |C| - |B \cap C|. \tag{10}$$

Clearly, $|A| = \lfloor n/2 \rfloor + 1$ and $|C| = 2^{n-1}$. Observe that $xf(x) = x\sum_{i=0}^{n-1} f_i x^i \in B$ if and only if $|\{0 \le i \le n-1 \mid f_i = 1\}|$ is even. Hence,

$$|B| = 2^{n-1}. \tag{11}$$

It is not difficult to see that

$$B \cap C = \{x^2 f(x) \mid f(x) \in \mathbb{F}_2[x]_{n-2},\ f(1) = 0\}, \tag{12}$$

and, hence,

$$|B \cap C| = 2^{n-2}. \tag{13}$$

Therefore,

$$\begin{aligned} T_{(2,n)} &= \left\lfloor \frac{n}{2} \right\rfloor + 1 + 2^{n-1} + 2^{n-1} - 2^{n-2} \\ &= 3 \cdot 2^{n-2} + \left\lfloor \frac{n}{2} \right\rfloor + 1 \end{aligned} \tag{14}$$

as desired. □

Next, we determine the set $\widehat{T}_{(2,n)}$ of good punctured polynomials of degree n over the binary field $\mathbb{F}_2$. Since $T_{(2,1)} = \mathbb{F}_2$, we have $\widehat{T}_{(2,1)} = \emptyset$. For $n \ge 2$, the set $\widehat{T}_{(2,n)}$ can be determined as follows.

Theorem 6. *If $n \ge 2$ is a positive integer, then*

$$\widehat{T}_{(2,n)} = \begin{cases} \widehat{A} \cup \widehat{B} \cup \widehat{C} & \text{if } n \text{ is even}, \\ \widehat{B} \cup \widehat{C} & \text{if } n \text{ is odd}, \end{cases} \tag{15}$$

where $\widehat{A} = \{\sum_{i=0}^{n} x^i\}$, $\widehat{B} = \{x(f(x) + x^{n-1}) \mid f(x) \in \mathbb{F}_2[x]_{n-2}$ and $f(1) = 1\}$, and $\widehat{C} = \{x^2(f(x) + x^{n-2}) \mid f(x) \in \mathbb{F}_2[x]_{n-3}\}$.

Proof. We prove the statement by determining the elements in $T_{(2,n)}$ of degree n. Let $A = \{\sum_{i=0}^{2k} x^i \mid 0 \le k \le n/2\}$, $B = \{xf(x) \mid f(x) \in \mathbb{F}_2[x]_{n-1}$ and $f(1) = 0\}$, and $C = \{x^2 f(x) \mid f(x) \in \mathbb{F}_2[x]_{n-2}\}$ be defined as in Theorem 4.

It is not difficult to see that the set of elements in B (resp., C) of degree n is $\widehat{B}$ (resp., $\widehat{C}$).

If n is even, then the set of elements in A of degree n is $\widehat{A}$. In the case where n is odd, the set of elements in A of degree n is empty.

By Theorem 4, the result, therefore, follows. □

Corollary 7. *If n is a positive integer, then*

$$|\widehat{T}_{(2,n)}| = \begin{cases} 0 & \text{if } n = 1, \\ 3 & \text{if } n = 2, \\ 3 \cdot 2^{n-3} & \text{if } n \ge 3 \text{ is odd}, \\ 3 \cdot 2^{n-3} + 1 & \text{if } n \ge 4 \text{ is even}. \end{cases} \tag{16}$$

Proof. By direct calculation, we have $\widehat{T}_{(2,1)} = \emptyset$ and

$$\widehat{T}_{(2,2)} = \{x^2, x^2 + x, x^2 + x + 1\}. \tag{17}$$

Hence, we have $|\widehat{T}_{(2,1)}| = 0$ and $\widehat{T}_{(2,2)} = 3$.

Next, assume that $n \ge 3$. By Theorem 6, we have

$$|\widehat{T}_{(2,n)}| = \begin{cases} |\widehat{A} \cup \widehat{B} \cup \widehat{C}| & \text{if } n \text{ is even}, \\ |\widehat{B} \cup \widehat{C}| & \text{if } n \text{ is odd}, \end{cases} \tag{18}$$

where $\widehat{A}$, $\widehat{B}$, and $\widehat{C}$ are defined as in Theorem 6. Since $\widehat{A}$ and $\widehat{B} \cup \widehat{C}$ are disjoint, by the inclusion-exclusion principle, we have

$$|\widehat{T}_{(2,n)}| = |\widehat{A} \cup \widehat{B} \cup \widehat{C}| = |\widehat{A}| + |\widehat{B}| + |\widehat{C}| - |\widehat{B} \cap \widehat{C}|. \tag{19}$$

We note that $|\widehat{A}| = 1$ and $|\widehat{C}| = 2^{n-2}$.

Since $x(f(x) + x^{n-1}) = x(\sum_{i=0}^{n-2} f_i x^i + x^{n-1}) \in \widehat{B}$ if and only if

$$|\{i \mid 0 \le i \le n-2,\ f_i = 1\}| \tag{20}$$

is odd, we have

$$|\widehat{B}| = 2^{n-2}. \tag{21}$$

TABLE 1: Punctured polynomials over $\mathbb{F}_2$.

n	1	2	3	4	5	6	7	8	9	10	11	12	13
$\lvert T_{(2,n)}\rvert$	2	5	8	15	27	52	100	197	389	774	1542	3079	6151
$\lvert \widehat{T}_{(2,n)}\rvert$	0	3	3	7	12	25	48	97	192	385	768	1537	3072

It is not difficult to see that

$$\widehat{B}\cap\widehat{C} = \left\{x^2\left(f(x)+x^{n-2}\right) \mid f(x)\in\mathbb{F}_2[x]_{n-3},\ f(1) = 1\right\}, \tag{22}$$

and, hence,

$$\left|\widehat{B}\cap\widehat{C}\right| = 2^{n-3}. \tag{23}$$

Therefore, by (18), we have

$$\begin{aligned}\left|\widehat{T}_{(2,n)}\right| &= \begin{cases}1+2^{n-2}+2^{n-2}-2^{n-3} & \text{if } n \text{ is even,}\\ 2^{n-2}+2^{n-2}-2^{n-3} & \text{if } n \text{ is odd}\end{cases}\\ &= \begin{cases}3\cdot 2^{n-3}+1 & \text{if } n \text{ is even,}\\ 3\cdot 2^{n-3} & \text{if } n \text{ is odd}\end{cases}\end{aligned} \tag{24}$$

as desired. □

Table 1 presents the numbers $|T_{(2,n)}|$ and $|\widehat{T}_{(2,n)}|$ for $n = 1,2,\ldots,13$. The relation $|T_{(2,n)}| = |T_{(2,n-1)}| + |\widehat{T}_{(2,n)}|$ in Remark 2 can be easily seen.

3. Punctured Polynomials over Nonbinary Finite Fields

In this section, we focus on punctured polynomials over nonbinary finite fields. Given a prime power $q > 2$, the characterization and enumeration of good punctured polynomials of degree less than or equal to 2 over $\mathbb{F}_q$ are completely determined. For $n \geq 3$, we construct subsets of $T_{(q,n)}$ and $\widehat{T}_{(q,n)}$ which lead to lower bounds of the cardinalities of $T_{(q,n)}$ and $\widehat{T}_{(q,n)}$, respectively.

Theorem 8. *If $q > 2$ is a prime power, then $T_{(q,1)} = \mathbb{F}_q$.*

Proof. By the definition, $\mathbb{F}_q \subseteq T_{(q,1)}$. Let $f(x) = ax+b \in T_{(q,1)}$. Since $f_{(0)}(x) = a$ has a root in $\mathbb{F}_q$, we have $a = 0$. Hence, $f(x) = b \in \mathbb{F}_q$ as desired. □

The next corollary follows immediately from Theorem 8.

Corollary 9. *If $q > 2$ is a prime power, then the following statements hold:*

(i) $|T_{(q,1)}| = q$.

(ii) $\widehat{T}_{(q,1)} = \emptyset$.

(iii) $|\widehat{T}_{(q,1)}| = 0$.

Theorem 10. *Let $q > 2$ be a prime power. Then*

$$\begin{aligned}T_{(q,2)} &= \left\{a_2x^2+a_1x+a_0 \mid (a_2,a_1,a_0)\in\mathbb{F}_q^*\times\mathbb{F}_q^*\times\mathbb{F}_q\right\}\\ &\quad\cup\left\{a_2x^2 \mid a_2\in\mathbb{F}_q^*\right\}\cup\mathbb{F}_q.\end{aligned} \tag{25}$$

Proof. Let $A = \{a_2x^2+a_1x+a_0 \mid (a_2,a_1,a_0)\in\mathbb{F}_q^*\times\mathbb{F}_q^*\times\mathbb{F}_q\}$ and $B = \{a_2x^2 \mid a_2\in\mathbb{F}_q^*\}$.

Let $f(x)\in A\cup B\cup\mathbb{F}_q$. We write $f(x) = a_2x^2+a_1x+a_0$ and consider the proof as two cases.

Case 1 ($f(x)\in A$). We have $f_{(0)}(x) = a_2x+a_1$, $f_{(1)}(x) = a_2x+a_0$, and $f_{(2)}(x) = a_1x+a_0$. Since a_2 and a_1 are nonzero, it follows that $-a_1a_2^{-1}$, $-a_0a_2^{-1}$, and $-a_0a_1^{-1}$ are roots of $f_{(0)}(x)$, $f_{(1)}(x)$, and $f_{(2)}(x)$, respectively. Hence, $f(x)\in T_{(q,2)}$.

Case 2 ($f(x)\in B$). Then $f(x) = a_2x^2$ for some $a_2\in\mathbb{F}_q^*$. It follows that 0 is a root of $f_{(0)}(x) = a_2x$, $f_{(1)}(x) = a_2x$, and $f_{(2)}(x) = 0$. Therefore, $f(x)\in T_{(q,2)}$.

Case 3 ($f(x)\in\mathbb{F}_q$). Then, by the definition, $f(x)\in T_{(q,2)}$.

On the other hand, let $f(x) = a_2x^2+a_1x+a_0\in T_{(q,2)}$. If $a_2 = 0$, then $f(x) = a_1x+a_0\in T_{(q,1)}$, and, hence, $f(x) = a_0\in\mathbb{F}_q$ by Theorem 8. Assume that $a_2\neq 0$. Then $f_{(0)}(x) = a_2x+a_1$, $f_{(1)}(x) = a_2x+a_0$, and $f_{(2)}(x) = a_1x+a_0$ have a root in $\mathbb{F}_q$.

Case 1 ($a_1 = 0$). We have that $f_{(2)}(x) = a_0$ has a root in $\mathbb{F}_q$ which implies that $a_0 = 0$. Therefore, $f(x) = a_2x^2\in B$.

Case 2 ($a_1\neq 0$). Since $f_{(0)}(x) = a_2x+a_1$ has a root in $\mathbb{F}_q$, we have $a_2\neq 0$. Hence, $f(x)\in A$.

From the two cases, it can be concluded that $f(x)\in A\cup B\cup\mathbb{F}_q$.

As desired, we have $T_{(q,2)} = A\cup B\cup\mathbb{F}_q$. □

Corollary 11. *If $q > 2$ is a prime power, then*

$$\left|T_{(q,2)}\right| = q\left(q^2-2q+3\right)-1. \tag{26}$$

Proof. Let $A = \{a_2x^2+a_1x+a_0 \mid (a_2,a_1,a_0)\in\mathbb{F}_q^*\times\mathbb{F}_q^*\times\mathbb{F}_q\}$ and $B = \{a_2x^2 \mid a_2\in\mathbb{F}_q^*\}$ be defined as in the proof of Theorem 10. It is not difficult to see that A, B, and $\mathbb{F}_q$ are disjoint. By Theorem 10, we have

$$\begin{aligned}\left|T_{(q,2)}\right| &= \left|A\cup B\cup\mathbb{F}_q\right| = |A|+|B|+\left|\mathbb{F}_q\right|\\ &= (q-1)(q-1)q+(q-1)+q\\ &= q\left(q^2-2q+3\right)-1\end{aligned} \tag{27}$$

as desired. □

Corollary 12. *If q is a prime power, then*

$$\widehat{T}_{(q,2)} = \left\{a_2x^2 + a_1x + a_0 \mid (a_2, a_1, a_0) \in \mathbb{F}_q^* \times \mathbb{F}_q^* \times \mathbb{F}_q\right\} \cup \left\{a_2x^2 \mid a_2 \in \mathbb{F}_q^*\right\}. \quad (28)$$

Proof. From Theorem 10, it is not difficult to see that the polynomials of degree less than 2 in $T_{(q,2)}$ are $f(x) = a \in \mathbb{F}_q$. Hence, the result follows. □

Corollary 13. *If q is a prime power, then*

$$\left|\widehat{T}_{(q,2)}\right| = (q-1)\left(q^2 - q + 1\right). \quad (29)$$

Proof. From Corollaries 11 and 12, it follows that

$$\begin{aligned}\left|\widehat{T}_{(q,2)}\right| &= \left|T_{(q,2)}\right| - q = q\left(q^2 - 2q + 3\right) - 1 - q \\ &= (q-1)\left(q^2 - q + 1\right).\end{aligned} \quad (30)$$

□

In the case where $n \geq 3$, determining the sets $T_{(q,n)}$ and $\widehat{T}_{(q,n)}$ is more tedious and complicated. For these cases, we give constructive lower bounds of $|T_{(q,n)}|$ and $|\widehat{T}_{(q,n)}|$.

The following results are important tools in constructing lower bounds of $|T_{(q,n)}|$ and $|\widehat{T}_{(q,n)}|$.

Theorem 14 (see [14, Page 588]). *Let n be a positive integer and let q be a prime power. Then the number of monic irreducible polynomials of degree n in $\mathbb{F}_q[x]$ is*

$$L(n,q) = \frac{1}{n}\sum_{d|n} \mu(d)\, q^{n/d}, \quad (31)$$

where

$$\mu(n) = \begin{cases} 1 & \text{if } n = 1, \\ 0 & \text{if } n \text{ contains a repeated prime factor}, \\ (-1)^r & \text{if } n \text{ is a product of } r \text{ distinct primes} \end{cases} \quad (32)$$

is the Möbius function.

Theorem 15 (see [3, Section 4.2, Theorem 1]). *Let $f(x)$ be a polynomial of degree 2 or 3 in $\mathbb{F}_q[x]$. Then $f(x)$ is reducible if and only if $f(x)$ has a root in $\mathbb{F}_q[x]$.*

Theorem 16. *If $q > 2$ is a prime power, then*

$$\left|\widehat{T}_{(q,3)}\right| \geq \frac{1}{2}(q-1)\,q\,(q+1). \quad (33)$$

Proof. Let $S := \{xf(x) \mid f(x) \in \widehat{\mathbb{F}}_q[x]_2$ and $f(x)$ has a root in $\mathbb{F}_q\}$.

First, we show that $S \subseteq \widehat{T}_{(q,3)}$. Let $g(x) = x(a_2x^2 + a_1x + a_0) \in S$, where $a_2 \in \mathbb{F}_q^*$ and $a_1, a_0 \in \mathbb{F}_q$. Then $g_{(0)}(x) = a_2x^2 + a_1x + a_0$ has a root in $\mathbb{F}_q$ and 0 is a root of

$$\begin{aligned} g_{(1)}(x) &= a_2x^2 + a_1x, \\ g_{(2)}(x) &= a_2x^2 + a_0x, \\ g_{(3)}(x) &= a_1x^2 + a_0x. \end{aligned} \quad (34)$$

Hence, $g(x) \in \widehat{T}_{(q,3)}$ is good punctured.

By Theorem 14, the number of monic irreducible polynomials of degree 2 over $\mathbb{F}_q$ is

$$\begin{aligned} L(q,2) &= \frac{1}{2}\sum_{d|2} \mu(d)\, q^{2/d} = \frac{1}{2}\left(\mu(1)\, q^2 + \mu(2)\, q^1\right) \\ &= \frac{1}{2}\left(q^2 - q\right). \end{aligned} \quad (35)$$

Hence, the number of irreducible polynomials of degree 2 over $\mathbb{F}_q$ is

$$\frac{1}{2}(q-1)\left(q^2 - q\right). \quad (36)$$

By Theorem 15, the number of polynomials of degree 2 over $\mathbb{F}_q$ having a root in $\mathbb{F}_q$ is

$$\begin{aligned} |S| &= (q-1)\,q^2 - \frac{1}{2}(q-1)\left(q^2 - q\right) \\ &= \frac{1}{2}(q-1)\,q\,(q+1). \end{aligned} \quad (37)$$

Since $S \subseteq \widehat{T}_{(q,3)}$, we have

$$\left|\widehat{T}_{(q,3)}\right| \geq |S| = \frac{1}{2}(q-1)\,q\,(q+1) \quad (38)$$

as desired. □

Corollary 17. *If $q > 2$ is a prime power, then*

$$\left|T_{(q,3)}\right| \geq \frac{3q^3 - 4q^2 + 5q - 2}{2}. \quad (39)$$

Proof. By Corollary 11 and Theorem 16, we have

$$\begin{aligned} \left|T_{(q,2)}\right| &= q\left(q^2 - 2q + 3\right) - 1, \\ \left|\widehat{T}_{(q,3)}\right| &\geq \frac{1}{2}(q-1)\,q\,(q+1). \end{aligned} \quad (40)$$

Hence, by Remark 2, we have the relation

$$\begin{aligned} \left|T_{(q,3)}\right| &= \left|T_{(q,2)}\right| + \left|\widehat{T}_{(q,3)}\right| \\ &\geq q\left(q^2 - 2q + 3\right) - 1 + \frac{1}{2}(q-1)\,q\,(q+1) \\ &= \frac{3q^3 - 4q^2 + 5q - 2}{2}. \end{aligned} \quad (41)$$

□

Theorem 18. *If $q > 2$ is a prime power, then*

$$\left|\widehat{T}_{(q,4)}\right| \geq \frac{1}{2}(q-1)q(q^2+1). \tag{42}$$

Proof. Let $S := \{xf(x) \mid f(x) \in \widehat{\mathbb{F}}_q[x]_3$ and $f(x)$ has a root in $\mathbb{F}_q\}$.

First, we show that $S \subseteq \widehat{T}_{(q,4)}$. Let $g(x) = x(a_3x^3 + a_2x^2 + a_1x + a_0) \in S$, where $a_3 \in \mathbb{F}_q^*$ and $a_2, a_1, a_0 \in \mathbb{F}_q$. Then $g_{(0)}(x) = a_3x^3 + a_2x^2 + a_1x + a_0$ has a root in $\mathbb{F}_q$ and 0 is a root of

$$\begin{aligned} g_{(1)}(x) &= a_3x^3 + a_2x^2 + a_1x, \\ g_{(2)}(x) &= a_3x^3 + a_2x^2 + a_0x, \\ g_{(3)}(x) &= a_3x^3 + a_1x^2 + a_0x, \\ g_{(4)}(x) &= a_2x^3 + a_1x^2 + a_0x. \end{aligned} \tag{43}$$

Therefore, $g(x) \in \widehat{T}_{(q,4)}$ is good punctured as desired.

By Theorem 14, the number of monic irreducible polynomials of degree 3 over $\mathbb{F}_q$ is

$$\begin{aligned} L(q,3) &= \frac{1}{3}\sum_{d|3}\mu(d)q^{3/d} = \frac{1}{2}\left(\mu(1)q^3 + \mu(3)q^1\right) \\ &= \frac{1}{2}\left(q^3 - q\right). \end{aligned} \tag{44}$$

Hence, the number of irreducible polynomials of degree 3 over $\mathbb{F}_q$ is

$$\frac{1}{2}(q-1)\left(q^3 - q\right). \tag{45}$$

By Theorem 15, the number of polynomials of degree 3 over $\mathbb{F}_q$ having a root in $\mathbb{F}_q$ is

$$\begin{aligned} |S| &= (q-1)q^3 - \frac{1}{2}(q-1)\left(q^3 - q\right) \\ &= \frac{1}{2}(q-1)q\left(q^2+1\right). \end{aligned} \tag{46}$$

Since $S \subseteq \widehat{T}_{(q,4)}$, we have

$$\left|\widehat{T}_{(q,4)}\right| \geq |S| = \frac{1}{2}(q-1)q\left(q^2+1\right) \tag{47}$$

as desired. □

Corollary 19. *If $q > 2$ is a prime power, then*

$$\left|T_{(q,4)}\right| \geq \frac{q^4 + 2q^3 - 3q^2 + 4q - 2}{2}. \tag{48}$$

Proof. By Corollary 17, we have

$$\left|T_{(q,3)}\right| \geq \frac{3q^3 - 4q^2 + 5q - 2}{2}. \tag{49}$$

From Theorem 18, we have

$$\left|\widehat{T}_{(q,4)}\right| \geq \frac{1}{2}(q-1)q\left(q^2+1\right). \tag{50}$$

Hence, by Remark 2, we have the relation

$$\begin{aligned} \left|T_{(q,4)}\right| &= \left|T_{(q,3)}\right| + \left|\widehat{T}_{(q,4)}\right| \\ &\geq \frac{3q^3 - 4q^2 + 5q - 2}{2} + \frac{1}{2}(q-1)q\left(q^2+1\right) \\ &= \frac{q^4 + 2q^3 - 3q^2 + 4q - 2}{2}. \end{aligned} \tag{51}$$

□

Theorem 20. *Let $q > 2$ be a prime power and let $n \geq 5$ be an integer. Then*

$$\left|T_{(q,n)}\right| \geq q^{n-1} + q^3 - 2q^2 + 2q. \tag{52}$$

Proof. Let $A = \{x^2 f(x) \mid f(x) \in \mathbb{F}_q[x]_{n-2}\}$.

First, we show that $A \cup T_{(q,2)} \subseteq T_{(q,n)}$. Clearly, $T_{(q,2)} \subseteq T_{(q,n)}$. Let $g(x) \in A$. Then 0 is a root of $g_{(i)}(x)$ for all $i = 0, 1, \ldots, \deg(g(x))$, and, hence, $g(x) \in T_{(q,n)}$. Note that $A \cap T_{(q,2)} = \{ax^2 \mid a \in \mathbb{F}_q^*\}$. By Corollary 11, we have

$$\left|T_{(q,2)}\right| = q\left(q^2 - 2q + 2\right). \tag{53}$$

Therefore, consider

$$\begin{aligned} \left|T_{(q,n)}\right| &\geq |A| + \left|T_{(q,2)}\right| - \left|A \cap T_{(q,2)}\right| \\ &= q^{n-1} + q\left(q^2 - 2q + 3\right) - 1 - (q-1) \\ &= q^{n-1} + q^3 - 2q^2 + 2q. \end{aligned} \tag{54}$$

□

Corollary 21. *Let $q > 2$ be a prime power and let $n \geq 5$ be an integer. Then $|\widehat{T}_{(q,n)}| \geq (q-1)q^{n-2}$.*

Proof. The set of elements in A of degree n in the proof of Theorem 20 is $\widehat{A} = \{x^2 f(x) \mid f(x) \in \widehat{\mathbb{F}}_q[x]_{n-2}\}$. By Theorem 20, we have

$$\left|\widehat{T}_{(q,n)}\right| \geq \left|\widehat{A}\right| \geq (q-1)q^{n-2}. \tag{55}$$

□

4. Conclusion and Open Problems

The concepts of punctured polynomials and good punctured polynomials are introduced. Over the finite field $\mathbb{F}_2$, the complete characterization and enumeration of such polynomials are given. Over nonbinary finite fields, the good punctured polynomials of degree less than or equal to 2 are completely determined. For $n \geq 3$, constructive lower bounds of the number of good punctured polynomials of degree n are given.

In general, the following related problems are also interesting:

(1) Determine the sets $\widehat{T}_{(q,n)}$ and $\widehat{T}_{(q,n)}$, where $q > 2$ is a prime power and $n \geq 3$ is an integer.

(2) Determine the exact values of $|\widehat{T}_{(q,n)}|$ and $|\widehat{T}_{(q,n)}|$, where $q > 2$ is a prime power and $n \geq 3$ is an integer.

(3) Improve lower bounds of $|\widehat{T}_{(q,n)}|$ and $|\widehat{T}_{(q,n)}|$, where $q > 2$ is a prime power and $n \geq 3$ is an integer.

(4) Characterize and enumerate the good punctured polynomials of degree n over the real numbers $\mathbb{R}$ or over the rational numbers $\mathbb{Q}$.

Competing Interests

The authors declare that there are no competing interests regarding the publication of this paper.

Acknowledgments

This research is supported by the Thailand Research Fund under Research Grant TRG5780065.

References

[1] B. Kalantari, *Polynomial Root-Finding and Polynomiography*, World Scientific Publishing, River Edge, NJ, USA, 2008.

[2] V. Y. Pan and A.-L. Zheng, "New progress in real and complex polynomial root-finding," *Computers & Mathematics with Applications*, vol. 61, no. 5, pp. 1305–1334, 2011.

[3] W. K. Nicholson, *Introduction to Abstract Algebra*, John Wiley & Sons, New York, NY, USA, 2012.

[4] C. C. Pinter, *A Book of Abstract Algebra*, Dover, Mineola, NY, USA, 2nd edition, 2010.

[5] R. Lidl and H. Niederreiter, *Finite Fields*, Cambridge University Press, Cambridge, UK, 1997.

[6] E. Sangwisut, S. Jitman, S. Ling, and P. Udomkavanich, "Hulls of cyclic and negacyclic codes over finite fields," *Finite Fields and their Applications*, vol. 33, pp. 232–257, 2015.

[7] A. Akbary and Q. Wang, "On some permutation polynomials over finite fields," *International Journal of Mathematics and Mathematical Sciences*, vol. 2005, no. 16, pp. 2631–2640, 2005.

[8] R. A. Mollin and C. Small, "On permutation polynomials over finite fields," *International Journal of Mathematics and Mathematical Sciences*, vol. 10, no. 3, pp. 535–543, 1987.

[9] A. Sahni and P. T. Sehgal, "Hermitian self-orthogonal constacyclic codes over finite fields," *Journal of Discrete Mathematics*, vol. 2014, Article ID 985387, 7 pages, 2014.

[10] E. R. Berlekamp, "Factoring polynomials over finite fields," *The Bell System Technical Journal*, vol. 46, pp. 1853–1859, 1967.

[11] M. C. R. Butler, "On the reducibility of polynomials over a finite field," *The Quarterly Journal of Mathematics*, vol. 5, pp. 102–107, 1954.

[12] J.-M. Couveignes and R. Lercier, "Fast construction of irreducible polynomials over finite fields," *Israel Journal of Mathematics*, vol. 194, no. 1, pp. 77–105, 2013.

[13] A. Knopfmacher and J. Knopfmacher, "Counting polynomials with a given number of zeros in a finite field," *Linear and Multilinear Algebra*, vol. 26, no. 4, pp. 287–292, 1990.

[14] D. S. Dummit and R. M. Foote, *Abstract Algebra*, John Wiley & Sons, New York, NY, USA, 3rd edition, 2003.

18

Generalized Bell Numbers and Peirce Matrix via Pascal Matrix

Eunmi Choi

Department of Mathematics, HanNam Univ., Daejeon, Republic of Korea

Correspondence should be addressed to Eunmi Choi; emc@hnu.kr

Academic Editor: Pentti Haukkanen

With the Stirling matrix S and the Pascal matrix T, we show that T^kS ($k \geq 0$) satisfies a type of generalized Stirling recurrence. Then, by expressing the sum of components of each row of T^kS as k-Bell number, we investigate properties of k-Bell numbers as well as k-Peirce matrix.

1. Introduction

The Stirling number $s_{m,n}$ of the second kind is the number of ways to partition an m elements set into n nonempty subsets for any $m, n \geq 0$. $s_{m,n}$ can be expressed by $n!s_{m,n} = \sum_{i=0}^{n}(-1)^i \binom{n}{i} (n-i)^m$, and the Stirling matrix $S = [s_{m,n}]$ $(m, n \geq 0)$ satisfies the Stirling recurrence $s_{m,n} = s_{m-1,n-1} + ns_{m-1,n}$ [1]. Some researches including [2–4] were devoted to investigating the Stirling matrix S with the Pascal matrix T and binomial expressions. On the other hand, the sum $B_m = \sum_{k=0}^{m} s_{m,k}$ of numbers of the mth row of S is called a Bell number, so $\{B_m \mid m \geq 0\} = \{1, 1, 2, 5, 15, \cdots\}$ counts the number of partitions of an m elements set. One effective way for generating Bell numbers is to use the Peirce matrix (often called Bell table or Aitken's array), on which Bell numbers appear along both borders [5–7].

A main purpose of work is to study T^kS for $k \geq 1$. We show that T^kS satisfies a kind of generalized Stirling recurrence, and then by expressing the sum of components of each row of T^kS as a k-Bell number we investigate k-Bell numbers as well as k-Peirce matrix. We discuss recurrence rules of a k-Peirce matrix and then interrelationships between each k-Peirce matrices.

2. Generalized k-Stirling Matrix and k-Bell Numbers

Throughout the work, we write T for the Pascal matrix and r_m for the mth row of T. And M^{tr} means a transpose matrix of M. Let the Stirling matrix be

$$S = \begin{bmatrix} 1 & & & & \\ 0 & 1 & & & \\ 0 & 1 & 1 & & \\ 0 & 1 & 3 & 1 & \\ 0 & 1 & 7 & 6 & 1 \\ & & \cdots & & \end{bmatrix}$$

$$\text{and } S^* = \begin{bmatrix} 1 & & & & \\ 1 & 1 & & & \\ 1 & 3 & 1 & & \\ 1 & 7 & 6 & 1 & \\ 1 & 15 & 25 & 10 & 1 \\ & \cdots & & & \end{bmatrix}. \quad (1)$$

$$\text{Clearly } S \begin{bmatrix} 1 \\ 1 \\ 1 \\ \cdots \end{bmatrix} = \begin{bmatrix} B_0 \\ B_1 \\ B_2 \\ \cdots \end{bmatrix}.$$

Theorem 1. *We have the following:*

(1) $S\begin{bmatrix}1\\2\\3\\\cdots\end{bmatrix} = \begin{bmatrix}B_1\\B_2\\B_3\\\cdots\end{bmatrix} = S^*\begin{bmatrix}1\\1\\1\\\cdots\end{bmatrix}$, $S^*\begin{bmatrix}1\\2\\3\\\cdots\end{bmatrix} = \begin{bmatrix}B_2-B_1\\B_3-B_2\\B_4-B_3\\\cdots\end{bmatrix}$ *and* $S^*\begin{bmatrix}2\\3\\4\\\cdots\end{bmatrix} = \begin{bmatrix}B_2\\B_3\\B_4\\\cdots\end{bmatrix}$.

(2) $T\begin{bmatrix}B_0\\B_1\\B_2\\\cdots\end{bmatrix} = \begin{bmatrix}B_1\\B_2\\B_3\\\cdots\end{bmatrix}$ *and* $T\begin{bmatrix}B_1\\B_2\\B_3\\\cdots\end{bmatrix} = S^*\begin{bmatrix}1\\2\\3\\\cdots\end{bmatrix}$.

Proof. The first two identities in (1) are easy to observe. And

$$S^*\begin{bmatrix}2\\3\\4\\\cdots\end{bmatrix} = S^*\begin{bmatrix}1\\1\\1\\\cdots\end{bmatrix} + S^*\begin{bmatrix}1\\2\\3\\\cdots\end{bmatrix} = \begin{bmatrix}B_1\\B_2\\B_3\\\cdots\end{bmatrix} + \begin{bmatrix}B_2-B_1\\B_3-B_2\\B_4-B_3\\\cdots\end{bmatrix} = \begin{bmatrix}B_2\\B_3\\B_4\\\cdots\end{bmatrix}. \tag{2}$$

Clearly $\begin{bmatrix}1&&\\1&1&\\1&2&1\end{bmatrix}\begin{bmatrix}1\\1\\2\end{bmatrix} = \begin{bmatrix}1\\2\\5\end{bmatrix} = \begin{bmatrix}B_1\\B_2\\B_3\end{bmatrix}$. And $B_m = \sum_{i=0}^{m-1}\binom{m-1}{i}B_i$ in [8] proves $r_m(B_0,\cdots,B_{m-1})^{tr} = B_m$ for $m \ge 1$, so we have $T\begin{bmatrix}B_0\\B_1\\B_2\\\cdots\end{bmatrix} = \begin{bmatrix}B_1\\B_2\\B_3\\\cdots\end{bmatrix}$. Moreover since

$$\begin{aligned} r_{m+1} &= \left(\binom{m-1}{0},\cdots,\binom{m-1}{m-1},0\right) \\ &\quad + \left(0,\binom{m-1}{0},\cdots,\binom{m-1}{m-1}\right) \\ &= (r_m;0) + (0;r_m), \end{aligned} \tag{3}$$

it follows that

$$\begin{aligned} B_{m+1} &= r_{m+1}(B_0,B_1,\cdots,B_m)^{tr} \\ &= (r_m;0)(B_0,B_1,\cdots,B_m)^{tr} \\ &\quad + (0;r_m)(B_0,B_1,\cdots,B_m)^{tr} \\ &= r_m(B_0,\cdots,B_{m-1})^{tr} + r_m(B_1,\cdots,B_m)^{tr} \\ &= B_m + r_m(B_1,\cdots,B_m)^{tr}. \end{aligned} \tag{4}$$

So $r_m(B_1,\cdots,B_m)^{tr} = B_{m+1} - B_m$ and $T\begin{bmatrix}B_1\\B_2\\B_3\\\cdots\end{bmatrix} = \begin{bmatrix}B_2-B_1\\B_3-B_2\\B_4-B_3\\\cdots\end{bmatrix} = S^*\begin{bmatrix}1\\2\\3\\\cdots\end{bmatrix}$. □

Hence the next corollary follows immediately.

Corollary 2 ($TS = S^*$). *As a generalization of $S^* = TS$, we shall consider T^kS for $k > 0$. Write $S^{(k)} = T^kS = [s_{m,n}^{(k)}]$ for $m, n \ge 0$, and let $S^{(0)} = S$. Clearly $S^{(1)} = S^*$.*

Theorem 3. *$S^{(k)}$ satisfies the recurrence $s_{m+1,n+1}^{(k)} = s_{m,n}^{(k)} + (n+k)s_{m,n+1}^{(k)}$.*

Proof. $S^{(1)} = [s_{m,n}^{(1)}] = S^*$ in Corollary 2 shows $s_{m+1,n+1}^{(1)} = s_{m,n}^{(1)} + (n+1)s_{m,n+1}^{(1)}$. And $S^{(2)} = \begin{bmatrix}1&&&&\\2&1&&&\\4&5&1&&\\8&19&9&1&\\16&65&55&14&1\end{bmatrix}$, $S^{(3)} = \begin{bmatrix}1&&&&\\3&1&&&\\9&7&1&&\\27&37&12&1&\\81&175&97&18&1\end{bmatrix}$ give identities $55 = 19 + (4)9$ and $97 = 37 + (5)12$, and it is easy to observe

$$s_{m+1,n+1}^{(k)} = s_{m,n}^{(k)} + (n+k)\,s_{m,n+1}^{(k)} \tag{5}$$

for $1 \le n, m \le 5$; $k = 2, 3$. Now we assume $s_{m+1,n+1}^{(k)} = s_{m,n}^{(k)} + (n+k)s_{m,n+1}^{(k)}$ for some k, m. Since

$$\left[s_{m,n}^{(k+1)}\right] = S^{(k+1)} = T^{k+1}S = TT^kS = TS^{(k)}, \tag{6}$$

any (m,n)th component $s_{m,n}^{(k+1)}$ comes from the mth row r_m of T and the nth column of $S^{(k)}$. So

$$\begin{aligned} s_{m,n}^{(k+1)} &= \Bigg(\underbrace{\binom{m-1}{0},\cdots,\binom{m-1}{n-2}}_{n-1},\binom{m-1}{n-1}, \\ &\quad \binom{m-1}{n},\cdots,\binom{m-1}{m-2},\binom{m-1}{m-1}\Bigg)\cdot\Big(\underbrace{0,\cdots,0}_{n-1},s_{n,n}^{(k)}, \\ &\quad s_{n+1,n}^{(k)},\cdots,s_{m-1,n}^{(k)},s_{m,n}^{(k)}\Big)^{tr} = \binom{m-1}{n-1}s_{n,n}^{(k)} \\ &\quad + \binom{m-1}{n}s_{n+1,n}^{(k)} + \cdots + \binom{m-1}{m-2}s_{m-1,n}^{(k)} \\ &\quad + \binom{m-1}{m-1}s_{m,n}^{(k)} \end{aligned} \tag{7}$$

and

$$\begin{aligned} s_{m,n+1}^{(k+1)} &= \Bigg(\underbrace{\binom{m-1}{0},\cdots,\binom{m-1}{n-1}}_{n},\binom{m-1}{n}, \\ &\quad \binom{m-1}{n+1},\cdots,\binom{m-1}{m-2},\binom{m-1}{m-1}\Bigg)\cdot\Big(\underbrace{0,\cdots,0}_{n}, \\ &\quad s_{n+1,n+1}^{(k)},s_{n+2,n+1}^{(k)},\cdots,s_{m-1,n+1}^{(k)},s_{m,n+1}^{(k)}\Big)^{tr} \\ &= \binom{m-1}{n}s_{n+1,n+1}^{(k)} + \binom{m-1}{n+1}s_{n+2,n+1}^{(k)} + \cdots \\ &\quad + \binom{m-1}{m-2}s_{m-1,n+1}^{(k)} + s_{m,n+1}^{(k)}. \end{aligned} \tag{8}$$

Therefore we have

$$s_{m,n}^{(k+1)} + (n+k+1)\, s_{m,n+1}^{(k+1)} = \left[\binom{m-1}{n-1} s_{n,n}^{(k)} + \binom{m-1}{n} s_{n+1,n}^{(k)} + \cdots + \binom{m-1}{m-2} s_{m-1,n}^{(k)} + s_{m,n}^{(k)}\right]$$
$$+ (n+k+1)\left[\binom{m-1}{n} s_{n+1,n+1}^{(k)} + \cdots + \binom{m-1}{m-2} s_{m-1,n+1}^{(k)} + s_{m,n+1}^{(k)}\right] = \left[\binom{m-1}{n-1} s_{n,n}^{(k)} + \binom{m-1}{n} s_{n+1,n}^{(k)} + \cdots + \binom{m-1}{m-2} s_{m-1,n}^{(k)} + s_{m,n}^{(k)}\right]$$
$$+ \left[\binom{m-1}{n} (n+k)\, s_{n+1,n+1}^{(k)} + \cdots + \binom{m-1}{m-2} (n+k)\, s_{m-1,n+1}^{(k)} + (n+k)\, s_{m,n+1}^{(k)}\right]$$
$$+ \left[\binom{m-1}{n} s_{n+1,n+1}^{(k)} + \cdots + \binom{m-1}{m-2} s_{m-1,n+1}^{(k)} + s_{m,n+1}^{(k)}\right] = \binom{m-1}{n-1} s_{n,n}^{(k)} + \binom{m-1}{n} \left[s_{n+1,n}^{(k)} + (n+k)\, s_{n+1,n+1}^{(k)}\right] + \cdots + \binom{m-1}{m-2} \left[s_{m-1,n}^{(k)} + (n+k)\, s_{m-1,n+1}^{(k)}\right] + \left[s_{m,n}^{(k)} + (n+k)\, s_{m,n+1}^{(k)}\right]$$
$$+ \left[\binom{m-1}{n} s_{n+1,n+1}^{(k)} + \cdots + \binom{m-1}{m-2} s_{m-1,n+1}^{(k)} + s_{m,n+1}^{(k)}\right] = \binom{m-1}{n-1} s_{n,n}^{(k)} + \binom{m-1}{n} s_{n+2,n+1}^{(k)} + \cdots + \binom{m-1}{m-2} s_{m,n+1}^{(k)} + s_{m+1,n+1}^{(k)} + \left[\binom{m-1}{n} s_{n+1,n+1}^{(k)} + \cdots + \binom{m-1}{m-2} s_{m-1,n+1}^{(k)} + s_{m,n+1}^{(k)}\right] \tag{9}$$

by the induction hypothesis on $S^{(k)}$. But since $s_{i,i}^{(k)} = 1$ for all i, it follows immediately from Pascal's rule that

$$s_{m,n}^{(k+1)} + (n+k+1)\, s_{m,n+1}^{(k+1)} = \left[\binom{m-1}{n-1} + \binom{m-1}{n}\right] s_{n,n}^{(k)} + \left[\binom{m-1}{n} + \binom{m-1}{n+1}\right] s_{n+2,n+1}^{(k)} + \cdots + \left[\binom{m-1}{m-2} + 1\right] s_{m,n+1}^{(k)} + s_{m+1,n+1}^{(k)}$$
$$= \binom{m}{n} s_{n,n}^{(k)} + \binom{m}{n+1} s_{n+2,n+1}^{(k)} + \cdots + \binom{m}{m-1} s_{m,n+1}^{(k)} + s_{m+1,n+1}^{(k)}$$
$$= \binom{m}{n} s_{n+1,n+1}^{(k)} + \binom{m}{n+1} s_{n+2,n+1}^{(k)} + \cdots + \binom{m}{m-1} s_{m,n+1}^{(k)} + s_{m+1,n+1}^{(k)}$$
$$= \left(\underbrace{\binom{m}{0}, \cdots, \binom{m}{n-1}}_{n}, \binom{m}{n}, \cdots, \binom{m}{m-1}, 1\right) \cdot \left(\underbrace{0, \cdots, 0}_{n}, s_{n+1,n+1}^{(k)}, \cdots, s_{m+1,n+1}^{(k)}\right)^{tr} = s_{m+1,n+1}^{(k+1)}. \tag{10}$$

□

Theorem 3 agrees with the Stirling rule if $k = 1$, so we may call $T^k S = S^{(k)}$ the **generalized** k**-Stirling matrix** for $k > 0$. Like the Bell numbers which are the sum of each row of S, we will take sum of each row of the k-Stirling matrix $S^{(k)}$ and call this the k-**Bell number** and denote it by $B^{(k)} = \{B_i^{(k)} \mid i \geq 0\}$. The first few terms of the k-Bell numbers are obtained by examining $S^{(k)}$.

$B^{(k)} = \{B_i^{(k)}\}$ (11)

i	0	1	2	3	4	5	6	7	8	
$B^{(0)}$	1	1	2	5	15	52	203	877	4140	$\cdots$
$B^{(1)}$	1	2	5	15	52	203	877	4140	21147	$\cdots$
$B^{(2)}$	1	3	10	37	151	674	3263	17007	94828	$\cdots$
$B^{(3)}$	1	4	17	77	372	1915	10481	60814	372939	$\cdots$
$B^{(4)}$	1	5	26	141	799	4736	29371	190497	1291020	$\cdots$
$B^{(5)}$	1	6	37	235	1540	10427	73013	529032	3967195	$\cdots$

Obviously $B_i^{(0)} = B_i$ the (original) Bell numbers, and $B_i^{(1)} = B_{i+1}^{(0)}$, so $B_i^{(2)} = B_{i+1}^{(1)} - B_i^{(1)} = B_{i+2}^{(0)} - B_{i+1}^{(0)}$ $(i \geq 0)$ by Theorem 1. Note that $B^{(2)}$ is listed in A005493-OEIS as the numbers of partitions with a distinguished block. And $B^{(k)}$ $(3 \leq k \leq 5)$ are coefficients of $\exp(kx + \exp(x) - 1)$ (A005494, A045379, and A196834). On the other hand, the k-Bell numbers $B^{(k)}$ were studied in [9] by using certain binomial expressions $B_m = \sum_{i=0}^{m} \left\{ {m \atop i} \right\}$ and $B_m^{(k)} = \sum_{i=0}^{m} \left\{ {m+k \atop i+k} \right\}$. A key feature of our study is to have these k-Bell numbers from multiplications of T and S explicitly. Hence it enables us to find relations between k-Bell numbers $B^{(k)}$ and $B^{(k+1)}$, as follows.

Theorem 4. $S^{(k)} \begin{bmatrix} 1 \\ 1 \\ 1 \\ \cdots \end{bmatrix} = T \begin{bmatrix} B_0^{(k-1)} \\ B_1^{(k-1)} \\ B_2^{(k-1)} \\ \cdots \end{bmatrix}$ *and* $S^{(k)} \begin{bmatrix} 1 \\ 2 \\ 3 \\ \cdots \end{bmatrix} = \begin{bmatrix} B_0^{(k+1)} \\ B_1^{(k+1)} \\ B_2^{(k+1)} \\ \cdots \end{bmatrix}$.

Proof. Since $S^{(k)} = T^k S = TS^{(k-1)}$, it is clear that

$$\begin{bmatrix} B_0^{(k)} \\ B_1^{(k)} \\ B_2^{(k)} \\ \cdots \end{bmatrix} = S^{(k)} \begin{bmatrix} 1 \\ 1 \\ 1 \\ \cdots \end{bmatrix} = TS^{(k-1)} \begin{bmatrix} 1 \\ 1 \\ 1 \\ \cdots \end{bmatrix} = T \begin{bmatrix} B_0^{(k-1)} \\ B_1^{(k-1)} \\ B_2^{(k-1)} \\ \cdots \end{bmatrix}. \quad (12)$$

And, due to Theorem 1, we have

$$S^{(0)} \begin{bmatrix} 1 \\ 2 \\ 3 \\ \cdots \end{bmatrix} = \begin{bmatrix} B_1^{(0)} \\ B_2^{(0)} \\ B_3^{(0)} \\ \cdots \end{bmatrix} = \begin{bmatrix} B_0^{(1)} \\ B_1^{(1)} \\ B_2^{(1)} \\ \cdots \end{bmatrix}$$

$$\text{and } S^{(1)} \begin{bmatrix} 1 \\ 2 \\ 3 \\ \cdots \end{bmatrix} = \begin{bmatrix} B_2^{(0)} - B_1^{(0)} \\ B_3^{(0)} - B_2^{(0)} \\ B_4^{(0)} - B_3^{(0)} \\ \cdots \end{bmatrix} = \begin{bmatrix} B_0^{(2)} \\ B_1^{(2)} \\ B_2^{(2)} \\ \cdots \end{bmatrix}. \quad (13)$$

So if we assume $S^{(k-1)} \begin{bmatrix} 1 \\ 2 \\ 3 \\ \cdots \end{bmatrix} = \begin{bmatrix} B_0^{(k)} \\ B_1^{(k)} \\ B_2^{(k)} \\ \cdots \end{bmatrix}$ for some k then we have

$$S^{(k)} \begin{bmatrix} 1 \\ 2 \\ 3 \\ \cdots \end{bmatrix} = TS^{(k-1)} \begin{bmatrix} 1 \\ 2 \\ 3 \\ \cdots \end{bmatrix} = T \begin{bmatrix} B_0^{(k)} \\ B_1^{(k)} \\ B_2^{(k)} \\ \cdots \end{bmatrix} = \begin{bmatrix} B_0^{(k+1)} \\ B_1^{(k+1)} \\ B_2^{(k+1)} \\ \cdots \end{bmatrix}. \quad (14)$$

□

Theorem 4 says $T \begin{bmatrix} B_0 \\ B_1 \\ B_2 \\ \cdots \end{bmatrix} = \begin{bmatrix} B_1 \\ B_2 \\ B_3 \\ \cdots \end{bmatrix}$, so $T \begin{bmatrix} B_1 \\ B_2 \\ B_3 \\ \cdots \end{bmatrix} = T^2 \begin{bmatrix} B_0 \\ B_1 \\ B_2 \\ \cdots \end{bmatrix}$. Thus the expression $r_m(B_1, \cdots, B_m)^{tr}$ in the proof of Theorem 1, which is the multiplication of the mth row of T with $(B_1, \cdots, B_m)^{tr}$, is equal to multiplication of the mth row of T^2 with $(B_0, \cdots, B_{m-1})^{tr}$. Note that T^2 equals the arithmetic table $T^{(2x+1)}$ of the polynomial $(2x+1)^n$, while Pascal's table T is the arithmetic table of $(x+1)^n$. Hence if we let $r_m^{(2x+1)}$ be the mth row of $T^{(2x+1)} = \begin{bmatrix} 1 & & & \\ 2 & 1 & & \\ 4 & 4 & 1 & \\ 8 & 12 & 6 & 1 \end{bmatrix}$ of $(2x+1)^n$ then

$$\begin{aligned} B_{m+1} - B_m &= \sum_{i=0}^{m} \binom{m-1}{i} B_{i+1} \\ &= \left(\binom{m-1}{0}, \binom{m-1}{1}, \cdots, \binom{m-1}{m-1} \right) \\ &\quad \cdot (B_1, \cdots, B_m)^{tr} = r_m^{(2x+1)} (B_0, \cdots, B_{m-1})^{tr}, \end{aligned} \quad (15)$$

which implies $T^2 \begin{bmatrix} B_0 \\ B_2 \\ \cdots \end{bmatrix} = \begin{bmatrix} B_2 - B_1 \\ B_3 - B_2 \\ \cdots \end{bmatrix} = \begin{bmatrix} B_0^{(2)} \\ B_1^{(2)} \\ \cdots \end{bmatrix} = S^{(2)} \begin{bmatrix} 1 \\ 1 \\ \cdots \end{bmatrix}$. Thus, owing to Theorem 4, it is clear to have generalizations to k-Bell numbers $(k \geq 0)$ that

$$\begin{aligned} T^u \begin{bmatrix} B_0^{(k)} \\ B_1^{(k)} \\ B_2^{(k)} \\ \cdots \end{bmatrix} &= T^u S^{(k)} \begin{bmatrix} 1 \\ 1 \\ 1 \\ \cdots \end{bmatrix} = S^{(u+k)} \begin{bmatrix} 1 \\ 1 \\ 1 \\ \cdots \end{bmatrix} \\ &= \begin{bmatrix} B_0^{(u+k)} \\ B_1^{(u+k)} \\ B_2^{(u+k)} \\ \cdots \end{bmatrix}. \end{aligned} \quad (16)$$

3. Matrix of General Bell Numbers

With all the j-Bell numbers $B^{(j)}$ $(j \geq 1)$, we make a table $\widehat{GB}$ in which each jth column is composed of j-Bell numbers. We call it a **matrix of general Bell numbers** and denote it by $\widehat{GB} = [b_{i,j}]$ $(i, j \geq 1)$.

$$\widehat{GB} = \left[B^{(1)} \mid \cdots \mid B^{(7)} \mid \cdots \right] \quad (17)$$

1						
2	1					
5	3	1				
15	10	4	1			
52	37	17	5	1		
203	151	77	26	6	1	
877	674	372	141	37	7	1

Theorem 5. *The entries of the* $(j+1)$*th column in* $\widehat{GB} = [b_{i,j}]$ *satisfy* $b_{i,j+1} = b_{i,j} - jb_{i-1,j}$. *So* $B_i^{(j+1)} = B_{i+1}^{(j)} - jB_i^{(j)}$.

Proof. Observe that each jth column $(2 \leq j \leq 5)$ satisfies the following:

2nd col	3rd col	4th col	5th col
$2-1=1$	$3-(2)1=1$	$4-(3)1=1$	$5-(4)1=1$
$5-2=3$	$10-(2)3=4$	$17-(3)4=5$	$26-(4)5=6$
$15-5=10$	$37-(2)10=17$	$77-(3)17=26$	$141-(4)26=37$
$52-15=37$	$151-(2)37=77$	$372-(3)77=141$	$799-(4)141=235$

(18)

Suppose $b_{i,j+1} = b_{i,j} - jb_{i-1,j}$ for some $i, j > 1$. By looking at

$$\left[b_{i,j}\right] = \begin{bmatrix} b_{1,1} & & & \\ b_{2,1} & b_{2,2} & & \\ b_{3,1} & b_{3,2} & b_{3,3} & \\ b_{4,1} & b_{4,2} & b_{4,3} & b_{4,4} \\ \cdots & & & \end{bmatrix} = \begin{bmatrix} B_0^{(1)} & & & \\ B_1^{(1)} & B_0^{(2)} & & \\ B_2^{(1)} & B_1^{(2)} & B_0^{(3)} & \\ B_3^{(1)} & B_2^{(2)} & B_1^{(3)} & B_0^{(4)} \\ \cdots & & & \end{bmatrix} \quad (19)$$

it is enough to prove the following two cases.

(i) For any $i \geq 0$, assume $B_i^{(j)} = B_{i+1}^{(j-1)} - (j-1)B_i^{(j-1)}$ in some jth column and show $B_i^{(j+1)} = B_{i+1}^{(j)} - jB_i^{(j)}$ is true at $(j+1)$th column.

(ii) For any $j \geq 1$, assume $B_{t-1}^{(j)} = B_t^{(j-1)} - (j-1)B_{t-1}^{(j-1)}$ at every tth row $(1 \leq t \leq i)$ and show $B_i^{(j)} = B_{i+1}^{(j-1)} - (j-1)B_i^{(j-1)}$ is true at $(i+1)$th row.

Using the ith row r_i of T, the induction hypothesis in (i) with (16) implies

$$\begin{aligned} B_i^{(j+1)} &= r_{i+1}\left(B_0^{(j)}, B_1^{(j)}, \cdots B_i^{(j)}\right)^{tr} = r_{i+1}\left(B_1^{(j-1)}\right. \\ &\quad \left. - (j-1)B_0^{(j-1)}, \cdots, B_{i+1}^{(j-1)} - (j-1)B_i^{(j-1)}\right)^{tr} \\ &= r_{i+1}\left(B_1^{(j-1)}, \cdots, B_{i+1}^{(j-1)}\right)^{tr} - (j-1) \\ &\quad \cdot r_{i+1}\left(B_0^{(j-1)}, \cdots, B_i^{(j-1)}\right)^{tr} = \left[\binom{i}{0}B_1^{(j-1)}\right. \\ &\quad \left. + \binom{i}{1}B_2^{(j-1)} + \cdots + \binom{i}{i}B_{i+1}^{(j-1)}\right] - (j-1) \\ &\quad \cdot \left[\binom{i}{0}B_0^{(j-1)} + \binom{i}{1}B_1^{(j-1)} + \cdots + \binom{i}{i}B_i^{(j-1)}\right] \\ &= \left(-(j-1)\binom{i}{0}, \binom{i}{0}\right. \\ &\quad \left. - (j-1)\binom{i}{1}, \cdots, \binom{i}{i-1} - (j-1)\binom{i}{i}, \binom{i}{i}\right) \\ &\quad \cdot \left(B_0^{(j-1)}, B_1^{(j-1)}, \cdots, B_{i+1}^{(j-1)}\right)^{tr}. \end{aligned} \quad (20)$$

On the other hand, we also have

$$\begin{aligned} B_{i+1}^{(j)} - jB_i^{(j)} &= r_{i+2}\left(B_0^{(j-1)}, B_1^{(j-1)}, \cdots, B_{i+1}^{(j-1)}\right)^{tr} \\ &\quad - jr_{i+1}\left(B_0^{(j-1)}, B_1^{(j-1)}, \cdots, B_i^{(j-1)}\right)^{tr} \\ &= \left[\binom{i+1}{0}B_0^{(j-1)} + \binom{i+1}{1}B_1^{(j-1)}\right. \\ &\quad \left. + \binom{i+1}{2}B_2^{(j-1)} + \cdots + \binom{i+1}{i+1}B_{i+1}^{(j-1)}\right] \\ &\quad - j\left[\binom{i}{0}B_0^{(j-1)} + \binom{i}{1}B_1^{(j-1)} + \cdots + \binom{i}{i}B_i^{(j-1)}\right] \\ &= \left(\binom{i+1}{0} - j\binom{i}{0}, \binom{i+1}{1} - j\binom{i}{1}, \binom{i+1}{2}\right. \\ &\quad \left. - j\binom{i}{2}, \cdots, \binom{i+1}{i} - j\binom{i}{i}, \binom{i+1}{i+1}\right) \\ &\quad \cdot \left(B_0^{(j-1)}, \cdots, B_i^{(j-1)}, B_{i+1}^{(j-1)}\right)^{tr} \\ &= \left(-(j-1)\binom{i}{0}, \binom{i}{0}\right. \\ &\quad \left. - (j-1)\binom{i}{1}, \cdots, \binom{i}{i-1} - (j-1)\binom{i}{i}, \binom{i}{i}\right) \\ &\quad \cdot \left(B_0^{(j-1)}, B_1^{(j-1)}, \cdots, B_{i+1}^{(j-1)}\right)^{tr}, \end{aligned} \quad (21)$$

for $\binom{i+1}{u} - j\binom{i}{u} = \binom{i}{u-1} + \binom{i}{u} - j\binom{i}{u} = \binom{i}{u-1} - (j-1) + \binom{i}{u}$ for all $0 \leq u \leq i$. Hence identities (20) and (21) prove (i).

Now, for (ii), the induction hypothesis $B_t^{(j-1)} + (j-2)B_t^{(j-2)} = B_{t+1}^{(j-2)}$ $(1 \leq t \leq i)$ in (ii) with again (16) implies

$$\begin{aligned} B_i^{(j)} + (j-1)B_i^{(j-1)} &= B_i^{(j)} + (j-2)B_i^{(j-1)} + B_i^{(j-1)} \\ &= r_{i+1}\left(B_0^{(j-1)}, \cdots, B_i^{(j-1)}\right)^{tr} + (j-2) \end{aligned}$$

$$\begin{aligned}&\cdot r_{i+1}\left(B_0^{(j-2)},\cdots,B_i^{(j-2)}\right)^{tr}\\&+r_{i+1}\left(B_0^{(j-2)},\cdots,B_i^{(j-2)}\right)^{tr}=r_{i-1}\left(B_0^{(j-1)}\right.\\&\left.+(j-2)B_0^{(j-2)},\cdots,B_i^{(j-1)}+(j-2)B_i^{(j-2)}\right)^{tr}\\&+r_{i+1}\left(B_0^{(j-2)},\cdots,B_i^{(j-2)}\right)^{tr}\\&=r_{i+1}\left(B_1^{(j-2)},\cdots,B_{i+1}^{(j-2)}\right)^{tr}\\&+r_{i+1}\left(B_0^{(j-2)},\cdots,B_i^{(j-2)}\right)^{tr}\end{aligned}\tag{22}$$

Moreover

$$\begin{aligned}B_{i+1}^{(j-1)}&=r_{i+2}\left(B_0^{(j-2)},B_1^{(j-2)},\cdots,B_{i+1}^{(j-2)}\right)^{tr}\\&=[(0;r_{i+1})+(r_{i+1};0)]\left(B_0^{(j-2)},B_1^{(j-2)},\cdots,B_{i+1}^{(j-2)}\right)^{tr}\\&=r_{i+1}\left(B_1^{(j-2)},\cdots,B_{i+1}^{(j-2)}\right)^{tr}\\&+r_{i+1}\left(B_0^{(j-2)},\cdots,B_i^{(j-2)}\right)^{tr}\end{aligned}\tag{23}$$

Then we complete the proof by comparing (22) and (23). □

Therefore Theorem 5 with (16) shows

$$\begin{aligned}T^u\begin{bmatrix}B_0^{(k)}\\B_1^{(k)}\\B_2^{(k)}\\\cdots\end{bmatrix}&=\begin{bmatrix}B_0^{(u+k)}\\B_1^{(u+k)}\\B_2^{(u+k)}\\\cdots\end{bmatrix}\\&=\begin{bmatrix}B_1^{(u+k-1)}-(k+u-1)B_0^{(u+k-1)}\\B_2^{(u+k-1)}-(k+u-1)B_1^{(u+k-1)}\\B_3^{(u+k-1)}-(k+u-1)B_2^{(u+k-1)}\\\cdots\end{bmatrix}.\end{aligned}\tag{24}$$

As an example, the 6-Bell numbers $B^{(6)}=\{B_i^{(6)}\}=\{1,7,50,365,2727,\cdots\}$ and then 7-Bell numbers $B^{(7)}=\{B_i^{(7)}\}=\{1,8,65,537,4516,\cdots\}$ are obtained sequentially from the 5-Bell numbers.

5th col.	6th col.	7th col.
6	$6-(5)1=1$	$7-(6)1=1$
37	$37-(5)6=7$	$50-(6)7=8$
235	$235-(5)37=50$	$365-(6)50=65$
1540	$1540-(5)235=365$	$2727-(6)365=537$

(25)

The next theorem gives another way to have k-Bell numbers $B^{(k)}$. Let $[0]_j$ denote the j-tuple $\{0,\cdots,0\}$.

Theorem 6. *Let* $X_j=([0]_{j-1};j,j+1,\cdots,n)^{tr}$ *and* $\rho_n=(b_{n,1},\cdots,b_{n,n})$ *be the nth row of* $\widehat{GB}$. *Then* $\rho_nX_j+1=b_{n+1,j}$ *for any* $1\le j\le n$.

Proof. Clearly $\widehat{GB}X_1+1=\widehat{GB}\begin{bmatrix}1\\2\\3\\\cdots\end{bmatrix}+1=\begin{bmatrix}2\\5\\15\\\cdots\end{bmatrix}$ and $\widehat{GB}X_2+1=\widehat{GB}\begin{bmatrix}0\\2\\3\\\cdots\end{bmatrix}+1=\begin{bmatrix}1\\3\\10\\\cdots\end{bmatrix}$. Similarly $\widehat{GB}X_j+1\ (j=3,4)$ equals $\begin{bmatrix}1\\1\\4\\\cdots\end{bmatrix}$ and $\begin{bmatrix}1\\1\\1\\\cdots\end{bmatrix}$. Thus, for instance, when $j=4$, with nth row ρ_n of $\widehat{GB}$, we have $\rho_3X_4+1=d_{4,4}=1$, $\rho_4X_4+1=d_{5,4}=5$, $\rho_5X_4+1=d_{6,4}=26$, $\rho_6X_4+1=d_{7,4}=141$, and so on. Hence we can say that $\rho_nX_4+1=b_{n+1,4}$ for $n\ge 3$.

Now assume $\rho_nX_j+1=b_{n+1,j}-1$ for $n\ge j-1$. Then since $X_{j+1}=X_j-([0]_{j-1};j,0,\cdots,0)^{tr}$, Theorem 5 implies

$$\begin{aligned}\rho_nX_{j+1}&=\rho_nX_j-\rho_n\left([0]_{j-1};j,0,\cdots,0\right)^{tr}\\&=b_{n+1,j}-1-jb_{n,j}=b_{n+1,j+1}-1.\end{aligned}\tag{26}$$

□

Theorem 6 can be restated in terms of k-Bell numbers.

Corollary 7. $(B_n^{(1)},B_{n-1}^{(2)},\cdots,B_1^{(n)})([0]_{j-1};j,j+1,\cdots,n)^{tr}=B_{n-j+2}^{(j)}-1$.

Hence $\widehat{GB}(1,2,\cdots,n)^{tr}=(B_{n+1}^{(1)},\cdots,B_{n+1}^{(n)})^{tr}$ and $\widehat{GB}([0]_{j-1};j,\cdots,n)^{tr}=([0]_{j-1};B_{n+1}^{(j)},\cdots,B_{n+1}^{(n)})^{tr}$. Precisely, with $\rho_6=(203,151,77,26,6,1)$, we have $\rho_6(0,0,0,0,0,6)^{tr}+1=7$, $\rho_6(0,0,0,0,5,6)^{tr}+1=37$, $\rho_6(0,0,0,4,5,6)^{tr}+1=141$, $\rho_6(0,0,3,4,5,6)^{tr}+1=372$, $\rho_6(0,2,3,4,5,6)^{tr}+1=674$, and $\rho_6(1,2,3,4,5,6)^{tr}+1=877$, where these yield the 7th row $\rho_7=(877,674,372,141,37,7,1)$.

4. k-Peirce Matrix

The Peirce matrix $[p_{i,j}]$ $(i,j\ge 0)$ was designed to generate Bell numbers.

$$P=\left[p_{i,j}\right]=\begin{bmatrix}1\\1&2\\2&3&5\\5&7&10&15\\15&20&27&37&52\end{bmatrix}\tag{27}$$

where it holds a recurrence $p_{i,j}=p_{i,j-1}+p_{i-1,j-1}$ with $p_{0,0}=B_0$, $p_{n,0}=p_{n-1,n-1}=B_n$. So the 6th row of P begins with 52 followed by 67, 87, 114, and 151 and then reach 203, the next Bell number of 52. Thus all Bell numbers are on both borders of P. The matrix is often called the Bell matrix or Aitken's array named after E. T. Bell and A. Aitken. Here we call it Peirce matrix after C. S. Peirce [10] to avoid confusion with the matrix of general Bell numbers in Section 3.

Now, for any $k\ge 1$, let $P^{(k)}=[p_{i,j}^{(k)}]$ $(i,j\ge 1)$ be a matrix satisfying the following two rules.

(i) Each ith row begins with the Bell number $p_{i,1}^{(k)}=B_{i-1}^{(0)}$.
(ii)

$$p_{i,j}^{(k)}=p_{i,j-1}^{(k)}+kp_{i-1,j-1}^{(k)}\quad\text{for all } i>1,\qquad\text{and } p_{i,j}^{(k)}=0\text{ for } j>i.\tag{28}$$

Obviously $P^{(1)}$ is the Peirce matrix, and $P^{(k)}$ $(k = 2, 3)$ are equal to

$$P^{(2)} = \begin{bmatrix} 1 & & & \\ 1 & 3 & & \\ 2 & 4 & 10 & \\ 5 & 9 & 17 & 37 \end{bmatrix}$$

$$\text{and } P^{(3)} = \begin{bmatrix} 1 & & & \\ 1 & 4 & & \\ 2 & 5 & 17 & \\ 5 & 11 & 26 & 77 \end{bmatrix} \tag{29}$$

Notice that the left border is always comprised of Bell numbers, while the right diagonals (r.diag.) of $P^{(2)}$ and $P^{(3)}$ are of 2 and 3-Bell numbers, respectively. And the second right diagonal (2nd r.diag.) of $P^{(2)}$ equals the r.diag. of $P^{(3)}$. The next theorem shows that the r.diag. of $P^{(k)} = [p_{i,j}^{(k)}]$ $(k \geq 1)$ is composed of k-Bell numbers $B_i^{(k)}$.

Theorem 8. *(1) The entries of r.diag. of $P^{(k)}$ are k-Bell numbers; i.e., $p_{i,i}^{(k)} = B_{i-1}^{(k)}$, while the entries of left border are Bell numbers.*

(2) The 2nd r.diag. is composed of the $(k+1)$-Bell numbers; i.e., $p_{i,i-1}^{(k)} = B_i^{(k+1)}$.

Proof. Following the rules in (28), $P^{(k)} = [p_{i,j}^{(k)}]$ is equal to

$$\begin{matrix} 1 \\ 1 & k+1 \\ 2 & k+2 & k^2+2k+2 \\ 5 & 2k+5 & k^2+4k+5 & k^3+3k^2+6k+5 \\ 15 & 5k+15 & 2k^2+10k+15 & k^3+6k^2+15k+15 & k^4+4k^3+12k^2+20k+15 \\ 52 & 15k+52 & 5k^2+30k+52 & 2k^3+15k^2+45k+52 & k^4+8k^3+30k^2+60k+52 & \cdots \end{matrix} \tag{30}$$

When $k = 1, 2, 3$, the r.diag. of $P^{(k)}$ yields $\{1, 2, 5, 15, \cdots\}$, $\{1, 3, 10, 37, \cdots\}$, and $\{1, 4, 17, 77, \cdots\}$, where these correspond to $B^{(1)}$, $B^{(2)}$, and $B^{(3)}$.

Assume the r.diag

$$\{1, k+1, k^2+2k+2, k^3+3k^2+6k+5, k^4+4k^3 + 12k^2+20k+15, \cdots\} \tag{31}$$

of $P^{(k)}$ equals the k-Bell numbers $B^{(k)}$. Then, due to (16), we have

$$\begin{bmatrix} B_0^{(k+1)} \\ B_1^{(k+1)} \\ B_2^{(k+1)} \\ B_3^{(k+1)} \\ B_4^{(k+1)} \end{bmatrix} = T \begin{bmatrix} B_0^{(k)} \\ B_1^{(k)} \\ B_2^{(k)} \\ B_3^{(k)} \\ B_4^{(k)} \end{bmatrix}$$

$$= T \begin{bmatrix} 1 \\ k+1 \\ k^2+2k+2 \\ k^3+3k^2+6k+5 \\ k^4+4k^3+12k^2+20k+15 \end{bmatrix}$$

$$= \begin{bmatrix} 1 & & & & \\ 1 & 1 & & & \\ 1 & 2 & 1 & & \\ 1 & 3 & 3 & 1 & \\ 1 & 4 & 6 & 4 & 1 \end{bmatrix} \begin{bmatrix} 1 & & & & \\ 1 & 1 & & & \\ 2 & 2 & 1 & & \\ 5 & 6 & 3 & 1 & \\ 15 & 20 & 12 & 4 & 1 \end{bmatrix} \begin{bmatrix} 1 \\ k \\ k^2 \\ k^3 \\ k^4 \end{bmatrix}$$

$$= \begin{bmatrix} 1 & & & & \\ 2 & 1 & & & \\ 5 & 4 & 1 & & \\ 15 & 15 & 6 & 1 & \\ 52 & 60 & 30 & 8 & 1 \end{bmatrix} \begin{bmatrix} 1 \\ k \\ k^2 \\ k^3 \\ k^4 \end{bmatrix} \tag{32}$$

And it is not hard to see that it corresponds to the r.diag. of $P^{(k+1)}$

$$\begin{matrix} 1 \\ 1 & k+2 \\ 2 & k+3 & k^2+4k+5 \\ 5 & 2k+7 & k^2+6k+10 & k^3+6k^2+15k+15 \\ 15 & 5k+20 & 2k^2+14k+27 & k^3+9k^2+30k+37 & k^4+8k^3+30k^2+60k+52 \\ 52 & 15k+67 & 5k^2+40k+87 & 2k^3+21k^2+81k+114 & & \cdots \end{matrix} \tag{33}$$

Moreover, by comparing $P^{(k)}$ with $P^{(k+1)}$, we also notice that the r.diag. of $P^{(k+1)}$ equals the 2nd r.diag. of $P^{(k)}$. □

Owing to Theorem 8, we may call $P^{(k)}$ the **Peirce k-matrix**. Write diag$[a, b, c, \cdots]$ for the diagonal matrix having diagonal entries $a, b, c, \cdots$. Every k-Bell number can be obtained by Bell numbers as follows.

Theorem 9. $B_n^{(k)} = r_n \text{diag}[B_0, B_1, \cdots, B_{n-1}](k^{n-1}, \cdots, k, 1)^{tr}$.

Proof. When $k = 1$, it is due to Theorem 1. And, by Theorem 8, we have

$$\begin{bmatrix} B_1^{(k)} \\ B_2^{(k)} \\ B_3^{(k)} \\ B_4^{(k)} \\ B_5^{(k)} \end{bmatrix} = \begin{bmatrix} 1 \\ 1+k \\ 2+2k+k^2 \\ 5+6k+3k^2+k^3 \\ 15+20k+12k^2+4k^3+k^4 \end{bmatrix} = \begin{bmatrix} 1 & & & & \\ 1 & 1 & & & \\ 2 & 2 & 1 & & \\ 5 & 6 & 3 & 1 & \\ 15 & 20 & 12 & 4 & 1 \end{bmatrix} \begin{bmatrix} 1 \\ k \\ k^2 \\ k^3 \\ k^4 \end{bmatrix}. \tag{34}$$

But we notice

$$\begin{bmatrix} 1 & & & & \\ 1 & 1 & & & \\ 1 & 2 & 2 & & \\ 1 & 3 & 6 & 5 & \\ 1 & 4 & 12 & 20 & 15 \end{bmatrix} = \begin{bmatrix} 1 & & & & \\ 1 & 1 & & & \\ 1 & 2 & 1 & & \\ 1 & 3 & 3 & 1 & \\ 1 & 4 & 6 & 4 & 1 \end{bmatrix} \begin{bmatrix} 1 & & & & \\ & 1 & & & \\ & & 2 & & \\ & & & 5 & \\ & & & & 15 \end{bmatrix} = T \,\text{diag}\,[B_0, B_1, \cdots, B_4]. \tag{35}$$

Hence, with the nth row r_n of T, the nth k-Bell number $B_n^{(k)}$ is equal to r_n diag$[B_0, \cdots, B_{n-1}](k^{n-1}, \cdots, k, 1)^{tr}$. □

For example, with $r_6 = (1, 5, 10, 10, 5, 1)$, the 6th$k$-Bell number $B_6^{(k)}$ equals

$$\begin{aligned} &r_6 \text{diag}\,[B_0, \cdots, B_5]\,(k^5, \cdots, 1)^{tr} \\ &= r_6 \,\text{diag}\,[1, 1, 2, 5, 15, 52]\,(k^5, \cdots, 1)^{tr} \\ &= k^5 + 5k^4 + 20k^3 + 50k^2 + 75k + 52. \end{aligned} \tag{36}$$

Conflicts of Interest

The author declares that there are no conflicts of interest regarding the publication of this paper.

Acknowledgments

This work was supported by 2018 HanNam University Research Fund.

References

[1] A. Tucker, *Applied combinatorics*, John Wiley, New York, NY, USA, 2nd edition, 1984.

[2] P. Maltais and T. A. Gulliver, "Pascal matrices and Stirling numbers," *Applied Mathematics Letters*, vol. 11, no. 2, pp. 7–11, 1998.

[3] M. Z. Spivey and A. M. Zimmer, "Symmetric polynomials, Pascal matrices, and Stirling matrices," *Linear Algebra and its Applications*, vol. 428, no. 4, pp. 1127–1134, 2008.

[4] S.-L. Yang and H. You, "On a connection between the Pascal, Stirling and Vandermonde matrices," *Discrete Applied Mathematics: The Journal of Combinatorial Algorithms, Informatics and Computational Sciences*, vol. 155, no. 15, pp. 2025–2030, 2007.

[5] M. Abbas and S. Bouroubi, "On new identities for Bell's polynomials," *Discrete Mathematics*, vol. 293, no. 1-3, pp. 5–10, 2005.

[6] M. Griffiths, "Families of sequences from a class of multinomial sums," *Journal of Integer Sequences*, vol. 15, no. 1, Article 12.1.8, 10 pages, 2012.

[7] S. Noschese and P. E. Ricci, "Differentiation of multivariable composite functions and Bell polynomials," *Journal of Computational Analysis and Applications*, vol. 5, no. 3, pp. 333–340, 2003.

[8] J. Shallit, "A triangle for the Bell numbers," in *A collection of manuscripts related to the Fibonacci sequence*, pp. 69–71, Fibonacci Assoc., Santa Clara, Calif., 1980.

[9] I. Mező, "TThe r-Bell numbers," *Journal of Integer Sequences*, vol. 14, no. 1, Article ID 11.1.1, 2011.

[10] C. S. Peirce, "On the algebra of logic," *American Journal of Mathematics*, vol. 3, no. 1, pp. 15–57, 1880.

19

On the Commutative Rings with At Most Two Proper Subrings

David E. Dobbs

Department of Mathematics, University of Tennessee, Knoxville, TN 37996-1320, USA

Correspondence should be addressed to David E. Dobbs; dobbs@math.utk.edu

Academic Editor: Kaiming Zhao

The commutative rings with exactly two proper (unital) subrings are characterized. An initial step involves the description of the commutative rings having only one proper subring.

1. Introduction

This paper is a sequel to [1]. All rings considered below are commutative with identity; all subrings, inclusions of rings, ring extensions, ring or algebra homomorphisms, modules, and submodules are unital. Our interest here is in characterizing rings that have at most two proper subrings. Observe that a ring R has no proper subrings if and only if either $R = 0$, $R \cong \mathbb{Z}$, or $R \cong \mathbb{Z}/n\mathbb{Z}$ for some integer $n \geq 2$ and that these three cases are mutually exclusive. We will answer the analogous questions for rings having exactly one proper subring (resp., for rings having exactly two proper subrings) in Theorems 5 and 13 (resp., in Theorems 6 and 15). In doing so, we may tacitly ignore the zero ring, as $R = 0$ does not have any proper subrings (or, for that matter, any proper ring extensions). Thus, we may organize our answers with a focus on the characteristic of a ring of interest, since every nonzero ring is a ring extension of an isomorphic copy of exactly one of the so-called prime rings (namely, $\mathbb{Z}$ and $\mathbb{Z}/n\mathbb{Z}$ for $n \geq 2$).

Notice that a ring R has exactly one proper subring if and only if R is a minimal ring extension (in the sense of [2]) of its prime (sub)ring. As the minimal ring extensions of a (commutative integral) domain were classified in [3], much of the technical work leading up to our main theorems will be concerned with determining the minimal ring extensions of the prime rings of positive characteristic. Then, by determining the minimal ring extensions of either *those* minimal ring extensions or the minimal ring extensions of $\mathbb{Z}$, we will be led to a list of candidates for the characterizations in Theorems 6 and 15. Determining which of those candidates survive will depend on the main result of [1] which, for convenience, is restated here as Theorem 2. Carrying out this program will require details about minimal ring extensions, most of which are summarized in the next paragraph.

Recall (cf. [2]) that a ring extension $A \subset B$ is a *minimal ring extension* if there does not exist a ring properly contained between A and B. A minimal ring extension $A \subset B$ is either integrally closed (in the sense that A is integrally closed in B) or integral. If $A \subset B$ is a minimal ring extension, it follows from [2, Théorème 2.2(i) and Lemme 1.3] that there exists a unique maximal ideal M of A (called the *crucial maximal ideal* of $A \subset B$) such that the canonical injective ring homomorphism $A_M \to B_M (:= B_{A\setminus M})$ can be viewed as a minimal ring extension while the canonical ring homomorphism $A_P \to B_P$ is an isomorphism for all prime ideals P of A except M. A minimal ring extension $A \subset B$ is integrally closed if and only if $A \hookrightarrow B$ is a flat epimorphism (in the category of commutative rings). If $A \subset B$ is an integral minimal ring extension with crucial maximal ideal M, there are three possibilities: $A \subset B$ is said to be, respectively, *inert*, *ramified*, or *decomposed* if $B/MB (= B/M)$ is isomorphic, as an algebra over the field $K := A/M$, to a minimal field extension of K, $K[X]/(X^2)$, or $K \times K$. (As usual, X denote an indeterminate over the ambient base ring.)

If A is a ring, then char(A) denotes the characteristic of A; Spec(A) (resp., Max(A)) denotes the set of prime (resp., maximal) ideals of A; and by the *dimension* of A, we mean the Krull dimension of A, denoted by dim(A). If $A \subseteq B$ are rings, then $[A, B]$ denotes the set of rings C such that $A \subseteq C \subseteq B$; and, as in [4], we say that $A \subseteq B$ satisfies FIP (for the "finitely many intermediate rings" property) if the set $[A, B]$ is finite. Following [5, page 28], we let INC, LO, and GU, respectively, denote the incomparable, lying-over, and going-up properties of ring extensions. Given rings $A \subseteq B$ with $P \in \mathrm{Spec}(A)$, then $B_P := B_{A\setminus P}$. As usual, $\subset$ denotes proper inclusion; $\mathbb{F}_q$ denotes the finite field of cardinality q; and $|H|$ denotes the cardinality of a set H. The reader may find it useful to have copies of [1, 6] at hand. Any otherwise unexplained material is standard, as in [5].

2. Results

Our work will be separated into individual approaches that depend on the characteristic of the ambient prime ring. Each of those approaches will make use of the next two results, which are the main results of [1, 6]. Prior to giving the main results for each of our approaches, we will provide most of the new technical results that will be needed to prove those main results. We will not need to make the cumbersome conditions from [6, Proposition 3.5] (ultimately, from [7, Theorem 5.18]) explicit which are mentioned in parts (xii) and (xiii) of Theorem 1.

Theorem 1 (see [6, Theorem 4.1]). *Let $R \subset S$ and $S \subset T$ be minimal ring extensions, with crucial maximal ideals M and N, respectively. Then $R \subset T$ satisfies FIP if and only if (exactly) one of the following conditions holds:*

(i) *Both $R \subset S$ and $S \subset T$ are integrally closed.*

(ii) *$R \subset S$ is integral and $S \subset T$ is integrally closed.*

(iii) *$R \subset S$ is integrally closed, $S \subset T$ is integral, and $N \cap R \not\subseteq M$.*

(iv) *Both $R \subset S$ and $S \subset T$ are integral and $N \cap R \neq M$.*

(v) *Both $R \subset S$ and $S \subset T$ are inert, $N \cap R = M$, and either R/M is finite or there exists $\gamma \in T_M$ such that $T_M = R_M[\gamma]$.*

(vi) *$R \subset S$ is decomposed, $S \subset T$ is inert, and $N \cap R = M$.*

(vii) *Both $R \subset S$ and $S \subset T$ are decomposed and $N \cap R = M$.*

(viii) *$R \subset S$ is inert, $S \subset T$ is decomposed, $N \cap R = M$.*

(ix) *$R \subset S$ is ramified, $S \subset T$ is decomposed, and $N \cap R = M$.*

(x) *$R \subset S$ is decomposed, $S \subset T$ is ramified, and $N \cap R = M$.*

(xi) *$R \subset S$ is ramified, $S \subset T$ is inert, and $N \cap R = M$.*

(xii) *$R \subset S$ is inert, $S \subset T$ is ramified, $N \cap R = M$, and the two conditions stated in [6, Proposition 3.5(a)] hold.*

(xiii) *Both $R \subset S$ and $S \subset T$ are ramified, $N \cap R = M$, and the two conditions stated in [6, Proposition 3.5(b)] hold.*

Theorem 2 (see [1, Theorem 2.9]). *Consider the 13 conditions, (i)–(xiii), in the statement of Theorem 1. Then, consider the following:*

(a) *If data satisfy condition (vi) or condition (xi), then $|[R,T]| = 3$.*

(b) *If data satisfy any of the seven conditions (iii), (iv), (vii), (viii), (ix), (x), and (xii), then $|[R,T]| > 3$ (and $|[R,T]| < \infty$).*

(c) *For each of the four conditions (i), (ii), (v), and (xiii), there exist data satisfying this condition for which $|[R,T]| = 3$ and there exist other data satisfying this condition for which $|[R,T]| > 3$ (and $|[R,T]| < \infty$).*

We will often use the fact [8, Theorem 25.1(3)] that if R is a ring and E is an R-module, then $\mathrm{Spec}(R(+)E) = \{P(+)E \mid P \in \mathrm{Spec}(R)\}$. As we now begin the approach that is specific to the context of characteristic 0, it is also convenient to record here another fact that will see frequent use below. To wit: the integrally closed minimal ring extensions of $\mathbb{Z}$ are, up to isomorphism, the minimal overrings of $\mathbb{Z}$ (inside $\mathbb{Q}$), namely, the rings $\mathbb{Z}[1/s]$, where s runs over the set of prime numbers. This fact seems to be well known and can be recovered from [9, Lemma V.2] (cf. also [3, Remark 2.8(a)]). More generally, the minimal overrings of an arbitrary principal ideal domain have been explicitly identified earlier: see, for instance, [10, Proposition 4.11].

Proposition 3. *Let p be a prime number and $A := \mathbb{Z}(+)\mathbb{F}_p$. Then, consider the following:*

(a) *Let q be a prime number that is distinct from p. Then, up to A-algebra isomorphism, $B := \mathbb{Z}[1/q](+)\mathbb{F}_p$ is the unique ring such that $A \subset B$ is an integrally closed minimal ring extension whose crucial maximal ideal is $q\mathbb{Z}(+)\mathbb{F}_p$.*

(b) *There does not exist a ring S such that $A \subset S$ is an integrally closed minimal ring extension whose crucial maximal ideal is $p\mathbb{Z}(+)\mathbb{F}_p$.*

Proof. We will show first that if q is a prime number that is distinct from p and we put $T := \mathbb{Z}[1/q](+)\mathbb{F}_p$, then $A \subset T$ is an integrally closed minimal ring extension whose crucial maximal ideal is $q\mathbb{Z}(+)\mathbb{F}_p$. The "minimal ring extension" assertion follows from the above comments because each member of $[A,T]$ takes the form $D(+)\mathbb{F}_p$ for some (uniquely determined) $D \in [\mathbb{Z}, \mathbb{Z}[1/q]]$. The remaining parts of the assertion follow via the lore of the idealization construction. Indeed, by [8, Theorem 25.5(1)], the total quotient ring of A is (canonically isomorphic to) $\mathbb{Z}_{p\mathbb{Z}}(+)(\mathbb{F}_p)_{\mathbb{Z}\setminus p\mathbb{Z}}$, that is, to $\mathbb{Z}_{p\mathbb{Z}}(+)\mathbb{F}_p$. As $\mathbb{Z}$ is integrally closed (in $\mathbb{Q}$ and, hence, in $\mathbb{Z}_{p\mathbb{Z}}$), it now follows from [8, Theorem 25.5(1)] that A coincides with its integral closure (in its total quotient ring). In particular, $A \subset T$ is an integrally closed ring extension. To show that the crucial maximal ideal of this extension is $I := q\mathbb{Z}(+)\mathbb{F}_p$, [2, Théorème 2.2(i)] reduces our task to showing that $A_I \subset T_I$. Using [8, Corollary 25.5(2)], we find canonical identifications

$$A_I = \mathbb{Z}_{q\mathbb{Z}}(+)\left(\mathbb{F}_p\right)_{\mathbb{Z}\setminus q\mathbb{Z}} = \mathbb{Z}_{q\mathbb{Z}}(+)0 = \mathbb{Z}_{q\mathbb{Z}} \quad (1)$$

and, similarly, $T_I = (\mathbb{Z}[1/q])_{\mathbb{Z}\setminus q\mathbb{Z}}(+)0 = \mathbb{Q}$.

It now suffices to show that if a ring B is such that $A \subset B$ is an integrally closed minimal ring extension, then up to A-algebra isomorphism, $B = \mathbb{Z}[1/q](+)\mathbb{F}_p$ for some prime number $q \neq p$. By [2, Théorème 2.2(iii)], the inclusion map $A \hookrightarrow B$ is a flat epimorphism (in the category of rings). We claim that, up to A-algebra isomorphism, B is an overring of A (inside the total quotient ring of A); that is, we claim that we can view $B \subseteq \mathbb{Z}_{p\mathbb{Z}}(+)\mathbb{F}_p$.

Since $A \hookrightarrow B$ is a flat epimorphism, we can view $B \subseteq M(A)$, where $M(A)$ is a certain universal object introduced by Lazard in [11]. Hence, to prove the above claim, it suffices to prove that $M(A)$ coincides with the total quotient ring of A. Therefore, by the proof of [11, Proposition 4.1, page 116], it suffices to show that the generization (inside Spec(A)) of the set of weak Bourbaki associated primes of A is the same as the canonical image inside Spec(A) of the set of all prime ideals of the total quotient ring of A, that is, the same as the two-element set whose members are $P_1 := p\mathbb{Z}(+)\mathbb{F}_p$ and $P_2 := 0(+)\mathbb{F}_p$. (Some authors translate "le générisé de" as "the generalization of"; we prefer the translation "the generization of" because it is more reminiscent of the relevant notion of a generic point.) By definition, a weak Bourbaki associated prime of A is a minimal prime (ideal of A) over the annihilator of some element of A. A routine case analysis shows that the set of such annihilators consists of $p\mathbb{Z} \times \{0\}$, $\{0\} \times \mathbb{F}_p$, and $\{(0,0)\}$. It follows that the set of weak Bourbaki associated primes of A is $\{P_1, P_2\}$. As this set is stable under generization, the above claim has been proved.

We have shown that $A \subset B \subseteq \mathbb{Z}_{p\mathbb{Z}}(+)\mathbb{F}_p$. It follows easily that $B = E(+)\mathbb{F}_p$ for some ring E such that $E \subseteq \mathbb{Z}_{p\mathbb{Z}}$ and $\mathbb{Z} \subset E$ is a minimal ring extension. By the above comments, $E = \mathbb{Z}[1/q]$ for some prime number q. In short, $B = \mathbb{Z}[1/q](+)\mathbb{F}_p$. Finally, $q \neq p$ since $1/q \in E \subseteq \mathbb{Z}_{p\mathbb{Z}}$. □

Proposition 4. *Let p be a prime number and $A := \mathbb{Z}(+)\mathbb{F}_p$. Then there does not exist a ring B such that $A \subset B$ is an inert extension whose crucial maximal ideal is $p\mathbb{Z}(+)\mathbb{F}_p$.*

Proof. Put $N := p\mathbb{Z}(+)\mathbb{F}_p(= p\mathbb{Z}(+)\mathbb{Z}/p\mathbb{Z})$. Then we can identify $A_N = \mathbb{Z}_{p\mathbb{Z}}(+)\mathbb{Z}/p\mathbb{Z}$ (cf. [8, Lemma 25.4 and Theorem 25.5(2)]). Now, suppose the assertion fails. Then $N \in \text{Max}(B)$; and by [12, Proposition 4.6], $A_N \subset B_N(:= B_{A\setminus N})$ is an inert extension. In particular, $B_N/NB_N \cong B/N$ is a minimal field extension of $A/N(\cong \mathbb{F}_p) \cong A_N/NA_N$. Hence, by a harmless change of notation, we can now take $A := \mathbb{Z}_{p\mathbb{Z}}(+)\mathbb{Z}/p\mathbb{Z}$, with $A \subset B$ being an inert extension, necessarily with crucial maximal ideal $N = p\mathbb{Z}(+)\mathbb{Z}/p\mathbb{Z}$. As we can view $\mathbb{F}_p \to B/N$ as a minimal field extension, the standard Galois theory of finite fields provides a prime number q (possibly equal to p) such that $B/N = \mathbb{F}_{p^q}$, whence $y^{p^q} = y$ for each element $y \in B/N$.

Pick $e \in B \setminus A$. Then $A[e] = B$ by the minimality of $A \subset B$. Also, since the integral extension $A \subset B$ must satisfy INC, LO, and GU (cf. [5, Theorem 44]), we see that (B, N) is quasi-local and, in fact, that Spec(A) = Spec(B) as sets. Since $x := (p, 1 + p\mathbb{Z}) \in N$ is a (nonzero) non-zero-divisor in A, it follows that x is a non-zero-divisor in B. Also, since $xe \in NB = N$, there exist $z, a \in \mathbb{Z}$ such that $xe = (pz, a + p\mathbb{Z})$. Thus, $b := a - z \in \mathbb{Z}$ satisfies

$$\begin{aligned} x(e - (z,0)) &= xe - x(z,0) \\ &= (pz, a + p\mathbb{Z}) - (pz, z + p\mathbb{Z}) \\ &= (0, b + p\mathbb{Z}). \end{aligned} \tag{2}$$

Taking p^qth powers, we get

$$\begin{aligned} x^{p^q}(e - (z,0))^{p^q} &= (x(e - (z,0)))^{p^q} = (0, b + p\mathbb{Z})^{p^q} \\ &= (0,0). \end{aligned} \tag{3}$$

As x^{p^q} is a non-zero-divisor in B, we now have $(e-(z,0))^{p^q} = (0,0) \in A \subset B$. Applying the canonical surjection $B \to B/N$ leads to

$$(e + N)^{p^q} - ((z,0) + N)^{p^q} = 0 \in B/N; \tag{4}$$

that is, $(e+N) - ((z^{p^q},0)+N) = 0 \in B/N$. Thus, $e \in (z^{p^q},0) + N \subseteq A$, the desired contradiction. □

Theorem 5. *Up to isomorphism, the rings R of characteristic zero that have exactly one proper subring can be classified as the rings satisfying (exactly) one of the following conditions:*

(a) *$R = \mathbb{Z}[1/p]$, where p is a prime number (which is uniquely determined by R);*

(b) *$R = \mathbb{Z} \times \mathbb{F}_p$, where p is a prime number (which is uniquely determined by R);*

(c) *$R = \mathbb{Z}(+)\mathbb{F}_p$, where p is a prime number (which is uniquely determined by R).*

Proof. We must identify, up to isomorphism, the rings R such that $\mathbb{Z} \subset R$ is a minimal ring extension which is either (a) integrally closed, (b) decomposed, (c) ramified, or (d) inert. Thanks to [3, Theorem 2.7] (or [13, Theorem 2.4]), we are led to the list of rings in the statement of this result, including the fact that there is no ring R such that $\mathbb{Z} \subset R$ is inert. To obtain the above formulation in (a), recall that the integrally closed minimal ring extensions of $\mathbb{Z}$ are, up to isomorphism, the minimal overrings of $\mathbb{Z}$ (inside $\mathbb{Q}$), namely, the rings $\mathbb{Z}[1/p]$, where p runs over the set of prime numbers.

The above-cited references show that conditions (a), (b), and (c) are mutually exclusive. It remains to establish the uniqueness assertions in those conditions. For (a), note that p is determined as the prime number that has a multiplicative inverse in R. The uniqueness assertions in (b) and (c) were established in [3, Theorem 2.7]. □

We next confront the much more arduous task of classifying the rings of characteristic zero that have exactly two proper subrings. Much of the work below will use the following well known fact. If $A := R_1 \times R_2$ is a nontrivial direct product of (nonzero) rings R_1 and R_2, then, up to isomorphism, the minimal ring extensions of A take one of the forms $E_1 \times R_2$ and $R_1 \times E_2$ where E_i is a minimal ring extension of R_i for $i = 1, 2$.

Theorem 6. *(1) Up to isomorphism, the rings R of characteristic zero that have exactly two proper subrings can be characterized as the rings satisfying (exactly) one of the following two conditions:*

(a) *$R := \mathbb{Z} \times \mathbb{F}_{p^q}$, where p and q are (possibly equal) prime numbers (which are uniquely determined by R);*

(b) *for some prime number p, there is a ramified extension* $B := \mathbb{Z}(+)\mathbb{F}_p \subset R$ *with crucial maximal ideal* $p\mathbb{Z}(+)\mathbb{F}_p$ *and, furthermore,* $\mathbb{Z}[u] = R$ *for all* $u \in R \setminus B$.

(2) For each prime number p, there exist (nonisomorphic) rings R_1 *and* R_2 *such that* $B := \mathbb{Z}(+)\mathbb{F}_p \subset R_i$ *is a ramified extension with crucial maximal ideal* $p\mathbb{Z}(+)\mathbb{F}_p$ *for each* $i \in \{1,2\}$, R_1 *has exactly two proper subrings, and* R_2 *has more than two proper subrings.*

Proof. (1) Thanks to Theorem 2, we need only address a ring R that can result from a construction described in one of six of the 13 conditions in the statement of Theorem 1, namely, conditions (i), (ii), (v), (vi), (xi), and (xiii). We will deal with these six conditions in the stipulated order.

Suppose, first, that R is a ring that results from condition (i) in the statement of Theorem 1. Then, as recalled prior to Proposition 3, there is a prime number p such that R is a minimal overring of the principal ideal domain $B := \mathbb{Z}[1/p]$. By [10, Proposition 4.11], R is then (up to isomorphism) $B[1/b]$ where b is an irreducible element of B. Such elements b take the form $b = \pm p^j q$, where $j \in \mathbb{Z}$ and q is a prime number that is distinct from p. Then

$$\begin{aligned} R = B\left[\frac{1}{b}\right] &= \mathbb{Z}\left[\frac{1}{p}\right]\left[\frac{1}{(p^j q)}\right] = \mathbb{Z}\left[\frac{1}{p}\right]\left[\frac{1}{q}\right] \\ &= \mathbb{Z}\left[\frac{1}{(pq)}\right]. \end{aligned} \tag{5}$$

However, $\mathbb{Z}[1/p]$ and $\mathbb{Z}[1/q]$ are distinct proper subrings of *this* R. In particular, this R satisfies $|[\mathbb{Z}, R]| \geq 4$. (In fact, $|[\mathbb{Z}, R]| = 4$.) Thus, condition (i) does not produce any contributions to the list that we are building/verifying.

Next, suppose that R results from condition (ii). Then, as noted in Theorem 5, there is a prime number p such that R is an integrally closed minimal ring extension of either $B_1 := \mathbb{Z} \times \mathbb{F}_p$ or $B_2 := \mathbb{Z}(+)\mathbb{F}_p$. We will first show that the first of these alternatives does not contribute to the list that we are building/verifying. As there is no integrally closed minimal ring extension of the form $\mathbb{F}_p \subset E_2$, it follows from the above comments that, up to isomorphism, $R = E_1 \times \mathbb{F}_p$, where E_1 is an integrally closed minimal ring extension of $\mathbb{Z}$. As recalled above, this means that there is a prime number q (possibly equal to p) such that we can take $E_1 = \mathbb{Z}[1/q]$. In short, $R = \mathbb{Z}[1/q] \times \mathbb{F}_p$. It will suffice to show that $\mathbb{Z} \subset R$ does not satisfy FIP. In fact, we will show that if $\alpha \in \mathbb{Z}[1/q] \setminus \mathbb{Z}$ (e.g., $\alpha = 1/q$) and $\beta \in \mathbb{F}_p$ (e.g., $\beta = 0$), then $u := (\alpha, \beta) \in R \setminus B_1$ is such that $\mathbb{Z} \subset \mathbb{Z}[u]$ does not satisfy FIP. This will follow from [14, Proposition 3.1] after we establish three facts (to show that the cited result is applicable here). The first of these facts is that $\mathbb{Z} \subset \mathbb{Z}[u]$ is not an integral extension; this is clear since condition (ii) ensures that the integral closure of $\mathbb{Z}$ in R is B_1. The second required fact (actually, slightly more than is needed) is that, for all nonzero elements $v \in R$, there exist integers a and b, with $a \neq 0$, such that $av = b$. (For a proof, pick a positive integer N such that $q^N v \in \mathbb{Z} \times \mathbb{F}_p$; note that $pq^N v = (c, 0) \in \mathbb{Z} \times \{0\}$ for some integer c; $pq^N v - c \in \{0\} \times \mathbb{F}_p$; and so $a := p^2 q^N$ and $b := pc$ satisfy $av = b$.) The last required fact is that $\mathbb{Z}$ is a residually finite ring. This completes the proof concerning B_1.

In the remaining subcase pertinent to condition (ii), R is an integrally closed minimal ring extension of $B := B_2 = \mathbb{Z}(+)\mathbb{F}_p$. Then by Proposition 3, there is a (uniquely determined) prime number $q \neq p$ such that, up to B-algebra isomorphism, $R = \mathbb{Z}[1/q](+)\mathbb{F}_p$. With modest changes to the above argument that applied [14, Proposition 3.1], we can show that $\mathbb{Z} \subset \mathbb{Z}[(1/q, 0)]$ does not satisfy FIP. Thus, B_2 also fails to lead to a contribution to the list that we are building. This completes the discussion relative to condition (ii).

Next, note that condition (v) cannot lead to a contribution to our developing list because there is no ring R such that $\mathbb{Z} \subset R$ is inert [3, Theorem 2.7]. We turn to condition (vi). This will lead to the first entry to our list. Indeed, up to isomorphism, the towers $\mathbb{Z} \subset B \subset R$ such that $\mathbb{Z} \subset B$ is decomposed with crucial maximal ideal M and $B \subset R$ is inert with crucial maximal ideal N such that $N \cap \mathbb{Z} = M$ can be characterized via $B := \mathbb{Z} \times \mathbb{F}_p$ and $R := \mathbb{Z} \times \mathbb{F}_{p^q}$, where p and q are (possibly equal) prime numbers. (This can be seen by combining [3, Theorem 2.7], [2, Lemme 1.2], and the classical Galois theory of finite fields. To verify the required behavior of the crucial maximal ideals, observe that $M = p\mathbb{Z}$ and $N = \mathbb{Z} \times 0$.) As for the uniqueness conclusions in the assertion (a), note first that p is the only prime number which, when viewed inside R, is a zero-divisor in R. It is slightly harder to establish the uniqueness of q without mentioning M and N explicitly, but it can be done as follows (now that the uniqueness of p has been shown). The factor ring R/M is isomorphic to (a unique direct product of the form) $\mathbb{F}_p \times F$, where F is a field (actually, $F = \mathbb{F}_{p^q}$) whose vector space dimension over $\mathbb{F}_p$ is q.

By Proposition 4, condition (xi) cannot lead to a contribution to our developing list. It remains only to describe the possible contributions from condition (xiii). Assume, then, that both $\mathbb{Z} \subset B$ and $B \subset R$ are ramified extensions whose crucial maximal ideals, denoted by M and N, respectively, satisfy $N \cap \mathbb{Z} = M$. By [3, Theorem 2.7], there exists a uniquely determined prime number p such that we can identify $B = \mathbb{Z}(+)\mathbb{F}_p$. Then, necessarily, $M = p\mathbb{Z}$ and $N = p\mathbb{Z}(+)\mathbb{F}_p$. The cumbersome conditions from [6, Proposition 3.5(b)] that were alluded to in the statement of condition (xiii) hold automatically in the present context. In other words, $\mathbb{Z} \subset R$ satisfies FIP. The easiest way to see this is to apply the noncumbersome part of [6, Proposition 3.5(b)], the point being that $\mathbb{Z}/M (\cong \mathbb{F}_p)$ is finite. However, our relative lack of understanding of ramified extensions (cf. the third paragraph of [1, Remark 2.11(a)]) has made us settle here for the nonspecificity in assertion (b). One should note that the part of assertion (b) which follows the word "furthermore" is what allows us to conclude that R has exactly two proper subrings. As we showed above that our data satisfy the conditions from [6, Proposition 3.5(b)], it is now clear that the earlier part of assertion (b) simply ensures that we are addressing data satisfying condition (xiii).

(2) We have $B := \mathbb{Z}(+)\mathbb{F}_p$, with $N := p\mathbb{Z}(+)\mathbb{F}_p \in \mathrm{Max}(B)$. Noting that B/N can be identified with $\mathbb{F}_p$, put

$$R_2 := B(+)B/N \left(= B(+)\mathbb{F}_p = \left(\mathbb{Z}(+)\mathbb{F}_p\right)(+)\mathbb{F}_p\right). \tag{6}$$

By [15], $A \subset R_2$ is a minimal ring extension. Its crucial maximal ideal is N; $N \cap \mathbb{Z} = p\mathbb{Z} =: M$; and this minimal ring extension is ramified (cf. [16, Lemma 2.1]). Consider the ring $D := (\mathbb{Z}(+)0)(+)\mathbb{F}_p$. It is clear that $D \in [\mathbb{Z}, R_2] \setminus \{\mathbb{Z}, B, R_2\}$, and so $|[\mathbb{Z}, R_2]| > 3$, as required. We turn next to the more difficult construction of a suitable R_1.

Once again, we are working with $B = \mathbb{Z}(+)\mathbb{F}_p$ and $N := p\mathbb{Z}(+)\mathbb{F}_p$. The B-algebra R_1 will be constructed so as to have essentially the same B-module structure as R_2. Specifically, as a set (and then as an additive group with the "×" symbols being viewed as "⊕"), let

$$R_1 := \mathbb{Z} \times \mathbb{F}_p x \times \mathbb{F}_p z, \quad (7)$$

where x and z each are generators of (distinct) one-dimensional vector spaces over $\mathbb{F}_p$. Next, we define the multiplication in R_1 by setting $(m_1 + \xi_1 x + \eta_1 z) \cdot (m_2 + \xi_2 x + \eta_2 z)$

$$:= m_1 m_2 + (m_1\xi_2 + m_2\xi_1 + \eta_1\eta_2)\, x + (m_1\eta_2 + m_2\eta_1)\, z \quad (8)$$

for all $m_1, m_2 \in \mathbb{Z}$ and all $\xi_1, \xi_2, \eta_1, \eta_2 \in \mathbb{F}_p$. (Of course, expressions such as $m_i\xi_j$ and $m_i\eta_j$ are interpreted by using the additive structure of $\mathbb{F}_p$.)

It is straightforward (albeit somewhat tedious) to verify that R_1 is a ring (having $1 \in \mathbb{Z}$ as its multiplicative identity element). Moreover, the function $B \to R_1$, given by $(m, \xi) \mapsto m + \xi x$ for all $m \in \mathbb{Z}$ and $\xi \in \mathbb{F}_p$, is an injective (unital) ring homomorphism, thus allowing us to view B as a subring of R_1. A more intuitive understanding of the structure of R_1 as a B-algebra is that we required the multiplication to satisfy the axiomatic restrictions and the additional relations $x^2 = 0$, $xz = 0$, and $z^2 = x$. (As $(0, 1) \in B$ has been identified with $1x = x$, we see that the last two relations differ from their analogues in R_2.)

Recall that $\mathbb{Z} \subset B$ is a ramified extension with crucial maximal ideal $M = p\mathbb{Z}$. We claim that $B \subset R_1$ is a ramified extension with crucial maximal ideal N (such that $N \cap \mathbb{Z} = M$). Consider the set (described additively as) $N_1 := p\mathbb{Z} \oplus \mathbb{F}_p x \oplus \mathbb{F}_p z \subseteq R_1$. It is straightforward to verify that N_1 is a prime ideal of R_1 such that $N_1 \cap B = N$. As R_1 is integral over B, the fact that N is a maximal ideal of B ensures that N_1 is a maximal ideal of R_1. It is also straightforward to verify that $(N_1)^2 \subseteq N \subset N_1$ and that the canonical map $(\mathbb{F}_p \cong \mathbb{Z}/M \cong) B/N \to R_1/N_1$ is an isomorphism. In addition, one can verify that N is an ideal of R_1 (since $p\mathbb{F}_p = 0$ and $xz = 0$) and R_1/N is a two-dimensional vector space over B/N. Therefore, according to [12, Theorem 4.2(c)], the claim has now been proved.

It remains to show that R_1 has only two proper subrings. It is enough to show that if $u \in R_1 \setminus B$, then $\mathbb{Z}[u] = R_1$. Without loss of generality, $u = \xi x + \eta z$ for some $\xi, \eta \in \mathbb{F}_p$. Note that $\eta \neq 0$ since $u \notin B$. Thus, the above rules for multiplication lead to $u^2 = \eta^2 x \neq 0$. In particular, $\mathbb{Z}[u]$ contains the element $\nu(\eta^2 x)$ for each integer ν. In other words, $\mathbb{F}_p(\eta^2 x) \subseteq \mathbb{Z}[u]$. But $\mathbb{F}_p(\eta^2 x) = (\eta^2 \mathbb{F}_p)x = \mathbb{F}_p x$. Now, since $\mathbb{F}_p x \subseteq \mathbb{Z}[u]$, we get that $\eta z = u - \xi x \in \mathbb{Z}[u]$. Since $\eta \neq 0$, we can see, by taking integer multiples as above, that $\mathbb{F}_p z = \mathbb{F}_p(\eta z) \subseteq \mathbb{Z}[u]$. It is now evident that $R_1 \subseteq \mathbb{Z}[u]$, which completes the proof. □

One can fairly state that the formulation of assertion (b) in part (1) of the preceding theorem means that we have given a characterization but not a classification of the rings R in question. As noted in the proof, this lack of a classification is due solely to our current inability to classify certain ramified extensions (up to isomorphism). This feature of having a characterization, but not a classification, will recur when we treat the corresponding problems for base rings of positive characteristic; there, too, the sole reason for a lack of a classification will be our incomplete information concerning certain ramified extensions.

It was shown in [9, Proposition V.1] that if a nonzero ring R has only finitely many subrings and $\text{char}(R) \neq 0$, then R is a finite ring. Having dispatched the case of characteristic 0 (and the trivial case of the zero ring), we henceforth devote our attention to nonzero finite rings. The first technical lemma leading to our main results on finite rings is given next. It will be applicable because any (nonzero) finite ring is zero-dimensional.

Proposition 7. *Let A be a (nonzero) zero-dimensional ring. Then there does not exist a ring B such that $A \subset B$ is an integrally closed minimal ring extension.*

Proof. Suppose the assertion fails. Let M denote the crucial maximal ideal of an integrally closed minimal ring extension $A \subset B$. As $A \subset B$ is a minimal ring extension which is not integral, it follows from [2, Théorème 2.2(ii)] that there does not exist a prime ideal Q of B such that $Q \cap A = M$. However, M is a minimal prime ideal of A (since $\dim(A) = 0$), and so by an application of Zorn's Lemma [5, Exercise 1, page 41], such a prime ideal Q *does* exist, the desired contradiction. □

Thanks to Proposition 7 and the comments that preceded it, we may now focus on integral minimal ring extensions of finite rings. The next result is in the same spirit as Proposition 3. To further motivate it, first recall from [2, Lemme 1.2] that if p is a prime number, the inert extensions of $\mathbb{Z}/p\mathbb{Z}$ (i.e., of $\mathbb{F}_p$) are the minimal field extensions of $\mathbb{F}_p$. The next result shows that if $\alpha > 1$ (and p is a prime number), then the base ring $\mathbb{Z}/p^\alpha\mathbb{Z}$ behaves differently. It will also be useful to note that the hypothesis of the next result accommodates base rings such as $\mathbb{F}_p[X]/(X^2)$. Indeed, it accommodates base rings that are arbitrary special principal ideal rings (other than fields).

Proposition 8. *Let (A, M) be a quasi-local zero-dimensional principal ideal ring which is not a field. Then there does not exist a ring B such that $A \subset B$ is an inert extension.*

Proof. Observe that M is the only prime ideal of A. As A is a principal ideal ring, there exists $x \in A$ such that $M = Ax$. As A is not a field, $M \neq 0$, and so $x \neq 0$. Suppose the assertion fails. Then M is the crucial maximal ideal of some inert extension $A \subset B$. The "inert" condition ensures that (B, M) is quasi-local. Next, choose any element $e \in B \setminus A$. Then $B = A[e]$, by the minimality of $A \subset B$. As $xe \in MB = M$, there exists $a \in A$ such that $xe = ax$, and so $x(a - e) = 0$. Since $x \neq 0$, $a - e$ cannot be a unit of B; that is, $a - e \in M$.

Hence, $e = a - (a - e) \in A + M = A$, the desired contradiction. □

Remark 9. Although Proposition 8 has been stated in a form that will be useful below, it may be of interest to record the following generalization of it. If A is a (nonzero) zero-dimensional ring such that A_M is a principal ideal ring and $MA_M \neq 0$ for each $M \in \operatorname{Max} A$, then there does not exist a ring B such that $A \subset B$ is an inert extension. For a proof by contradiction, take M to be the crucial maximal ideal of an inert extension $A \subset B$. Next, apply [12, Proposition 4.6] to replace $A \subset B$ with $(A_M, MA_M) \subset B_M$. Then repeat the proof of Proposition 8 to get the desired contradiction.

In contrast to Propositions 7 and 8, Proposition 10 shows that there does exist a ring B with the relevant property (and that it is unique up to isomorphism). As an application, it also shows a way in which the base ring $\mathbb{Z}/p^{\alpha}\mathbb{Z}$ behaves regardless of whether the positive integer α exceeds 1. In view of Proposition 8, this highlights an important difference in the behavior of the "inert" and "decomposed" concepts.

Proposition 10. *Let (A, M) be a quasi-local zero-dimensional principal ideal ring. Then, up to A-algebra isomorphism, the unique ring B such that $A \subset B$ is a decomposed extension is $B = A \times A/M$.*

Proof. By [3, Lemma 2.2] (or the first paragraph of the proof of [13, Corollary 2.5]), $A \subset A \times A/M$ is a minimal ring extension. As this minimal ring extension is integral, it follows from [2, Théorème 2.2(ii)] that its crucial maximal ideal is its conductor, namely, M (viewed as $\mathscr{M} := M \times 0 \subset A \times A/M$). Since $(A \times A/M)/\mathscr{M} \cong A/M \times A/M$, we see that $A \subset A \times A/M$ is a decomposed extension.

Conversely, suppose that $A \subset B$ is a decomposed extension. We will prove that B is A-algebra isomorphic to $A \times A/M$. By [2, Lemme 1.2], we may assume that A is not a field. In particular, $M \neq 0$. Since R is a principal ideal ring, there exists $x \in A$ such that $M = Ax$. As $M \neq 0$, we get $x \neq 0$. As M is the only prime ideal of A, it consists of all the nilpotent elements of A. Let $k(\geq 2)$ be the index of nilpotence of x. Clearly, M is a nilpotent ideal whose index of nilpotence is k. Next, notice that A is an Artinian ring, since it is a zero-dimensional Noetherian ring. Since B is finitely generated as an A-module, it follows that B is also an Artinian ring. By a fundamental structure theorem [17, Theorem 3, page 205], B is (essentially uniquely expressible as) the direct product of finitely many (nonzero Artinian quasi-) local rings, say, $B = \prod_{i=1}^{n} C_i$. In fact, $n = 2$ since the "decomposed" hypothesis ensures that B has exactly two maximal ideals, say N_1 and N_2. The "decomposed" hypothesis also ensures that $N_1 \cap N_2 = M$; and the canonical ring (field) homomorphism $A/M \to B/N_i$ is an isomorphism for $i = 1, 2$. With $\mathscr{N}_i$ denoting the unique maximal ideal of C_i, there is no harm in taking $N_1 = \mathscr{N}_1 \times C_2$ and $N_2 = C_1 \times \mathscr{N}_2$. It will be useful later to notice that $C_i/\mathscr{N}_i \cong B/N_i (\cong A/M)$ for $i = 1, 2$.

Next, note that A is a special principal ideal ring (in short, an SPIR), in the sense of [17, page 245]. (The literature contains a variety of definitions of an SPIR, but those definitions are all equivalent.) Consequently, each nonzero proper ideal of R is of the form M^j for some integer j such that $1 \leq j \leq k - 1$ [17, page 245]. Then, for each $i = 1, 2$, the First Isomorphism Theorem shows that the composition of the inclusion map $R \hookrightarrow S$ with the canonical projection map $S \to C_i$ has an image which is a subring of C_i that is isomorphic as an A-algebra to A/M^{λ_i} for some nonnegative integer $\lambda_i \leq k$. There is no harm in viewing A/M^{λ_i} as a subring of C_i for $i = 1, 2$. Observe that the canonical map $\varphi : A \to A/M^{\lambda_1} \times A/M^{\lambda_2}$ is an injection (since the composition of φ with the inclusion map $A/M^{\lambda_1} \times A/M^{\lambda_2} \hookrightarrow C_1 \times C_2$ is the inclusion map $A \hookrightarrow B$). For each $a \in A$, we have $\varphi(a) = (a + M^{\lambda_1}, a + M^{\lambda_2})$. Put $\lambda := \max(\lambda_1, \lambda_2)$. Then $\varphi(x^{\lambda}) = (0, 0)$. Since φ is an injection, $\lambda \geq k$. Thus, either λ_1 or λ_2 is k. Without loss of generality $\lambda_1 = k$. Hence, we can view $A(\cong A/0 = A/M^k = A/M^{\lambda_1})$ as a subring of C_1.

We claim that A is not a proper subring of C_1. To see this, suppose, on the contrary, that $A \subset C_1$. Using φ to identify A with its image in B, we then get $A \subsetneq A \times C_2 \subset C_1 \times C_2 = B$, and so the minimality of $A \subset B$ yields that $A = A \times C_2$. However, this gives a contradiction, since A has only one prime ideal while $A \times C_2$ has two distinct prime ideals. This contradiction proves the claim. Thus, $B = A \times C_2$. It remains to show that $C_2 \cong A/M$ as algebras over A/M.

We have $N_1 = M \times C_2$ and $N_2 = A \times \mathscr{N}_2$. It follows that

$$\begin{aligned} M &= N_1 \cap N_2 = (M \cap A) \times (C_2 \cap \mathscr{N}_2) = M \times \mathscr{N}_2 \\ &\subsetneq A \times C_2 = B. \end{aligned} \tag{9}$$

Of course, when viewed inside $A \times C_2$, each element y of M is identified with (y, z_y), where $z_y \in C_2$ is the image of y under the canonical projection map $S \to C_2$. Thus, viewing matters in $A \times C_2$, we have

$$\{(y, z_y) \in S \mid y \in M\} = M = M \times \mathscr{N}_2. \tag{10}$$

Each $y \in M$ occurs as the first coordinate of only one element in the last-displayed set, and so $\mathscr{N}_2$ must be a singleton set. Hence, $\mathscr{N}_2 = 0$; that is, C_2 is a field. Therefore, $C_2 \cong C_2/0 = C_2/\mathscr{N}_2 \cong A/M$, as desired. □

The next result classifies certain ramified extensions. Its proof served us as motivation for the second construction that was used in the proof of Theorem 6(2).

Proposition 11. *Let $B := \mathbb{F}_p[X]/(X^2)$, where p is a prime number. As usual, write $B = \mathbb{F}_p \oplus \mathbb{F}_p x$, where $x^2 = 0 \neq x := X + (X^2)$. Then, up to B-algebra isomorphism, there are exactly two rings R_i $(i = 1, 2)$ such that $B \subset R_i$ is a ramified extension. One of these rings, say R_1, can be constructed as the B-algebra whose additive structure as a vector space over $\mathbb{F}_p$ is given by $R_1 = B \oplus \mathbb{F}_p z$ for some nonzero element z and whose multiplication is determined by the relations $xz = 0$ and $z^2 = x$ (and $x^2 = 0$). The other ring R_2 can be taken as $B(+)\mathbb{F}_p$. Moreover, R_1 has exactly two proper subrings, while R_2 has more than two proper subrings (and so R_1 is not isomorphic to R_2).*

Proof. To shape the constructions of the rings R_i, consider any ramified extension R of B. The crucial maximal ideal of $B \subset R$ is necessarily then the unique prime ideal of B; namely, $N := \mathbb{F}_p x$. As recalled in the proof of Proposition 10, the fact that $B \subset R$ is an integral minimal ring extension gives that $N = (B : R)$. In particular, N is an ideal of R. Moreover, since $B \subset R$ is ramified and R is quasi-local, it follows that R must be quasi-local, say with maximal ideal Q, such that $Q^2 \subseteq N \subset Q$, $R/Q \cong B/N$ ($\cong \mathbb{F}_p$), and $R/N \cong (B/N)[X]/(X^2) \cong \mathbb{F}_p[X]/(X^2)$. Hence $\dim_{\mathbb{F}_p}(R) = \dim_{\mathbb{F}_p}(R/N) + \dim_{\mathbb{F}_p}(N) = 2 + 1 = 3$, and so $\dim_{\mathbb{F}_p}(Q) = \dim_{\mathbb{F}_p}(R) - \dim_{\mathbb{F}_p}(R/Q) = 3 - 1 = 2$. Therefore, taking z to be any element of $Q \setminus N$, we have $Q = N \oplus \mathbb{F}_p z = \mathbb{F}_p x \oplus \mathbb{F}_p z$. Observe that $z^2 \in Q^2 \subseteq N$.

We claim that $xz = 0$. In fact, $xz \in NB = N$ and so $xz = ax$ for some $a \in \mathbb{F}_p$. Thus $x(a - z) = 0$. Since $x \neq 0$, it follows that $a - z$ is not invertible in R. As $z^4 = (z^2)^2 \in (Q^2)^2 \subseteq N^2 = 0$, we see that z is nilpotent. Hence, a is not invertible (in R). Since a is a member of the field $\mathbb{F}_p$, we get $a = 0$, whence $xz = 0$, thus proving the above claim. Note also that $z^3 = 0$ since $z^3 = z^2 z \in Q^2 z \subseteq Nz = \mathbb{F}_p xz = 0$.

Since $\mathbb{F}_p \cap Q = 0$ and $\dim_{\mathbb{F}_p}(R) = \dim_{\mathbb{F}_p}(\mathbb{F}_p) + \dim_{\mathbb{F}_p}(Q)(< \infty)$, it follows that the linear structure of R must be given by $R = \mathbb{F}_p \oplus Q = \mathbb{F}_p \oplus \mathbb{F}_p x \oplus \mathbb{F}_p z$. Given this explicit linear structure and the specific facts that we have found about multiplication in R, there are now two general ways to describe a set of necessary conditions for the B-algebra R. One would impose the following set of relations: $x^2 = 0$, $xz = 0$, and $z^2 = 0$. This set of conditions determines the B-algebra $R_2 := B(+)\mathbb{F}_p$. It is known that $B \subset R_2$ is ramified (see [15] and [16, Lemma 2.1]), and so this given set of relations should not be augmented in a possible search for other possible R. To show that R_2 has more than two proper subrings, one need only tweak an argument that was given in the proof of Theorem 6(2): consider $D := \mathbb{F}_p \oplus 0 \oplus \mathbb{F}_p z$.

However, there is one other general way to prescribe the multiplication of $R = \mathbb{F}_p \oplus \mathbb{F}_p x \oplus \mathbb{F}_p z$, namely, as follows: $x^2 = 0$, $xz = 0$, and $z^2 = bx$, where b is some nonzero element of $\mathbb{F}_p$. We next explain that any such prescription does give a B-algebra and that any two such prescriptions (using different values of the nonzero element b) give isomorphic B-algebras. The first of these tasks can be handled classically, as we have merely stipulated how to multiply the basis elements of a vector space. (Lacking that vectorial context led us to the more complicated argumentation given for the somewhat similar construction in the proof of Theorem 6(2).) Let R_1 denote the B-algebra that is constructed when we use $b := 1$. The second task is now handled by noticing that any other nonzero element b, say $\beta \in \mathbb{F}_p \setminus \{0\}$, gives a B-algebra that is isomorphic to R_1. What it *does* give is a multiplication that is determined on the elements of the $\mathbb{F}_p$-basis $\{1, x, z\}$ of the $\mathbb{F}_p$-algebra by $x^2 = 0$, $xz = 0$, and $z^2 = \beta x$. Consider the induced multiplication on the basis consisting of $1, x_1 := \beta^3 x$, and $z_1 := \beta z$. One easily finds that $x_1^2 = 0$, $x_1 z_1 = 0$, and $z_1^2 = \beta^2 z^2 = \beta^2 \beta x = \beta^3 x = x_1$. Thus, the construction using β has produced an algebra that is isomorphic (via the change of bases from $\{1, x, z\}$ to $\{1, x_1, z_1\}$) to the algebra that was constructed using $b := 1$.

Only two tasks remain: to prove that $B \subset R_1$ is ramified and that R_1 has only two proper subrings. To accomplish the first of these, we will invoke [12, Theorem 4.2]. For that, it suffices (since $\mathrm{Max}(B) = \{N\}$) to show the following: $Q := N + \mathbb{F}_p z (= \mathbb{F}_p x \oplus \mathbb{F}_p z)$ is a maximal ideal of R such that $Q^2 \subseteq N \subset Q$, $\dim_{B/N}(R/N) = 2$, and the canonical map $B/N \to R/Q$ is an isomorphism. All of the preceding assertions are easily verified and so we omit those details. Finally, to show that R_1 has only two proper subrings, it suffices to show that if $u \in R \setminus B$, then $\mathbb{F}_p[u] = R$. Without loss of generality, $u = \xi x + \eta z$ for some $\xi, \eta \in \mathbb{F}_p$. To complete this proof, it suffices to repeat the argument in the final paragraph of the proof of Theorem 6, with that proof's coefficient ring $\mathbb{Z}$ now being modified to $\mathbb{F}_p$ for the purposes of this proof. □

Note that the idealization used in Proposition 11 can also be described as $R_2 = \mathbb{F}_p[X, Z]/(X^2, XZ, Z^2)$. In the same spirit, one could describe the ring R_1 from that result as $\mathbb{F}_p[X, Z]/(X^2, XZ, Z^2 - X)$. This description of R_1 is admittedly more compact than the earlier one, but that earlier description seemed more suitable in presenting the proof of Proposition 11.

Part (a) of the next result was developed in correspondence with Jay Shapiro and is included here with his kind permission.

Proposition 12. *Let $A := \mathbb{Z}/p^\alpha\mathbb{Z}$, where p is a prime number and $\alpha \geq 2$ is an integer. Let $M := pA$, the maximal ideal of A. Then, consider the following:*

(a) *Let B be a ring such that $A \subset B$ is a ramified extension. Let Q denote the maximal ideal of B. Then there exists $y \in Q \setminus M$ such that $B = A[y]$, $y^2 \in M$, and $py \in M$. In fact, $p^{\alpha-1} y = 0$. Moreover, there exist units u and v of A and uniquely determined integers i and j such that $y^2 = up^i$, $py = vp^j$, and $1 \leq i, j \leq \alpha$. Also, $(u - v^2)p^{2j} = 0$. If $i < \alpha - 2$, then i is even and $j = (i+2)/2$ (so $2 \leq j < \alpha/2$). If $i \geq \alpha - 2$, then $j \geq \alpha/2$.*

(b) *Let B be a ring such that $A \subset B$ is a ramified extension. Let $y \in B$ be as in (a). Let $H := \{ry \in B \mid 0 \leq r \leq p - 1 \text{ in } \mathbb{Z}\}$. Then H is a subset (not necessarily an additive subgroup) of B of cardinality p. Moreover, the assignment $(a, ky) \mapsto a + ky$ gives a bijection $A \times H \to B$ (which does not necessarily preserve any algebraic structure).*

(c) *If B is a ring such that $A \subset B$ is a ramified extension, then $|B| = p^{\alpha+1}$.*

(d) *$A \subset A(+)\mathbb{F}_p$ is a ramified extension.*

(e) *Up to isomorphism, $\mathbb{Z}/4\mathbb{Z}$ has exactly two ramified extensions, namely, the ring $B_1 := \mathbb{Z}/4\mathbb{Z}(+)\mathbb{F}_2$ given by (d) and the ring $B_2 := (\mathbb{Z}/4\mathbb{Z})[X]/(X^2 - 2, 2X)$.*

(f) *Suppose that p is an odd prime number. Then, up to isomorphism, $A := \mathbb{Z}/p^2\mathbb{Z}$ has exactly three ramified extensions, namely, the ring $B_a := A(+)\mathbb{F}_p$ given by (d), the ring $B_b := A[X]/(pX, X^2 - p)$, and the ring*

$B_c := A[X]/(pX, X^2 - dp)$, where d is a(ny) quadratic nonresidue of p.

Proof. (a) By [13, Proposition 2.12], there exists $y \in Q \setminus M$ such that $B = A[y]$, $y^2 \in M$, $y^3 \in M$, and $(A : B) = M$. In particular, $py \in MB = M$. As A is an SPIR whose unique prime ideal is generated by p, it follows from [17, page 245] that there exist uniquely determined integers i and j and units u and v of A such that $y^2 = up^i$, $py = vp^j$, and $1 \le i, j \le \alpha$. Thus

$$v^2p^{2j} = \left(vp^j\right)^2 = (py)^2 = p^2y^2 = p^2\left(up^i\right) = up^{i+2}. \quad (11)$$

If $i < \alpha - 2$, then the right-hand side of the displayed equation is nonzero (again, by [17, page 245]), and thus so is the left-hand side, whence $2j = i + 2$ (again, by [17, page 245]), so that $i = 2j - 2$ and $j = (i + 2)/2$. In particular, if $i < \alpha - 2$, then i is even, $2 \le j < \alpha/2$, and $(u - v^2)p^{2j} = up^{2j} - v^2p^{2j} = up^{i+2} - v^2p^{2j} = 0$. On the other hand, if $i \ge \alpha - 2$, then the right-hand side of the displayed equation is 0, hence so is its left-hand side, whence $2j \ge \alpha$ and $j \ge \alpha/2$.

It remains only to prove that $p^{\alpha-1}y = 0$. First, note that y is nilpotent since y^2 is a member of the only prime ideal of A. (In general, $y^{2\alpha} = (y^2)^\alpha \in M^\alpha = 0$.) Recall that $py = vp^j$. As $p^{\alpha-1}y = p^{\alpha-2}(py)$ and $p^\alpha = 0$, it suffices to show that $j \ge 2$. Suppose, on the contrary, that $py = vp$. Then $p(v - y) = 0$. As the nilpotence of y implies that $v - y$ is a unit of B, it follows that $p = 0$, contradicting $\alpha > 1$.

(b) Let $k \in \mathbb{Z}$ with $1 \le k < p^\alpha$. By the division algorithm, there exist nonnegative integers q and r such that $k = qp + r$ and $r < p^\alpha$. Thus, $ky = q(py) + ry$ can be expressed as the sum of an element of A with an element of H. We will show that this expression is unique.

We claim that there cannot exist nonnegative integers $\lambda < \mu < p$ such that $(\mu - \lambda)y \in A$. For a proof, suppose, on the contrary, that $v := \mu - \lambda$ satisfies $vy \in A$. Since $1 \le v \le p - 1$, v and p are relatively prime, and so there exist integers n_1, n_2 such that $n_1v + n_2p = 1$. Hence, $n_1(vy) + n_2(py) = y$. As vy and py are elements of A, it follows that $y \in A$, the desired contradiction. This proves the above claim.

Now suppose that $a_1 + k_1y = a_2 + k_2y$, where $a_1, a_2 \in A$ and $0 \le k_1 \le k_2 < p$. As $(k_2 - k_1)y = a_1 - a_2 \in A$, it follows from the above claim that $k_1 = k_2$ (hence $k_1y = k_2y$) and $a_1 = a_2$. This proves the above uniqueness assertion. It also shows that $|H| = p$ and establishes the asserted bijection.

(c) One proof of (c) is via (b), for $|B| = |A \times H| = |A| \cdot |H| = p^\alpha p = p^{\alpha+1}$. We next give an alternate proof. Let Q denote the maximal ideal of B. Since $A \subset B$ is ramified, the canonical map $A/M \to B/Q$ is an isomorphism and $B/M \cong (A/M)[X]/(X^2) (\cong \mathbb{F}_p[X]/(X^2))$. Therefore, by Lagrange's Theorem, $|B| = |B/M| \cdot |M| = p^2p^{\alpha-1} = p^{\alpha+1}$.

(d) By [15], $B := A(+)\mathbb{F}_p$ is a minimal ring extension of A; and this extension is ramified (i.e., subintegral) by [16, Lemma 2.1].

(e), (f) Let B be a ramified extension of $\mathbb{Z}/p^2\mathbb{Z}$ and let Q denote the maximal ideal of B. We adopt the other data from (a) (now with $A = \mathbb{Z}/p^2\mathbb{Z}$). Recall from (a) that $p^{\alpha-1}y = 0$. As $\alpha = 2$ here, $py = 0$. Next, since $y^3 \in M$, it follows as above via [17, page 245] that $y^3 = wp^k$ for some unit w of A and some (uniquely determined) positive integer $k \le 2$. It cannot be the case that $i = k = 1$. (Otherwise, $y^2 = up$ and $y^3 = wp$, with u and w suitable units of A as above, and so $p = w^{-1}(wp) = w^{-1}y^3 = w^{-1}y^2y = w^{-1}(up)y = (w^{-1}u) \cdot 0 = 0$, a contradiction since $\alpha > 1$.) On the other hand, if $i = 2$, then $k = 2$ (the point being that if $y^2 = 0$, then $y^3 = 0$). Thus, (i, k) is either $(2, 2)$ or $(1, 2)$. In any event, $k = 2$, and so $y^3 = 0$. In summary, we have $py = 0$, $y^3 = 0$, and y^2 is either 0 or of the form up (for some unit u of A).

We first consider (e), where $p = 2$. Then, since the set of units of A is $\{1, -1\}$, we have that y^2 is either 0 or $up = \pm p = \pm 2 = 2$. By (b) and (c), $B = \{0, 1, 2, 3, y, y+1, y+2, y+3\}$ has cardinality 8. Since the algebraic structure of B is determined by the values of y^2 and $2y (= 0)$, it follows that (up to isomorphism) there are at most two ramified extensions B of A. The first of these candidates is $B_1 := (\mathbb{Z}/4\mathbb{Z})[X]/(X^2, 2X)$, which arises from the identities $y^2 = 0$ and $2y = 0$. Note that B_1 is isomorphic to $\mathbb{Z}/4\mathbb{Z}(+)\mathbb{F}_2$, which we know, by (d), is a ramified extension of $\mathbb{Z}/4\mathbb{Z}$ (where $y \in 0(+)\mathbb{F}_2$ can be taken to be $(0, 1)$). Hence, B_1 is a ramified extension of $\mathbb{Z}/4\mathbb{Z}$.

The remaining candidate is $B_2 := (\mathbb{Z}/4\mathbb{Z})[X]/(X^2 - 2, 2X)$, which arises from the identities $y^2 = 2$ and $2y = 0$. While it was rather obvious that $A \subset B_1$, we next include a proof that $A \subset B_2$. To view $A \subseteq B_2$, we will show that the canonical A-algebra homomorphism $A \to B_2$ is an injection; in other words, we claim that $A \cap (X^2 - 2, 2X) = 0$. Suppose, on the contrary, that $a = (X^2 - 2)f + 2Xg$ for some $a \in A \setminus \{0\}$ and some $f, g \in A[X]$. Equating constant terms shows that $a \in M (= 2A)$. As $a \ne 0$, it follows that the constant term of f must be a unit of A. Hence, the coefficient of X^2 in $(X^2 - 2)f + 2Xg$ is a unit of A. This is a contradiction (since $a = a + 0X^2$). This proves the above claim. We leave to the reader a similar argument (focusing on the coefficient of X) which reveals that $X \notin A + (X^2 - 2, 2X)$. Thus, $y := X + (X^2 - 2, 2X)$ satisfies $A \subset B_2 = A[y] = A + Ay$, with $y^2 = 2$ and $2y = 0$. Next, note that the set of nilpotent elements of B_2, namely, its unique prime ideal, is $Q = \{0, 2, y, y + 2\}$. (The same could be said of B_1.) By verifying the conditions in [12, Theorem 4.2(c)], it is straightforward to verify that B_2 is a ramified extension of $\mathbb{Z}/4\mathbb{Z}$. (In detail, $|B_2/Q| = |B_2|/|Q| = 8/4 = 2$, so $B_2/Q \cong \mathbb{F}_2 \cong A/M$; $\dim_{A/M}(B/M) = 2$ since $|B/M| = |B|/|M| = 8/2 = 4 = |A/M|^2$; and $Q^2 = \{0, 2\} \subseteq M \subset Q$.)

It remains only to verify that B_1 and B_2 are not isomorphic. For each of these rings B_i, the only elements of B_i that can play the role of y in (a) are y and $y + 2$; and $Q \setminus M = \{y, y + 2\}$. In B_1 (resp., B_2), each element of $Q \setminus M$ has index of nilpotence 2 (resp., 3). This completes the proof of (e).

We turn to (f), where p is an odd prime and $A := \mathbb{Z}/p^2\mathbb{Z}$. The first candidate for a ramified extension of A is $B_a := A[X]/(X^2, pX)$, which arises from the identities $y^2 = 0$ and $py = 0$. As above, we see via (d) that B_a is a ramified extension of A since $B_a \cong A(+)\mathbb{F}_p$. Any other (nonisomorphic) candidate for a ramified extension of A must arise from the identities $y^2 = up$ and $py = 0$ for some unit u of A. This candidate can be viewed as $C_u := A[X]/(X^2 - up, pX)$. Ostensibly, u could be the coset of $p^2\mathbb{Z}$ in

$\mathbb{Z}$ that is represented by any positive integer $m \leq p^2 - 1$ which is not divisible by p. However, it suffices to further restrict $m \leq p-1$, for if the division algorithm gives $m = qp+r$, then $p^2 = 0 \in A$ leads to $C_m = C_r$ since $X^2 - mp = (X^2 - rp) + qp^2 = X^2 - rp \in A[X]$. As the above illustrates, it will be convenient (and harmless) to blur the distinction between u and its coset representative m, so we will view $u \in \mathcal{S} := \{1, \ldots, p-1\}$. As is well known (cf. [18, Theorem 9.1]), exactly half of the elements in the set $\mathcal{S}$ are quadratic nonresidues of p (and the other half are quadratic residues of p). To finish the proof of (f), it suffices to establish the following five facts (where $u, u_1, u_2 \in \mathcal{S}$): each C_u is a ramified extension of A; C_u is not isomorphic to B_a; if u_1 and u_2 each are quadratic residues of p, then $C_{u_1} \cong C_{u_2}$; if u_1 and u_2 each are quadratic nonresidues of p, then $C_{u_1} \cong C_{u_2}$; and if u_1 is a quadratic residue of p and u_2 is a quadratic nonresidue of p, then C_{u_1} and C_{u_2} are not isomorphic.

Let $u \in \mathcal{S}$. By reworking the above analysis of B_2 in the proof of (e), we can view $A \subset C_u$ and see that $y := X + (X^2 - up, pX)$ satisfies $C_2 = A[y] = A + Ay$, with $y^2 = up$ and $py = 0$. Next, by reworking the proof of (b), one can show that each element of C_u can be uniquely expressed as the sum of an element of A and an element of the form vy where $v \in \mathcal{S} \cup \{0\}$. It follows easily that $|C_u| = p^3$. Moreover, the set of nilpotent elements of C_u is the ideal $Q := M + Ay = M + (\mathcal{S} \cup \{0\})y$ of cardinality p^2. It follows that Q is the unique maximal ideal of C_u; $Q \cap A = M$; $Q^2 = M \subset Q$; and the canonical ring homomorphism $A/M \to C_u/Q$ is an isomorphism (since it is an injective map of fields both of which have cardinality p). In addition, since $C_u = A + Ay$ and $My = 0$, we see that M is an ideal of C_u. Then C_u/M must be a two-dimensional vector space over $\mathbb{F}_p (\cong A/M)$ since $|C_u/M| = |C_u|/|M| = p^2$. We have verified all the criteria in [12, Theorem 4.2(c)], and so $A \subset C_u$ is a ramified extension.

Next, we show that there cannot exist an A-algebra isomorphism $\varphi : B_a \to C_u$. Suppose, on the contrary, that such φ exists. View $B_a = A(+)\mathbb{F}_p = A[\gamma]$ with $\gamma := (0,1)$. As $\gamma^2 = 0$, we see by applying φ that $z := \varphi(\gamma) \in C_u$ must satisfy $A[z] = C_u$ and $z^2 = 0$. Consequently, $z \in Q \setminus M$. Therefore, as noted in the preceding paragraph (and using its notation), $z = m + sy$ for some $m \in M$ and some $s \in \mathcal{S}$. (Note that $s \neq 0$ since $z \notin M$.) Then $0 = z^2 = (m+sy)^2 = m^2 + 2s(my) + s^2y^2 = 0 + 2s \cdot 0 + s^2up = s^2up \neq 0$ (since s^2u is not divisible by p), the desired contradiction. Hence, B_a is not isomorphic to C_u.

Next, to show that $C_{u_1} \cong C_{u_2}$ whenever u_1 and u_2 are quadratic residues of p, it is enough to show that if $C := C_1$ and u is a quadratic residue of p, then $C_u \cong C$. It suffices to prove that there exists $z \in C$ such that $C = A[z]$, $z^2 = up$, and $pz = 0$ (for then $C = A[z] \cong C_u$). By hypothesis, there exists $a \in \mathcal{S}$ such that $a^2 \equiv u (\text{mod}\, p)$. By the definition of C_1, there exists $y \in C$ such that $C = A[y]$, $y^2 = p$, and $py = 0$. Put $z := ay$. Since a is a unit of A, we have $Az = (Aa)y = Ay$, and so $A + Az = A + Ay = C$. It follows that $A[z] = C$. It is easy to check that $pz = 0$. Moreover, since $a^2 - u = tp$ for some "integer" t and $p^2 = 0$, we get $z^2 = a^2y^2 = (u+tp)p = up + t \cdot 0 = up$, as required.

Next, we will show that $C_{u_1} \cong C_{u_2}$ whenever u_1 and u_2 each are quadratic nonresidues of p. By the well known multiplicative property of the Legendre symbols [18, Theorem 9.3(ii)], u_1u_2 is a quadratic residue of p. Pick $a \in \mathcal{S}$ such that $u_1u_2 \equiv a^2 (\text{mod}\, p)$. Let b be the element of $\mathcal{S}$ such that $(a + p\mathbb{Z})/(u_2 + p\mathbb{Z}) = b + p\mathbb{Z} \in \mathbb{F}_p$. Then, working in $\mathbb{F}_p$, we have $(u_1 + p\mathbb{Z})/(u_2 + p\mathbb{Z})$

$$= \frac{(a^2 + p\mathbb{Z})}{(u_2^2 + p\mathbb{Z})} = \left[\frac{(a + p\mathbb{Z})}{(u_2 + p\mathbb{Z})}\right]^2 = (b + p\mathbb{Z})^2, \qquad (12)$$

and so $u_1 \equiv b^2u_2 (\text{mod}\, p)$. Given z such that $C_{u_2} = A[z]$, $z^2 = u_2p$, and $pz = 0$, put $y := bz \in C_{u_2}$. As b is a unit in A, we see as above that $A + Ay = A + Az$, and so $A[y] = C_{u_2}$. It is easy to check that $py = 0$. Moreover, for some "integer" t,

$$\begin{aligned} y^2 &= b^2z^2 = \left(b^2u_2\right)p = (u_1 + tp)\,p = u_1p + t \cdot 0 \\ &= u_1p, \end{aligned} \qquad (13)$$

and so $C_{u_2} = A[y] \cong C_{u_1}$, as required.

Finally, to complete the proof, it suffices to show that if u is a quadratic nonresidue of p, then $C := C_1$ is not isomorphic to C_u. Suppose, on the contrary, that there exists an isomorphism $\psi : C \to C_u$. By hypothesis, there exists $y \in C$ such that $C = A[y]$, $y^2 = p$, and $py = 0$; and there exists $z \in C_u$ such that $C_u = A[z]$, $z^2 = up$, and $pz = 0$. Put $y^* := \psi(y)$. Since $A \subset A[y] = C$, we see by applying the isomorphism ψ that $A \subset A[y^*] = C_u$. Therefore, by (b), $y^* = a + \lambda z$ for some (uniquely determined) $a \in A$ and $\lambda \in \mathcal{S}$. Since y is nilpotent, so is y^*. But z is also nilpotent. Thus, a is nilpotent; that is, $a \in M$. Hence $a^2 = 0$. In addition, $ay^* = \psi(ay) \in \psi(My) = \psi(0) = 0$; that is, $ay^* = 0$. Also, $(y^*)^2 = \psi(y^2) = \psi(p) = p$.

Consider $y^* - a = \lambda z \neq 0$. Squaring both sides leads to the equality of $\lambda^2z^2 = \lambda^2up$ with

$$(y^*)^2 - 2ay^* + a^2 = p - 2 \cdot 0 + 0 = p. \qquad (14)$$

Thus $p = \lambda^2up$; that is, $p(1 - \lambda^2u) = 0$. Hence, $1 - \lambda^2u \equiv 0 (\text{mod}\, p)$, and so $\lambda^2u \equiv 1 (\text{mod}\, p)$. As λ^2 is obviously a quadratic residue of p and u is assumed to be a quadratic nonresidue of p, it follows from the multiplicative property of the Legendre symbols [18, Theorem 9.3(ii)] that λ^2u is a quadratic nonresidue of p. Therefore, so is 1, the desired contradiction. The proof is complete. □

The five rings featured in parts (e) and (f) of Proposition 12 were identified in [19] as being, up to isomorphism, the (commutative unital) rings of cardinality 8 and (when p is an odd prime number) p^3. We provided full details in proving those parts of Proposition 12 for two reasons: [19] provided few details relevant to this context and, unfortunately, there were some errors in [19]. In his Mathematical Review of [19], Kruse [20] listed the correct number (up to isomorphism) of (not necessarily commutative or unital) rings of cardinality p^3 and reported that a correct list of isomorphism class representatives for these rings had been given in [21]. See also the later treatment in [22] and the discussion of other relevant literature in Mal'tsev's Mathematical Review [23] of [22].

Proposition 12(e) gives some evidence concerning the essentiality of the role of the parameters given in Proposition 12(a). Indeed, for the case $p^\alpha = 4$, we found that B_1 featured $i = 2 = j$, with u and v being nonuniquely determined units of A, while B_2 featured $i = 1$, $j = 2$, and $u = 1$, with v a nonuniquely determined unit of A. Moreover, not every ordered pair (i, j) that fit the (necessary) conditions in Proposition 12(a) actually led to a viable candidate in Proposition 12(e). For readers who may wish to see the extent to which the behavior in parts (e) and (f) may be exhibited when $\alpha \geq 3$, we would suggest, rather than continuing to do computations by hand with the generator-and-relations information from Proposition 12(a), the use of appropriate computer software. We will comment further on such studies in the final remark of this paper.

We can now state our main results involving rings R of positive characteristic. If the prime ring of R is (isomorphic to) $\mathbb{Z}/n\mathbb{Z}$ and n has the prime-power factorization $n = \prod_{i=1}^k p_i^{\alpha_i}$, then by the Chinese Remainder Theorem, the prime ring of R is $\prod_{i=1}^k \mathbb{Z}/p_i^{\alpha_i}\mathbb{Z}$. Accordingly, R can be expressed as $R = \prod_{i=1}^k E_i$, where each E_i is a suitable ring extension of the corresponding $\mathbb{Z}/p_i^{\alpha_i}\mathbb{Z}$. The forays into classification in the next two theorems are built from this point of view, featuring a lack of specificity only when a chain of minimal ring extensions from some $\mathbb{Z}/p_i^{\alpha_i}\mathbb{Z}$ to its corresponding E_i includes a ramified extension of $\mathbb{Z}/p_i^{\alpha_i}\mathbb{Z}$, where $\alpha_i \geq 2$. As usual, if E is any ring, the direct product of E with an empty direct product is being viewed as $(E \times \{0\} \cong)E$.

Theorem 13. *Up to isomorphism, the rings R of positive characteristic that have exactly one proper subring can be characterized as follows. The prime ring of R is (isomorphic to) the direct product $\prod_{i=1}^k \mathbb{Z}/p_i^{\alpha_i}\mathbb{Z}$, where $p = p_1, p_2, \ldots, p_k$ are pairwise distinct prime numbers for some positive integer k and $\alpha = \alpha_1, \alpha_2, \ldots, \alpha_k$ are positive integers (and possibly $\alpha_i = \alpha_j$ for some $i \neq j$). Then (up to isomorphism), R is the direct product $E \times \prod_{i=2}^k \mathbb{Z}/p_i^{\alpha_i}\mathbb{Z}$, where E is a ring satisfying (exactly) one of the following five conditions:*

(a) *($\alpha = 1$ and) $E = \mathbb{F}_{p^q}$, where q is a prime number (which is possibly equal to p);*

(b) *($\alpha = 1$ and) $E = \mathbb{F}_p \times \mathbb{F}_p$;*

(c) *($\alpha = 1$ and) $E = \mathbb{F}_p[X]/(X^2)$;*

(d) *$\alpha \geq 2$ and $E = \mathbb{Z}/p^\alpha\mathbb{Z} \times \mathbb{F}_p$;*

(e) *$\alpha \geq 2$ and $\mathbb{Z}/p^\alpha\mathbb{Z} \subset R$ is a ramified extension.*

Moreover, in (a)–(e), the parameters $p_1, p_2, \ldots, p_k$, $\alpha_1, \alpha_2, \ldots, \alpha_k$ are determined by R.

Proof. By the above comments, the rings R in question are the direct products of the asserted form in which E is a minimal ring extension of $\mathbb{Z}/p^\alpha\mathbb{Z}$. If $\alpha = 1$, then [2, Lemme 1.2] provides the classification given by (a), (b), and (c). If $\alpha \geq 2$, then Propositions 7 and 8 give that the suitable rings E are either decomposed extensions of $\mathbb{Z}/p^\alpha\mathbb{Z}$ or ramified extensions of $\mathbb{Z}/p^\alpha\mathbb{Z}$. By Proposition 10, the former are given (of course, up to isomorphism) in (d); and the latter are stipulated in (e). As for the "Moreover" assertion, note that when R (being a finite, hence an Artinian, ring) is expressed as a direct product $\prod_{i=1}^k E_i$ of nonzero local rings E_i, there is exactly one index i such that E_i is not a prime ring. This fact determines the prime number that plays the role of p_1, and the uniqueness of the other parameters is now clear. □

By Proposition 12(d), condition (e) of Theorem 13 is nonvacuous for any prime number p and any integer $\alpha \geq 2$. In case $\alpha = 2$ and p is any prime number, parts (e) and (f) of Proposition 12 serve to complete the classification of the rings R figuring in condition (e) of Theorem 13. It would be interesting to know if such a classification could be completed for arbitrary prime-powers p^α when $\alpha \geq 3$.

We pause to isolate a result of some independent interest. Lemma 14 will be used often in the proof of Theorem 15.

Lemma 14. *Let A and B be rings, with E a ring extension of A. Put $C := A \times B$ and $D := E \times B$. Let M be an ideal of A and put $\mathcal{M} := M \times B$. Then, consider the following:*

(a) *$C \subsetneq D$ is a minimal ring extension if and if $A \subsetneq E$ is a minimal ring extension.*

(b) *$C \subsetneq D$ is an integral extension if and only if $A \subsetneq E$ is an integral extension.*

(c) *$\mathcal{M} \in \mathrm{Max}(C)$ if and only if $M \in \mathrm{Max}(A)$.*

(d) *$C \subsetneq D$ is an integral minimal ring extension with crucial maximal ideal $\mathcal{M}$ if and only if $A \subsetneq E$ is an integral minimal ring extension with crucial maximal ideal M. Moreover, when these equivalent conditions hold, $C \subsetneq D$ and $A \subsetneq E$ are the same type of integral minimal ring extension (i.e., to say, inert, decomposed, or ramified).*

Proof. (a) This is evident since $[C, D] = \{R \times B \mid R \in [A, E]\}$.

(b) Let $e \in E$ and $b \in B$. Straightforward calculations show that (e, b) is integral over C if and only if e is integral over A.

(c) This is evident since $C/\mathcal{M} \cong A/M$.

(d) Note that $\mathcal{M}D = \mathcal{M}$ if and only if $ME = M$. Assume that these equivalent conditions hold. In view of (a), (b), and (c), the assertion now follows by considering the commutative diagram formed by the inclusion maps $C/\mathcal{M} \hookrightarrow D/\mathcal{M}$ and $A/M \hookrightarrow E/M$ and the isomorphisms $C/\mathcal{M} \to A/M$ and $D/\mathcal{M} \to E/M$. □

Theorem 15. *Up to isomorphism, the rings R of positive characteristic that have exactly two proper subrings can be characterized as follows. The prime ring of R is (isomorphic to) the direct product $\prod_{i=1}^k \mathbb{Z}/p_i^{\alpha_i}\mathbb{Z}$, where $p = p_1, p_2, \ldots, p_k$ are pairwise distinct prime numbers for some positive integer k and $\alpha = \alpha_1, \alpha_2, \ldots, \alpha_k$ are positive integers (and possibly $\alpha_i = \alpha_j$ for some $i \neq j$). Then (up to isomorphism), R is the direct product $E \times \prod_{i=2}^k \mathbb{Z}/p_i^{\alpha_i}\mathbb{Z}$, where E is a ring satisfying (exactly) one of the following five conditions:*

(a) $E = \mathbb{F}_{p^{q^2}}$, *where q is a prime number (which is possibly equal to p);*

(b) $E = \mathbb{F}_{p^q} \times \mathbb{F}_p$, *where q is a prime number (which is possibly equal to p);*

(c) $E = R_1$, *the ring that was constructed (in terms of any given prime number p) in Proposition 11;*

(d) $\alpha \geq 2$ *and* $E = \mathbb{Z}/p^\alpha\mathbb{Z}\times\mathbb{F}_{p^q}$, *where q is a prime number (which is possibly equal to p);*

(e) $\alpha \geq 2$ *and there exists a ring B such that both* $\mathbb{Z}/p^\alpha\mathbb{Z} \subset B$ *and* $B \subset R$ *are ramified extensions and* $(\mathbb{Z}/p^\alpha\mathbb{Z})[u] = R$ *for all* $u \in R \setminus B$.

Proof. Let S be a prime ring that is expressed as a direct product as in the statement of this result. A ring extension, say R, of S can be expressed (up to isomorphism) as $\prod_{i=1}^{k} D_i$, where D_i is a ring extension of $\mathbb{Z}/p_i^{\alpha_i}\mathbb{Z}$ for each $i = 1, \ldots, k$. If we wish to consider only rings R such that $|[S, R]| = 3$, then there cannot exist distinct indices i and j such that $\mathbb{Z}/p_i^{\alpha_i}\mathbb{Z} \subset D_i$ and $\mathbb{Z}/p_j^{\alpha_j}\mathbb{Z} \subset D_j$. Indeed, if one assumes the contrary, with $\{i, j\} = \{1, 2\}$ for simplicity of notation, then

$$V := D_1 \times \prod_{i=2}^{k} \mathbb{Z}/p_i^{\alpha_i}\mathbb{Z},$$

$$W := \mathbb{Z}/p^\alpha\mathbb{Z} \times D_2 \times \prod_{i=3}^{k} \mathbb{Z}/p_i^{\alpha_i}\mathbb{Z} \qquad (15)$$

are unequal members of $[S, R] \setminus \{S, R\}$, whence $|[S, R]| \geq 4$. This proves the assertion that we need only consider candidates R expressible in the form $E \times \prod_{i=2}^{k} \mathbb{Z}/p_i^{\alpha_i}\mathbb{Z}$ as in the statement of this result. Therefore, by Lemma 14, our task translates to determining (up to isomorphism) those rings E with prime ring $\mathbb{Z}/p^\alpha\mathbb{Z}$ such that E has exactly two proper subrings. Building on Theorem 13, we will find many candidates. Those exhibiting the desired behavior will be deemed "good" (they will form the five mutually exclusive collections in the statement of this result), while the candidates that do not exhibit the desired behavior will be deemed "bad." The decision as to whether a given candidate is "good" or "bad" will be based on the conclusion in Theorem 2 concerning the condition from Theorem 1 that is satisfied by a tower of minimal ring extensions leading from $\mathbb{Z}/p^\alpha\mathbb{Z}$ to E. Any candidate for which Theorem 2(c) is applicable will be deemed "ambiguous." By Proposition 7, conditions (v) and (xiii) are the only conditions from Theorem 1 that can lead to the "ambiguous" designation. The candidates will be developed by taking as the "first step" of any such tower one of the rings identified in Theorem 13.

Our first candidates for suitable E will come by building towers whose first step (from the prime ring $\mathcal{P} = \mathbb{F}_p$) arrives at the ring identified in Theorem 13(a), namely, $B = \mathbb{F}_{p^q}$, where q is a prime number (which is possibly equal to p). For these candidates, the extension $B \subset E$ may be inert, decomposed, or ramified. The resulting towers $\mathcal{P} \subset B \subset E$, respectively, satisfy conditions (v), (viii), and (xii) from Theorem 1. According to Theorem 2, these candidates are, respectively, ambiguous, bad, and bad. By [2, Lemme 1.2], that ambiguous candidate is $E = \mathbb{F}_{p^{qr}}$, where r is a prime number (which is possibly equal to p or q). The scrutiny of this candidate reveals two subcases. Indeed, if $q = r$, it follows from the classical Galois theory of finite fields that B is the only ring (field, in this case) that is properly contained between $\mathbb{F}_p$ and $\mathbb{F}_{p^{q^2}}$. However, if $q \neq r$, then $\mathbb{F}_{p^q}$ and $\mathbb{F}_{p^r}$ are distinct fields that are properly contained between $\mathbb{F}_p$ and $\mathbb{F}_{p^{qr}}$. Accordingly, candidates from the first subcase are good (and this fact is recorded in part (a) of the statement of this result), while candidates from the second subcase are bad.

The next candidates for suitable E will come via Theorem 13(b), building on the prime ring $\mathcal{P} = \mathbb{F}_p$ and the intermediate ring $B = \mathbb{F}_p \times \mathbb{F}_p$. For these candidates, the extension $B \subset E$ may be inert, decomposed, or ramified. The resulting towers $\mathcal{P} \subset B \subset E$, respectively, satisfy conditions (vi), (vii), and (x) from Theorem 1. According to Theorem 2, these candidates are, respectively, good, bad, and bad. By combining Lemma 14(d) with [2, Lemme 1.2], we find that good candidate to be (up to isomorphism) of the form $\mathbb{F}_{p^q}\times\mathbb{F}_p$, where q is a prime number (which is possibly equal to p). This fact is recorded in part (b) of the statement of this result.

Next, we consider the candidates for E coming via Theorem 13(c), building on the intermediate ring $B = \mathbb{F}_p[X]/(X^2)$ (and its prime ring $\mathcal{P} = \mathbb{F}_p$). For these candidates, Proposition 8 shows that the extension $B \subset E$ may only be decomposed or ramified. The resulting towers $\mathcal{P} \subset B \subset E$, respectively, satisfy conditions (ix) and (xiii) from Theorem 1. Accordingly, these candidates are, respectively, bad and ambiguous. In fact, an isomorphic copy of this bad candidate was already designated as "bad" in the preceding paragraph. As for the ambiguous candidate, there are two subcases. These were completely assessed in Proposition 11. The final assertion in the statement of that result produces part (c) of the statement of this result.

The next candidates for E come via Theorem 13(d), with $\mathcal{P} = \mathbb{Z}/p^\alpha\mathbb{Z}$, where p is a prime number and the integer $\alpha \geq 2$, and $B = \mathbb{Z}/p^\alpha\mathbb{Z}\times\mathbb{F}_p$. By Lemma 14 and Proposition 8, the relevant candidates E are of two kinds: the "type 1" candidates are those of the form $E_1 \times \mathbb{F}_p$, where $\mathbb{Z}/p^\alpha\mathbb{Z} \subset E_1$ is either decomposed or ramified; the "type 2" candidates are those of the form $\mathbb{Z}/p^\alpha\mathbb{Z} \times E_2$, where $\mathbb{F}_p \subset E_2$ is inert, decomposed, or ramified. By Lemma 14, the type 1 candidates, respectively, satisfy conditions (vii) and (x) from Theorem 1 and, hence, are deemed to be, respectively, bad and bad. The type 2 candidates, respectively, satisfy conditions (vi), (vii), and (x) from Theorem 1 and so are, respectively, good, bad, and bad. Thus (up to isomorphism), the sole good candidate emanating from Theorem 13(d) is (thanks to [2, Lemme 1.2]) given by $\mathbb{Z}/p^\alpha\mathbb{Z}\times\mathbb{F}_{p^q}$, where q is a prime number (which is possibly equal to p). This fact is recorded in part (d) of the statement of this result.

Finally, we must address the candidates for E that come via Theorem 13(e). These involve the prime ring $\mathcal{P} = \mathbb{Z}/p^\alpha\mathbb{Z}$, where p is a prime number and the integer $\alpha \geq 2$. However, as we have classified the ramified extensions of the form $\mathbb{Z}/p^\alpha\mathbb{Z} \subset B$ only when $\alpha \leq 2$, we must settle for the formulation of part (e) of the statement of this result. □

Remark 16. (a) We next prove that condition (e) of Theorem 15 is nonvacuous for any p^α (where, as above, p is a prime number and α is an integer exceeding 1). Consider the tower $(A, M) := \mathbb{Z}/p^\alpha\mathbb{Z} \subset B := A(+)M^{\alpha-1} \subset R := A(+)M^{\alpha-2}$. (As usual, $M^0 := A$.) Since

$$M^{\alpha-1} = (pA)^{\alpha-1} = \frac{p^{\alpha-1}\mathbb{Z}}{p^\alpha\mathbb{Z}} \cong \frac{\mathbb{Z}}{p\mathbb{Z}} \cong \frac{A}{M}, \qquad (16)$$

the usual combination of [15] and [16, Lemma 2.1] shows that $A \subset B$ is ramified (necessarily with maximal ideal M). Note that the unique prime ideal of B is $N := M(+)M^{\alpha-1}$. Identifying A with $A(+)0 \subset B$ as usual via $a \mapsto (a, 0)$ for each $a \in A$, we see that $N \cap A = M$. Next, to show that $B \subset R$ is ramified, it is straightforward to check the criteria given in [12, Theorem 4.2(c)]. (In detail, check that $(B : R) = N$; the unique prime ideal of R is $Q := M(+)M^{\alpha-2}$ and the canonical ring homomorphism $B/N \to R/Q$ is an isomorphism of fields; and R/N is a two-dimensional vector space over $B/N(\cong \mathbb{F}_p)$, since as a vector space, $R/N \cong A/M \oplus M^{\alpha-2}/M^{\alpha-1} \cong \mathbb{F}_p \oplus \mathbb{F}_p \cong B/N \oplus B/N$.) Of course, $Q \cap B = N$. Next, to show that the tower $A \subset B \subset R$ satisfies condition (e) of Theorem 15 (so that R is a ring of characteristic p^α that has only two proper subrings), it suffices to show that if $u \in R \setminus B$, then $A[u]$ contains at least one element of $B \setminus A$. It will suffice to show that $pu \in B \setminus A$ (where pu denotes the sum $u + \cdots + u$ of p summands each of which is u). We can write $u = (a, a^*)$ where $a \in A$ and $a^* \in M^{\alpha-2} \setminus M^{\alpha-1} = p^{\alpha-2}A \setminus p^{\alpha-1}A$. It remains only to observe that $pu = (pa, pa^*) \in B \setminus A$, the critical point being that $pa^* \neq 0 \in A$ since $a^* \notin p^{\alpha-1}A$.

(b) Not only is condition (e) of Theorem 15 nonvacuous (as we just proved in (a)), this condition is also not redundant. More precisely, one cannot delete the requirement "$(\mathbb{Z}/p^\alpha\mathbb{Z})[u] = R$ for all $u \in R \setminus B$" in the statement of condition (e) of Theorem 15. This can be shown for any p^α (where, as above, p is a prime number and α is an integer exceeding 1) as follows. As above, take $(A, M) := \mathbb{Z}/p^\alpha\mathbb{Z}$. Next, take $B := A(+)\mathbb{F}_p$ (recalling that $A/M \cong \mathbb{F}_p$). Observe that the only prime ideal of B is $N := M(+)\mathbb{F}_p$; $N \cap A = M$; and the canonical ring homomorphism $A/M \to B/N$ is an isomorphism of fields. Put $R := B(+)\mathbb{F}_p(\cong B(+)B/N)$. Then the usual combination of [15] and [16, Lemma 2.1] shows that $A \subset B$ and $B \subset R$ are ramified extensions, with crucial maximal ideals M and N, respectively. Of course, R has only finitely many subrings because R is finite. However, as we have seen in analyzing the rings called D in the proofs of Theorem 6(2) and Proposition 11, such iterated idealizations always produce at least three proper subrings. For the present data, we have $A(+)0(+)\mathbb{F}_p \in [A, R] \setminus \{A, B, R\}$.

(c) It may be useful to note that condition (e) of Theorem 15 is equivalent to the following condition, (e)$'$: $\alpha \geq 2$, $|R| = p^{\alpha+2}$, $\text{char}(R) = p^\alpha$, R is subintegral over its prime ring, and there exists a proper subring B of R such that each element of $R \setminus B$ generates R as a (unital) ring. By reasoning as in the second proof of Proposition 12(c), it is easy to show that (e) $\Rightarrow$ (e)$'$. Conversely, assume (e)$'$. Then there is no harm in viewing $A := \mathbb{Z}/p^\alpha\mathbb{Z} \subset R$. To see that this is not a minimal ring extension, combine Propositions 7 and 8 with Proposition 12(c). This leads to a ring $B \in [A, R]$ such that $A \subset B$ is ramified, and then (e) follows easily.

The above formulation of (e)$'$ makes it clear that further progress on this program will depend on knowing the structure of certain local rings of prime-power cardinality p^n when $n \geq 4$. According to the authors of [24], "All the commutative rings (of cardinality p^n) are known" because Wilson [25] has characterized the local ones. In our opinion, that characterization is somewhat noncommutative in spirit, inasmuch as it features matrix rings (over Galois rings). In a subsequent work, we hope to make that work sufficiently explicit so as to convert condition (e) of Theorem 15 into an explicit list that would be in the same spirit as the statements of the other conditions in Theorem 15.

(d) Despite the information in (a) and (b), one may wish for a fuller implementation of condition (e) of Theorem 15 that would require the classification (up to isomorphism) of all the ramified extensions of $\mathbb{Z}/p^\alpha\mathbb{Z}$ for all integers $\alpha \geq 2$ (and all prime numbers p). This was done only for $\alpha = 2$ in Proposition 12 (e), (f). To investigate whether the behavior that was established there extends to the case $\alpha \geq 3$, we can suggest (in addition to pursuing the matrical methods of Wilson that were mentioned in (c)) a computer-aided study of the kind that was recommended following the proof of Proposition 12.

(e) Let R be a ring with prime ring A. The above analysis has benefitted from the fact that if R has at most two proper subrings, then $[A, R]$ is linearly ordered by inclusion. Any study that would go beyond the present work by seeking to classify (or even to nontrivially characterize) those R having three proper subrings must deal with the loss of such linear order. For instance, it follows from [9, Lemma V.2] (and the second paragraph of the proof of Theorem 6) that the integral domains C such that $\text{char}(C) = 0$ and C has exactly three proper subrings are characterized, up to isomorphism, as the rings having the form $C = \mathbb{Z}[1/(pq)]$, where p and q are distinct prime numbers. As we showed in the above-mentioned proof, $[\mathbb{Z}, C]$ is not linearly ordered.

The above-noted increased level of complexity continues for positive characteristic. In the spirit of condition (a) in Theorem 15 (see also the second paragraph of [6, Remark 3.4(b)]), we will illustrate this by using field-theoretic examples. Let p be a prime number. Then (up to isomorphism), the finite fields of characteristic p that have exactly three proper subrings (subfields) are the fields having one of the forms $k_1 = \mathbb{F}_{p^{q^3}}$ and $k_2 = \mathbb{F}_{p^{qr}}$ where q and r are distinct prime numbers (one of which may be p). (The smallest such fields are $k_1 = \mathbb{F}_{256}$ and $k_2 = \mathbb{F}_{64}$.) In general, we have that $[\mathbb{F}_p, k_1]$ is linearly ordered and $[\mathbb{F}_p, k_2]$ is not linearly ordered.

The ring R_1 in the proof of Theorem 6(2) is isomorphic to the 2-trivial extension $\mathbb{Z} \ltimes_2 \mathbb{F}_p z \ltimes \mathbb{F}_p x$ (in the sense of [26]). The theory developed in [26] can illuminate such constructions. For instance, the assertion that N_1 is a prime ideal of R_1 in the above-mentioned proof can be seen via [26, Theorem 4.7].

Competing Interests

The author declares that there is no conflict of interests in the publication of this paper.

References

[1] D. E. Dobbs, "When the juxtaposition of two minimal ring extensions produces no new intermediate rings," *Palestine Journal of Mathematics*, vol. 6, no. 1, 2017.

[2] D. Ferrand and J.-P. Olivier, "Homomorphismes minimaux d'anneaux," *Journal of Algebra*, vol. 16, pp. 461–471, 1970.

[3] D. E. Dobbs and J. Shapiro, "A classification of the minimal ring extensions of an integral domain," *Journal of Algebra*, vol. 305, no. 1, pp. 185–193, 2006.

[4] D. D. Anderson, D. E. Dobbs, and B. Mullins, "The primitive element theorem for commutative algebras," *Houston Journal of Mathematics*, vol. 25, no. 4, pp. 603–623, 1999.

[5] I. Kaplansky, *Commutative Rings*, University of Chicago Press, Chicago, Ill, USA, 1974.

[6] D. E. Dobbs and J. Shapiro, "When only finitely many intermediate rings result from juxtaposing two minimal ring extensions," *Palestine Journal of Mathematics*, vol. 5, pp. 13–31, 2016.

[7] D. E. Dobbs, G. Picavet, and M. Picavet-L'Hermitte, "Characterizing the ring extensions that satisfy FIP or FCP," *Journal of Algebra*, vol. 371, pp. 391–429, 2012.

[8] J. A. Huckaba, *Commutative rings with zero divisors*, vol. 117 of *Monographs and Textbooks in Pure and Applied Mathematics*, Marcel Dekker, New York, NY, USA, 1988.

[9] D. E. Dobbs, B. Mullins, G. Picavet, and M. Picavet-L'Hermitte, "On the FIP property for extensions of commutative rings," *Communications in Algebra*, vol. 33, no. 9, pp. 3091–3119, 2005.

[10] M. S. Gilbert, *Extensions of commutative rings with linearly ordered intermediate rings [Ph.D. dissertation]*, University of Tennessee, Knoxville, Tenn, USA, 1996.

[11] D. Lazard, "Autour de la platitude," *Bulletin de la Société Mathématique de France*, vol. 97, pp. 81–128, 1969.

[12] D. E. Dobbs, G. Picavet, M. Picavet-L'Hermitte, and J. Shapiro, "On intersections and composites of minimal ring extensions," *JP Journal of Algebra, Number Theory and Applications*, vol. 26, no. 2, pp. 103–158, 2012.

[13] D. E. Dobbs and J. Shapiro, "A classification of the minimal ring extensions of certain commutative rings," *Journal of Algebra*, vol. 308, no. 2, pp. 800–821, 2007.

[14] D. E. Dobbs, B. Mullins, and M. Picavet-L'Hermitte, "The singly generated unital rings with only finitely many unital subrings," *Communications in Algebra*, vol. 36, no. 7, pp. 2638–2653, 2008.

[15] D. E. Dobbs, "Every commutative ring has a minimal ring extension," *Communications in Algebra*, vol. 34, no. 10, pp. 3875–3881, 2006.

[16] G. Picavet and M. Picavet-L'Hermitte, "Modules with finitely many submodules," *International Electronic Journal of Algebra*, vol. 19, pp. 119–131, 2016.

[17] O. Zariski and P. Samuel, *Commutative algebra, Volume I*, The University Series in Higher Mathematics, D. Van Nostrand, Princeton, NJ, USA, 1958.

[18] K. H. Rosen, *Elementary Number Theory and its Applications*, Addison-Wesley, Reading, Mass, USA, 3rd edition, 1993.

[19] R. Gilmer and J. Mott, "Associative rings of order P^3," *Proceedings of the Japan Academy*, vol. 49, pp. 795–799, 1973.

[20] R. L. Kruse, Article ID 0369422, MR 0369422 (51 # 5655), Mathematical Reviews Division of the American Mathematical Society, Providence, 1976.

[21] R. Ballieu, "Anneaux finis; systèmes hypercomplexes de rang trois sur un corps commutatif," *Annales de la Société Scientifique de Bruxelles. Série I*, vol. 61, pp. 222–227, 1947.

[22] V. G. Antipkin and V. P. Elizarov, "Rings of order p^3," *Sibirskii Matematicheskii Zhurnal*, vol. 23, no. 4, pp. 9–18, 1982.

[23] Y. N. Mal'tsev, *Mathematical Reviews*, American Mathematical Society, Providence, RI, USA, 1984.

[24] J. B. Derr, G. F. Orr, and P. S. Peck, "Noncommutative rings of order p^4," *Journal of Pure and Applied Algebra*, vol. 97, no. 2, pp. 109–116, 1994.

[25] R. S. Wilson, "Representations of finite rings," *Pacific Journal of Mathematics*, vol. 53, pp. 647–649, 1974.

[26] D. D. Anderson, D. Bennis, B. Fahid, and A. Shaiea, "n-trivial extensions of rings," https://arxiv.org/abs/1604.01486.

20

Structure of n-Lie Algebras with Involutive Derivations

Ruipu Bai,[1] **Shuai Hou**,[2] **and Yansha Gao**[2]

[1]College of Mathematics and Information Science, Hebei University, Key Laboratory of Machine Learning and Computational Intelligence of Hebei Province, Baoding 071002, China
[2]College of Mathematics and Information Science, Hebei University, Baoding 071002, China

Correspondence should be addressed to Ruipu Bai; bairuipu@hbu.edu.cn

Academic Editor: Kaiming Zhao

We study the structure of n-Lie algebras with involutive derivations for $n \geq 2$. We obtain that a 3-Lie algebra A is a two-dimensional extension of Lie algebras if and only if there is an involutive derivation D on $A = A_1 + A_{-1}$ such that $\dim A_1 = 2$ or $\dim A_{-1} = 2$, where A_1 and A_{-1} are subspaces of A with eigenvalues 1 and -1, respectively. We show that there does not exist involutive derivations on nonabelian n-Lie algebras with $n = 2s$ for $s \geq 1$. We also prove that if A is a $(2s+2)$-dimensional $(2s+1)$-Lie algebra with $\dim A^1 = r$, then there are involutive derivations on A if and only if r is even, or r satisfies $1 \leq r \leq s+2$. We discuss also the existence of involutive derivations on $(2s+3)$-dimensional $(2s+1)$-Lie algebras.

1. Introduction

Derivation is an important tool in studying the structure of n-Lie algebras [1]. The derivation algebra $Der(A)$ of an n-Lie algebra A over the field of real numbers is the Lie algebra of the automorphism group $Aut(A)$, which is a Lie group if $\dim A < \infty$ [2]. Any n-Lie algebra-module (V, ρ) is a module of the inner derivation algebra $ad(A)$, which is a linear Lie algebra [3]. Also, derivations have close relationship with extensions of n-Lie algebras.

The concept of 3-Lie classical Yang-Baxter equations is introduced in [4]. It is known that if there is an involutive derivation D on A, then $(A, \{,,\}_D)$ is a 3-pre-Lie algebra, where $\{x, y, z\}_D = D(ad(x, y)D(z))$, $\forall x, y, z \in A$, and the 3-Lie algebra A is a subadjacent 3-Lie algebra of $(A, \{,,\}_D)$, and $r = \sum_i e_i^* \otimes D(e_i) - D(e_i) \otimes e_i^*$ is a skew-symmetric solution of the 3-Lie classical Yang-Baxter equation in the 3-Lie algebra $A \ltimes_{ad^*} A^*$, where $\{e_1, \cdots, e_m\}$ is a basis of A and $\{e_1^*, \cdots, e_m^*\}$ is the dual basis of A^*.

Due to this importance of involutive derivations on 3-Lie algebras, we investigate in this paper the existence of involutive derivations on finite dimensional n-Lie algebras. More specifically, in Section 2, we discuss the properties of involutive derivations on n-Lie algebras. In Section 3, we study the existence of involutive derivations on $(2s+2)$-dimensional $(2s+1)$-Lie algebras. In Section 4, we consider the existence of involutive derivations on $(2s+3)$-dimensional $(2s+1)$-Lie algebras. In Section 5, we investigate a class of 3-Lie algebras with involutive derivations which are two-dimensional extension of Lie algebras.

In the following, we assume that all algebras are over an algebraically closed field $\mathbb{F}$ with characteristic zero, Id is the identity mapping, and $\mathbb{Z}$ is the set of integers. For $\lambda \in \mathbb{F}$ and an $\mathbb{F}$-linear mapping D on a vector space A, A_λ denotes the subspace $\{x \in A \mid D(x) = \lambda x\}$.

2. n-Lie Algebras with Involutive Derivations

An n-Lie algebra [1] is a vector space A over a field $\mathbb{F}$ equipped with a linear multiplication $[, \cdots,] : \wedge^n A \longrightarrow A$ satisfying, for all $x_1, \cdots, x_n, y_2, \cdots, y_n \in A$,

$$[[x_1, \cdots, x_n], y_2, \cdots, y_n] = \sum_{i=1}^{n} [x_1, \cdots, [x_i, y_2, \cdots, y_n], \cdots, x_n]. \tag{1}$$

Equation (1) is usually called *the generalized Jacobi identity*, or *Filippov identity*.

The derived algebra of an n-Lie algebra A is a subalgebra of A generated by $[x_1,\cdots,x_n]$ for all $x_1,\cdots,x_n \in A$, and is denoted by A^1. We use $Z(A)$ to denote the center of A; that is, $Z(A)=\{x \mid x\in A, [x,A,\cdots,A]=0\}$.

A derivation of A is an endomorphism of A satisfying

$$D([x_1,\cdots,x_n]) = \sum_{i=1}^{n} [x_1,\cdots,D(x_i),\cdots,x_n], \quad \forall x_1,\cdots,x_n \in A. \tag{2}$$

If a derivation D satisfies that $D^2 = Id$, then D is called an *involutive derivation* on A. Der(A) denotes the derivation algebra of A.

For $x_1,\cdots,x_{n-1}\in A$, map $\mathrm{ad}(x_1,\cdots,x_{n-1})\colon A\longrightarrow A$,

$$\mathrm{ad}(x_1,\cdots,x_{n-1})(x) = [x_1,\cdots,x_{n-1},x], \quad \forall x\in A \tag{3}$$

is called *a left multiplication* defined by elements $x_1,\cdots,x_{n-1}$. From (1), left multiplications are derivations.

The following lemma can be easily verified.

Lemma 1. *Let V be a finite dimensional vector space over $\mathbb{F}$ and D be an endomorphism of V with $D^2 = Id$. Then V can be decomposed into the direct sum of subspaces $V = V_1 \dotplus V_{-1}$, where $V_1 = \{v\in V \mid Dv = v\}$ and $V_{-1} = \{v\in V \mid Dv = -v\}$.*

If A is a finite dimensional n-Lie algebra with an involutive derivation D, then we have

$$A = A_1 \dotplus A_{-1}. \tag{4}$$

Lemma 2. *Let A be an n-Lie algebra over $\mathbb{F}$. If $D\in Der(A)$ is an involutive derivation, then, for all $x_1,\cdots,x_n\in A$,*

$$[x_1,\cdots,x_n] = \frac{-2}{n-1}\sum_{i<j}[x_1,\cdots,x_{i-1},D(x_i),x_{i+1},\cdots, x_{j-1},D(x_j),x_{j+1},\cdots,x_n], \tag{5}$$

$$[D(x_1),\cdots,D(x_n)] = \frac{-2}{n-1}\sum_{i<j}[Dx_1,\cdots,D(x_{i-1}),x_i, D(x_{i+1}),\cdots,D(x_{j-1}),x_j,D(x_{j+1}),\cdots,D(x_n)]. \tag{6}$$

Proof. If D is an involutive derivation on A, then, for all $x_1,\cdots,x_n\in A$,

$$\begin{aligned}[x_1,\cdots,x_n] &= D^2([x_1,\cdots,x_n]) \\ &= D\left(\sum_{i=1}^{n}[x_1,\cdots,D(x_i),\cdots,x_n]\right) \\ &= n[x_1,\cdots,x_n] \\ &\quad + 2\sum_{1\le i<j\le n}[x_1,\cdots,D(x_i),\cdots,D(x_j),\cdots,x_n].\end{aligned} \tag{7}$$

Equation (5) follows. Equation (6) follows from (4) and $D^2 = Id$. □

Theorem 3. *Let A be a finite dimensional n-Lie algebra with $n=2s$, $s\ge 1$. Then there is an involutive derivation D on A if and only if A is abelian.*

Proof. If A is abelian, then the result is trivial.

Conversely, let D be an involutive derivation on A. By Lemma 1, $A = A_1 \dotplus A_{-1}$. Then, for any $i\in\mathbb{Z}$, $1\le i\le n$, $x_1,\cdots,x_n\in A_1$, and $y_1,\cdots,,y_n\in A_{-1}$,

$$\begin{aligned}D([x_1,\cdots,x_i,y_1,\cdots,y_{n-i}]) &= i[x_1,\cdots,x_i,y_1,\cdots,y_{n-i}] \\ &\quad -(n-i)[x_1,\cdots,x_i,y_1,\cdots,y_{n-i}] \\ &= (2i-2s)[x_1,\cdots,x_i,y_1,\cdots,y_{n-i}]\in A_{2i-2s}.\end{aligned} \tag{8}$$

$$D([x_1,\cdots,x_n]) = 2s[x_1,\cdots,x_n],$$

$$D([y_1,\cdots,y_n]) = -2s[y_1,\cdots,y_n].$$

Thanks to $\pm 2s\neq\pm 1$ and $2i-2s\neq\pm 1$, $A_{2i-n} = A_{\pm 2s} = 0$. Therefore, A is abelian. □

Theorem 4. *Let A be a finite dimensional n-Lie algebra with $n=2s+1$, $s\ge 1$, and D be an involutive derivation on A. Then A_1 and A_{-1} are abelian subalgebras, and*

$$\Big[\underbrace{A_1,\cdots,A_1}_{i},\underbrace{A_{-1},\cdots,A_{-1}}_{2s+1-i}\Big] = 0, \quad \forall 1\le i\le 2s,\ i\neq s,\ s+1, \tag{9}$$

$$\Big[\underbrace{A_1,\cdots,A_1}_{s},\underbrace{A_{-1},\cdots,A_{-1}}_{s+1}\Big] \subseteq A_{-1}, \tag{10}$$

$$\Big[\underbrace{A_1,\cdots,A_1}_{s+1},\underbrace{A_{-1},\cdots,A_{-1}}_{s}\Big] \subseteq A_1. \tag{11}$$

Proof. Since $D\in DerA$, $[\underbrace{A_1,\cdots,A_1}_{i},\underbrace{A_{-1},\cdots,A_{-1}}_{2s+1-i}] \subseteq A_{2i-2s-1}, 0\le i\le 2s+1$. If $[\underbrace{A_1,\cdots,A_1}_{i},\underbrace{A_{-1},\cdots,A_{-1}}_{2s+1-i}]\neq 0$, then $2i-2s-1=\pm 1$, that is, $i=s+1$, or $i=s$. Therefore, $[A_1,\cdots,A_1] = [A_{-1},\cdots,A_{-1}] = 0$. The result follows. □

Theorem 5. *Let A be an m-dimensional n-Lie algebra with $n=2s+1$, $s\ge 1$. Then there is an involutive derivation on A if and only if A has the decomposition $A = B\dotplus C$ (as direct sum of subspaces), and*

$$\Big[\underbrace{B,\cdots,B}_{i},\underbrace{C,\cdots,C}_{2s+1-i}\Big] = 0, \quad 0\le i\le 2s+1,\ i\neq s, s+1, \tag{12}$$

$$\Big[\underbrace{B,\cdots,B}_{s},\underbrace{C,\cdots,C}_{s+1}\Big] \subseteq C, \qquad \Big[\underbrace{B,\cdots,B}_{s+1},\underbrace{C,\cdots,C}_{s}\Big] \subseteq B. \tag{13}$$

Proof. If there is an involutive derivation D on A, then, by Theorem 4, $B = A_1$ and $C = A_{-1}$ satisfy (12) and (13).

Conversely, define an endomorphism D of A by $D(x) = x, D(y) = -y, \forall x \in B, y \in C$. Then $D^2 = Id, B = A_1$ and $C = A_{-1}$. By (12) and (13), D is a derivation. □

Corollary 6. *Let A be a $(2s+1)$-dimensional $(2s+1)$-Lie algebra with the multiplication $[e_1, \cdots, e_{2s+1}] = e_1$, where $\{e_1, \cdots, e_{2s+1}\}$ is a basis of A. Then the linear mapping $D: A \longrightarrow A$ defined by $D(e_i) = e_i$, $1 \le i \le s+1$, $D(e_j) = -e_j$, $s+2 \le j \le 2s+1$ is an involutive derivation on A.*

Proof. The result follows from a direct computation. □

3. Involutive Derivations on $(n+1)$-Dimensional n-Lie Algebras with $n = 2s+1$

In this section, we study involutive derivations on $(n+1)$-dimensional n-Lie algebras over $\mathbb{F}$. From Theorem 3, we only need to discuss the case of $n = 2s+1$, $s \ge 1$.

Lemma 7 (see [5]). *Let A be an $(n+1)$-dimensional non-abelian n-Lie algebra over $\mathbb{F}$, $n \ge 3$. Then up to isomorphisms A is one and only one of the following possibilities:*

$$(b_1)\ [e_2, e_3, \cdots, e_{n+1}] = e_1.$$

$$(b_2)\ [e_1, e_2, \cdots, e_n] = e_1.$$

$$(c_1)\ \begin{cases} [e_2, \cdots, e_{n+1}] = e_1, \\ [e_1, e_3, \cdots, e_{n+1}] = e_2. \end{cases}$$

$$(c_2)\ \begin{cases} [e_2, \cdots, e_{n+1}] = \alpha e_1 + e_2, \\ [e_1, e_3, \cdots, e_{n+1}] = e_2, \end{cases} \quad \alpha \in \mathbb{F},\ \alpha \ne 0. \tag{14}$$

$$(c_3)\ \begin{cases} [e_1, e_3, \cdots, e_{n+1}] = e_1, \\ [e_2, \cdots, e_{n+1}] = e_2. \end{cases}$$

$$(d_r)\ [e_1, \cdots, \hat{e}_i, \cdots, e_{n+1}] = e_i, \quad 1 \le i \le r,$$

where $\{e_1, \cdots, e_{n+1}\}$ is a basis of A, $3 \le r \le n+1$, and $\hat{e}_i$ means that e_i is omitted.

Theorem 8. *Let A be a $(2s+2)$-dimensional $(2s+1)$-Lie algebra over $\mathbb{F}$ and $\dim A^1 = r$. Then there exists an involutive derivation D on A if and only if r is even, or $0 \le r \le s+2$.*

Proof. If $\dim A^1 = r \le s+2$, then, by Lemma 7, and a direct computation, the linear mapping $D: A \longrightarrow A$ defined by $D(e_i) = e_i, D(e_j) = -e_j, 1 \le i \le s+2, s+3 \le j \le 2s+2$, is an involutive derivation on A.

Now we discuss the case $\dim A^1 = r \ge s+3$. Let $\{e_1, \cdots, e_{2s+2}\}$ be a basis of A and the multiplication in the basis be as follows:

$$e^i = (-1)^{2s+2+i}[e_1, \cdots, \hat{e}_i, \cdots, e_{2s+2}] = \sum_{l=1}^{2s+2} \beta_{il} e_l, \quad \beta_{il} \in F,\ 1 \le i \le 2s+2, \tag{15}$$

where $\beta_{il} \in \mathbb{F}, 1 \le i, l \le 2s+2$. Thanks to Theorem 3 in [1], A is a 3-Lie algebra if and only if the $(2s+2) \times (2s+2)$-matrix $B = (\beta_{il})$ is symmetric.

If $r = 2t > 3$, $2 \le t \le s+1$, then define the multiplication on A by

$$[e_1, \cdots, \hat{e}_i, \cdots, e_{2s+2}] = (-1)^i e_{2s+3-i}, \quad 1 \le i \le t,\ 3-t < i-2s \le 2, \tag{16}$$

$$[e_1, \cdots, \hat{e}_j, \cdots, e_{2s+2}] = 0, \quad t < j \le 2s+3-t,$$

that is, $\beta_{i,2s+3-i} = \beta_{2s+3-i,i} = 1$ for $1 \le i \le t$, or $2s+3-t < i \le 2s+2$, and others are zero. Then, $B = (\beta_{il})$ is symmetric. Therefore, A is a $(2s+1)$-Lie algebra with the multiplication (16).

Define an endomorphism D of A by $De_i = e_i, 1 \le i \le s+1$, and $De_j = -e_j$ for $s+2 \le j \le 2s+2$. Then D is an involutive derivation on A.

For the case $\dim A^1 = r = 2t+1 \ge s+3$. Suppose $l = \dim A_1, l' = \dim A_{-1}$.

If there is an involutive derivation D on A, then, by Theorem 4, $l + l' = 2s+2$, $s \le l \le s+2$ and $s \le l' \le s+2$. Since $\dim A^1 = r = 2t+1 \ge s+3$, $A^1 \cap A_1 \ne 0$ and $A^1 \cap A_{-1} \ne 0$. Therefore, $\dim A_1 = \dim A_{-1} = s+1$. Without loss of generality, we can suppose $\{e_1, \cdots, e_{s+1}\} \subseteq A_1$, and $\{e_{s+2}, \cdots, x_{2s+2}\} \subseteq A_{-1}$. By (10) and (11), the $(2s+2)\times(2s+2)$-matrix $B = (\beta_{il})$ defined by (15) is nonsymmetric, which is a contradiction. Therefore, if $\dim A^1 = r = 2t+1 \ge s+3$, then there do not exist involutive derivations on A. □

By Theorem 8, if A is a 10-dimensional 9-Lie algebra with $\dim A^1 = 7$, or 9, then there does not exist involutive derivation on A. If $1 \le \dim A^1 = r \le 10$ and $r \ne 7, 9$, then there are involutive derivations on A.

4. Involutive Derivations on $(n+2)$-Dimensional n-Lie Algebras with $n = 2s+1$

By Theorem 3, we only need to discuss the case where n is odd. So we suppose that A is a $(2s+3)$-dimensional $(2s+1)$-Lie algebra over $\mathbb{F}$, $s \ge 1$, and that $E_t = \mathrm{Diag}(1, \cdots 1)$ is the $(t \times t)$-unit matrix.

Lemma 9 (see [6]). *Let A be a $(2s+3)$-dimensional $(2s+1)$-Lie algebra over $\mathbb{F}$ with a basis $\{e_1, \cdots, e_{2s+3}\}$. Then A is isomorphic to one and only one of the following possibilities:*

(*a*) *A is an abelian.*

(*b*) $\dim A^1 = 1$:

(b^1) $[e_2, \cdots, e_{2s+2}] = e_1$;

(b^2) $[e_1, \cdots, e_{2s+1}] = e_1$.

(*c*) $\dim A^1 = 2$:

$$(c^1)\ \begin{cases} [e_2, \cdots, e_{2s+2}] = e_1, \\ [e_3, \cdots, e_{2s+3}] = e_2; \end{cases}$$

$$(c^2)\ \begin{cases} [e_2, \cdots, e_{2s+2}] = e_1, \\ [e_2, e_4, \cdots, e_{2s+3}] = e_2, \\ [e_1, e_4, \cdots, e_{2s+3}] = e_1; \end{cases}$$

$$(c^3)\ \begin{cases} [e_2, \cdots, e_{2s+2}] = \alpha e_1 + e_2, \\ [e_1, e_3, \cdots, e_{2s+2}] = e_2, \\ [e_2, e_4, \cdots, e_{2s+3}] = e_2, \\ [e_1, e_4 \cdots, e_{2s+3}] = e_1; \end{cases}$$

$$(c^4)\ \begin{cases} [e_2, \cdots, e_{2s+2}] = e_1, \\ [e_2, e_3, \cdots, e_{2s+2}] = e_2, \\ [e_2, e_4, \cdots, e_{2s+3}] = e_2, \\ [e_1, e_4, \cdots, e_{2s+3}] = e_1; \end{cases}$$

$$(c^5)\ \begin{cases} [e_2, \cdots, e_{2s+2}] = e_1, \\ [e_1, e_3, \cdots, e_{2s+2}] = e_2; \end{cases}$$

$$(c^6)\ \begin{cases} [e_2, \cdots, e_{2s+2}] = \alpha e_1 + e_2, \\ [e_1, e_3, \cdots, e_{2s+2}] = e_2; \end{cases} \qquad \alpha \in \mathbb{F},\ \alpha \neq 0.$$

$$(c^7)\ \begin{cases} [e_1, e_3, \cdots, e_{2s+2}] = e_1, \\ [e_2, e_3, \cdots, e_{2s+2}] = e_2; \end{cases}$$

(*d*) $\dim A^1 = 3$:

$$(d^1)\ \begin{cases} [e_2, \cdots, e_{2s+2}] = e_1, \\ [e_2, e_4, \cdots, e_{2s+3}] = e_2, \\ [e_3, \cdots, e_{2s+3}] = e_3; \end{cases}$$

$$(d^2)\ \begin{cases} [e_2, \cdots, e_{2s+2}] = e_1, \\ [e_3, \cdots, e_{2s+3}] = e_3 + \alpha e_2, \\ [e_2, e_4, \cdots, e_{2s+3}] = e_3, \\ [e_1, e_4, \cdots, e_{2s+3}] = e_1; \end{cases}$$

$$(d^3)\ \begin{cases} [e_2, \cdots, e_{2s+2}] = e_1, \\ [e_3, e_4, \cdots, e_{2s+3}] = e_3, \\ [e_2, e_4, \cdots, e_{2s+3}] = e_2, \\ [e_1, e_4, \cdots, e_{2s+3}] = 2e_1; \end{cases}$$

$$(d^4)\ \begin{cases} [e_2, \cdots, e_{2s+2}] = e_1, \\ [e_1, e_3, \cdots, e_{2s+2}] = e_2, \\ [e_1, e_2, e_4, \cdots, e_{2s+2}] = e_3; \end{cases}$$

$$(d^5)\ \begin{cases} [e_1, e_4, \cdots, e_{2s+3}] = e_1, \\ [e_2, e_4, \cdots, e_{2s+3}] = e_3, \\ [e_3, e_4, \cdots, e_{2s+3}] = \beta e_2 + (1 + \beta) e_3,; \end{cases}$$

$$(d^6)\ \begin{cases} [e_1, e_4, \cdots, e_{2s+3}] = e_1, \\ [e_2, e_4, \cdots, e_{2s+3}] = e_2, \\ [e_3, e_4, \cdots, e_{2s+3}] = e_3; \end{cases}$$

$$(d^7)\ \begin{cases} [e_1, e_4, \cdots, e_{2s+3}] = e_2, \\ [e_2, e_4, \cdots, e_{2s+3}] = e_3, \\ [e_3, e_4, \cdots, e_{2s+3}] = se_1 + te_2 + ue_3, \end{cases}$$

$$\beta, s, t, u \in \mathbb{F},\ \beta \neq 0, 1,\ s \neq 0. \tag{17}$$

And n-Lie algebras corresponding to the case (d^7) with coefficients s, t, u and s', t', u' are isomorphic if and only if there exists a nonzero element $\lambda \in \mathbb{F}$ such that $s = \lambda^3 s', t = \lambda^2 t'$, $u = \lambda u'$, $s, s', t, t', u, u' \in \mathbb{F}$.

(*r*) $\dim A^1 = r$,

$$4 \le r < 2s + 3,\ \textit{for}\ 2 \le j \le r,\ 1 \le i \le r,$$

$$(r^1)\ \begin{cases} [e_2, \cdots, e_{2s+2}] = e_1, \\ [e_2, \cdots, \widehat{e}_j, \cdots, e_{2s+3}] = e_j; \end{cases} \tag{18}$$

$$(r^2)\ [e_1, \cdots, \widehat{e}_i, \cdots, e_{2s+2}] = e_i.$$

Theorem 10. *If A is a $(2s+3)$-dimensional $(2s+1)$-Lie algebra over $\mathbb{F}$ with $\dim A^1 = r < s + 3$, then there are involutive derivations on A.*

Proof. Define linear mappings $D_j : A \longrightarrow A$, $1 \le j \le 6$ by

$$D_1(e_i) = \begin{cases} e_i, & 1 \le i \le s + 2, \text{ or } i = 2s + 3, \\ -e_i, & \text{otherwise}; \end{cases}$$

$$D_2(e_i) = \begin{cases} e_i, & 1 \le i \le s + 2, \\ -e_i, & \text{otherwise}; \end{cases}$$

$$D_3(e_i) = \begin{cases} -e_i, & s+2 \le i \le 2s+1, \\ e_i, & \text{otherwise}; \end{cases}$$

$$D_4(e_i) = \begin{cases} e_i, & 1 \le i \le s+1, \text{ or } i = 2s+2, \\ -e_i, & \text{otherwise}; \end{cases}$$

$$D_5(e_i) = \begin{cases} e_i, & 1 \le i \le s+1, \text{ or } i = 2s+3, \\ -e_i, & \text{otherwise}; \end{cases}$$

$$D_6(e_i) = \begin{cases} e_i, & 1 \le i \le s+3, \\ -e_i, & \text{otherwise}. \end{cases} \tag{19}$$

Since $\dim A^1 = r \le s+2$, it is easy to verify that D_1 is an involutive derivation on the 3-Lie algebras of the cases of (b^1), (c^i), (d^j), and (r^k), where $1 \le i \le 7$, $1 \le j \le 4$ and $1 \le k \le 2$. D_2 is an involutive derivation on the 3-Lie algebras of the cases of (b^1), (d^4), and (c^i), where $5 \le i \le 7$. D_3, D_4, and D_5 are involutive derivations on the 3-Lie algebras of the case of (b^2). And D_6 is an involutive derivation on the 3-Lie algebras of the cases of (d^5), (d^6), and (d^7). Also D_i are involutive derivations on abelian algebras for $1 \le i \le 6$. □

Next, we discuss the case of $\dim A^1 = r \ge s+3$. Let D be an endomorphism of A,

$$De_i = \sum_{j=1}^{2s+3} b_{ij} e_j, \quad b_{ij} \in \mathbb{F}, \ 1 \le i \le 2s+3, \tag{20}$$

and $B = (b_{ij})$ be the $(2s+3) \times (2s+3)$-matrix. Then

$$\begin{aligned} D(e_1, \cdots, e_{2s+3})^T &= B(e_1, \cdots, e_{2s+3})^T \\ &= \begin{pmatrix} B_1 & B_0 \\ B_2 & B_3 \end{pmatrix} (e_1, \cdots, e_{2s+3})^T, \end{aligned} \tag{21}$$

where $\begin{pmatrix} B_1 & B_0 \\ B_2 & B_3 \end{pmatrix}$ is the block matrix of B. First we discuss $(2s+3)$-dimensional $(2s+1)$-Lie algebras of the case (r^1) in Lemma 9.

Lemma 11. *If A is a $(2s+3)$-dimensional $(2s+1)$-Lie algebra of the case (r^1) with $\dim A^1 = r \ge s+3, s \ge 1$. Then the linear mapping D is an involutive derivation on A if and only if the block matrix $B = \begin{pmatrix} B_1 & B_0 \\ B_2 & B_3 \end{pmatrix}$ satisfies that $B_0 = O_{r\times(2s+3-r)}$ (which is the zero $(r \times (2s+3-r))$-matrix), and*

$$\begin{gathered} B_1^2 = E_r, \\ B_3^2 = E_{2s+3-r}, \\ B_2B_1 + B_3B_2 = 0, \end{gathered} \tag{22}$$

$$\begin{gathered} \sum_{j=2}^{2s+2} b_{jj} = b_{11}, \\ \sum_{j=2, j\ne i}^{2s+3} b_{jj} = b_{ii}, \quad 2 \le i \le r, \\ b_{2s+3,i} = (-1)^{i+1} b_{i,1}, \quad 2 \le i \le r, \\ b_{j,i} = (-1)^{j-i-1} b_{ij}, \quad 2 \le i, j \le r, \ i \ne j. \end{gathered} \tag{23}$$

Proof. By (2), and a direct computation, D is a derivation of A if and only if matrix B has the property:

$$\begin{gathered} \sum_{j=2}^{2s+2} b_{jj} = b_{11}, \\ \sum_{j=2, j\ne i}^{2s+3} b_{jj} = b_{ii}, \quad 2 \le i \le r, \\ b_{1l} = 0, \quad 2 \le l \le 2s+3, \\ b_{2s+3,i} = (-1)^{i+1} b_{i,1}, \quad b_{il} = 0, \ 2 \le i \le r, \ l \ge r+1, \\ b_{j,i} = (-1)^{j-i-1} b_{ij}, \quad 2 \le i, j \le r, \ i \ne j. \end{gathered} \tag{24}$$

Therefore, matrix B satisfies (23) and $B_0 = O_{r\times(2s+3-r)}$. And $D^2 = Id$ if and only if

$$\begin{aligned} B^2 &= \begin{pmatrix} B_1^2 & B_1B_0 + B_0B_3 \\ B_2B_1 + B_3B_2 & B_2B_0 + B_3^2 \end{pmatrix} \\ &= \begin{pmatrix} E_r & O \\ O & E_{2s+3-r} \end{pmatrix}. \end{aligned} \tag{25}$$

Thanks to $B_0 = O_{r\times(2s+3-r)}$, (22) holds. □

Theorem 12. *Let A be a $(2s+3)$-dimensional $(2s+1)$-Lie algebra of the case (r^1) with $\dim A^1 = r \ge s+3$, $s \ge 1$. If r is odd, then there are involutive derivations on A.*

Proof. Let $r = 2t+1 \ge s+3$. Then $t \ge 2$ and $r \ge 5$. Suppose D is an endomorphism of A and the matrix of D with respect to the basis $\{e_1, \cdots e_{2s+3}\}$ is $B = (b_{ij}) = \begin{pmatrix} B_1 & B_0 \\ B_2 & B_3 \end{pmatrix}$ which satisfies (22) and (23), and $B_0 = O_{r\times(2s+3-r)}$. Then

$$B_1 = \begin{pmatrix} b_{11} & 0 & 0 & \cdots & 0 & 0 \\ b_{21} & b_{22} & b_{23} & \cdots & b_{2,r-1} & b_{2,r} \\ b_{31} & b_{23} & b_{33} & \cdots & b_{3,r-1} & b_{3,r} \\ \vdots & \vdots & \vdots & \ddots & \vdots \\ b_{r-1,1} & (-1)^r b_{2,r-1} & (-1)^{r-1} b_{3,r-1} & \cdots & b_{r-1,r-1} & b_{r-1,r} \\ b_{r,1} & (-1)^{r+1} b_{2,r} & (-1)^r b_{3,r} & \cdots & b_{r-1,r} & b_{r,r} \end{pmatrix},$$

$$B_2 = \begin{pmatrix} b_{r+1,1} & b_{r+1,2} & b_{r+1,3} & \cdots & b_{r+1,r} \\ \vdots & \vdots & \vdots & \ddots & \vdots \\ b_{2s+2,1} & b_{2s+2,2} & b_{2s+2,3} & \cdots & b_{2s+2,r} \\ b_{2s+3,1} & -b_{2,1} & b_{3,1} & \cdots & (-1)^{r+1} b_{r,1} \end{pmatrix}. \quad (26)$$

Since $\sum_{j=2}^{2s+2} b_{jj} = b_{11}, \sum_{j=2,j\neq i}^{2s+3} b_{jj} = b_{ii}, 2 \le i \le r$, we have $-b_{11} + 2b_{22} - b_{2s+3,2s+3} = 0, (r-3)b_{22} + \sum_{2s+3}^{r+1} b_{ii} = 0, b_{22} = b_{ii}, 3 \le i \le r$. Therefore,

$$\begin{aligned} b_{11} &= \frac{-1}{r-3}\left((r-1)k_1 + 2\sum_{j=2}^{2s+3-r} k_j\right), \\ b_{ii} &= \frac{-1}{r-3}\sum_{j=1}^{2s+3-r} k_j, \quad 2 \le i \le r, \\ b_{jj} &= k_{2s+3-j+1}, \quad r+1 \le j \le 2s+3, \ k_{2s+3-j+1} \in \mathbb{F}. \end{aligned} \quad (27)$$

Suppose

$$B_1^2 = \begin{pmatrix} c_{11} & c_{12} & \cdots & c_{1i} & \cdots & c_{1j} & \cdots & c_{1r} \\ \vdots & \vdots & \ddots & \vdots & \ddots & \vdots & \ddots & \vdots \\ c_{i1} & c_{i2} & \cdots & c_{ii} & \cdots & c_{ij} & \cdots & c_{ir} \\ \vdots & \vdots & \ddots & \vdots & \ddots & \vdots & \ddots & \vdots \\ c_{j1} & c_{j2} & \cdots & c_{ji} & \cdots & c_{jj} & \cdots & c_{jr} \\ \vdots & \vdots & \ddots & \vdots & \ddots & \vdots & \ddots & \vdots \\ c_{r1} & c_{r2} & \cdots & c_{ri} & \cdots & c_{rj} & \cdots & c_{rr} \end{pmatrix}. \quad (28)$$

By (23),

$$\begin{aligned} c_{11} &= b_{11}^2, \\ c_{1l} &= 0, \quad 2 \le l \le r, \\ c_{ii} &= \sum_{l=1}^{i} (-1)^{i+l-1} b_{li}^2 + \sum_{l=i+1}^{r} (-1)^{i+l-1} b_{il}^2, \quad 2 \le i \le r, \\ c_{ij} &= \sum_{l=1}^{i} (-1)^{l+i+1} b_{li} b_{lj} + \sum_{l=i+1}^{j} b_{il} b_{lj} \\ &\quad + \sum_{l=j+1}^{r} (-1)^{l+j+1} b_{il} b_{jl}, \quad 2 \le i < j \le r, \\ c_{ij} &= \sum_{l=1}^{j} (-1)^{l+i+1} b_{lj} b_{li} + \sum_{l=j+1}^{i} (-1)^{i+j} b_{jl} b_{li} \\ &\quad + \sum_{l=i+1}^{r} (-1)^{l+j+1} b_{jl} b_{il}, \quad 1 \le j < i \le r. \end{aligned} \quad (29)$$

Therefore, the endomorphisms D of A, which are defined by

$$\begin{aligned} De_1 &= e_1 \\ \text{or } De_1 &= -e_1, \\ De_2 &= \sum_{k=3}^{r} e_k, \\ De_i &= (-1)^{i-1} e_2 + \sum_{k=3}^{i-1} (-1)^i e_k + (-1)^{i-1} \sum_{k=i+1}^{r} e_k, \\ &\qquad 3 \le i \le r, \\ De_j &= (-1)^j e_j, \quad r+1 \le j \le 2s+3, \\ De_1 &= e_1 \\ \text{or } De_1 &= -e_1, \\ De_2 &= \sum_{k=3}^{r} e_k, \\ De_i &= (-1)^{i-1} e_2 + \sum_{k=3}^{i-1} (-1)^i e_k + (-1)^{i-1} \sum_{k=i+1}^{r} e_k, \\ &\qquad 3 \le i \le r, \\ De_j &= (-1)^{j-1} e_j, \quad r+1 \le j \le 2s+3, \end{aligned} \quad (30)$$

are involutive derivations on A. □

Theorem 13. *Let A be a $(2s+3)$-dimensional $(2s+1)$-Lie algebra of the case (r^1) with $\dim A^1 = r = 2s+2$ $(s \ge 1)$, then there does not exist an involutive derivation on A.*

Proof. If D is an involutive derivation on A, then, by Lemma 11 and (23),

$$\begin{aligned} b_{11} &= \frac{-(2s+1)k_1}{2s-1}, \\ b_{2s+3,2s+3} &= k_1, \\ b_{ii} &= \frac{-k_1}{2s-1}, \\ &\qquad 2 \le i \le r, \ k_1 \in \mathbb{F}. \end{aligned} \quad (31)$$

Thanks to (22), $b_{2s+3,2s+3}^2 = b_{11}^2 = k_1^2 = 1$. Therefore, $(-(2s+1)k_1/(2s-1))^2 = ((2s+1)/(2s-1))^2 = 1$, which is a contradiction. □

Now we discuss case (r^2).

Theorem 14. *Let A be a $(2s+3)$-dimensional $(2s+1)$-Lie algebra of the case (r^2) with $\dim A^1 = r \ge s+3$. Then there exist involutive derivations on A if and only if r is even.*

Proof. By Lemma 7, $A = A_1 \dotplus \mathbb{F}e_{2s+3}$, where $e_{2s+3} \in Z(A)$, and A_1 is a $(2s+2)$-dimensional $(2s+1)$-Lie subalgebra of A with $\dim A^1 = \dim A_1^1 = r$. Then there exist involutive derivations on A if and only if there exist involutive derivations on A_1.

By Theorem 3 in [1], there is a basis $\{e_1, \cdots, e_{2s+2}\}$ of A_1 such that

$$\begin{aligned} &\left[e_1, \cdots, \widehat{e}_j, \cdots, e_{2s+2}\right] = 0, \quad t < j \le 2s+3-t, \\ &\left[e_1, \cdots, \widehat{e}_i, \cdots, e_{2s+2}\right] = (-1)^i e_{2s+3-i}, \\ &\qquad 1 \le i \le t, \text{ or } 2s+3-t < i \le 2s+2. \end{aligned} \tag{32}$$

If r is even, then $r = 2t \ge 4$. By Theorem 8 and (32), endomorphism D_1 of A_1 defined by

$$D_1(e_i) = \begin{cases} e_i, & i = 1, \cdots, s+1, \\ -e_i, & i = s+2, \cdots, 2s+2 \end{cases} \tag{33}$$

is an involutive derivation on A_1. Therefore, the endomorphism D of A defined by

$$\begin{aligned} D(e_i) &= e_i, \quad 1 \le i \le s+1, \\ D(e_j) &= -e_j, \quad s+2 \le j \le 2s+2, \\ D(e_{2s+3}) &= \pm e_{2s+3} \end{aligned} \tag{34}$$

is involutive derivation on A.

If $\dim A^1 = r$ is odd and endomorphism D of A is an involutive derivation on A, then $r = 2t+1 > 4$. Suppose $D(e_i) = \sum_{j=1}^{2s+3} a_{ij} e_j$, $1 \le i \le 2s+3$. Then

$$\begin{aligned} &\left[D(e_{2s+3}), e_{i_1}, \cdots, e_{i_{2s}}\right] \\ &\quad = D\left[e_{2s+3}, e_{i_1}, \cdots, e_{i_{2s}}\right] \\ &\quad - \sum_{j=1}^{2s} \left[e_{2s+3}, \cdots, De_{i_j}, \cdots, e_{i_{2s}}\right] = 0. \end{aligned} \tag{35}$$

We get $D(e_{2s+3}) \in Z(A) = \mathbb{F}e_{2s+3}$. Since $D^2 = Id$, $D(e_{2s+3}) = \pm e_{2s+3}$. By (32), $A^1 = \mathbb{F}e_1 + \cdots + \mathbb{F}e_t + \mathbb{F}e_{2s+1} + \cdots + \mathbb{F}e_{2s+3-t}$, and

$$\begin{aligned} &(-1)^i D(e_{2s+3-i}) \\ &\quad = \sum_{k=1, k\ne i}^{2s+2} \left[e_1, \cdots, \widehat{e}_i, \cdots, D(e_k), \cdots, e_{2s+2}\right], \end{aligned} \tag{36}$$

where $1 \le i \le t$, and $2s+3-t < i \le 2s+2$. Then $a_{i,2s+3} = 0$, for $1 \le i \le t$, or $2s+3-t < i \le 2s+2$, and $DA^1 \subseteq A_1$. Then the endomorphism D_2 of A_1 defined by

$$\begin{aligned} &D_2(e_i) = D(e_i), \\ &\qquad 1 \le i \le t, \text{ or } 2s+3-t < i \le 2s+2, \\ &D_2(e_j) = D(e_j) - a_{j,2s+3} e_{2s+3} = \sum_{j=1}^{2s+2} a_{ij} e_j, \\ &\qquad t < j \le 2s+3-t \end{aligned} \tag{37}$$

is an involutive derivation on the $(2s+2)$-dimensional $(2s+1)$-Lie algebra A_1, contradiction (Theorem 8). Therefore, there does not exist involutive derivation on A. □

5. Structure of 3-Lie Algebras with Involutive Derivations

Let $(L, [,])$ be a Lie algebra over $\mathbb{F}$, and p be an element which is not contained in L. Then $A = L \dotplus \mathbb{F}p$ is a 3-Lie algebra in the multiplication

$$\begin{aligned} &[x, y, z] = 0, \\ &[p, x, y] = [x, y], \\ &\qquad \text{for all } x, y, z \in L. \end{aligned} \tag{38}$$

And the 3-Lie algebra $(A, [,,])$ is called *one-dimensional extension of L*.

Theorem 15. *Let A be a 3-Lie algebra, then A is one-dimensional extension of a Lie algebra if and only if there exists an involutive derivation D on A such that $\dim A_1 = 1$, or $\dim A_{-1} = 1$.*

Proof. If A is an one-dimensional extension of a Lie algebra L, then $A = L \dotplus \mathbb{F}p$. Define the endomorphism D of A by $D(p) = -p$ (or p), and $D(x) = x$ (or $-x$), $\forall x \in L$. Thanks to (38), $D^2 = Id$, and $D([x, y, z]) = 0 = [Dx, y, z] + [x, Dy, z] + [x, y, Dz]$, $D([p, x, y]) = [p, x, y] = [Dp, x, y] + [p, Dx, y] + [p, x, Dy]$, for all $x, y, z \in L$. Therefore, D is an involutive derivation on A, and $\dim A_{-1} = 1$ (or $\dim A_1 = 1$).

Conversely, let D be an involutive derivation on a 3-Lie algebra A, and $\dim A_{-1} = 1$ (or $\dim A_1 = 1$). Let $A_{-1} = \mathbb{F}p$, and $A_1 = L$ (or $A_{-1} = L$, $A_1 = \mathbb{F}p$), where $p \in A - L$. Thanks to Theorem 3, L is a Lie algebra with the multiplication $[x, y] = [p, x, y]$, for all $x, y \in L$, and A is one-dimensional extension of L. □

Let $(L, [,]_1)$ and $(L, [,]_2)$ be Lie algebras and $\{x_1, \cdots, x_m\}$ be a basis of L. For convenience, denote Lie algebras $(L, [,]_k)$ by L_k, $k = 1, 2$, respectively. Suppose p_1 and p_2 are two distinct elements which are not contained in L, and 3-Lie algebras $(B, [,,]_1)$ and $(C, [,,]_2)$ are one-dimensional extensions of Lie algebras L_1 and L_2, respectively, where $B = L \dotplus \mathbb{F}p_1$, $C = L \dotplus \mathbb{F}p_2$. Then $Der(L_1)$ and $Der(L_2)$ are subalgebras of $gl(L)$.

Definition 16. Let $L_1 = (L, [,]_1)$ and $L_2 = (L, [,]_2)$ be two Lie algebras, and p_1, p_2 be two distinct elements which are not contained in L, and $A = L \dotplus \mathbb{F}p_1 \dotplus \mathbb{F}p_2$. Then 3-algebra $(A, [,,])$ is called a two-dimensional extension of Lie algebras L_k, $k = 1, 2$, where $[,,] : A \wedge A \wedge A \longrightarrow A$ defined by

$$\begin{aligned} &[x, y, p_1] = [x, y]_1, \\ &[x, y, p_2] = [x, y]_2, \\ &[x, y, z] = 0, \\ &[p_1, p_2, x] = \lambda_x p_1 + \mu_x p_2, \\ &\qquad \forall x, y, z \in L, \ \lambda_x, \mu_x \in \mathbb{F}. \end{aligned} \tag{39}$$

If A is a 3-Lie algebra, then A is called a two-dimensional extension 3-Lie algebra of Lie algebras L_k, $k = 1, 2$.

Let $A = L \dotplus W$ be a two-dimensional extension of Lie algebras L_k, $k = 1, 2$, where $W = \mathbb{F}p_1 \dotplus \mathbb{F}p_2$. Define linear mappings $D_1, D_2 : L \longrightarrow End(L)$ and $D : L \longrightarrow W$ by

$$\begin{aligned} D_1(x) &= ad(p_1, x), \\ D_2(x) &= ad(p_2, x), \\ D(x) &= ad(p_1, p_2)(x), \\ &\qquad \forall x \in L, \end{aligned} \tag{40}$$

that is, for all $y \in L$, $D_1(x)(y) = [p_1, x, y] = [x, y]_1$, $D_2(x)(y) = [p_2, x, y] = [x, y]_2$, $D(x) = [p_1, p_2, x]$. We have the following result.

Theorem 17. *Let 3-algebra A be a two-dimensional extension of Lie algebras L_1 and L_2. Then A is a 3-Lie algebra if and only if linear mappings D_1, D_2, and D satisfy that $D_1 : L_1 \longrightarrow Der(L_1)$, $D_2 : L_2 \longrightarrow Der(L_2)$ are Lie homomorphisms, and*

$$\begin{aligned} D_1(x_3)([x_1, x_2]_2) = {} & [D_1(x_3)(x_1), x_2]_2 \\ & + [x_1, D_1(x_3)(x_2)]_2 \\ & - \lambda_{x_3}[x_1, x_2]_1 - \mu_{x_3}[x_1, x_2]_2, \end{aligned} \tag{41}$$

$$\begin{aligned} D_2(x_3)([x_1, x_2]_1) = {} & [D_2(x_3)(x_1), x_2]_1 \\ & + [x_1, D_2(x_3)(x_2)]_1 \\ & + \lambda_{x_3}[x_1, x_2]_1 + \mu_{x_3}[x_1, x_2]_2, \end{aligned} \tag{42}$$

$$\begin{aligned} D([x_1, x_2]_1) &= \left(\mu_{x_1}\lambda_{x_2} - \lambda_{x_1}\mu_{x_2}\right)p_1, \\ D([x_1, x_2]_2) &= \left(\mu_{x_1}\lambda_{x_2} - \lambda_{x_1}\mu_{x_2}\right)p_2, \end{aligned} \tag{43}$$

$$\begin{aligned} \lambda_{[x_1,x_2]_1} &= \mu_{[x_1,x_2]_2} = \mu_{x_1}\lambda_{x_2} - \lambda_{x_1}\mu_{x_2}, \\ \mu_{[x_1,x_2]_1} &= \lambda_{[x_1,x_2]_2} = 0, \end{aligned} \tag{44}$$

$$\begin{aligned} D_k(x_1)(x_2) &= -D_k(x_2)(x_1), \\ &\qquad \textit{for all } x_1, x_2 \in L,\ k = 1, 2, \end{aligned} \tag{45}$$

where $x_1, x_2, x_3 \in L$, $D(x_i) = \lambda_{x_i}p_1 + \mu_{x_i}p_2$, $i = 1, 2, 3$.

Proof. If A is a two-dimensional extension 3-Lie algebra, then, by Definition 16, linear mappings D_k satisfy that $D_k(L_k) \subseteq Der(L_k)$, and D_k are Lie homomorphisms, $k = 1, 2$. Thanks to (39),

$$\begin{aligned} D_1(x_3)([x_1, x_2]_2) = {} & [p_2, [p_1, x_3, x_1], x_2] \\ & + [p_2, x_1, [p_1, x_3, x_2]] \\ & + [[p_1, x_3, p_2], x_1, x_2] \\ = {} & [D_1(x_3)(x_1), x_2]_2 \\ & + [x_1, D_1(x_3)(x_2)]_2 \\ & - \lambda_{x_3}[x_1, x_2]_1 - \mu_{x_3}[x_1, x_2]_2, \end{aligned} \tag{46}$$

for all $x_1, x_2, x_3 \in L$, (41) holds. Similarly, we have (42).

Thanks to (39) and (40),

$$\begin{aligned} D([x_1, x_2]_1) &= ad(p_1, p_2)[x_1, x_2]_1 \\ &= \lambda_{[x_1,x_2]_1}p_1 + \mu_{[x_1,x_2]_1}p_2 \\ &= \left(\mu_{x_1}\lambda_{x_2} - \lambda_{x_1}\mu_{x_2}\right)p_1, \\ D([x_1, x_2]_2) &= ad(p_1, p_2)[x_1, x_2]_2 \\ &= \lambda_{[x_1,x_2]_2}p_1 + \mu_{[x_1,x_2]_2}p_2 \\ &= \left(\mu_{x_1}\lambda_{x_2} - \lambda_{x_1}\mu_{x_2}\right)p_2, \end{aligned} \tag{47}$$

Equations (43) and (44) hold. Equation (45) follows from (39) and (40), directly.

Conversely, by (39), $\forall x_1, x_2, x_3, x \in L$,

$$\begin{aligned} [x_1, x_2, x_3] &= 0, \\ [p_1, x_1, x_2] &= D_1(x_1)(x_2) = [x_1, x_2]_1, \\ [p_2, x_1, x_2] &= D_2(x_1)(x_2) = [x_1, x_2]_2, \\ [p_1, p_2, x] &= D(x) = \lambda(x)p_1 + \mu_x p_2. \end{aligned} \tag{48}$$

Since $D_k(L) \subseteq Der(L_k)$ and D_k are Lie homomorphisms, $B = L \dotplus \mathbb{F}p_1$ and $C = L \dotplus \mathbb{F}p_2$ are 3-Lie algebras, which are one-dimensional extension 3-Lie algebras of Lie algebras L_k, $k = 1, 2$, respectively.

Next we only need to prove that the multiplication on A defined by (39) satisfies (1). For all $x_i \in L$, $1 \le i \le 5$, that products $[[x_1, x_2, x_3], x_4, x_5]$, $[[p_j, x_2, x_3], x_4, x_5]$, $[[x_1, x_2, x_3], x_4, p_j]$ and $[[x_1, x_2, p_j], x_4, p_j]$ satisfy (1), $j = 1, 2$ follow from that B and C are one-dimensional extension 3-Lie algebras of L_k and (39), directly.

From (41) and (42), it follows that products $[[p_i, x_1, x_2], p_j, x_3]$, $1 \le i \ne j \le 2$, satisfy (1). It follows from (43)–(45) that products $[p_1, p_2, [p_i, x_1, x_2]]$, $[x_1, x_2, [p_i, p_2, x_3]]$, and $[p_i, x_1, [p_1, p_2, x_2]$, $i = 1, 2$, satisfy (1). We omit the computation process. □

Theorem 18. *Let $(A, [,,])$ be a 3-Lie algebra. Then A is a two-dimensional extension 3-Lie algebra of Lie algebras if and only if there is an involutive derivation T on A such that* $\dim A_1 = 2$ *or* $\dim A_{-1} = 2$.

Proof. If A is a two-dimensional extension 3-Lie algebra of Lie algebras. Then by Theorem 15, there are Lie algebras $L_1 = (L, [,]_1)$ and $L_2 = (L, [,]_2)$, such that $A = L \dotplus W$ and the multiplication of A is defined by (39), where $W = \mathbb{F}p_1 + \mathbb{F}p_2$.

Define the endomorphism T of A by $T(x) = x, T(p_1) = -p_1, T(p_2) = -p_2$, or $T(x) = -x, T(p_1) = p_1, T(p_2) = p_2, \forall x \in L$. Then $T^2 = Id$, and $A_1 = L, A_{-1} = W$, or $A_{-1} = L$, $A_1 = W$. Thanks to (38) and (41)-(45), T is a derivation of A.

Conversely, if there is an involutive derivation T on the 3-Lie algebra A such that $\dim A_{-1} = 2$ (or $\dim A_1 = 2$). By Theorem 4, $[A_1, A_1, A_1] = 0$, $[A_1, A_1, A_{-1}] \subseteq A_1$, $[A_1, A_{-1}, A_{-1}] \subseteq A_{-1}$. Let $L = A_1$ and $A_{-1} = \mathbb{F}p_1 \dotplus \mathbb{F}p_2$. Then $[L, L, p_1] \subseteq L, [L, L, p_2] \subseteq L$, and $(L, [,]_1)$ and

$(L, [,]_2)$ are Lie algebras, where $[x, y]_1 = [x, y, p_1]$, $[x, y]_2 = [x, y, p_2]$, $\forall x, y \in L$. Thanks to Theorem 17, the 3-Lie algebra A is a two-dimensional extension 3-Lie algebra of Lie algebras L_1 and L_2. □

Conflicts of Interest

The authors declare that they have no conflicts of interest.

Acknowledgments

The first named author was supported in part by the Natural Science Foundation (11371245) and the Natural Science Foundation of Hebei Province (A2018201126).

References

[1] V. Filippov, "n-Lie algebras," *Siberian Mathematical Journal*, vol. 26, no. 6, pp. 126–140, 1985.

[2] R. P. Bai and P. P. Jia, "The real compact n-Lie algebras and invariant bilinear forms," *Acta Mathematica Scientia. Series A. Shuxue Wuli Xuebao. Chinese Edition*, vol. 27A, no. 6, pp. 1074–1081, 2007.

[3] S. Kasymov, "On a theory of n-Lie algebras," *Algebra and Logic*, vol. 26, no. 3, pp. 277–297, 1987.

[4] C. Bai, L. Guo, and Y. Sheng, "Bialgebras, the classical Yang-Baxter equation and main triples for 3-Lie algebras," *Mathematical Physics*, 2016.

[5] R. Bai and G. Song, "The classification of six-dimensional 4-Lie algebras," *Journal of Physics A: Mathematical and General*, vol. 42, no. 3, 035207, 17 pages, 2009.

[6] R. Bai, G. Song, and Y. Zhang, "On classification of n-Lie algebras," *Frontiers of Mathematics in China*, vol. 6, no. 4, pp. 581–606, 2011.

21

On Killing Forms and Invariant Forms of Lie-Yamaguti Superalgebras

Patricia L. Zoungrana[1,2] and A. Nourou Issa[1,2]

[1]*Département de Mathématiques de la Décision, Université Ouaga 2, 12 BP 412, Ouagadougou 12, Burkina Faso*
[2]*Département de Mathématiques, Université d'Abomey-Calavi, 01 BP 4521, Cotonou, Benin*

Correspondence should be addressed to Patricia L. Zoungrana; patibffr@yahoo.fr

Academic Editor: Hernando Quevedo

The notions of the Killing form and invariant form in Lie algebras are extended to the ones in Lie-Yamaguti superalgebras and some of their properties are investigated. These notions are also $\mathbb{Z}_2$-graded generalizations of the ones in Lie-Yamaguti algebras.

1. Introduction

A *Lie-Yamaguti algebra* is a triple $(T, *, [\cdot,\cdot,\cdot])$ consisting of a vector space T, a bilinear map $* : T \times T \to T$, and a trilinear map $[\cdot,\cdot,\cdot] : T \times T \times T \to T$ such that

(LY1) $x * y = -y * x$,

(LY2) $[x, y, z] = -[y, x, z]$,

(LY3) $\circlearrowleft_{x,y,z}\{((x * y) * z) + [x, y, z]\} = 0$,

(LY4) $\circlearrowleft_{x,y,z}[x * y, z, u] = 0$,

(LY5) $[x, y, u * v] = [x, y, u] * v + u * [x, y, v]$,

(LY6) $[u, v, [x, y, z]] = [[u, v, x], y, z] + [x, [u, v, y], z] + [x, y, [u, v, z]]$,

for all u, v, x, y, z, in T, where $\circlearrowleft_{x,y,z}$ denotes the sum over cyclic permutation of x, y, z. The bilinear map $*$ sometimes will be denoted by juxtaposition. If $x * y = 0$, $\forall x, y \in T$, one gets a *Lie triple system* $(T, [\cdot,\cdot,\cdot])$, while $[x, y, z] = 0$ in $(T, *, [\cdot,\cdot,\cdot])$ induces a Lie algebra $(T, *)$.

Lie-Yamaguti algebras were introduced by Yamaguti [1] (who formerly called them "generalized Lie triple systems") in an algebraic study of the characteristic properties of the torsion and curvature of a homogeneous space with canonical connection [2]. Later on, these algebraic objects were called "Lie triple algebras" [3] and the terminology of "Lie-Yamaguti algebras" is introduced in [4] for these algebras. For further development of the theory of Lie-Yamaguti algebras one may refer, for example, to [5–8]. From the standard enveloping Lie algebra of a given Lie-Yamaguti algebra, the notions of the Killing-Ricci form and the invariant form of a Lie-Yamaguti algebra are introduced and studied in [9]. Further properties of invariant forms of Lie-Yamaguti algebras were considered in [10].

Lie superalgebras as a $\mathbb{Z}_2$-graded generalization of Lie algebras are considered in [11, 12] while a $\mathbb{Z}_2$-graded generalization of Lie triple systems (called Lie supertriple systems) was first considered in [13]. For an application of Lie supertriple systems in physics, one may refer to [14]. Next, *Lie-Yamaguti superalgebras* as a $\mathbb{Z}_2$-graded generalization of Lie-Yamaguti algebras were first considered in [15].

Definition 1 (see [16]). A Lie-Yamaguti superalgebra is a $\mathbb{Z}_2$-graded vector space $T = T_0 \oplus T_1$ with a binary operation denoted by juxtaposition satisfying $T_iT_j \subseteq T_{i+j}$ and a ternary operation $[\cdot,\cdot,\cdot]$ satisfying $[T_i, T_j, T_k] \subseteq T_{i+j+k}$ $(i, j, k \in \mathbb{Z}_2)$ such that

(LYS1) $xy = -(-1)^{\overline{x}\overline{y}}yx$,

(LYS2) $[x, y, z] = -(-1)^{\overline{x}\overline{y}}[y, x, z]$,

(LYS3) $\circlearrowleft_{x,y,z}(-1)^{\overline{x}\overline{z}}\{((xy)z) + [x, y, z]\} = 0$,

(LYS4) $\circlearrowleft_{x,y,z}(-1)^{\overline{x}\overline{z}}[xy, z, u] = 0$,

(LYS5) $[x, y, uv] = [x, y, u]v + (-1)^{\overline{u}(\overline{x}+\overline{y})}u[x, y, v]$,

(LYS6) $[u, v, [x, y, z]] = [[u, v, x], y, z] + (-1)^{\overline{x}(\overline{u}+\overline{v})}[x, [u, v, y], z] + (-1)^{(\overline{x}+\overline{y})(\overline{u}+\overline{v})}[x, y, [u, v, z]]$,

for all u, v, x, y, z, in T.

Observe that T_0 is a Lie-Yamaguti algebra.

As a part of the general theory of superalgebras, the notion of the Killing form of Lie algebras is extended to the one of Lie triple systems (see [17] and references therein), Lie superalgebras [12], and next Lie supertriple systems [18] (see also [19]).

In this paper we define and study the Killing form and invariant form of Lie-Yamaguti superalgebras as a generalization of the ones of both of Lie-Yamaguti algebras [9] and Lie supertriple systems [18, 19] including Lie superalgebras [12].

The paper is organized as follows. In Section 2 we record some useful results on Lie-Yamaguti superalgebras (see [16]). In Section 3 the Killing form of a Lie-Yamaguti superalgebra is defined (see Theorem 10 and Definition 11) and some of its properties are investigated (Proposition 13, Theorem 14, and Corollary 15). In Section 4 the invariant form of a Lie-Yamaguti superalgebra is defined (Definition 16) and, under some conditions, it is shown (Theorem 21) that the Killing form of a Lie-Yamaguti superalgebra T is nondegenerate if and only if the standard enveloping Lie superalgebra of T is semisimple.

All vector spaces and algebras are finite-dimensional over a fixed ground field $\mathbb{K}$ of characteristic 0.

2. Some Basics on Lie-Yamaguti Superalgebras

We give here some definitions and results which can be found in [11, 12, 16].

A superalgebra over $\mathbb{K}$ is a $\mathbb{Z}_2$-graded algebra $A = A_0 \oplus A_1$, where $A_iA_j \subseteq A_{i+j}$, $(i, j \in \mathbb{Z}_2)$. The subspaces A_0 and A_1 are called the even and the odd parts of the superalgebra and so are called the elements from A_0 and from A_1, respectively. Below, all the elements are assumed to be homogeneous, that is, either even or odd, and for a homogeneous element $x \in A_i$, $i = 0, 1$, the notation $\overline{x} = i$ is used and means the parity of x.

Let G be the *Grassmann algebra* over $\mathbb{K}$ generated by the elements $1, e_1, \ldots, e_n$ such that $e_i^2 = 0$, $e_ie_j = -e_je_i$ for $i \neq j$. The elements 1, $e_{i_1}e_{i_2}\cdots e_{i_r}$, $i_1 < i_2 < \cdots < i_r$ form a basis of G. Denote by G_0 (resp., G_1) the span of the products of even length (resp., odd length) in the generators. The product of zero e_i's is by convention equal to 1. Then $G = G_0 \oplus G_1$ is an associative and supercommutative superalgebra; that is, $g_1g_2 = (-1)^{\overline{g_1}\overline{g_2}}g_2g_1$, where $g_1, g_2 \in G_0 \cup G_1$. Let $A = A_0 \oplus A_1$ be a superalgebra. Consider the graded tensor product $G \otimes A$ which becomes a superalgebra with the product given by $(x \otimes g_1)(y \otimes g_2) = (-1)^{\overline{x}\overline{g_2}}xy \otimes g_1g_2$, for homogeneous elements $g_1, g_2 \in G$, $x, y \in A$ and grading given by $(G \otimes A)_0 = G_0 \otimes A_0 \oplus G_1 \otimes A_1$, $(G \otimes A)_1 = G_0 \otimes A_1 \oplus G_1 \otimes A_0$. The subalgebra $G(A) = (G \otimes A)_0 = G_0 \otimes A_0 \oplus G_1 \otimes A_1$ is called the *Grassmann envelope* of the superalgebra A.

Having in mind that if V is a homogeneous variety of algebras [20], a superalgebra $A = A_0 \oplus A_1$ is called a V-superalgebra, if its Grassmann envelope $G(A)$ belongs to V, we can state the following proposition.

Proposition 2. *A superalgebra $T = T_0 \oplus T_1$ equipped with bilinear and trilinear products verifying $T_iT_j \subseteq T_{i+j}$ and $[T_i, T_j, T_k] \subseteq T_{i+j+k}$ is a Lie-Yamaguti superalgebra if its Grassmann envelope $G(T) = G_0 \otimes T_0 \oplus G_1 \otimes T_1$ is a Lie-Yamaguti algebra under the following products:*

$$\begin{aligned}(x \otimes g_1)(y \otimes g_2) &= (-1)^{\overline{x}\overline{y}}xy \otimes g_1g_2;\\ [x \otimes g_1, y \otimes g_2, z \otimes g_3] &\\ &= (-1)^{\overline{x}\overline{y}+\overline{y}\overline{z}+\overline{x}\overline{z}}[x, y, z] \otimes g_1g_2g_3.\end{aligned} \tag{1}$$

Proof. The proof is straightforward by using the fact that, for any element $x \otimes g$ in $G(T)$, we have $\overline{x} = \overline{g}$. □

Example 3. (1) Lie superalgebras are Lie-Yamaguti superalgebras with $[x, y, z] = 0$.

(2) If $xy = 0$ for any $x, y \in T_0 \cup T_1$ then (LYS2), (LYS3), and (LYS6) define a Lie supertriple system.

(3) Let $M = M_0 \oplus M_1$ be a Malcev superalgebra; that is, for any x, y, z, t in T,

$$\begin{aligned}xy &= -(-1)^{\overline{x}\overline{y}}yx;\\ -(-1)^{\overline{y}\overline{z}}(xz)(yt) &= ((xy)z)t\\ &\quad + (-1)^{\overline{x}(\overline{y}+\overline{z}+\overline{t})}((yz)t)x\\ &\quad + (-1)^{(\overline{x}+\overline{y})(\overline{z}+\overline{t})}((zt)x)y\\ &\quad + (-1)^{\overline{t}(\overline{x}+\overline{y}+\overline{z})}((tx)y)z.\end{aligned} \tag{2}$$

It is shown in [16] that M becomes a Lie-Yamaguti superalgebra if we set $[x, y, z] = x(yz) - (-1)^{\overline{x}\overline{y}}y(xz) + (xy)z$. Conversely, if on a Malcev superalgebra $(M, \cdot)$ we define a trilinear operation by $[x, y, z] = x \cdot yz - (-1)^{\overline{x}\overline{y}}y \cdot xz + xy \cdot z$ then $(M, \cdot, [\cdot, \cdot, \cdot])$ is a Lie-Yamaguti superalgebra.

Definition 4. Let $T = T_0 \oplus T_1$ be a Lie-Yamaguti superalgebra. A graded subspace $H = H_0 \oplus H_1$ of T is a graded Lie-Yamaguti subalgebra of T if $H_iH_j \subseteq H_{i+j}$ and $[H_i, H_j, H_k] \subseteq H_{i+j+k}$ for any $i, j, k \in \mathbb{Z}_2$.

Definition 5. A graded subalgebra of a Lie-Yamaguti superalgebra T is an invariant graded subalgebra (resp., an ideal) of T if $[T, T, H] \subseteq H$ (resp., $TH \subseteq H$ and $[T, H, T] \subseteq H$).

If H is an ideal of T, it is an invariant graded subalgebra of T. Obviously the center $Z(T)$ of a Lie-Yamaguti superalgebra T defined by $Z(T) = \{x \in T,\ xy = 0 \text{ and } [x, y, z] = 0,\ \forall y, z \in T\}$ is an ideal of T.

Definition 6. Let $T = T_0 \oplus T_1$ and $T' = T_0' \oplus T_1'$ be Lie-Yamaguti superalgebras. A linear map $f : T \to T'$ is said to be of degree r if $f(T_i) \subseteq T'_{r+i}$ for all $r, i \in \mathbb{Z}_2$.

Definition 7. Let $T = T_0 \oplus T_1$ and $T' = T_0' \oplus T_1'$ be Lie-Yamaguti superalgebras. A linear map $f : T \to T'$ is called a homomorphism of Lie-Yamaguti superalgebras if

(1) f preserves the grading, that is, $f(T_i) \subseteq T_i'$, $i \in \mathbb{Z}_2$;

(2) $f(xy) = f(x)f(y)$;

(3) $f([x, y, z]) = [f(x), f(y), f(z)]$ for any $x, y, z \in T_0 \cup T_1$.

Recall [11] that if $V = V_0 \oplus V_1$ is a $\mathbb{Z}_2$-graded vector space then, if we set $\mathrm{End}_r(V) = \{f \in \mathrm{End}(V)/f(V_i) \subseteq V_{r+i}\}$, we obtain an associative superalgebra $\mathrm{End}(V) = \mathrm{End}_0(V) \oplus \mathrm{End}_1(V)$; $\mathrm{End}_r(V)$ consists of the linear mappings of V into itself which are homogeneous of degree r. The bracket $[f, g] = fg - (-1)^{\overline{f}\overline{g}} gf$ makes $\mathrm{End}(V)$ into a Lie superalgebra which we denote by $l(V)$ or $l(m,n)$ where $m = \dim V_0$ and $n = \dim V_1$. Let $e_1, \ldots, e_m, e_{m+1}, \ldots, e_{m+n}$ be a basis of V. In this basis the matrix of $a \in l(m,n)$ is expressed as $\left(\begin{smallmatrix} \alpha & \beta \\ \gamma & \delta \end{smallmatrix}\right)$, α being an $(m \times m)$-, δ an $(n \times n)$-, β an $(m \times n)$-, and γ an $(n \times m)$-matrix. The matrices of even elements have the form $\left(\begin{smallmatrix} \alpha & 0 \\ 0 & \delta \end{smallmatrix}\right)$ and those of odd ones $\left(\begin{smallmatrix} 0 & \beta \\ \gamma & 0 \end{smallmatrix}\right)$. For $a = \left(\begin{smallmatrix} \alpha & \beta \\ \gamma & \delta \end{smallmatrix}\right)$, the supertrace of a is defined by $\mathrm{str}(a) = \mathrm{tr}\,\alpha - \mathrm{tr}\,\delta$ and does not depend on the choice of a homogeneous basis. We have $\mathrm{str}([a, b]) = 0$ that is $\mathrm{str}(ab) = (-1)^{\overline{a}\overline{b}} \mathrm{str}(ba)$ and $\mathrm{str}(aba^{-1}) = \mathrm{str}(b)$.

Definition 8. Let $T = T_0 \oplus T_1$ be a Lie-Yamaguti superalgebra; $D \in \mathrm{End}_r(T)$ is a superderivation of T if, for any $x, y, z \in T_0 \cup T_1$,

$$D(xy) = D(x)\,y + (-1)^{r\overline{x}} xD(y);$$
$$D([x, y, z]) = [D(x), y, z] + (-1)^{r\overline{x}} [x, D(y), z] + (-1)^{r(\overline{x}+\overline{y})} [x, y, D(z)]. \tag{3}$$

Let $D_r(T)$ consist of all the superderivations of degree r and $D(T) = D_0(T) \oplus D_1(T)$. It is easy to check that $D(T)$ is a graded subalgebra of $\mathrm{End}(T)$ called the Lie superalgebra of superderivations of T.

Let $T = T_0 \oplus T_1$ be a Lie-Yamaguti superalgebra. For any $x, y \in T_0 \cup T_1$, denote by $D_{x,y}$ the endomorphism of T defined by $D_{x,y}(z) = [x, y, z]$ for any $z \in T$. We have, for any $x, y \in T_0 \cup T_1$, $r \in \mathbb{Z}_2$, $D_{x,y}(T_r) \subseteq T_{r+\overline{x}+\overline{y}}$; that is, $D_{x,y}$ is a linear map of degree $\overline{x} + \overline{y}$. Moreover, it comes from (LYS5) and (LYS6) that

$$D_{x,y}(zw) = D_{x,y}(z)\,w + (-1)^{\overline{z}(\overline{x}+\overline{y})} zD_{x,y}(w); \tag{4}$$
$$D_{x,y}([z, v, w]) = [D_{x,y}(z), v, w] + (-1)^{\overline{z}(\overline{x}+\overline{y})} [z, D_{x,y}(v), w] + (-1)^{(\overline{z}+\overline{v})(\overline{x}+\overline{y})} [z, v, D_{x,y}(w)] \tag{5}$$

for any $x, y, z, v, w \in T_0 \cup T_1$. It follows that $D_{x,y}$ is a superderivation of T called an *inner superderivation* of T.

Let $D(T, T)$ be the vector space spanned by all $D_{x,y}$ $(x, y \in T)$.

We can define naturally a $\mathbb{Z}_2$-gradation by setting $D(T, T) = D_0(T, T) \oplus D_1(T, T)$, where $D_r(T, T)$ consists of the superderivation $D_{x,y}$ of degree r. From (5) we also have that, for any $x, y, z, v, w \in T_0 \cup T_1$,

$$[D_{x,y}, D_{z,v}] = D_{[x,y,z],v} + (-1)^{\overline{z}(\overline{x}+\overline{y})} D_{z,[x,y,v]}. \tag{6}$$

It is clear from (6) that $D(T, T)$ is a $\mathbb{Z}_2$-graded Lie subalgebra of $D(T)$ called the *Lie superalgebra of all inner superderivations of T.*

Now, let $(T, \cdot, [\cdot, \cdot, \cdot])$ be a Lie-Yamaguti superalgebra.

Set $L_i(T) = T_i \oplus D_i(T, T)$, $i = 0, 1$, and define a new bracket operation in $L(T) = L_0(T) \oplus L_1(T) = T \oplus D(T, T)$ as follows: for any $x, y \in T_0 \cup T_1$, $D_1, D_2 \in D_0(T, T) \cup D_1(T, T)$,

$$[x, y] = xy + D_{x,y};$$
$$[D, x] = -(-1)^{\overline{x}\overline{D}} [x, D] = D(x); \tag{7}$$
$$[D_1, D_2] = D_1 D_2 - (-1)^{\overline{D_1}\overline{D_2}} D_2 D_1.$$

Theorem 9. *Let $T = T_0 \oplus T_1$ be a Lie-Yamaguti superalgebra. Then*

(1) *$L(T)$ is a Lie superalgebra called the standard enveloping Lie superalgebra of T and $D(T, T)$ becomes a graded subalgebra of $L(T)$.*

(2) *If H is an ideal of T then $H \oplus D(T, H)$ is an ideal of $T \oplus D(T, T)$.*

Proof. The bracket $[\cdot, \cdot]$ is bilinear by definition and $XY = -(-1)^{\overline{X}\overline{Y}} YX$ for any $X, Y \in L(T)$ by (LYS1) and (LYS2). Jacobi's superidentity follows from (LTS3–6).

(2) is obvious. □

3. Killing Forms of Lie-Yamaguti Superalgebras

The definition of the Killing form given here for Lie-Yamaguti superalgebras stems from [9] in the case of Lie-Yamaguti algebras and extends the one given in [18] for Lie supertriple systems. Let $T = T_0 \oplus T_1$ be an n-dimensional Lie-Yamaguti superalgebra. Denote by α the Killing form of the standard enveloping Lie superalgebra $L(T) = (T_0 \oplus D_0(T, T)) \oplus (T_1 \oplus D_1(T, T))$. Consider the bilinear form β of T obtained by restricting α to $T \times T$. For any x, y, z in T, define the endomorphisms L_x and $R_{x,y}$ of the vector space T by $L_x(y) = xy$ and $R_{x,y}(z) = (-1)^{\overline{z}(\overline{x}+\overline{y})}[z, x, y] = (-1)^{\overline{z}(\overline{x}+\overline{y})} D_{z,x}(y)$. It is clear that $R_{x,y}$ is of degree $\overline{x} + \overline{y}$ and $[D_{z,t}, R_{x,y}] = R_{[z,t,x],y} + (-1)^{(\overline{z}+\overline{t})\overline{x}} R_{x,[z,t,y]}$.

Theorem 10. *For $x, y \in T$, we have*

$$\beta(x, y) = \mathrm{str}(L_x L_y) + \mathrm{str}(R_{x,y} + (-1)^{\overline{x}\overline{y}} R_{y,x}). \tag{8}$$

Proof. Let $\{a_i\}$, $\{b_i\}$, $\{u_i\}$, $\{v_i\}$ be bases of T_0, T_1, $D_0(T,T)$, $D_1(T,T)$, respectively. For these bases, we express the operations of T and $D(T,T)$ as follows:

$$\begin{aligned}
a_ia_j &= \sum_l S^l_{ij}a_l; \quad a_ia_j \in T_0,\\
a_ib_j &= \sum_l T^l_{ij}b_l; \quad a_ib_j \in T_1,\\
b_ib_j &= \sum_l R^l_{ij}a_l; \quad b_ib_j \in T_0,\\
D_{a_i,a_j} &= \sum_\alpha D^\alpha_{ij}u_\alpha; \quad D_{a_i,a_j} \in D_0(T,T),\\
D_{a_i,b_j} &= \sum_\alpha C^\alpha_{ij}v_\alpha; \quad D_{a_i,b_j} \in D_1(T,T),\\
D_{b_i,b_j} &= \sum_\alpha X^\alpha_{ij}u_\alpha; \quad D_{b_i,b_j} \in D_0(T,T),\\
[u_\alpha, a_i] &= u_\alpha(a_i) = \sum_j K^j_{\alpha i}a_j;\\
[v_\alpha, a_i] &= v_\alpha(a_i) = \sum_j L^j_{\alpha i}b_j;\\
[u_\alpha, b_i] &= u_\alpha(b_i) = \sum_j H^j_{\alpha i}b_j;\\
[v_\alpha, b_i] &= v_\alpha(b_i) = \sum_j Q^j_{\alpha i}a_j.
\end{aligned} \tag{9}$$

To prove the theorem, it suffices to show that $\beta(a_i,a_j) = \alpha(a_i,a_j)$, $\beta(a_i,b_j) = \alpha(a_i,b_j)$ and $\beta(b_i,b_j) = \alpha(b_i,b_j)$. Since $(L_{a_i}L_{b_j})(T_0) \subseteq T_1$ and $(L_{a_i}L_{b_j})(T_1) \subseteq T_0$, we have $\operatorname{str}(L_{a_i}L_{b_j}) = 0$. Also, $R_{a_i,b_j}(T_1) \subseteq T_0$ and $R_{a_i,b_j}(T_0) \subseteq T_1$ give $\operatorname{str}(R_{a_i,b_j} + R_{b_j,a_i}) = 0$ and then $\beta(a_i,b_j) = 0 = \alpha(a_i,b_j)$ because of the consistency property of α ($a_i \in T_0 \oplus D_0(T,T)$, $b_j \in T_1 \oplus D_1(T,T)$). Hence, it remains to show that $\beta(a_i,a_j) = \alpha(a_i,a_j)$ and $\beta(b_i,b_j) = \alpha(b_i,b_j)$. The operations in T and the identities (7) imply the following:

$$\begin{aligned}
[a_i,[a_j,a_k]] &= [a_i, a_ja_k + D_{a_j,a_k}]\\
&= \left[a_i, \sum_m S^m_{jk}a_m + \sum_\alpha D^\alpha_{jk}u_\alpha\right]\\
&= \sum_m S^m_{jk}(a_ia_m + D_{a_i,a_m}) - \sum_\alpha D^\alpha_{jk}\sum_l K^l_{\alpha i}a_l\\
&= \sum_m S^m_{jk}\left(\sum_l S^l_{im}a_l + \sum_\alpha D^\alpha_{im}u_\alpha\right)\\
&\quad - \sum_{\alpha,l} D^\alpha_{jk}K^l_{\alpha i}a_l\\
&= \sum_{m,l} S^m_{jk}S^l_{im}a_l + \sum_{m,\alpha} S^m_{jk}D^\alpha_{im}u_\alpha\\
&\quad - \sum_{\alpha,l} D^\alpha_{jk}K^l_{\alpha i}a_l.
\end{aligned} \tag{10}$$

In a similar way, we get

$$\begin{aligned}
[a_i,[a_j,b_k]] &= \sum_{m,l} T^m_{jk}T^l_{im}b_l + \sum_{m,\alpha} T^m_{jk}C^\alpha_{im}v_\alpha\\
&\quad - \sum_{\alpha,l} C^\alpha_{jk}L^l_{\alpha i}b_l,\\
[a_i,[a_j,u_\alpha]] &= -[a_i, u_\alpha(a_j)] = -\left[a_i, \sum_m K^m_{\alpha j}a_m\right]\\
&= -\sum_{m,l} K^m_{\alpha j}S^l_{im}a_l - \sum_{m,\beta} K^m_{\alpha j}D^\beta_{im}u_\beta,\\
[a_i,[a_j,v_\alpha]] &= -[a_i, v_\alpha(a_j)] = -\left[a_i, \sum_m L^m_{\alpha j}b_m\right]\\
&= -\sum_{m,l} L^m_{\alpha j}T^l_{im}b_l - \sum_{m,\beta} L^m_{\alpha j}C^\beta_{im}v_\beta.
\end{aligned} \tag{11}$$

Also,

$$\begin{aligned}
L_{a_i}L_{a_j}(a_k) &= a_i(a_ja_k) = a_i\left(\sum_p S^p_{jk}a_p\right)\\
&= \sum_p S^p_{jk}(a_ia_p) = \sum_{p,l} S^p_{jk}S^l_{ip}a_l,\\
L_{a_i}L_{a_j}(b_k) &= a_i(a_jb_k) = a_i\left(\sum_p T^p_{jk}b_p\right)\\
&= \sum_p T^p_{jk}(a_ib_p) = \sum_{p,l} T^p_{jk}T^l_{ip}b_l,\\
R_{a_i,a_j}(a_k) &= [a_k,a_i,a_j] = D_{a_k,a_i}(a_j) = \sum_\alpha D^\alpha_{ki}u_\alpha(a_j)\\
&= \sum_\alpha D^\alpha_{ki}\sum_m K^m_{\alpha j}a_m = \sum_{\alpha,m} D^\alpha_{ki}K^m_{\alpha j}a_m;\\
R_{a_i,a_j}(b_k) &= [b_k,a_i,a_j] = D_{b_k,a_i}(a_j)\\
&= -\sum_\alpha C^\alpha_{ik}v_\alpha(a_j) = -\sum_\alpha C^\alpha_{ik}\sum_m L^m_{\alpha j}b_m\\
&= -\sum_{\alpha,m} C^\alpha_{ik}L^m_{\alpha j}b_m.
\end{aligned} \tag{12}$$

By interchanging i and j, we have

$$\begin{aligned}
R_{a_j,a_i}(a_k) &= \sum_{\alpha,m} D^\alpha_{kj}K^m_{\alpha i}a_m;\\
R_{a_j,a_i}(b_k) &= -\sum_{\alpha,m} C^\alpha_{jk}L^m_{\alpha i}b_m.
\end{aligned} \tag{13}$$

Therefore,

$$\begin{aligned}\beta(a_i,a_j) &= \alpha(a_i,a_j)\\ &= \sum_{m,k} S^m_{jk}S^k_{im} - \sum_{\alpha,k} D^\alpha_{jk}K^k_{\alpha i}a_k - \sum_{m,k} T^m_{jk}T^k_{im} + \sum_{\alpha,k} C^\alpha_{jk}L^k_{\alpha i}\\ &\quad - \sum_{m,\alpha} K^m_{\alpha j}D^\alpha_{im} + \sum_{m,\alpha} L^m_{\alpha j}C^\alpha_{im},\\ \operatorname{str}(L_{a_i}L_{a_j}) &+ \operatorname{str}(R_{a_j,a_i} + R_{a_j,a_i})\\ &= \sum_{p,k} S^p_{jk}S^k_{ip} - \sum_{p,k} T^p_{jk}T^k_{ip} + \sum_{\alpha,k} D^\alpha_{ki}K^k_{\alpha j} + \sum_{\alpha,k} D^\alpha_{kj}K^k_{\alpha i}\\ &\quad + \sum_{\alpha,k} C^\alpha_{ik}L^k_{\alpha j} + \sum_{\alpha,k} C^\alpha_{jk}L^k_{\alpha i} = \beta(a_i,a_j).\end{aligned} \tag{14}$$

It remains to show that $\beta(b_i,b_j) = \operatorname{str}(L_{b_i}L_{b_j}) + \operatorname{str}(R_{b_j,b_i} + R_{b_j,b_i})$,

$$\begin{aligned}[b_i,[b_j,b_k]] &= [b_i, b_jb_k + D_{b_j,b_k}]\\ &= \left[b_i, \sum_m R^m_{jk}a_m + \sum_\alpha X^\alpha_{jk}u_\alpha\right]\\ &= -\sum_{m,l} R^m_{jk}T^l_{mi}b_l - \sum_{m,\alpha} R^{m\alpha}_{jkmi}v_\alpha\\ &\quad - \sum_{\alpha,l} X^\alpha_{jk}H^l_{\alpha i}b_l.\end{aligned} \tag{15}$$

Likewise, we have

$$\begin{aligned}[b_i,[b_j,a_k]] &= -\sum_{m,l} T^m_{kj}R^l_{im}a_l - \sum_{m,\alpha} T^m_{kj}X^\alpha_{im}u_\alpha\\ &\quad - \sum_{\alpha,l} C^\alpha_{kj}Q^l_{\alpha i}a_l,\\ [b_i,[b_j,u_\alpha]] &= -[b_i,u_\alpha(b_j)] = -\left[b_i,\sum_l H^l_{\alpha j}b_l\right]\\ &= -\sum_l H^l_{\alpha j}R^m_{il}a_m - \sum_{l,\beta} H^l_{\alpha j}X^\beta_{il}u_\beta,\\ [b_i,[b_j,v_\alpha]] &= [b_i,v_\alpha(b_j)] = \left[b_i,\sum_l Q^l_{\alpha j}a_l\right]\\ &= -\sum_{l,m} Q^l_{\alpha j}T^m_{li}b_m - \sum_{l,\beta} Q^l_{\alpha j}C^\beta_{li}v_\beta.\end{aligned} \tag{16}$$

Therefore,

$$\begin{aligned}\beta(a_i,a_j) &= \alpha(a_i,a_j)\\ &= -\sum_{m,k} T^m_{kj}R^k_{im} - \sum_{\alpha,k} C^\alpha_{kj}Q^k_{\alpha i} + \sum_{m,k} R^m_{jk}T^k_{mi}\\ &\quad + \sum_{\alpha,k} X^\alpha_{jk}H^k_{\alpha i} - \sum_{l,\alpha} H^l_{\alpha j}X^\alpha_{il} + \sum_{l,\alpha} Q^l_{\alpha j}C^\alpha_{li}.\end{aligned} \tag{17}$$

Now,

$$\begin{aligned}L_{b_i}L_{b_j}(a_k) &= b_i(b_ja_k) = -b_i\left(\sum_m T^m_{jk}b_m\right)\\ &= -\sum_m T^m_{jk}(b_ib_m) = -\sum_{m,l} T^m_{jk}R^l_{im}a_l;\\ L_{b_i}L_{b_j}(b_k) &= b_i(b_jb_k) = b_i\left(\sum_m R^m_{jk}a_m\right)\\ &= -\sum_m R^m_{jk}(a_mb_i) = -\sum_{m,l} R^m_{jk}T^l_{mi}b_l;\\ R_{b_i,b_j}(a_k) &= [a_k,b_i,b_j] = D_{a_k,b_i}(b_j) = \sum_\alpha C^\alpha_{ki}v_\alpha(b_j)\\ &= \sum_\alpha C^\alpha_{ki}\sum_l Q^l_{\alpha j}a_l = \sum_{\alpha,l} C^\alpha_{ki}Q^l_{\alpha j}a_l;\\ R_{b_i,b_j}(b_k) &= [b_k,b_i,b_j] = D_{b_k,b_i}(b_j) = \sum_\alpha X^\alpha_{ki}u_\alpha(b_j)\\ &= \sum_\alpha X^\alpha_{ki}\sum_l H^l_{\alpha j}b_l = \sum_{\alpha,l} X^\alpha_{ki}H^l_{\alpha j}b_l.\end{aligned} \tag{18}$$

By interchanging i and j, we have

$$\begin{aligned}R_{b_j,b_i}(a_k) &= \sum_{\alpha,l} C^\alpha_{kj}Q^l_{\alpha i}a_l;\\ R_{b_j,b_i}(b_k) &= \sum_{\alpha,l} X^\alpha_{kj}H^l_{\alpha i}b_l,\\ \operatorname{str}(L_{b_i}L_{b_j}) &+ \operatorname{str}(R_{b_j,b_i} + R_{b_j,b_i})\\ &= -\sum_{m,k} T^m_{jk}R^k_{im} + \sum_{m,k} R^m_{jk}T^k_{mi} + \sum_{\alpha,k} C^\alpha_{ki}Q^k_{\alpha j} - \sum_{\alpha,k} X^\alpha_{ki}H^k_{\alpha j}\\ &\quad - \sum_{\alpha,k} C^\alpha_{kj}Q^k_{\alpha i} + \sum_{\alpha,k} X^\alpha_{kj}H^k_{\alpha i}b_l = \beta(b_i,b_j).\end{aligned} \tag{19}$$

Hence the theorem is proved. □

Definition 11. The bilinear form β defined on the Lie-Yamaguti superalgebra $T = T_0 \oplus T_1$ by

$$\beta(x,y) = \operatorname{str}(L_xL_y) + \operatorname{str}\left(R_{x,y} + (-1)^{\bar{x}\bar{y}}R_{y,x}\right) \tag{20}$$

for $x, y \in T$ is called the Killing form of T.

Remark 12. Recall that if T is a Lie superalgebra, then the Killing form β on T is defined as $\beta(x,y) = \operatorname{str}(L_xL_y)$, $x, y \in T$. Likewise, if T is a Lie supertriple system (resp., a Lie-Yamaguti algebra), the Killing form on T is defined as $\beta(x,y) = \operatorname{str}(R_{x,y} + (-1)^{\bar{x}\bar{y}}R_{y,x})$ (resp., $\beta(x,y) = \operatorname{tr}(L_xL_y) + \operatorname{tr}(R_{x,y} + R_{y,x})$) with L_u and $R_{u,v}$ defined according to the considered structure on T. So if a Lie-Yamaguti superalgebra T is reduced to a Lie superalgebra (resp., a Lie supertriple system, a Lie-Yamaguti algebra), then β as defined in Definition 11 is the Killing form of the Lie superalgebra (resp., the Lie supertriple system, the Lie-Yamaguti algebra) T.

Proposition 13. *Let $T = T_0 \oplus T_1$ be a Lie-Yamaguti superalgebra with a Killing form denoted by β. Then,*

(1) $\beta(T_0, T_1) = 0$ *(consistence),*

(2) $\beta(x, y) = (-1)^{\overline{x}\,\overline{y}}\beta(y, x)$ *(supersymmetry),*

(3) $\beta(A(x), A(y)) = \beta(x, y)$, $A \in \mathrm{Aut}(T)$.

Proof. As $L_{T_0}L_{T_1}(T_1) \subseteq T_0$, $L_{T_0}L_{T_1}(T_0) \subseteq T_1$, $R_{T_0,T_1}(T_0) \subseteq T_1$ and $R_{T_0,T_1}(T_1) \subseteq T_0$ we can state that $\beta(T_0, T_1) = 0$.

(2) comes from the definition of β.

Now, for any A in Aut(T), x in T, $\overline{A(x)} = \overline{x}$, and

$$\beta(A(x), A(y)) = \operatorname{str}\left(L_{A(x)}L_{A(y)}\right) + \operatorname{str}\left(R_{A(x),A(y)} + (-1)^{\overline{x}\,\overline{y}} R_{A(y),A(x)}\right). \tag{21}$$

As $A(xy) = A(x)A(y)$ then $AL_x(y) = L_{A(x)}A(y)$; that is, $AL_x = L_{A(x)}A$ and $AL_xA^{-1} = L_{A(x)}$.

Hence, $\operatorname{str}(L_{A(x)}L_{A(y)}) = \operatorname{str}(AL_xA^{-1}AL_yA^{-1}) = \operatorname{str}(AL_xL_yA^{-1}) = \operatorname{str}(L_xL_y)$. Also, $A[x, y, z] = [A(x), A(y), A(z)]$ gives $AR_{y,z}(x) = R_{A(y),A(z)}(A(x))$; that is, $AR_{y,z}A^{-1} = R_{A(y),A(z)}$. Then,

$$\begin{aligned}\beta(A(x), A(y)) &= \operatorname{str}\left(L_{A(x)}L_{A(y)}\right) + \operatorname{str}\left(R_{A(x),A(y)} + (-1)^{\overline{x}\,\overline{y}} R_{A(y),A(x)}\right)\\ &= \operatorname{str}\left(L_xL_y\right) + \operatorname{str}\left(AR_{x,y}A^{-1} + (-1)^{\overline{x}\,\overline{y}} AR_{y,x}A^{-1}\right)\\ &= \operatorname{str}\left(L_xL_y\right) + \operatorname{str}\left(A\left(R_{x,y} + (-1)^{\overline{x}\,\overline{y}} R_{y,x}\right)A^{-1}\right)\\ &= \operatorname{str}\left(L_xL_y\right) + \operatorname{str}\left(R_{x,y} + (-1)^{\overline{x}\,\overline{y}} R_{y,x}\right)\\ &= \beta(x, y).\end{aligned} \tag{22}$$

□

Now, let γ be a trilinear form in T given by $\gamma(x, y, z) = \operatorname{str}(D_{x,y}L_z)$ for any $x, y, z \in T$. We can easily see that, for any $x, y, z \in T$, $\gamma(x, y, z) = -(-1)^{\overline{x}\,\overline{y}}\gamma(y, x, z)$ and that γ vanishes identically if T is reduced to Lie superalgebra or Lie supertriple system.

Theorem 14. *Let $T = T_0 \oplus T_1$ be a Lie-Yamaguti superalgebra with a Killing form denoted by β. Then, β satisfies the identities*

$$\beta(xy, z) + (-1)^{\overline{x}\,\overline{y}} \beta(y, xz) = (-1)^{\overline{x}\,\overline{y}} \gamma(y, x, z) + (-1)^{(\overline{x}+\overline{y})\overline{z}} \gamma(z, x, y); \tag{23}$$

$$\beta(x, [y, z, w]) + (-1)^{\overline{w}(\overline{y}+\overline{z})} \beta([x, w, y], z) = (-1)^{\overline{w}(\overline{y}+\overline{z})} \gamma(x, w, yz) - (-1)^{\overline{x}(\overline{y}+\overline{z})} \gamma(y, z, xw) \tag{24}$$

for all $x, y, z \in T$.

Proof. The Killing form α of $L = (T_0 \oplus D_0(T, T)) \oplus (T_1 \oplus D_1(T, T))$ satisfies $\alpha(y, [x, z]) + (-1)^{\overline{x}\,\overline{y}}\alpha([x, y], z) = 0$; that is, $\alpha(y, [x, z]) = \alpha([y, x], z)$. But, using (7), we have

$$\begin{aligned}\alpha([x, y], z) &= \alpha\left(xy + D_{x,y}, z\right)\\ &= \alpha(xy, z) + \alpha\left(D_{x,y}, z\right)\\ &= \beta(xy, z) + \operatorname{str}\left(D_{x,y}L_z\right)\\ &= \beta(xy, z) + \gamma(x, y, z),\\ \alpha(y, [x, z]) &= \alpha\left(y, xz + D_{x,z}\right)\\ &= \alpha(y, xz) + \operatorname{str}\left(L_yD_{x,z}\right)\\ &= \beta(y, xz) + (-1)^{\overline{y}(\overline{x}+\overline{z})} \operatorname{str}\left(D_{x,z}L_y\right)\\ &= \beta(y, xz) + (-1)^{\overline{y}(\overline{x}+\overline{z})} \gamma(x, z, y).\end{aligned} \tag{25}$$

Then the identity $\alpha(y, [x, z]) + (-1)^{\overline{x}\,\overline{y}}\alpha([x, y], z) = 0$ gives $\beta(xy, z) + \gamma(x, y, z) + (-1)^{\overline{x}\,\overline{y}}\beta(y, xz) + (-1)^{\overline{y}\,\overline{z}}\gamma(x, z, y) = 0$ that is $\beta(xy, z) + (-1)^{\overline{x}\,\overline{y}}\beta(y, xz) = -\gamma(x, y, z) - (-1)^{\overline{y}\,\overline{z}}\gamma(x, z, y) = (-1)^{\overline{x}\,\overline{y}}\gamma(y, x, z) + (-1)^{(\overline{x}+\overline{y})\overline{z}}\gamma(z, x, y)$ and (23) is obtained.

From $\alpha([y, x], z) = \alpha(y, [x, z])$ we deduce $\alpha(x, [w, [y, z]]) = \alpha([x, w], [y, z]) = \alpha([[x, w], y], z)$ that is $-(-1)^{\overline{w}(\overline{y}+\overline{z})}\alpha(x, [[y, z], w]) = \alpha([[x, w], y], z)$ and

$$\alpha(x, [[y, z], w]) + (-1)^{\overline{w}(\overline{y}+\overline{z})} \alpha([[x, w], y], z) = 0. \tag{26}$$

Then, using (7) again and developing (26), we have $\alpha(x, [yz + D_{y,z}, w]) + (-1)^{\overline{w}(\overline{y}+\overline{z})}\alpha([xw + D_{x,w}, y], z) = 0$ and we get

$$\alpha\left(x, (yz)w + D_{yz,w} + [y, z, w]\right) + (-1)^{\overline{w}(\overline{y}+\overline{z})} \alpha\left((xw)y + D_{xw,y} + [x, w, y], z\right) = 0. \tag{27}$$

This gives

$$\begin{aligned}&\beta(x, (yz)w) + \alpha\left(x, D_{yz,w}\right) + \beta(x, [y, z, w]) + (-1)^{\overline{w}(\overline{y}+\overline{z})} \beta((xw)y, z)\\ &\quad + (-1)^{\overline{w}(\overline{y}+\overline{z})} \alpha\left(D_{xw,y}, z\right) + (-1)^{\overline{w}(\overline{y}+\overline{z})} \beta([x, w, y], z) = 0,\\ &\beta(x, (yz)w) + \operatorname{str}\left(L_xL_{D_{yz,w}}\right) + \beta(x, [y, z, w]) + (-1)^{\overline{w}(\overline{y}+\overline{z})} \beta((xw)y, z)\\ &\quad + (-1)^{\overline{w}(\overline{y}+\overline{z})} \operatorname{str}\left(L_{D_{xw,y}}L_z\right) + (-1)^{\overline{w}(\overline{y}+\overline{z})} \beta([x, w, y], z) = 0.\end{aligned} \tag{28}$$

Thus,

$$\begin{aligned}\beta(x,(yz)w) &+ (-1)^{\overline{x}(\overline{y}+\overline{z}+\overline{w})}\operatorname{str}\left(L_{D_{yz,w}}L_x\right)\\ &+\beta(x,[y,z,w]) + (-1)^{\overline{w}(\overline{y}+\overline{z})}\beta((xw)y,z)\\ &+(-1)^{\overline{w}(\overline{y}+\overline{z})}\operatorname{str}\left(L_{D_{xw,y}}L_z\right)\\ &+(-1)^{\overline{w}(\overline{y}+\overline{z})}\beta([x,w,y],z)=0;\end{aligned}\tag{29}$$

that is,

$$\begin{aligned}\beta(x,(yz)w) &+ (-1)^{\overline{x}(\overline{y}+\overline{z}+\overline{w})}\gamma(yz,w,x)\\ &+\beta(x,[y,z,w]) + (-1)^{\overline{w}(\overline{y}+\overline{z})}\beta((xw)y,z)\\ &+(-1)^{\overline{w}(\overline{y}+\overline{z})}\gamma(xw,y,z)\\ &+(-1)^{\overline{w}(\overline{y}+\overline{z})}\beta([x,w,y],z)=0.\end{aligned}\tag{30}$$

This implies

$$\begin{aligned}\beta(x,[y,z,w]) &+ (-1)^{\overline{w}(\overline{y}+\overline{z})}\beta([x,w,y],z)\\ &= -\beta(x,(yz)w) - (-1)^{\overline{w}(\overline{y}+\overline{z})}\beta((xw)y,z)\\ &\quad-(-1)^{\overline{x}(\overline{y}+\overline{z}+\overline{w})}\gamma(yz,w,x)\\ &\quad-(-1)^{\overline{w}(\overline{y}+\overline{z})}\gamma(xw,y,z).\end{aligned}\tag{31}$$

But (23) gives

$$\begin{aligned}(-1)^{\overline{w}\,\overline{x}}\beta(x,w(yz)) &+ \beta(wx,yz)\\ &= (-1)^{\overline{w}\,\overline{x}}\gamma(x,w,yz)\\ &\quad+(-1)^{(\overline{w}+\overline{x})(\overline{y}+\overline{z})}\gamma(yz,w,x),\\ (-1)^{\overline{w}\,\overline{x}}\beta((xw)y,z) &+ \beta(wx,yz)\\ &= \gamma(wx,y,z) + (-1)^{(\overline{w}+\overline{x})(\overline{y}+\overline{z})+\overline{y}\,\overline{z}}\gamma(z,y,wx).\end{aligned}\tag{32}$$

Then,

$$\begin{aligned}-\beta(x,w(yz)) &- (-1)^{\overline{w}(\overline{y}+\overline{z})}\beta((xw)y,z)\\ &= (-1)^{\overline{w}(\overline{y}+\overline{z})}\gamma(x,w,yz)\\ &\quad+(-1)^{\overline{x}(\overline{w}+\overline{y}+\overline{z})}\gamma(yz,w,x)\\ &\quad-(-1)^{\overline{w}(\overline{x}+\overline{y}+\overline{z})}\gamma(wx,y,z)\\ &\quad-(-1)^{\overline{x}(\overline{w}+\overline{y}+\overline{z})+\overline{y}\,\overline{z}}\gamma(z,y,wx).\end{aligned}\tag{33}$$

Hence, $\beta(x,[y,z,w])+(-1)^{\overline{w}(\overline{y}+\overline{z})}\beta([x,w,y],z)= (-1)^{\overline{w}(\overline{y}+\overline{z})}\gamma(x, w, yz) - (-1)^{\overline{x}(\overline{y}+\overline{z})}\gamma(z, y, wx)$ and (24) is proved. □

Corollary 15. *Let $T = T_0 \oplus T_1$ be a Lie-Yamaguti superalgebra with a Killing form denoted by β. Then, β satisfies the following for $x, y, z \in T$:*

$$([y,z,x],w) + (-1)^{\overline{x}(\overline{y}+\overline{z})}\beta(x,[y,z,w]) = 0;\tag{34}$$

$$\begin{aligned}\beta(R_{w,y}(x),z) &- (-1)^{(\overline{w}+\overline{y})\overline{x}+\overline{w}\,\overline{y}}\beta(x,R_{y,w}(z))\\ &= (-1)^{\overline{x}(\overline{w}+\overline{y})}\gamma(x,w,yz)\\ &\quad+(-1)^{\overline{w}(\overline{y}+\overline{z})+\overline{x}\,\overline{z}}\gamma(y,z,xw).\end{aligned}\tag{35}$$

Proof. Using (24) we have

$$\begin{aligned}\beta([y,z,x],w) &= (-1)^{(\overline{w}+\overline{x})(\overline{y}+\overline{z})}\beta([x,w,y],z)\\ &\quad-(-1)^{(\overline{w}+\overline{x})(\overline{y}+\overline{z})}\gamma(x,w,yz)\\ &\quad-(-1)^{\overline{w}\,\overline{x}}\gamma(y,z,wx);\end{aligned}\tag{36}$$

$$\begin{aligned}(-1)^{\overline{x}(\overline{y}+\overline{z})}\beta(x,[y,z,w]) &= -(-1)^{(\overline{w}+\overline{x})(\overline{y}+\overline{z})}\beta([x,w,y],z)\\ &\quad+(-1)^{(\overline{w}+\overline{x})(\overline{y}+\overline{z})}\gamma(x,w,yz)\\ &\quad+(-1)^{\overline{w}\,\overline{x}}\gamma(y,z,wx).\end{aligned}\tag{37}$$

By adding memberwise (36) and (37) we obtain the identity (34).

Also, the identity (37) is equivalent to $-(-1)^{\overline{y}\,\overline{z}}\beta(x, [z, y, w]) + (-1)^{\overline{w}(\overline{y}+\overline{z})}\beta([x, w, y], z) = (-1)^{\overline{w}(\overline{y}+\overline{z})}\gamma(x, w, yz) - (-1)^{\overline{x}(\overline{y}+\overline{z})}\gamma(y, z,xw)$. Then, we obtain $(-1)^{\overline{w}\,\overline{y}+\overline{x}(\overline{w}+\overline{y})}\beta(R_{w,y}(x), z, z) - \beta(x, R_{y,w}(z)) = (-1)^{\overline{w}\,\overline{y}}\gamma(w,x, yz) - (-1)^{\overline{x}(\overline{y}+\overline{z})+\overline{w}\,\overline{z}}\gamma(y, z, xw)$ that is $\beta(R_{w,y}(x), z) - (-1)^{(\overline{w}+\overline{y})\overline{x}+\overline{w}\,\overline{y}}\beta(x, R_{y,w}(z)) = (-1)^{\overline{x}(\overline{w}+\overline{y})}\gamma(x, w, yz) + (-1)^{\overline{w}(\overline{y}+\overline{z})+\overline{x}\,\overline{z}}\gamma(y, z, xw)$ and the remaining assertion is proved. □

4. Invariant Forms of Lie-Yamaguti Superalgebras

In this section we introduced the concept of invariant forms of Lie-Yamaguti superalgebras as generalizations of those of Lie superalgebras and Lie supertriple systems.

Definition 16. An invariant form b of a Lie-Yamaguti superalgebra $T = T_0 \oplus T_1$ is a supersymmetric bilinear form on T satisfying the identities

$$b(xy,z) + (-1)^{\overline{x}\,\overline{y}}b(y,xz) = 0;\tag{38}$$

$$b([x,w,y],z) + (-1)^{\overline{w}(\overline{y}+\overline{z})}b(x,[y,z,w]) = 0\tag{39}$$

for all x, y, z, w in T.

Remark 17. (1) If $\gamma = 0$, the Killing form of T is an invariant form of T.

(2) If T is reduced to a Lie supertriple system (resp., a Lie superalgebra, a Lie-Yamaguti algebra), then b is reduced to an invariant form of a Lie supertriple system [19] (resp., a Lie superalgebra [12], a Lie-Yamaguti algebra [10]).

Definition 18. Let b be an invariant form of a Lie-Yamaguti superalgebra T and S a subset of T. The orthogonal $S^{\perp}$ of S with respect to b is defined by $S^{\perp} = \{x \in T,\ b(x, y) = 0,\ \forall y \in S\}$. The invariant form b is nondegenerate if $T^{\perp} = \{0\}$.

Lemma 19. *Let b be an invariant form of a Lie-Yamaguti superalgebra T. Then, for any x, y, z, w in T, we have*

$$b([x, y, w], z) + (-1)^{(\overline{x}+\overline{y})\overline{z}}\, b(w, [x, y, z]) = 0. \qquad (40)$$

Proof. By interchanging y and w in (39) we have

$$b([x, y, w], z) + (-1)^{\overline{y}(\overline{w}+\overline{z})}\, b(x, [w, z, y]) = 0 \qquad (41)$$

that is $(-1)^{(\overline{x}+\overline{y})\overline{z}} b(z, [x, y, w]) + (-1)^{\overline{y}(\overline{w}+\overline{z})+\overline{w}\,\overline{z}} b(x, [w, z, y]) = 0$ by supersymmetry. Also by switching z and w in (41), we obtain $b([x, y, z], \overline{w}) + (-1)^{\overline{y}(\overline{w}+\overline{z})} b(x, [z, w, y]) = 0$ that is

$$b([x, y, z], w) - (-1)^{\overline{y}(\overline{w}+\overline{z})+\overline{w}\,\overline{z}}\, b(x, [w, z, y]) = 0. \qquad (42)$$

Thus adding (41) and (42) we get (40) whence the lemma. □

Lemma 20. *Let b be an invariant form of a Lie-Yamaguti superalgebra T. Then,*

(1) *$(T + [T, T, T])^{\perp} = Z(T)$ if b is nondegenerate;*

(2) *if H is an ideal of T then $H^{\perp}$ is an ideal of T. In particular, $T^{\perp}$ is an ideal of T.*

Proof. Consider x in $(T + [T, T, T])^{\perp}$. Then, for any $u, v, w \in T$, we have $b(x, uv) = 0$ and $b(x, [u, v, w]) = 0$. This implies, by (38) and (39), that $(-1)^{\overline{u}\,\overline{x}} b(ux, v) = 0$ and $(-1)^{\overline{w}(\overline{u}+\overline{v})} b([x, w, u], v) = 0$ that is $b(ux, v) = 0$ and $b([x, w, u], v) = 0$. As b is nondegenerate, we get $ux = 0$ and $[x, w, u] = 0$ for any $u, v, w \in T$. This gives $x \in Z(T)$.

Conversely, if $x \in Z(T)$, we have, for any $u, v, u', v', w' \in T$, $b(x, uv + [u', v', w']) = b(x, uv) + b(x, [u', v', w']) = 0$ and $x \in (T + [T, T, T])^{\perp}$ whence $(T + [T, T, T])^{\perp} = Z(T)$.

Now, suppose that H is an ideal of T that is $TH \subseteq H$ and $[T, H, T] \subseteq H$; then for any $x, y \in T$, $u \in H^{\perp}$, and $h \in H$, we have $b(xu, h) = -(-1)^{\overline{x}\,\overline{u}} b(u, xh) = 0$ and $b([x, u, y], h) = -(-1)^{\overline{x}\,\overline{u}} b([u, x, y], h) = (-1)^{\overline{x}(\overline{y}+\overline{h})+\overline{x}\,\overline{u}} b(u, [y, h, x]) = 0$. Then $TH^{\perp} \subseteq H^{\perp}$ and $[T, H^{\perp}, T] \subseteq H^{\perp}$ which proves (2). □

We are now ready to prove the following theorem.

Theorem 21. *Let $T = T_0 \oplus T_1$ be a Lie-Yamaguti superalgebra with $\gamma = 0$. Then the Killing form β is nondegenerate if and only if the standard enveloping Lie superalgebra $L(T) = T \oplus D(T, T)$ is a semisimple Lie superalgebra.*

Proof. Let α be the Killing form of the Lie superalgebra $L(T)$. If $\gamma = 0$, we have, for any $x, y, z \in T$, $\gamma(x, y, z) = \operatorname{str}(D_{x,y} L_z) = 0$ and

$$\alpha\left(D_{x,y}, z\right) = 0. \qquad (43)$$

Then, using the invariance of α and (43), we have, for any $x, y, z, w \in T$: $\alpha([x, y], D_{z,w}) = \alpha(x, [y, D_{z,w}])$; that is, by (7),

$\alpha(xy + D_{x,y}, D_{z,w}) = -(-1)^{\overline{y}(\overline{z}+\overline{w})} \alpha(x, [z, w, y])$ and $\alpha(D_{x,y}, D_{z,w}) = -(-1)^{\overline{y}(\overline{z}+\overline{w})} \alpha(x, [z, w, y])$. This gives

$$\alpha\left(D_{x,y}, D_{z,w}\right) = -(-1)^{\overline{y}(\overline{z}+\overline{w})}\, \beta(x, [z, w, y]). \qquad (44)$$

Thus, if β is nondegenerate, the restriction of α on $D(T, T) \times D(T, T)$ is nondegenerate and α is nondegenerate.

Now, suppose that β is degenerate. Then by the lemma above, $T^{\perp}$ is an ideal of T so $T^{\perp} \oplus D(T, T^{\perp})$ is a nonzero ideal of T.

Using the identities (43) and (44) we get

$$\begin{aligned} &\alpha\left(T^{\perp} \oplus D(T, T^{\perp}), T \oplus D(T, T)\right) \\ &= \alpha(T^{\perp}, T) + \alpha(T^{\perp}, D(T, T)) + \alpha(D(T, T^{\perp}), T) \\ &\quad + \alpha(D(T, T^{\perp}), D(T, T)) \\ &= \alpha(T^{\perp}, T) + \beta(T, [T, T^{\perp}, T]) = 0. \end{aligned} \qquad (45)$$

It comes that α is degenerate and $T \oplus D(T, T)$ is not semisimple which proves the theorem. □

The results of this paper could be used for a study of the structure of a pair consisting of a semisimple Lie superalgebra and its semisimple graded subalgebra.

Competing Interests

The authors declare that they have no competing interests.

References

[1] K. Yamaguti, "On the Lie triple system and its generalization," *Journal of Science of the Hiroshima University, Series A*, vol. 21, pp. 107–113, 1957/1958.

[2] K. Nomizu, "Invariant affine connections on homogeneous spaces," *American Journal of Mathematics*, vol. 76, pp. 33–65, 1954.

[3] M. Kikkawa, "Geometry of homogeneous Lie loops," *Hiroshima Mathematical Journal*, vol. 5, no. 2, pp. 141–179, 1975.

[4] M. K. Kinyon and A. Weinstein, "Leibniz algebras, Courant algebroids, and multiplications on reductive homogeneous spaces," *American Journal of Mathematics*, vol. 123, no. 3, pp. 525–550, 2001.

[5] P. Benito, A. Elduque, and F. Martín-Herce, "Irreducible Lie-Yamaguti algebras," *Journal of Pure and Applied Algebra*, vol. 213, no. 5, pp. 795–808, 2009.

[6] P. Benito, A. Elduque, and F. Martín-Herce, "Irreducible Lie-Yamaguti algebras of generic type," *Journal of Pure and Applied Algebra*, vol. 215, no. 2, pp. 108–130, 2011.

[7] J. Lin, L. Y. Chen, and Y. Ma, "On the deformation of Lie-Yamaguti algebras," *Acta Mathematica Sinica (English Series)*, vol. 31, no. 6, pp. 938–946, 2015.

[8] T. Zhang and J. Li, "Deformations and extensions of Lie-Yamaguti algebras," *Linear and Multilinear Algebra*, vol. 63, no. 11, pp. 2212–2231, 2015.

[9] M. Kikkawa, "On Killing-Ricci forms of Lie triple algebras," *Pacific Journal of Mathematics*, vol. 96, no. 1, pp. 153–161, 1981.

[10] M. Kikkawa, "Remarks on invariant forms of Lie triple algebras," *Memoirs of the Faculty of Science and Engineering, Shimane University*, vol. 16, pp. 23–27, 1982.

[11] V. G. Kac, "Lie superalgebras," *Advances in Mathematics*, vol. 26, no. 1, pp. 8–96, 1977.

[12] M. Scheunert, *The Theory of Lie Superalgebras*, Springer, Berlin, Germany, 1979.

[13] H. Tilgner, "A graded generalization of Lie triples," *Journal of Algebra*, vol. 47, no. 1, pp. 190–196, 1977.

[14] S. Okubo, "Parastatistics as Lie-supertriple systems," *Journal of Mathematical Physics*, vol. 35, no. 6, pp. 2785–2803, 1994.

[15] M. F. Ouédraogo, *Sur les Superalgèbres triples de Lie, Thèse de Doctorat 3è cycle Mathématiques [Ph.D. thesis]*, Université de Ouagadougou, Ouagadougou, Burkina Faso, 1999.

[16] P. L. Zoungrana, "A note on Lie-Yamaguti superalgebras," *Far East Journal of Mathematical Sciences*, vol. 100, no. 1, pp. 1–18, 2016.

[17] T. S. Ravisankar, "Some remarks on Lie triple systems," *Kumamoto Journal of Science (Mathematics)*, vol. 11, pp. 1–8, 1974.

[18] S. Okubo and N. Kamiya, "Quasi-classical Lie superalgebras and Lie supertriple systems," *Communications in Algebra*, vol. 30, no. 8, pp. 3825–3850, 2002.

[19] Z. Zhixue and J. Peipei, "The Killing forms and decomposition theorems of Lie supertriple systems," *Acta Mathematica Scientia. Series B*, vol. 29, no. 2, pp. 360–370, 2009.

[20] I. P. Shestakov, "Prime Mal'tsev superalgebras," *Matematicheskii Sbornik*, vol. 182, no. 9, pp. 1357–1366, 1991 (Russian).

The Noncentral Version of the Whitney Numbers: A Comprehensive Study

Mahid M. Mangontarum,[1] Omar I. Cauntongan,[2] and Amila P. Macodi-Ringia[1]

[1]*Department of Mathematics, Mindanao State University, Main Campus, 9700 Marawi City, Philippines*
[2]*Department of Natural Sciences and Mathematics, Mindanao State University, Maigo School of Arts and Trades, 9206 Maigo, Philippines*

Correspondence should be addressed to Mahid M. Mangontarum; mmangontarum@yahoo.com

Academic Editor: Palle E. Jorgensen

This paper is a comprehensive study of a certain generalization of Whitney-type and Stirling-type numbers which unifies the classical Whitney numbers, the translated Whitney numbers, the classical Stirling numbers, and the noncentral Stirling (or r-Stirling) numbers. Several identities, applications, and occurrences are also presented.

1. Introduction

For a finite group G of order $m > 0$, the *Dowling lattice* of rank n, denoted by $Q_n(G)$, associated with G is defined by Dowling [1] as a class of geometric lattices and is known to generalize the partition lattice. Following this, Benoumhani [2] defined the *Whitney numbers of the first and second kind of* $Q_n(G)$, denoted by $w_m(n,k)$ and $W_m(n,k)$, respectively, as coefficients in the expansions of the relations

$$m^n (x)_n = \sum_{k=0}^{n} w_m(n,k)(mx+1)^k,$$
$$(mx+1)^n = \sum_{k=0}^{n} m^k W_m(n,k)(x)_k, \tag{1}$$

where

$$(x)_k = x(x-1)(x-2)\cdots(x-k+1). \tag{2}$$

Fundamental properties of these numbers were mostly established by Benoumhani [2–4].

In a recent paper, Belbachir and Bousbaa [5] defined the *translated Whitney numbers of the first kind* $\left[{n \atop k}\right]^{(\alpha)}$ *and second kind* $\left\{{n \atop k}\right\}^{(\alpha)}$ to be the number of permutations (resp., partitions) of n elements with k cycles (resp., parts) such that the elements of each cycle (resp., parts) can mutate in α ways except for the dominant one. These numbers are known to obey the following relations:

$$(x \mid -\alpha)_n = \sum_{k=0}^{n} \begin{bmatrix} n \\ k \end{bmatrix}^{(\alpha)} x^k, \tag{3}$$

$$x^n = \sum_{k=0}^{n} \begin{Bmatrix} n \\ k \end{Bmatrix}^{(\alpha)} (x \mid \alpha)_k, \tag{4}$$

where

$$(x \mid \alpha)_n = \prod_{i=0}^{n-1} (x - i\alpha). \tag{5}$$

Further study of these numbers was done due to Mangontarum et al. [6] and Mangontarum and Dibagulun [7].

It is important to note that the numbers in (1), (3), and (4) are known to be generalizations of the well-known *Stirling numbers* [8]. To be more precise, if G is the trivial group ($m = 1$), it can be easily verified that

$$w_1(n,k) = \begin{bmatrix} n+1 \\ k+1 \end{bmatrix}, \quad W_1(n,k) = \begin{Bmatrix} n+1 \\ k+1 \end{Bmatrix}. \tag{6}$$

Similarly, when $\alpha = -1$ in (3) and $\alpha = 1$ in (4),

$$\begin{bmatrix} n \\ k \end{bmatrix}^{(-1)} = \begin{bmatrix} n \\ k \end{bmatrix}, \quad \begin{Bmatrix} n \\ k \end{Bmatrix}^{(1)} = \begin{Bmatrix} n \\ k \end{Bmatrix}, \tag{7}$$

where $\left[{n \atop k} \right]$ and $\left\{ {n \atop k} \right\}$ denote the classical Stirling numbers of the first and second kind, respectively. Moreover, we have

$$\begin{bmatrix} n \\ k \end{bmatrix}^{(\alpha)} = (-\alpha)^{n-k} \begin{bmatrix} n \\ k \end{bmatrix}, \quad \begin{Bmatrix} n \\ k \end{Bmatrix}^{(\alpha)} = \alpha^{n-k} \begin{Bmatrix} n \\ k \end{Bmatrix}. \tag{8}$$

In line with this, Mező [9] defined the *r-Whitney numbers* $w_{m,r}(n,k)$ *and* $W_{m,r}(n,k)$ *of the first and second kind* via expressions

$$m^n (x)_n = \sum_{k=0}^{n} w_{m,r}(n,k)(mx+r)^k, \tag{9}$$

$$(mx+r)^n = \sum_{k=0}^{n} m^k W_{m,r}(n,k)(x)_k, \tag{10}$$

respectively, to obtain remarkable formulas related to the Bernoulli and Harmonic polynomials. Notice that by suitable assignments of parameters, the previously mentioned numbers appear to be particular cases of (9) and (10). Further combinatorial and algebraic properties of these numbers were later on studied by Cheon and Jung [10] and Mező and Ramirez [11].

The motivation of this paper is partly influenced by the pair of generalizations of the numbers $\left[{n \atop k} \right]$ and $\left\{ {n \atop k} \right\}$ earlier considered by Koutras [12] which were defined by the relations

$$(t)_n = \sum_{k=0}^{n} \frac{1}{k!} \left[\frac{d^k}{dt^k} (t)_n \right]_{t=a} (t-a)^k, \tag{11}$$

$$(t-a)^n = \sum_{k=0}^{n} \frac{1}{k!} \left[\Delta^k (t-a)^n \right]_{t=0} (t)_k, \tag{12}$$

where the following notations are used:

$s_a(n,k) = (1/k!)[(d^k/dt^k)(t)_n]_{t=a}$:= noncentral Stirling numbers of the first kind,

$S_a(n,k) = (1/k!)[\Delta^k(t-a)^n]_{t=0}$:= noncentral Stirling numbers of the second kind.

Note that these numbers can be shown to be equivalent to Broder's [13] r-Stirling numbers. However, the methods by which the former were defined appear to be of distinct motivation (cf. [13, equations (3) and (4)]). Keeping this in mind, we propose "noncentral" versions for the classical Whitney numbers parallel to the work of Koutras as seen in (11) and (12). These will serve as unified generalizations of all the abovementioned sequences of special numbers. In this comprehensive study, we present fundamental combinatorial properties such as recurrence relations, generating functions and explicit formulas, and derive more results such as the orthogonality and the inverse relations, matrix decompositions, Hankel transform, and other notable identities. Several conjectures and questions are also mentioned for further research.

2. Definitions and Basic Properties

2.1. Noncentral Whitney Numbers of the First Kind. For any real numbers a, t, nonnegative integer n, and positive integer m, the expansion of $(t \mid m)_n$ in a Taylor series gives

$$(t \mid m)_n = \sum_{k=0}^{n} \frac{1}{k!} \left[\frac{d^k}{dt^k} (t \mid m)_n \right]_{t=a} (t-a)^k. \tag{13}$$

We define the noncentral Whitney numbers of the first kind, denoted by $\widetilde{w}_{m,a}(n,k)$, as

$$\widetilde{w}_{m,a}(n,k) = \frac{1}{k!} \left[\frac{d^k}{dt^k} (t \mid m)_n \right]_{t=a} \tag{14}$$

with the initial conditions $\widetilde{w}_{m,a}(0,0) = 1$ and $\widetilde{w}_{m,a}(n,k) = 0$ if $n < k$ or $n, k < 0$. Obviously, we have the next identity.

Proposition 1. *The numbers* $\widetilde{w}_{m,a}(n,k)$ *satisfy the horizontal generating function*

$$(t \mid m)_n = \sum_{k=0}^{n} \widetilde{w}_{m,a}(n,k)(t-a)^k. \tag{15}$$

Notice that (15) is equivalent to

$$m^n (x)_n = \sum_{k=0}^{n} \widetilde{w}_{m,a}(n,k)(mx-a)^k \tag{16}$$

when t is replaced with mx. Hence, we can see that from (9) and (11)

$$\widetilde{w}_{m,-r}(n,k) = w_{m,r}(n,k), \quad \widetilde{w}_{1,a}(n,k) = s_a(n,k). \tag{17}$$

Moreover,

$$\widetilde{w}_{-\alpha,0}(n,k) = \begin{bmatrix} n \\ k \end{bmatrix}^{(\alpha)}, \quad \widetilde{w}_{1,0}(n,k) = \begin{bmatrix} n \\ k \end{bmatrix}. \tag{18}$$

From (15), one has

$$\begin{aligned}\sum_{k=0}^{n+1}\widetilde{w}_{m,a}(n+1,k)(mx-a)^k &= \sum_{k=0}^{n}\widetilde{w}_{m,a}(n,k)\\ &\cdot(mx-a)^k(mx-nm)\\ &= \sum_{k=0}^{n+1}\{\widetilde{w}_{m,a}(n,k-1)+(a-nm)\widetilde{w}_{m,a}(n,k)\}\\ &\cdot(mx-a)^k.\end{aligned} \tag{19}$$

Comparing coefficients of $(mx-a)^k$ yields the next identity useful in finding the values of $\widetilde{w}_{m,a}(n,k)$.

Proposition 2. *The noncentral Whitney numbers of the first kind satisfy the following triangular recurrence relation:*

$$\begin{aligned}\widetilde{w}_{m,a}(n+1,k) &= \widetilde{w}_{m,a}(n,k-1)\\ &+(a-nm)\widetilde{w}_{m,a}(n,k).\end{aligned} \tag{20}$$

It can be seen from this recurrence relation that

$$\widetilde{w}_{m,a}(n,0) = (a\mid m)_n, \tag{21}$$

$$\widetilde{w}_{m,a}(n,k) = 1, \quad k=n. \tag{22}$$

The next corollary can be obtained by successive application of (20).

Corollary 3. *The noncentral Whitney numbers of the first kind satisfy the following recurrence relations:*

$$\begin{aligned}\widetilde{w}_{m,a}(n+1,k+1) &= \sum_{j=k}^{n}\frac{(a\mid m)_{n+1}}{(a\mid m)_{j+1}}\widetilde{w}_{m,a}(j,k),\\ \widetilde{w}_{m,a}(n,k) &= \sum_{j=0}^{n-k}(mn-a)^j\,\widetilde{w}_{m,a}(n+1,k+j+1).\end{aligned} \tag{23}$$

Note that (15) can be written as

$$m^n n!\binom{\frac{x+a}{m}}{n} = \sum_{k=0}^{n}\widetilde{w}_{m,a}(n,k)x^k \tag{24}$$

when x is replaced with $x+a$. We are now ready to state the following proposition.

Proposition 4. *The exponential generating function of the sequence $\{\widetilde{w}_{m,a}(n,k)\}$ is given by*

$$\sum_{n=k}^{\infty}\widetilde{w}_{m,a}(n,k)\frac{z^n}{n!} = (1+mz)^{a/m}\frac{[\log(1+mz)]^k}{m^k k!}. \tag{25}$$

Proof. Multiplying both sides of (24) by $z^n/n!$ and summing over n gives us

$$\sum_{n=0}^{\infty}\binom{\frac{x+a}{m}}{n}m^n z^n = \sum_{n=0}^{\infty}\sum_{k=0}^{n}\widetilde{w}_{m,a}(n,k)x^k\frac{z^n}{n!}. \tag{26}$$

Since the left-hand side is just

$$\begin{aligned}\sum_{n=0}^{(x+a)/m}\binom{\frac{x+a}{m}}{n}(mz)^n &= (1+mz)^{a/m}\sum_{k=0}^{\infty}\left(\frac{x}{m}\right)\frac{[\log(1+mz)]^k}{k!},\end{aligned} \tag{27}$$

then we have

$$\begin{aligned}\sum_{k=0}^{\infty}\left\{\sum_{n=k}^{\infty}\widetilde{w}_{m,a}(n,k)\frac{z^n}{n!}\right\}x^k &= \sum_{k=0}^{\infty}\left\{\frac{(1+mz)^{a/m}}{k!}\cdot\left[\frac{\log(1+mz)}{m}\right]^k\right\}x^k.\end{aligned} \tag{28}$$

Comparing the coefficients of x^k completes the proof. □

Theorem 5. *The noncentral Whitney numbers $\widetilde{w}_{m,a}(n,k)$ satisfy the following relations:*

$$\widetilde{w}_{m,a}(n,k) = \sum_{i=k}^{n}\binom{i}{k}a^{i-k}m^{n-i}\begin{bmatrix}n\\i\end{bmatrix}, \tag{29}$$

$$\widetilde{w}_{m,a}(n,k) = \sum_{i=k}^{n}\binom{i}{k}a^{i-k}\begin{bmatrix}n\\i\end{bmatrix}^{(m)}. \tag{30}$$

Proof. Notice that (30) is an obvious consequence of (29). Hence, we only choose to prove (29). Note that multiplying by m^n the defining relation for the Stirling numbers $\left[{n \atop k}\right]$ given by

$$(x)_n = \sum_{i=0}^{n}\begin{bmatrix}n\\i\end{bmatrix}x^i \tag{31}$$

yields

$$\begin{aligned}m^n(x)_n &= m^n\sum_{i=0}^{n}\begin{bmatrix}n\\i\end{bmatrix}\left(\frac{mx-a+a}{m}\right)^i\\ &= m^n\sum_{i=0}^{n}\begin{bmatrix}n\\i\end{bmatrix}\sum_{k=0}^{i}\binom{i}{k}a^{i-k}(mx-a)^k\frac{1}{m^i}\\ &= \sum_{k=0}^{n}\left\{\sum_{i=k}^{n}m^{n-i}\begin{bmatrix}n\\i\end{bmatrix}\binom{i}{k}a^{i-k}\right\}(mx-a)^k.\end{aligned} \tag{32}$$

Comparing the coefficients of $(mx-a)^k$ with (16) gives the desired result. □

It is known that there is no simple method in expressing *first kind Stirling-type numbers* explicitly. In the next theorem, we express the numbers $\widetilde{w}_{m,a}(n,k)$ in elementary symmetric polynomial form by induction.

Theorem 6. *The numbers $\widetilde{w}_{m,a}(n,k)$ satisfy the explicit formula*

$$\widetilde{w}_{m,a}(n,k) = \sum_{0\le i_1<i_2<\cdots<i_{n-k}\le n-1}\prod_{j=1}^{n-k}(a-i_j m). \tag{33}$$

Proof. Note that the theorem yields $\widetilde{w}_{m,a}(0,0) = 1$, which is in line with the initial value of $\widetilde{w}_{m,a}(n,k)$ stated earlier in this section. Now, suppose the theorem holds up to n for $k = 0, 1, 2, \ldots, n$. Then by (20),

$$\begin{aligned}\widetilde{w}_{m,a}(n+1,k) &= \sum_{0\le i_1<i_2<\cdots<i_{n+1-k}\le n-1} \prod_{j=1}^{n+1-k}(a-i_jm) \\ &\quad + (a-nm)\sum_{0\le i_1<i_2<\cdots<i_{n-k}\le n-1}\prod_{j=1}^{n-k}(a-i_jm) \\ &= \sum_{0\le i_1<i_2<\cdots<i_{n+1-k}\le n}\prod_{j=1}^{n+1-k}(a-i_jm).\end{aligned} \tag{34}$$

Finally (33) yields $\widetilde{w}_{m,a}(n+1,k) = 1$ when $k = n+1$. This is in accordance with (22). □

2.2. Noncentral Whitney Numbers of the Second Kind. Analogous to what is being done in (12), we define the noncentral Whitney numbers of the second kind by

$$m^k k! \widetilde{W}_{m,a}(n,k) = \left[\Delta^k (mt-a)^n\right]_{t=0}, \tag{35}$$

for any real numbers a and t, nonnegative integer n, and positive integer m. Now, let $f(x) = (mx-a)^n$. The known difference operator

$$\Delta^k f(x) = \sum_{j=0}^{k} (-1)^{k-j}\binom{k}{j} f(x+j) \tag{36}$$

yields the explicit formula

$$\left[\Delta^k (mx-a)^n\right]_{x=0} = \sum_{j=0}^{k}(-1)^{k-j}\binom{k}{j}(mj-a)^n. \tag{37}$$

Hence we propose the following combinatorial properties of the numbers $\widetilde{W}_{m,a}(n,k)$.

Proposition 7. *An explicit formula for $\widetilde{W}_{m,a}(n,k)$ is given by*

$$\widetilde{W}_{m,a}(n,k) = \frac{1}{m^k k!}\sum_{j=0}^{k}(-1)^{k-j}\binom{k}{j}(mj-a)^n. \tag{38}$$

Moreover, the exponential generating function of the sequence $\{\widetilde{W}_{m,a}(n,k)\}$ is given by

$$\sum_{n=k}^{\infty}\widetilde{W}_{m,a}(n,k)\frac{z^n}{n!} = \frac{e^{-az}}{m^k k!}\left(e^{mz}-1\right)^k \tag{39}$$

while the horizontal generating function is

$$(t-a)^n = \sum_{k=0}^{n}\widetilde{W}_{m,a}(n,k)(t\mid m)_k. \tag{40}$$

Replacing t with mx in (40) yields

$$(mx-a)^k = \sum_{k=0}^{n} m^k \widetilde{W}_{m,a}(n,k)(x)_k. \tag{41}$$

Hence,

$$\begin{aligned}\widetilde{W}_{m,-r}(n,k) &= W_{m,r}(n,k), \\ \widetilde{W}_{1,a}(n,k) &= S_a(n,k), \\ \widetilde{W}_{\alpha,0}(n,k) &= \left\{ {n \atop k} \right\}^{(\alpha)}, \\ \widetilde{W}_{1,0}(n,k) &= \left\{ {n \atop k} \right\}.\end{aligned} \tag{42}$$

The next identity is also immediately obtained.

Proposition 8. *The noncentral Whitney numbers of the second kind satisfy the following triangular recurrence relation:*

$$\begin{aligned}\widetilde{W}_{m,a}(n+1,k) &= \widetilde{W}_{m,a}(n,k-1) \\ &\quad + (km-a)\widetilde{W}_{m,a}(n,k).\end{aligned} \tag{43}$$

Obviously,

$$\begin{aligned}\widetilde{W}_{m,a}(n,0) &= (-a)^n, \\ \widetilde{W}_{m,a}(n,k) &= 1\end{aligned} \tag{44}$$

when $k = n$. Also, from (23), we deduce the following.

Corollary 9. *The noncentral Whitney numbers of the second kind satisfy the following recurrence relations:*

$$\begin{aligned}\widetilde{W}_{m,a}(n+1,k+1) &= \sum_{j=k}^{n}\left[m(k+1)-a\right]^{n-j}\widetilde{W}_{m,a}(j,k); \\ \widetilde{W}_{m,a}(n,k) &= \sum_{j=0}^{n-k}(-1)^j\frac{(-a\mid m)_{n+1}}{(-a\mid m)_{n-j+1}}\widetilde{W}_{m,a}(n+1,k+j+1).\end{aligned} \tag{45}$$

It is also possible to express $\widetilde{W}_{m,a}(n,k)$ in terms of the classical Stirling numbers of the second kind. To do so, note that

$$(mx-a)^n = \sum_{j=0}^{n}\binom{n}{j}(-a)^{n-j}m^j x^j. \tag{46}$$

Using the defining relation

$$x^n = \sum_{k=0}^{n}\left\{ {n \atop k} \right\}(x)_k \tag{47}$$

of the Stirling numbers of the second kind, we obtain

$$(mx-a)^n = \sum_{j=0}^{n}\binom{n}{j}(-a)^{n-j}m^j\sum_{k=0}^{j}\left\{{j \atop k}\right\}(x)_k = \sum_{k=0}^{n}\left\{\sum_{j=k}^{n}\binom{n}{j}(-a)^{n-j}m^j\left\{{j \atop k}\right\}\right\}(x)_k. \quad (48)$$

Comparing the coefficients of $(x)_k$ with (40) yields

$$\widetilde{W}_{m,a}(n,k) = \sum_{j=k}^{n}\binom{n}{j}(-a)^{n-j}m^{j-k}\left\{{j \atop k}\right\}. \quad (49)$$

Let us formally state this in the next theorem.

Theorem 10. *The numbers $\widetilde{W}_{m,a}(n,k)$ satisfy the following identities:*

$$\widetilde{W}_{m,a}(n,k) = \sum_{j=k}^{n}\binom{n}{j}(-a)^{n-j}m^{j-k}\left\{{j \atop k}\right\}, \quad (50)$$

$$\widetilde{W}_{m,a}(n,k) = \sum_{j=k}^{n}\binom{n}{j}(-a)^{n-j}\left\{{j \atop k}\right\}^{(m)}. \quad (51)$$

Proof. The other equality follows directly. □

The next theorem can be proved by similar method used to prove (33).

Theorem 11. *The numbers $\widetilde{W}_{m,a}(n,k)$ satisfy the explicit formula in complete symmetric polynomial form given by*

$$\widetilde{W}_{m,a}(n,k) = (-1)^{n-k}\sum_{0\le i_1\le i_2\le\cdots\le i_{n-k}\le k}\prod_{j=1}^{n-k}\left(a-i_jm\right). \quad (52)$$

2.3. Application to the Bernoulli Polynomials. The well-known *Bernoulli polynomials* $B_n(x)$ defined by the exponential generating function [9]

$$\sum_{n=0}^{\infty}B_n(x)\frac{z^n}{n!} = \frac{ze^{zx}}{e^z-1}, \quad (53)$$

where $B_n(0) = B_n$ are the *Bernoulli numbers.* In relation to this, Mező [9] obtained some identities showing interesting relationships between the r-Whitney numbers and the Bernoulli polynomials. The said identities are as follows:

$$\binom{n+1}{l}B_{n-l+1} = \frac{n+1}{m^{n-l+1}}\sum_{k=0}^{n}W_{m,r}(n,k)\frac{w_{m,r}(k+1,l)}{k+1}, \quad (54)$$

$$\binom{n+1}{l}B_{n-l+1}\left(\frac{r}{m}\right) = \frac{n+1}{m^n}\sum_{k=0}^{n}\frac{m^k}{k+1}W_{m,r}(n,k)\left[{k+1 \atop l}\right]. \quad (55)$$

Note that when $m = 1$ and $r = 0$ in both (54) and (55), we obtain the classical identity [14]

$$\binom{n+1}{l}B_{n-l+1} = (n+1)\sum_{k=0}^{n}\left\{{n \atop k}\right\}\left[{k+1 \atop l}\right]\frac{1}{k+1}. \quad (56)$$

Following the same method used by Mező [9] and through the aid of the exponential generating function in (25) and the identity

$$\sum_{n=k}^{\infty}\left[{n \atop k}\right]\frac{z^n}{n!} = \frac{\left[\log(1+z)\right]^l}{l!}, \quad (57)$$

we propose an analogous relationship between the Bernoulli polynomials and the noncentral Whitney numbers of both kinds as follows.

Proposition 12. *The noncentral translated Whitney numbers $\widetilde{W}_{m,a}(n,k)$ and $\widetilde{w}_{m,a}(k+1,l)$ satisfy the following identities:*

$$\binom{n+1}{l}B_{n-l+1} = \frac{n+1}{m^{n-l+1}}\sum_{k=0}^{n}\widetilde{W}_{m,a}(n,k)\frac{\widetilde{w}_{m,a}(k+1,l)}{k+1},$$

$$\binom{n+1}{l}B_{n-l+1}\left(-\frac{a}{m}\right) = \frac{n+1}{m^n}\sum_{k=0}^{n}\frac{m^k}{k+1}\widetilde{W}_{m,a}(n,k)\left[{k+1 \atop l}\right]. \quad (58)$$

3. Matrix Relations for the Noncentral Whitney Numbers

3.1. Orthogonality and Inverse Relations

Proposition 13. *The noncentral Whitney numbers of the first and second kind satisfy the following orthogonality relations:*

$$\sum_{k=j}^{n}\widetilde{W}_{m,a}(n,k)\widetilde{w}_{m,a}(k,j) = \sum_{k=j}^{n}\widetilde{w}_{m,a}(n,k)\widetilde{W}_{m,a}(k,j) = \delta_{nj}, \quad (59)$$

where

$$\delta_{nj} = \begin{cases}0, & if\ j\neq n\\ 1, & if\ j=n\end{cases} \quad (60)$$

is the Kronecker delta.

This proposition can be easily proved by combining (16) and (40).

Now, since $\widetilde{W}_{m,a}(n,k) = \widetilde{w}_{m,a}(n,k) = 0$ when $n < k$, then we have

$$\sum_{k=0}^{\infty} \widetilde{W}_{m,a}(n,k)\,\widetilde{w}_{m,a}(k,j) = \sum_{k=0}^{\infty} \widetilde{w}_{m,a}(n,k)\,\widetilde{W}_{m,a}(k,j) = \delta_{nj}. \tag{61}$$

If we define $\mathscr{N}_{m,a} = (\widetilde{W}_{m,a}(i,l))$ to be the infinite matrix with $\widetilde{W}_{m,a}(i,l)$ as the (i,l)th entries for $i,l = 0,1,2,3,\ldots$ and $\mathscr{M}_{m,a} = (\widetilde{w}_{m,a}(i,l))$ as similar matrix for $\widetilde{w}_{m,a}(i,l)$ then we have

$$\mathscr{N}_{m,a} \cdot \mathscr{M}_{m,a} = \mathscr{M}_{m,a} \cdot \mathscr{N}_{m,a} = \mathscr{I}, \tag{62}$$

where $\mathscr{I}$ is the infinite-dimensional identity matrix. Thus, $\mathscr{M}_{m,a} = \mathscr{N}_{m,a}^{-1}$, where $\mathscr{N}_{m,a}^{-1}$ is the inverse of $\mathscr{N}_{m,a}$.

Corollary 14. *The following inverse relations hold:*

$$\begin{aligned} f_n &= \sum_{k=0}^{n} \widetilde{w}_{m,a}(n,k)\, g_k \Longleftrightarrow \\ g_n &= \sum_{k=0}^{n} \widetilde{W}_{m,a}(n,k)\, f_k, \\ f_k &= \sum_{n=k}^{\infty} \widetilde{w}_{m,a}(n,k)\, g_n \Longleftrightarrow \\ g_k &= \sum_{n=k}^{\infty} \widetilde{W}_{m,a}(n,k)\, f_n. \end{aligned} \tag{63}$$

3.2. Matrix Decomposition of the Noncentral Whitney Numbers. In a recent paper, Pan [15] introduced a remarkable matrix decomposition to give an explicit and nonrecursive way of computing the Unified Generalized Stirling numbers of Hsu and Shiue [16]. The said result is as follows [15, Theorem 7]:

$$\mathscr{S}_{\alpha,\beta,\gamma} = \mathscr{S}_{\alpha,0,0} \cdot \mathscr{S}_{0,0,\gamma} \cdot \mathscr{S}_{0,\beta,0}. \tag{64}$$

Here, the matrix $\mathscr{S}_{\alpha,\beta,\gamma}$ is given by

$$\mathscr{S}_{\alpha,\beta,\gamma} = \left(S(n,k;\alpha,\beta,\gamma)\right), \tag{65}$$

where $S(n,k;\alpha,\beta,\gamma)$ is Hsu and Shiue's [16] generalized Stirling numbers defined by

$$(x \mid \alpha)_n = \sum_{k=0}^{\infty} S(n,k;\alpha,\beta,\gamma)\,(x-\gamma \mid \beta)_k. \tag{66}$$

It can be verified that the (n,k)th entries of the matrices $\mathscr{S}_{\alpha,0,0}$, $\mathscr{S}_{0,\beta,0}$, and $\mathscr{S}_{0,0,\gamma}$ are

$$\begin{aligned} &\alpha^{n-k}\begin{bmatrix} n \\ k \end{bmatrix}, \\ &\beta^{n-k}\begin{Bmatrix} n \\ k \end{Bmatrix}, \\ &\gamma^{n-k}\binom{n}{k}, \end{aligned} \tag{67}$$

respectively.

Although the noncentral Whitney numbers can be written as

$$\begin{aligned} S(n,k;-a,m,0) &= \widetilde{w}_{m,a}(n,k), \\ S(n,k;0,m,-a) &= \widetilde{W}_{m,a}(n,k), \end{aligned} \tag{68}$$

it is not wise to assume that

$$\begin{aligned} \mathscr{M}_{m,a} &:= \mathscr{S}_{-a,m,0} = \mathscr{S}_{-a,0,0} \cdot \mathscr{S}_{0,0,0} \cdot \mathscr{S}_{0,m,0}, \\ \mathscr{N}_{m,a} &:= \mathscr{S}_{0,m,m} = \mathscr{S}_{0,0,0} \cdot \mathscr{S}_{0,0,-a} \cdot \mathscr{S}_{0,m,0}. \end{aligned} \tag{69}$$

Hence, it is justifiable to establish the matrix decomposition for the noncentral Whitney numbers of both kinds. For convenience, we will refer to the matrices $\mathscr{M}_{m,a}$ and $\mathscr{N}_{m,a}$ as *noncentral Whitney matrix of the first and second kind*, respectively. Also, we let

$$\mathscr{V}_{\alpha}(x) = \left(1, x, (x \mid \alpha)_2, (x \mid \alpha)_3, \ldots, (x \mid \alpha)_n, \ldots\right)^T \tag{70}$$

be an infinite column vector.

Remark 15. In reference to the relations in (15) and (40), the following identities seem natural:

$$\mathscr{V}_m(x) = \mathscr{M}_{m,a} \cdot \mathscr{V}_0(x-a), \tag{71}$$

$$\mathscr{V}_0(x-a) = \mathscr{N}_{m,a} \cdot \mathscr{V}_m(x). \tag{72}$$

Since $\widetilde{w}_{m,0}(n,k) = \left[{n \atop k}\right]^{(-m)}$ and $\widetilde{W}_{m,0}(n,k) = \left\{{n \atop k}\right\}^{(m)}$, then we have the following matrices:

$$\begin{aligned} \mathscr{M}_{m,0} &= \left(\begin{bmatrix} n \\ k \end{bmatrix}^{(-m)}\right), \\ \mathscr{N}_{m,0} &= \left(\begin{Bmatrix} n \\ k \end{Bmatrix}^{(m)}\right). \end{aligned} \tag{73}$$

By using the "signed" translated Whitney numbers $w_{(m)}^{*}(n,k)$ [7], the matrix $\mathscr{M}_{m,0}$ can also be rewritten as

$$\mathscr{M}_{m,0} = \left(w_{(m)}^{*}(n,k)\right). \tag{74}$$

Now, from (15),

$$0 = \sum_{k=0}^{n} \widetilde{w}_{0,a}(n,k)\,(-a)^k \tag{75}$$

while the binomial theorem yields

$$0 = (-a+a)^n = \sum_{k=0}^{n} \binom{n}{k} a^{n-k} (-a)^k. \tag{76}$$

Thus, we have

$$\widetilde{w}_{0,a}(n,k) = a^{n-k}\binom{n}{k}. \tag{77}$$

It can be shown using a similar manner that

$$\widetilde{W}_{0,a}(n,k) = (-a)^{n-k}\binom{n}{k}. \tag{78}$$

It then follows that

$$\mathcal{M}_{0,a} = \left(a^{n-k}\binom{n}{k}\right),$$
$$\mathcal{N}_{0,a} = \left((-a)^{n-k}\binom{n}{k}\right). \tag{79}$$

We are now ready to state the next theorem.

Theorem 16. *The decomposition formulas of the matrices $\mathcal{M}_{m,a}$ and $\mathcal{N}_{m,a}$ are*

$$\mathcal{M}_{m,a} = \mathcal{M}_{m,0}\cdot\mathcal{M}_{0,a}, \tag{80}$$

$$\mathcal{N}_{m,a} = \mathcal{N}_{0,a}\cdot\mathcal{N}_{m,0}. \tag{81}$$

Proof. Since

$$\mathcal{V}_m(x) = \mathcal{M}_{m,0}\cdot\mathcal{V}_0(x),$$
$$\mathcal{V}_0(x) = \mathcal{M}_{0,a}\cdot\mathcal{V}_0(x-a) \tag{82}$$

then

$$\mathcal{V}_m(x) = \mathcal{M}_{m,0}\mathcal{M}_{0,a}\cdot\mathcal{V}_0(x-a). \tag{83}$$

Combining this with (71) gives us

$$(\mathcal{M}_{m,a} - \mathcal{M}_{m,0}\cdot\mathcal{M}_{0,a})\mathcal{V}_0(x-a) = \mathbf{0}, \tag{84}$$

where $\mathbf{0}$ denotes an infinite-dimensional zero matrix. Consequently, because x is an arbitrary real or complex number and $\mathcal{V}_0(x-a)$ is a nonzero vector, then

$$\mathcal{M}_{m,a} = \mathcal{M}_{m,0}\cdot\mathcal{M}_{0,a}. \tag{85}$$

Equation (81) can be shown similarly. □

4. Noncentral Dowling and Noncentral Tanny-Dowling Polynomials

Benoumhani [2, 3] was the first to introduce the following familiar polynomials:

$$D_m(n;x) = \sum_{k=0}^{n} W_m(n,k)\,x^k, \tag{86}$$

$$\mathcal{F}_m(n;x) = \sum_{k=0}^{n} k!W_m(n,k)\,x^k. \tag{87}$$

$D_m(n;x)$ and $\mathcal{F}_m(n;x)$ are known as the Dowling and the Tanny-Dowling polynomials. Moreover, when $m = 1$ in (87), the resulting polynomial

$$\mathcal{F}_1(n;x) = \sum_{k=0}^{n} k!\begin{Bmatrix} n \\ k \end{Bmatrix} x^k \tag{88}$$

is called geometric polynomials [17] and was earlier studied by Tanny [18].

Denoted by $\widetilde{\mathcal{D}}_{m,a}(n;x)$ and $\widetilde{\mathcal{F}}_{m,a}(n;x)$, the noncentral Dowling and the noncentral Tanny-Dowling polynomials can be defined as

$$\widetilde{\mathcal{D}}_{m,a}(n;x) = \sum_{k=0}^{n} \widetilde{W}_{m,a}(n,k)\,x^k, \tag{89}$$

$$\widetilde{\mathcal{F}}_{m,a}(n;x) = \sum_{k=0}^{n} k!\widetilde{W}_{m,a}(n,k)\,x^k. \tag{90}$$

For brevity, we also call $\widetilde{\mathcal{D}}_{m,a}(n;1) \equiv \widetilde{\mathcal{D}}_{m,a}(n)$ and $\widetilde{\mathcal{F}}_{m,a}(n;1) \equiv \widetilde{\mathcal{F}}_{m,a}(n)$ as the noncentral Dowling and the noncentral Tanny-Dowling numbers, respectively. Notice that through the use of the exponential generating function in (39) and the explicit formula (38), the noncentral Dowling polynomials can be defined alternatively by (91) or explicitly by the Dobinski-type identity (92) in the following proposition.

Proposition 17. *The following identities hold:*

$$\sum_{n=0}^{\infty} \widetilde{\mathcal{D}}_{m,a}(n;x)\frac{z^n}{n!} = e^{-az+(e^{mz}-1)(x/m)}, \tag{91}$$

$$\widetilde{\mathcal{D}}_{m,a}(n;x) = e^{-x/m}\sum_{i=0}^{\infty}\left(\frac{x}{m}\right)^i\frac{(mi-a)^n}{i!}. \tag{92}$$

These identities are actually equivalent to those of the r-Whitney polynomials when $a = -r$ and are generalizations of the translated Dowling polynomials (cf. [6, equations (22) and (25)]). As for the noncentral Tanny-Dowling polynomials, since (39) can be rewritten as

$$\sum_{n=k}^{\infty} k!\widetilde{W}_{m,a}(n,k)\frac{z^n}{n!} = e^{-az}\left(\frac{e^{mz}-1}{m}\right)^k, \tag{93}$$

then we get

$$\begin{aligned}\sum_{n=k}^{\infty}\widetilde{\mathcal{F}}_{m,a}(n;x)\frac{z^n}{n!} &= \sum_{n=0}^{\infty}\sum_{k=0}^{n} k!\widetilde{W}_{m,a}(n,k)\frac{z^n}{n!} \\ &= \sum_{k=0}^{\infty} e^{-az}\left(\frac{e^{mz}-1}{m}\right)^k \\ &= e^{-az}\left(\frac{1}{1-(e^{mz}-1)x/m}\right).\end{aligned} \tag{94}$$

This is equivalent to (95) in the next theorem.

Theorem 18. *The polynomials $\widetilde{\mathcal{F}}_{m,a}(n;x)$ satisfy the exponential generating function*

$$\sum_{n=k}^{\infty}\widetilde{\mathcal{F}}_{m,a}(n;x)\frac{z^n}{n!} = \frac{me^{-az}}{m-x(e^{mz}-1)}. \tag{95}$$

The case in (95), where $x = my$, immediately yields

$$\sum_{n=k}^{\infty}\widetilde{\mathcal{F}}_{m,a}(n;my)\frac{z^n}{n!} = \frac{e^{-az}}{1-y(e^{mz}-1)}. \tag{96}$$

Equations (95) and (96) reduce to Benoumhani's results (cf. [3, Theorem 3]) when $a = -1$. The next theorem presents the explicit formula for the polynomials $\widetilde{\mathscr{F}}_{m,a}(n; x)$.

Theorem 19. *The following explicit formula holds:*

$$\widetilde{\mathscr{F}}_{m,a}(n; x) = \frac{m}{x+m}\sum_{k=0}^{\infty}\left(\frac{x}{x+m}\right)^k (mk-a)^n. \quad (97)$$

Proof. To prove this theorem, we first show that

$$\widetilde{\mathscr{F}}_{m,a}(n; my) = \frac{1}{1+y}\sum_{k=0}^{\infty}\left(\frac{y}{1+y}\right)^k (mk-a)^n. \quad (98)$$

Note that by algebraic manipulation, one readily gets

$$\begin{aligned}\sum_{n=0}^{\infty}&\left(\frac{1}{1+y}\sum_{k=0}^{\infty}\left(\frac{y}{1+y}\right)^k (mk-a)^n\right)\frac{z^n}{n!}\\ &=\sum_{n=0}^{\infty}\left(\frac{1}{1+y}\right.\\ &\quad\left.\cdot\sum_{k=0}^{\infty}\left(\frac{y}{1+y}\right)^k\cdot\sum_{i=0}^{n}\binom{n}{i}(km)^{n-i}(-a)^i\right)\frac{z^n}{n!}\\ &=\frac{1}{1+y}\sum_{i=0}^{\infty}\frac{(-az)^i}{i!}\sum_{k=0}^{\infty}\left(\frac{y}{1+y}\right)^k\cdot\sum_{n=i}^{\infty}\frac{(km)^{n-i}}{(n-i)!}z^{n-i}.\end{aligned} \quad (99)$$

Reindexing the third sum and using (96), we get

$$\begin{aligned}&\frac{1}{1+y}\sum_{i=0}^{\infty}\frac{(-az)^i}{i!}\sum_{k=0}^{\infty}\left(\frac{y}{1+y}\right)^k\sum_{n=i}^{\infty}\frac{(kmz)^{n-i}}{(n-i)!}\\ &\quad=\frac{1}{1+y}e^{-az}\sum_{k=0}^{\infty}\left(\frac{y}{1+y}\right)^k\sum_{\ell=0}^{\infty}\frac{(kmz)^{\ell}}{\ell!}\\ &\quad=\frac{1}{1+y}e^{-az}\sum_{k=0}^{\infty}\left(\frac{y}{1+y}\right)^k e^{kmz}\\ &\quad=\frac{1}{1+y}\cdot\frac{e^{-az}}{1-(y/(1+y))e^{mz}}\\ &\quad=\sum_{n=0}^{\infty}\widetilde{\mathscr{F}}_{m,a}(n; my)\frac{z^n}{n!}.\end{aligned} \quad (100)$$

Comparing the coefficients of z^n yields (98). The proof is then completed when my is replaced with x in (98). □

Identity (98) used in the proof of the previous theorem is a generalization of Benoumhani's [3, Theorem 4]. On the other hand, when $x = 1$,

$$\widetilde{\mathscr{F}}_{m,a}(n) = \sum_{k=0}^{\infty}\frac{1}{2^{k+1}}(mk-a)^n. \quad (101)$$

Moreover, we get the familiar representation of $\mathscr{F}_m(n; 1)$ due to Rota [19] given by

$$\widetilde{\mathscr{F}}_{m,-1}(n; 1) = \sum_{k=0}^{\infty}\frac{1}{2^{k+1}}(mk+1)^n \quad (102)$$

when $a = -1$ in (101).

It is well-known that the binomial coefficients $\binom{n}{k}$ satisfy the binomial inversion formula

$$\begin{aligned}f_k &= \sum_{j=0}^{k}\binom{k}{j}g_j \Longleftrightarrow\\ g_k &= \sum_{j=0}^{k}(-1)^{k-j}\binom{k}{j}f_j.\end{aligned} \quad (103)$$

The rest of this section contains corollaries which are obtained through the use of this.

Theorem 20. *The noncentral Whitney numbers of the second kind satisfy the following recursion formula:*

$$\widetilde{W}_{m,a+1}(n,k) = \sum_{j=0}^{n}(-1)^{n-j}\binom{n}{j}\widetilde{W}_{m,a}(j,k). \quad (104)$$

Proof. Using the explicit formula (38) gives us

$$\begin{aligned}\widetilde{W}_{m,a+1}(n,k) &= \frac{1}{m^k k!}\sum_{i=0}^{k}(-1)^{k-i}\binom{k}{i}(mi-a-1)^n\\ &=\frac{1}{m^k k!}\sum_{i=0}^{k}(-1)^{k-i}\binom{k}{i}\sum_{j=0}^{n}\binom{n}{j}(mi-a)^{n-j}(-1)^j\\ &=\sum_{j=0}^{n}(-1)^j\binom{n}{j}\\ &\quad\cdot\frac{1}{m^k k!}\sum_{i=0}^{k}\binom{k}{i}(-1)^{k-i}(mi-a)^{n-j}\\ &=\sum_{j=0}^{n}(-1)^j\binom{n}{j}\widetilde{W}_{m,a}(n-j,k).\end{aligned} \quad (105)$$

Reindexing the summation yields (104). □

Applying the abovementioned binomial inversion formula for $g_k = \widetilde{W}_{m,a+1}(n,k)$ and $f_j = \widetilde{W}_{m,a}(j,k)$, we get the following corollary.

Corollary 21. *The noncentral Whitney numbers of the second kind satisfy the relation given by*

$$\widetilde{W}_{m,a}(n,k) = \sum_{j=0}^{n}\binom{n}{j}\widetilde{W}_{m,a+1}(j,k). \quad (106)$$

Using (104), we have

$$\begin{aligned}&\sum_{k=0}^{n}\widetilde{W}_{m,a+1}(n,k)x^k\\ &\quad=\sum_{k=0}^{n}\sum_{j=0}^{n}(-1)^j\binom{n}{j}\widetilde{W}_{m,a}(n-j,k)x^k\\ &\quad=\sum_{j=0}^{n}(-1)^{n-j}\binom{n}{j}\sum_{k=0}^{n}\widetilde{W}_{m,a}(j,k)x^k.\end{aligned} \quad (107)$$

Similarly, we have

$$\begin{aligned}\sum_{k=0}^{n} k!\widetilde{W}_{m,a+1}(n,k)\,x^k &= \sum_{k=0}^{n}\sum_{j=0}^{n}(-1)^j\binom{n}{j}k!\widetilde{W}_{m,a}(n-j,k)\,x^k \\ &= \sum_{j=0}^{n}(-1)^{n-j}\binom{n}{j}\sum_{k=0}^{n}k!\widetilde{W}_{m,a}(j,k)\,x^k.\end{aligned}\tag{108}$$

Thus, we have the following theorem.

Theorem 22. *The noncentral Dowling and the noncentral Tanny-Dowling polynomials satisfy the following recursions:*

$$\widetilde{\mathscr{D}}_{m,a+1}(n;x) = \sum_{j=0}^{n}(-1)^{n-j}\binom{n}{j}\widetilde{\mathscr{D}}_{m,a}(j;x),\tag{109}$$

$$\widetilde{\mathscr{F}}_{m,a+1}(n;x) = \sum_{j=0}^{n}(-1)^{n-j}\binom{n}{j}\widetilde{\mathscr{F}}_{m,a}(j;x).\tag{110}$$

Consequently, we have the following.

Corollary 23. *The noncentral Dowling and Tanny-Dowling polynomials satisfy the relations given by*

$$\widetilde{\mathscr{D}}_{m,a}(n;x) = \sum_{j=0}^{n}\binom{n}{j}\widetilde{\mathscr{D}}_{m,a+1}(j;x),\tag{111}$$

$$\widetilde{\mathscr{F}}_{m,a}(n;x) = \sum_{j=0}^{n}\binom{n}{j}\widetilde{\mathscr{F}}_{m,a+1}(j;x).\tag{112}$$

Theorem 24. *For nonnegative real number m, the noncentral Whitney numbers of the second kind satisfy*

$$\widetilde{W}_{m+1,a}(n,k) = \frac{1}{(m+1)^k m^{n-k}}\sum_{j=0}^{n}\binom{n}{j}a^{n-j}(m+1)^j\widetilde{W}_{m,a}(j,k).\tag{113}$$

Proof. We can rewrite the explicit formula (38) as

$$\begin{aligned}\widetilde{W}_{m+1,a}(n,k) &= \frac{(m+1)^{n-k}}{k!}\cdot\sum_{i=0}^{k}(-1)^{k-i}\binom{k}{i}\left(i-\frac{a}{m+1}\right)^n = \frac{(m+1)^{n-k}}{k!}\\ &\cdot\sum_{i=0}^{k}(-1)^{k-i}\binom{k}{i}\left(i-\frac{a}{m}+\frac{a}{m(m+1)}\right)^n\\ &= \frac{(m+1)^{n-k}}{k!}\sum_{i=0}^{k}(-1)^{k-i}\binom{k}{i}\sum_{j=0}^{n}\binom{n}{j}\left(i-\frac{a}{m}\right)^{n-j}\\ &\cdot\left(\frac{a}{m(m+1)}\right)^j = (m+1)^n\\ &\cdot\sum_{j=0}^{n}\binom{n}{j}\frac{m^k}{(m+1)^k m^{n-j}}\left(\frac{a}{m(m+1)}\right)^j\cdot\frac{1}{m^k k!}\\ &\cdot\sum_{i=0}^{k}\binom{k}{i}(-1)^{k-i}(mi-a)^{n-j} = \frac{1}{(m+1)^k m^{n-k}}\\ &\cdot\sum_{j=0}^{n}\binom{n}{j}a^j(m+1)^{n-j}\widetilde{W}_{m,a}(n-j,k).\end{aligned}\tag{114}$$

Reindexing the summation yields (113). □

The next corollary is easily obtained by applying binomial inversion formula to (113).

Corollary 25. *The noncentral Whitney numbers of the second kind satisfy the relation given by*

$$\widetilde{W}_{m,a}(n,k) = \frac{1}{(m+1)^{n-k}}\sum_{j=0}^{n}\binom{n}{j}(-a)^{n-j}(m)^{j-k}\widetilde{W}_{m+1,a}(j,k).\tag{115}$$

Theorem 26. *The noncentral Dowling and the noncentral Tanny-Dowling polynomials satisfy the following recursions:*

$$\begin{aligned}\widetilde{\mathscr{D}}_{m+1,a}(n;x) &= \frac{1}{m^n}\sum_{j=0}^{n}a^{n-j}\binom{n}{j}(m+1)^j\widetilde{\mathscr{D}}_{m,a}\left(j;\frac{m}{m+1}x\right),\\ \widetilde{\mathscr{F}}_{m+1,a}(n;x) &= \frac{1}{m^n}\sum_{j=0}^{n}a^{n-j}\binom{n}{j}(m+1)^j\widetilde{\mathscr{F}}_{m,a}\left(j;\frac{m}{m+1}x\right).\end{aligned}\tag{116}$$

Proof. Combining (89) and (113) gives us

$$\begin{aligned}\widetilde{\mathscr{D}}_{m+1,a}(n;x) &= \sum_{k=0}^{n}\widetilde{W}_{m+1,a}(n,k)\,x^k\\ &= \sum_{k=0}^{n}\frac{1}{(m+1)^k m^{n-k}}\\ &\cdot\sum_{j=0}^{n}\binom{n}{j}a^j(m+1)^{n-j}\widetilde{W}_{m,a}(n-j,k)\,x^k = \frac{1}{m^n}\\ &\cdot\sum_{j=0}^{n}a^j\binom{n}{j}(m+1)^{n-j}\sum_{k=0}^{n}\widetilde{W}_{m,a}(n-j,k)\\ &\cdot\left(\frac{mx}{m+1}\right)^k = \frac{1}{m^n}\sum_{j=0}^{n}a^{n-j}\binom{n}{j}(m+1)^j\end{aligned}$$

$$\cdot \sum_{k=0}^{n} \widetilde{W}_{m,a}(j,k)\left(\frac{mx}{m+1}\right)^{k} = \frac{1}{m^{n}} \cdot \sum_{j=0}^{n} a^{n-j}\binom{n}{j}(m+1)^{j}\widetilde{\mathscr{D}}_{m,a}\left(j, \frac{mx}{m+1}\right). \tag{117}$$

Similarly, (90) and (113) give us

$$\begin{aligned}\widetilde{\mathscr{F}}_{m+1,a}(n;x) &= \sum_{k=0}^{n} k!\widetilde{W}_{m+1,a}(n,k)x^{k} \\ &= \sum_{k=0}^{n} \frac{k!}{(m+1)^{k} m^{n-k}} \sum_{j=0}^{n}\binom{n}{j} a^{j}(m+1)^{n-j}\widetilde{W}_{m,a} \\ &\cdot (n-j,k)x^{k} = \frac{1}{m^{n}}\sum_{j=0}^{n} a^{j}\binom{n}{j}(m+1)^{n-j} \\ &\cdot \sum_{k=0}^{n} k!\widetilde{W}_{m,a}(n-j,k)\left(\frac{mx}{m+1}\right)^{k} = \frac{1}{m^{n}} \\ &\cdot \sum_{j=0}^{n} a^{n-j}\binom{n}{j}(m+1)^{j}\sum_{k=0}^{n} k!\widetilde{W}_{m,a}(j,k)\left(\frac{mx}{m+1}\right)^{k} \\ &= \frac{1}{m^{n}}\sum_{j=0}^{n} a^{n-j}\binom{n}{j}(m+1)^{j}\widetilde{\mathscr{F}}_{m,a}\left(j, \frac{mx}{m+1}\right).\end{aligned} \tag{118}$$

□

Remark 27. When $a = -r$, we get

$$\begin{aligned}D_{m+1,r}(n;x) &= \frac{1}{m^{n}}\sum_{j=0}^{n}(-r)^{n-j}\binom{n}{j}(m+1)^{j}D_{m,r}\left(j;\frac{m}{m+1}x\right), \\ \widetilde{\mathscr{F}}_{m+1,-r}(n;x) &= \frac{1}{m^{n}}\sum_{j=0}^{n}(-r)^{n-j}\binom{n}{j}(m+1)^{j}\widetilde{\mathscr{F}}_{m,-r}\left(j;\frac{m}{m+1}x\right),\end{aligned} \tag{119}$$

where $D_{m,r}(n;x)$ are the r-Dowling polynomials [10, 17]. Moreover, the case where $a = -1$ yields [17, Theorems 2 and 3]

$$\begin{aligned}D_{m+1}(n;x) &= \frac{1}{m^{n}}\sum_{j=0}^{n}(-1)^{n-j}\binom{n}{j}(m+1)^{j}D_{m}\left(j;\frac{m}{m+1}x\right), \\ \mathscr{F}_{m+1}(n;x) &= \frac{1}{m^{n}}\sum_{j=0}^{n}(-r)^{n-j}\binom{n}{j}(m+1)^{j}\mathscr{F}_{m}\left(j;\frac{m}{m+1}x\right).\end{aligned} \tag{120}$$

The binomial inversion formula readily yields the following.

Corollary 28. *The noncentral Dowling and Tanny-Dowling polynomials satisfy the relations given by*

$$\begin{aligned}\widetilde{\mathscr{D}}_{m,a}\left(n;\frac{m}{m+1}x\right) &= \frac{1}{(m+1)^{n}}\sum_{j=0}^{n}(-a)^{n-j}\binom{n}{j}m^{j}\widetilde{\mathscr{D}}_{m+1,a}(j;x), \\ \widetilde{\mathscr{F}}_{m,a}\left(n;\frac{m}{m+1}x\right) &= \frac{1}{(m+1)^{n}}\sum_{j=0}^{n}(-a)^{n-j}\binom{n}{j}m^{j}\widetilde{\mathscr{F}}_{m+1,a}(j;x).\end{aligned} \tag{121}$$

The next theorem and corollary can be obtained by similar method as the previous ones. The proof is left as exercise.

Theorem 29. *The following recursion formulas hold:*

$$\begin{aligned}\widetilde{W}_{m+1,a+1}(n,k) &= \frac{1}{(m+1)^{k} m^{n-k}}\sum_{j=0}^{n}\binom{n}{j}(a-m)^{n-j} \\ &\cdot (m+1)^{j}\widetilde{W}_{m,a}(j,k), \\ \widetilde{\mathscr{D}}_{m+1,a+1}(n;x) &= \frac{1}{m^{n}}\sum_{j=0}^{n}(a-m)^{n-j}\binom{n}{j}(m+1)^{j} \\ &\cdot \widetilde{\mathscr{D}}_{m,a}\left(j;\frac{m}{m+1}x\right), \\ \widetilde{\mathscr{F}}_{m+1,a+1}(n;x) &= \frac{1}{m^{n}}\sum_{j=0}^{n}(a-m)^{n-j}\binom{n}{j}(m+1)^{j} \\ &\cdot \widetilde{\mathscr{F}}_{m,a}\left(j;\frac{m}{m+1}x\right).\end{aligned} \tag{122}$$

Corollary 30. *The following recursion formulas hold:*

$$\begin{aligned}\widetilde{W}_{m,a}(n,k) &= \frac{1}{(m+1)^{n-k} m^{k}}\sum_{j=0}^{n}\binom{n}{j}(m-a)^{n-j} \\ &\cdot m^{j}\widetilde{W}_{m+1,a+1}(j,k), \\ \widetilde{\mathscr{D}}_{m,a}\left(n;\frac{m}{m+1}x\right) &= \frac{1}{(m+1)^{n}}\sum_{j=0}^{n}(m-a)^{n-j} \\ &\cdot \binom{n}{j}m^{j}\widetilde{\mathscr{D}}_{m+1,a+1}(j;x), \\ \widetilde{\mathscr{F}}_{m,a}\left(n;\frac{m}{m+1}x\right) &= \frac{1}{(m+1)^{n}}\sum_{j=0}^{n}(m-a)^{n-j} \\ &\cdot \binom{n}{j}m^{j}\widetilde{\mathscr{F}}_{m+1,a+1}(j;x).\end{aligned} \tag{123}$$

The nth Bell polynomial

$$\phi_{n}(x) = \sum_{k=0}^{n}\begin{Bmatrix} n \\ k \end{Bmatrix}x^{k} \tag{124}$$

is known to satisfy the explicit formula

$$\phi_n(x) = \left(\frac{1}{e}\right)^x \sum_{i=0}^{\infty} \frac{i^n}{i!} x^i. \quad (125)$$

Now, from (50), we have

$$\begin{aligned}\widetilde{\mathscr{D}}_{m,a}(n;x) &= \sum_{k=0}^{n}\sum_{j=0}^{n}\binom{n}{j}(-a)^{n-j} m^{j-k}\left\{ {j \atop k} \right\} x^k \\ &= \sum_{j=0}^{n}\binom{n}{j}(-a)^{n-j} m^{j}\sum_{k=0}^{n}\left\{ {j \atop k} \right\}\left(\frac{x}{m}\right)^k.\end{aligned} \quad (126)$$

Thus we have the next theorem.

Theorem 31. *The noncentral Dowling polynomials satisfy*

$$\widetilde{\mathscr{D}}_{m,a}(n;x) = \sum_{j=0}^{n}(-a)^{n-j}\binom{n}{j} m^j \phi_j\left(\frac{x}{m}\right). \quad (127)$$

Corollary 32. *The Bell polynomials satisfy the following identity:*

$$\phi_n\left(\frac{x}{m}\right) = \frac{1}{m^n}\sum_{j=0}^{n} a^{n-j}\binom{n}{j}\widetilde{\mathscr{D}}_{m,a}(j;x). \quad (128)$$

The case where $a = -1$ in (127) and (128) is due to Rahmani (cf. [17, Theorem 4 and Corollary 1]). Obviously, $\widetilde{\mathscr{D}}_{m,a}(n;x)$ is the binomial transform of $a^{n-j}m^j\phi_j(x/m)$. The curious identity

$$\begin{aligned}&\sum_{k=0}^{\ell}\binom{\ell}{k}\binom{n+k}{s}\alpha_{n+k-s} \\ &= \sum_{k=0}^{\ell}\binom{n}{k}\binom{\ell+k}{s}(-1)^{n-k}\beta_{\ell+k-s}\end{aligned} \quad (129)$$

is due to Chen [20, Theorem 3.2]. When $\ell = s = n$, and if we let $\alpha_k = a^{n-k}m^k\phi_k(x/m)$ and $\beta_k = \widetilde{\mathscr{D}}_{m,a}(k;x)$, then we have the following.

Corollary 33. *The noncentral Dowling polynomials and the Bell polynomials can be related as follows:*

$$\begin{aligned}&\sum_{k=0}^{n}\binom{n}{k}\binom{n+k}{s}a^{n-k}m^k\phi_k\left(\frac{x}{m}\right) \\ &= \sum_{k=0}^{n}\binom{n}{k}\binom{n+k}{s}(-1)^{n-k}\widetilde{\mathscr{D}}_{m,a}(k;x).\end{aligned} \quad (130)$$

5. The Hankel Transform of Noncentral Dowling Numbers

Hankel matrices had been studied by several mathematicians because of their connections in some areas of mathematics, physics, and computer science. Further theories and applications of these matrices have been established including the Hankel determinant and Hankel transform. The Hankel transform was first introduced in Sloane's sequence *A*055878 [21] and was later on studied by Layman [22]. Layman [22] first defined the Hankel transform of an integer sequence as the sequence of Hankel determinants of order n of a given sequence. Among the remarkable properties established by Layman [22] is the property that any integer sequence has the same Hankel transform as its binomial transform, as well as its invert transform. In this section, we thoroughly investigate the Hankel transform of the noncentral Dowling numbers using this property.

Let $\Gamma = (b_{n,k})$ be the infinite lower triangular matrix defined recursively by

$$b_{n,k} = b_{n-k,k-1} + (mk+1)\,b_{n-1,k} + m(k+1)\,b_{n-1,k+1}, \quad (131)$$

where $n \geq 1$, $b_{0,0} = 1$, $b_{0,k} = 0$ if $k > 0$ and $b_{n,k} = 0$ if $n < k$.

The next proposition shows that (131) is a recurrence relation of the infinite lower triangular matrix $\Gamma = (b_{n,k})$, where the entries in the 0-column are the numbers $\widetilde{D}_{m,0}(n)$.

Proposition 34. *Let $\Phi_k(z)$ be the exponential generating function of the kth column of matrix Γ. Then*

$$\Phi_k(z) = e^{m^{-1}(e^{mz}-1)}\frac{(e^{mz}-1)^k}{m^k k!}, \quad (132)$$

where $k \geq 0$ and the 0-column entries of Γ are the numbers $\widetilde{D}_{m,0}(n)$.

To obtain the Hankel transform of the noncentral Dowling numbers, the next lemma which may be proved by induction is essential.

Lemma 35. *Let c_n be the nth row of $\Gamma = (b_{n,k})$. Define*

$$c_n \circ c_p = \sum_{k\geq 0} b_{n,k}b_{p,k}m^k k!. \quad (133)$$

Then for all nonnegative integers n and p

$$c_n \circ c_p = b_{n+p,0} = \widetilde{D}_{m,0}(n+p). \quad (134)$$

Before stating the next theorem, we first let

$$\mathfrak{D}_{m,a} = \begin{pmatrix} \widetilde{\mathscr{D}}_{m,a}(0) & \widetilde{\mathscr{D}}_{m,a}(1) & \cdots & \widetilde{\mathscr{D}}_{m,a}(n) \\ \widetilde{\mathscr{D}}_{m,a}(1) & \widetilde{\mathscr{D}}_{m,a}(2) & \cdots & \widetilde{\mathscr{D}}_{m,a}(n+1) \\ \vdots & \vdots & \cdots & \vdots \\ \widetilde{\mathscr{D}}_{m,a}(n) & \widetilde{\mathscr{D}}_{m,a}(n+1) & \cdots & \widetilde{\mathscr{D}}_{m,a}(2n) \end{pmatrix} \quad (135)$$

and $H\{\mathfrak{D}_{m,a}\}$ be the Hankel transform of the numbers $\widetilde{\mathscr{D}}_{m,a}(n)$.

Theorem 36. *The Hankel transform of the numbers $\widetilde{\mathscr{D}}_{m,0}(n)$ is given by*

$$H\{\mathfrak{D}_{m,0}\} = m^{\binom{n+1}{2}}\prod_{j\geq 0} j!. \quad (136)$$

Proof. Suppose Γ_n is the lower triangular submatrix of Γ consisting of the rows and columns numbered from 0 to n. Then $\det \Gamma_n^T = 1$. Let $\overline{\Gamma}_n = (m^j j! b_{i,j})_{0\le i,i\le n}$. It implies that

$$\det \overline{\Gamma}_n = \prod_{j=0}^{n} m^j j!. \tag{137}$$

On the other hand, by (134), we have $\overline{\Gamma}_n \cdot \Gamma_n^T = (c_{i,j})_{0\le i,j\le n}$, where

$$c_{i,j} = \sum_{k\ge 0} b_{i,j} b_{j,k} m^j j! = b_{i+j,0} = \widetilde{D}_{m,0}(i+j). \tag{138}$$

That is, $\overline{\Gamma}_n \cdot \Gamma_n^T = (\widetilde{D}_{m,0}(i+j))_{0\le i,j\le n}$. Thus, we have

$$\begin{aligned} \det\left(\overline{\Gamma}_n \cdot \Gamma_n^T\right) &= \left(\det \overline{\Gamma}_n\right)\left(\det \Gamma_n^T\right) = \prod_{j=0}^{n} m^j j! \\ &= m^{n(n+1)/2} \prod_{j=0}^{n} j! = m^{\binom{n+1}{2}} \prod_{j=0}^{n} j!. \end{aligned} \tag{139}$$

This is the desired result. □

Notice that by (109) and (111), $\widetilde{D}_{m,a+1}(n)$ is the binomial transform of $\widetilde{D}_{m,a}(n)$. Hence by the abovementioned property of Layman's [22], $\widetilde{D}_{m,0}(n)$ and $\widetilde{D}_{m,a}(n)$ have the same Hankel transform. Finally, we have the following theorem.

Theorem 37. *The Hankel transform of the noncentral Dowling numbers is given by*

$$H\{\mathfrak{D}_{m,a}\} = m^{\binom{n+1}{2}} \prod_{j\ge 0} j!. \tag{140}$$

6. More Theorems on $\widetilde{\mathscr{D}}_{m,a}(n;x)$

Proposition 38. *The noncentral Whitney numbers of the second kind satisfy the rational generating function*

$$\sum_{n=k}^{\infty} \widetilde{W}_{m,a}(n,k) z^{n-k} = \frac{1}{\prod_{i=0}^{k}(1-(mi-a)z)}. \tag{141}$$

It is easy to express this identity as

$$\begin{aligned} \sum_{n=k}^{\infty} \widetilde{W}_{m,a}(n,k) z^n &= \frac{1}{m^k(1+az)} \\ &\quad \cdot \frac{(-1)^k}{\langle((m-a)z-1)/mz\rangle_k}, \end{aligned} \tag{142}$$

where $\langle x\rangle_k = x(x+1)(x+2)\cdots(x+k-1)$. Hence, we have

$$\begin{aligned} &\sum_{k=0}^{\infty}\left(\sum_{n=k}^{\infty} \widetilde{W}_{m,a}(n,k) z^n\right) x^k \\ &= \frac{1}{1+az}\sum_{k=0}^{\infty} \frac{\langle 1\rangle_k}{\langle((m-a)z-1)/mz\rangle_k} \cdot \frac{(-x/m)^k}{k!}. \end{aligned} \tag{143}$$

Since the left-hand side is just

$$\sum_{n=0}^{\infty}\left(\sum_{k=0}^{\infty} \widetilde{W}_{m,a}(n,k) x^k\right) z^n = \sum_{n=0}^{\infty} \widetilde{\mathscr{D}}_{m,a}(n;x) z^n, \tag{144}$$

then by using the hypergeometric function defined by

$$\begin{aligned} {}_pF_q\left(\begin{matrix} a_1, a_2, \ldots, a_p \\ b_1, b_2, \ldots, b_q \end{matrix} \mid t\right) &= \sum_{k=0}^{\infty} \frac{\langle a_1\rangle_k \langle a_2\rangle_k \cdots \langle a_p\rangle_k}{\langle b_1\rangle_k \langle b_2\rangle_k \cdots \langle b_q\rangle_k} \frac{t^k}{k!}, \end{aligned} \tag{145}$$

we get

$$\begin{aligned} &\sum_{n=0}^{\infty} \widetilde{\mathscr{D}}_{m,a}(n;x) z^n \\ &= \frac{1}{1+az}\, {}_1F_1\left(\begin{matrix} 1 \\ \dfrac{(m-a)z-1}{mz} \end{matrix} \mid -\frac{x}{m}\right). \end{aligned} \tag{146}$$

Theorem 39. *The noncentral Dowling polynomials satisfy the generating function*

$$\begin{aligned} &\sum_{n=0}^{\infty} \widetilde{\mathscr{D}}_{m,a}(n;x) z^n \\ &= \frac{1}{1+az}\left(\frac{1}{e}\right)^{x/m} {}_1F_1\left(\begin{matrix} \dfrac{-az-1}{mz} \\ \dfrac{(m-a)z-1}{mz} \end{matrix} \mid \frac{x}{m}\right). \end{aligned} \tag{147}$$

Proof. Applying Kummer's formula [23, page 505] given by

$$e^{-x}\, {}_1F_1\left(\begin{matrix} \alpha \\ \beta \end{matrix} \mid x\right) = {}_1F_1\left(\begin{matrix} \beta-\alpha \\ \beta \end{matrix} \mid -x\right) \tag{148}$$

to (146) with $\alpha = (-az-1)/mz$ and $\beta = ((m-a)z-1)/mz$ yields the desired result. □

Remark 40. When $m = 1$ and $a = -r$, we get [24, Theorem 3.2]

$$\begin{aligned} &\frac{-1}{rz-1}\left(\frac{1}{e}\right)^{x} {}_1F_1\left(\begin{matrix} \dfrac{rz-1}{z} \\ \dfrac{rz+z-1}{mz} \end{matrix} \mid \frac{x}{m}\right) \\ &= \sum_{n=0}^{\infty} \phi_{n,r}(x) z^n, \end{aligned} \tag{149}$$

where $\phi_{n,r}(x)z^n$ denotes the r-Bell polynomials. In a similar manner, we get [7, page 10]

$$\left(\frac{1}{e}\right)^{x/\alpha} {}_1F_1\left(\begin{matrix} -\dfrac{1}{\alpha z} \\ \dfrac{\alpha z-1}{\alpha z} \end{matrix} \mid \frac{x}{\alpha}\right) = \sum_{n=0}^{\infty} \widetilde{D}_{(\alpha)}(n;x) z^n, \tag{150}$$

where $\widetilde{D}_{(\alpha)}(n;x)$ denotes the translated Dowling polynomials if we set $m=\alpha$ and $a=0$.

An equivalent of the result in the previous theorem is due to R. B. Corcino and C. B. Corcino [25, Theorem 4.1] obtained when $m=\beta$ and $a=-r$. This identity is established for the (r,β)-Bell polynomials $G_{n,\beta,r}(x)$.

Definition 41 (see [26, page 268]). A real sequence v_k, $k=0,1,2,\dots$ is called convex on an interval $[a,b]$, where $[a,b]$ contains at least 3 consecutive integers, if

$$v_k \le \frac{1}{2}\left(v_{k-1}+v_{k+1}\right), \quad k\in[a+1,b-1]. \tag{151}$$

The above inequality is often referred to as the *convexity property*.

Theorem 42. *For $a,m\ge 0$, the sequence of noncentral Dowling polynomials $\widetilde{\mathscr{D}}_{m,a}(n;x)$, $x>0$, satisfies the convexity property.*

Proof. Suppose $mk-a\ge 0$. Then

$$\begin{aligned} 0 &\le [1-(mk-a)]^2,\\ 0 &\le 1-2(mk-a)+(mk-a)^2,\\ mk-a &\le \frac{1}{2}\left[1+(mk-a)^2\right],\\ (mk-a)^{n+1} &\le \frac{1}{2}\left[(mk-a)^n+(mk-a)^{n+2}\right]. \end{aligned} \tag{152}$$

Multiplying the above inequality by $(x/m)^i(1/i!)$ and summing over i give

$$\widetilde{\mathscr{D}}_{m,a}(n+1;x) \le \frac{1}{2}\left[\widetilde{\mathscr{D}}_{m,a}(n;x)+\widetilde{\mathscr{D}}_{m,a}(n+2;x)\right]. \tag{153}$$

This is precisely the desired result. □

Cesàro [27] obtained an integral representation of the Bell numbers $\phi_n := \phi_n(1)$, namely,

$$\phi_n = \frac{2n!}{\pi e}\operatorname{Im}\int_0^\pi e^{e^{e^{i\theta}}}\sin(n\theta)\,d\theta. \tag{154}$$

Several generalizations of this remarkable representation were presented by Mező [24], Mangontarum et al. [6], and R. B. Corcino and C. B. Corcino [25]. To establish an analogous representation for $\widetilde{\mathscr{D}}_{m,a}(n;x)$, we take the explicit formula in (38) and substitute it to the right-hand side of Callan's [28] integral identity given by

$$\operatorname{Im}\int_0^\pi e^{je^{i\theta}}\sin(n\theta)\,d\theta = \frac{\pi}{2}\frac{j^n}{n!}. \tag{155}$$

That is, we obtain

$$\begin{aligned} \widetilde{W}_{m,a}(n,k) &= \frac{2n!}{m^k k!\pi}\sum_{j=0}^{k}(-1)^{k-j}\binom{k}{j}\operatorname{Im}\int_0^\pi e^{(mj-a)e^{i\theta}}\\ &\quad\cdot\sin(n\theta)\,d\theta = \frac{2n!}{m^k k!\pi}\\ &\quad\cdot\operatorname{Im}\int_0^\pi\left[\sum_{j=0}^{k}(-1)^{k-j}\binom{k}{j}\left(e^{me^{i\theta}}\right)^j\right]e^{-ae^{i\theta}}\\ &\quad\cdot\sin(n\theta)\,d\theta = \frac{2n!}{\pi}\\ &\quad\cdot\operatorname{Im}\int_0^\pi\frac{\left(\left(e^{me^{i\theta}}-1\right)/m\right)^k}{k!}e^{-ae^{i\theta}}\sin(n\theta)\,d\theta. \end{aligned} \tag{156}$$

Multiplying both sides by x^k and summing over k give

$$\begin{aligned} \sum_{k=0}^{\infty}\widetilde{W}_{m,a}(n,k)\,x^k &= \frac{2n!}{\pi}\\ &\quad\cdot\operatorname{Im}\int_0^\pi\left\{\sum_{k=0}^{\infty}\frac{\left(\left(\left(e^{me^{i\theta}}-1\right)/m\right)x\right)^k}{k!}\right\}e^{-ae^{i\theta}}\\ &\quad\cdot\sin(n\theta)\,d\theta = \frac{2n!}{\pi}\operatorname{Im}\int_0^\pi e^{(e^{me^{i\theta}}-1)x/m-ae^{i\theta}}\\ &\quad\cdot\sin(n\theta)\,d\theta. \end{aligned} \tag{157}$$

Hence,

$$\widetilde{\mathscr{D}}_{m,a}(n;x) = \frac{2n!}{\pi e^{x/m}}\operatorname{Im}\int_0^\pi e^{x(e^{me^{i\theta}}/m)-ae^{i\theta}}\sin(n\theta)\,d\theta \tag{158}$$

and we have the following theorem.

Theorem 43. *The noncentral Dowling polynomials have the integral representation*

$$\begin{aligned} &\widetilde{\mathscr{D}}_{m,a}(n;x)\\ &\quad = \frac{2n!}{\pi e^{x/m}}\operatorname{Im}\int_0^\pi e^{xm^{-1}e^{me^{i\theta}}}e^{-ae^{i\theta}}\sin(n\theta)\,d\theta. \end{aligned} \tag{159}$$

Remark 44. The earlier mentioned results due to Mező, Mangontarum et al., and R. B. Corcino and C. B. Corcino can be obtained from this theorem by carefully assigning values to the parameters m, a, and x (cf. [24, Theorem 6.1], [6, Theorems 10], and [25, Equation (4.12)]).

Another interesting fact on the Bell polynomials is the relation

$$E_\lambda\left[X^n\right] = \phi_n(\lambda), \tag{160}$$

where $E_\lambda[X^n]$ denotes the nth moment of a Poisson random variable X with mean λ. From this point on, we use X to denote such random variable. Another identity in line with this was obtained by Privault [29], namely,

$$E_\lambda\left[(X+y-\lambda)^n\right]=\phi_n(y,-\lambda), \tag{161}$$

where $\phi_n(y,\lambda)$ is an extension of Bell polynomials satisfying

$$\sum_{k=0}^{\infty}\phi_n(y,\lambda)\frac{t^k}{k!}=e^{ty-\lambda(e^t-t-1)}, \tag{162}$$

$$\phi_n(y,-\lambda)=\sum_{k=0}^{n}\binom{n}{k}(y-\lambda)^{n-k}\sum_{j=0}^{k}\left\{{k \atop j}\right\}\lambda^j. \tag{163}$$

It is also known that the nth factorial moment of X is

$$E_\lambda\left[(X)_n\right]=\lambda^n. \tag{164}$$

Now, if we take the expectation of (40), we get

$$\begin{aligned}E_\lambda\left[(mX-a)^n\right]&=\sum_{k=0}^{\infty}m^k\widetilde{W}_{m,a}(n,k)E_\lambda\left[(X)_k\right]\\&=\sum_{k=0}^{\infty}\frac{1}{k!}\sum_{j=0}^{k}(-1)^j\binom{k}{j}(m(j-k)-a)^n\lambda^k\\&=\sum_{j=0}^{\infty}\sum_{k=j}^{\infty}\frac{(-1)^j(m(j-k)-a)^n\lambda^k}{j!(k-j)!}\end{aligned} \tag{165}$$

through the aid of (38). Reindexing the sum yields

$$\begin{aligned}E_\lambda\left[(mX-a)^n\right]&=\sum_{j=0}^{\infty}\frac{(-1)^j}{j!}\sum_{i=0}^{\infty}\frac{(mi-a)^n}{i!}\lambda^i\\&=e^{-\lambda}\sum_{i=0}^{\infty}\frac{(mi-a)^n}{i!}\lambda^i.\end{aligned} \tag{166}$$

Clearly, when λ is replaced with λ/m,

$$E_{\lambda/m}\left[(mX-a)^n\right]=\widetilde{\mathscr{D}}_{m,a}(n;\lambda). \tag{167}$$

On the other hand, using the binomial theorem,

$$\begin{aligned}E_\lambda\left[(mX-a)^n\right]&=\sum_{k=0}^{n}\binom{n}{k}(-a)^{n-k}m^kE_\lambda\left[X^k\right]\\&=\sum_{k=0}^{n}\binom{n}{k}(-a)^{n-k}m^k\phi_k(\lambda).\end{aligned} \tag{168}$$

The above results are compiled in the next theorem.

Theorem 45. *The following identities hold:*

$$E_\lambda\left[(mX-a)^n\right]=e^{-\lambda}\sum_{i=0}^{\infty}\frac{(mi-a)^n}{i!}\lambda^i, \tag{169}$$

$$E_\lambda\left[(mX-a)^n\right]=\sum_{k=0}^{n}\binom{n}{k}(-a)^{n-k}m^k\phi_k(\lambda), \tag{170}$$

$$E_{\lambda/m}\left[(mX-a)^n\right]=\widetilde{\mathscr{D}}_{m,a}(n;\lambda). \tag{171}$$

Remark 46. When $m=1$ and $a=0$ in (169), we get the classical identity (160). Similarly, if $m=1$ and $a=\lambda-y$ in (170), we recover Privault's identities in (161) and (163). Some results reported by Mangontarum and Corcino [30, Remarks 1, 2, and 3] can also be obtained from this theorem.

7. Some Questions and Conjectures

There are a number of further applications and possible extensions of the numbers introduced in this paper. The authors would like to direct the attention of the readers to some questions and conjectures.

Several studies regarding the identification of the index for which certain Stirling-type numbers attain their maximum value were conducted earlier by some mathematicians, for instance, Mező [24, 31] for the r-Stirling and r-Bell numbers, R. B. Corcino and C. B. Corcino [32] for the generalized Stirling numbers, and recently, Corcino et al. [33] for the noncentral Stirling numbers of the first kind.

Question 1. Is it possible to identify the maximizing index of the noncentral Whitney numbers of both kinds? Will these be different from the said earlier results?

Perhaps this question may be answered by the so-called "Erdős and Stone Theorem" mentioned in [33].

The study of asymptotic estimates/approximations and asymptotic formulas for Stirling-type numbers (such as the (r,β)-Stirling numbers and the r-Whitney numbers of the second kind) has been the interest of several mathematicians, especially C. B. Corcino and R. B. Corcino [34, 35], Corcino et al. [36], and Corcino et al. [37]. The next question is an interesting motivation for further study.

Question 2. It is compelling to study asymptotic approximations and obtain formulas for the noncentral Whitney numbers. However, will these formulas be distinct from those results found in [34–37] or will they be equivalent?

Corcino et al. [38] defined distinct "*q-analogues*" for the noncentral Stirling numbers of the second kind. Using Definition 47, the said q-noncentral Stirling numbers of the second kind were given a combinatorial interpretation in the context of A-tableaux. A more general study can actually be seen in the work of de Médicis and Leroux [39].

Definition 47 (see [39]). An A-tableau is a list Φ of columns c of Ferrer's diagram of a partition Λ (by decreasing order of length) such that the lengths $|c|$ are part of the sequence $A=(a_i)_{i\geq 0}$, a strictly increasing sequence of nonnegative integers.

In line with this, denoting by $T^A(x,y)$ the set of A-tableaux with $A=\{0,1,2,\ldots,x\}$ and exactly y columns and letting $\omega_A(\Phi) = \prod_{c\in\Phi}\omega(|c|)$, $\Phi \in T^A(x,y)$, we have the following conjecture.

Conjecture 48. *For complex numbers m and a, and $\omega : N \to K$ a function from the set of nonnegative integers N to a ring K, defined by*

$$\omega(|c|) = m|c| - a, \tag{172}$$

where $|c|$ is the length of column c of an A-tableaux in $T^A(k, n-k)$, one has

$$\widetilde{W}_{m,a}(n,k) = \sum_{\Phi\in T^A(k,n-k)} \prod_{c\in\Phi} \omega(|c|). \tag{173}$$

The key to proving this remarkable observation might be achieved using the explicit formula in (52).

A "multiparameter version" of the noncentral Stirling numbers of Koutras [12] was introduced by El-Desouky [40]. This extended the number of parameters from the usual one that is a to a sequence a_i, $i = 0, 1, \ldots, a_{n-1}$. q-analogues of these numbers were then defined by Corcino and Mangontarum [41].

Conjecture 49. *It is possible to establish a multiparameter version of the noncentral Whitney numbers of both kinds (may be called multiparameter noncentral Whitney numbers) either by means of a triangular recurrence relation or by a certain generating function.*

Before ending this section, let us first note that Mangontarum and Katriel [42] investigated a connection of the defining relations of the r-Whitney numbers of both kinds in (9) and (10) and the *Boson operators* a and $a^\dagger$ known to satisfy the commutation relation

$$\left[a, a^\dagger\right] \equiv aa^\dagger - a^\dagger a = 1. \tag{174}$$

Using their observations in this matter, they were able to define a remarkable q-deformation of the r-Whitney numbers using the q-Boson operators of Arik and Coon [43] which satisfies

$$\left[a, a^\dagger\right]_q \equiv aa^\dagger - qa^\dagger a = 1. \tag{175}$$

Competing Interests

The authors declare that no competing interests exist regarding the publication of this paper.

Acknowledgments

This study is supported by the Office of the Vice Chancellor for Academic Affairs and the Office of the Vice Chancellor for Research and Extension of Mindanao State University, Main Campus, Marawi City, the Office of the Campus Head of Mindanao State University, Maigo School of Arts and Trades, and the Office of the President, Mindanao State University, Main Campus, Marawi City.

References

[1] T. A. Dowling, "A class of geometric lattices based on finite groups," *Journal of Combinatorial Theory. Series B*, vol. 15, pp. 61–86, 1973.

[2] M. Benoumhani, "On Whitney numbers of Dowling lattices," *Discrete Mathematics*, vol. 159, no. 1–3, pp. 13–33, 1996.

[3] M. Benoumhani, "On some numbers related to Whitney numbers of Dowling lattices," *Advances in Applied Mathematics*, vol. 19, no. 1, pp. 106–116, 1997.

[4] M. Benoumhani, "Log-concavity of Whitney numbers of Dowling lattices," *Advances in Applied Mathematics*, vol. 22, no. 2, pp. 186–189, 1999.

[5] H. Belbachir and I. E. Bousbaa, "Translated Whitney and r-Whitney numbers: a combinatorial approach," *Journal of Integer Sequences*, vol. 16, article 13.8.6, 2013.

[6] M. M. Mangontarum, A. P. Macodi-Ringia, and N. S. Abdulcarim, "The translated Dowling polynomials and numbers," *International Scholarly Research Notices*, vol. 2014, Article ID 678408, 8 pages, 2014.

[7] M. M. Mangontarum and A. Dibagulun, "On the translated Whitney numbers and their combinatorial properties," *British Journal of Applied Science and Technology*, vol. 11, no. 5, pp. 1–15, 2015.

[8] D. Callan, *Cesàro's Integral Formula for the Bell Numbers (Corrected)*, 2005, http://www.stat.wisc.edu/~callan/notes/cesaro/cesaro.pdf.

[9] I. Mező, "A new formula for the Bernoulli polynomials," *Results in Mathematics*, vol. 58, no. 3-4, pp. 329–335, 2010.

[10] G.-S. Cheon and J.-H. Jung, "r-Whitney numbers of Dowling lattices," *Discrete Mathematics*, vol. 312, no. 15, pp. 2337–2348, 2012.

[11] I. Mező and J. L. Ramirez, "The linear algebra of the r-Whitney matices," *Integral Transforms and Special Functions*, vol. 26, no. 3, pp. 213–225, 2015.

[12] M. Koutras, "Non-central Stirling numbers and some applications," *Discrete Mathematics*, vol. 42, no. 1, pp. 73–89, 1982.

[13] A. Z. Broder, "The r-Stirling numbers," *Discrete Mathematics*, vol. 49, no. 3, pp. 241–259, 1984.

[14] R. L. Graham, D. E. Knuth, and O. Patashnik, *Concrete Mathematics*, Addison-Wesley, Reading, Mass, USA, 1989.

[15] J. Pan, "Matrix decomposition of the unified generalized Stirling numbers and inversion of the generalized factorial matrices," *Journal of Integer Sequences*, vol. 15, article 12.6.6, 2012.

[16] L. C. Hsu and P. J. Shiue, "A unified approach to generalized Stirling numbers," *Advances in Applied Mathematics*, vol. 20, no. 3, pp. 366–384, 1998.

[17] M. Rahmani, "Some results on Whitney numbers of Dowling lattices," *Arab Journal of Mathematical Sciences*, vol. 20, no. 1, pp. 11–27, 2014.

[18] S. Tanny, "On some numbers related to the Bell numbers," *Canadian Mathematical Bulletin*, vol. 17, pp. 733–738, 1975.

[19] G.-C. Rota, "The number of partitions of a set," *The American Mathematical Monthly*, vol. 71, pp. 498–504, 1964.

[20] K.-W. Chen, "Identities from the binomial transform," *Journal of Number Theory*, vol. 124, no. 1, pp. 142–150, 2007.

[21] N. J. Sloane, "Least positive sequence with Hankel transform $\{1, 1, 1, 1, 1, \ldots\}$," The On-Line Encyclopedia of Integer Sequences, July 2000, http://oeis.org.

[22] J. W. Layman, "The Hankel transform and some of its properties," *Journal of Integer Sequences*, vol. 4, no. 1, Article ID 01.1.5, 2001.

[23] M. Abramowitz and I. A. Stegun, Eds., *Handbook of Mathematical Functions with Formulas, Graphs, and Mathematical Tables*, Dover, New York, NY, USA, 9th edition, 1972.

[24] I. Mező, "The r-Bell numbers," *Journal of Integer Sequences*, vol. 14, no. 1, Article ID 11.1.1, 2011.

[25] R. B. Corcino and C. B. Corcino, "On generalized Bell polynomials," *Discrete Dynamics in Nature and Society*, vol. 2011, Article ID 623456, 21 pages, 2011.

[26] L. Comtet, *Advanced Combinatorics*, D. Reidel, Dordrecht, Netherlands, 1974.

[27] M. Cesàro, "Sur une équation aux différences melées," *Nouvelles Annales de Mathématiques*, vol. 4, pp. 36–40, 1883.

[28] J. Stirling, *Methodus Differentialissme Tractus de Summatione et Interpolatione Serierum Infinitarum*, London, UK, 1730, (English translation by F. Holliday with the title: The Differential Method, London, UK, 1749).

[29] N. Privault, "Generalized Bell polynomials and the combinatorics of Poisson central moments," *Electronic Journal of Combinatorics*, vol. 18, no. 1, pp. 1–10, 2011.

[30] M. M. Mangontarum and R. B. Corcino, "The generalized factorial moments in terms of a Poisson random variable," *GSTF Journal of Mathematics, Statistics and Operations Research*, vol. 2, no. 1, pp. 64–67, 2013.

[31] I. Mező, "On the maximum of r-Stirling numbers," *Advances in Applied Mathematics*, vol. 41, no. 3, pp. 293–306, 2008.

[32] R. B. Corcino and C. B. Corcino, "On the maximum of generalized Stirling numbers," *Utilitas Mathematica*, vol. 86, pp. 241–256, 2011.

[33] R. B. Corcino, C. B. Corcino, and P. B. Aranas, "The peak of noncentral Stirling numbers of the first kind," *International Journal of Mathematics and Mathematical Sciences*, vol. 2015, Article ID 982812, 7 pages, 2015.

[34] C. B. Corcino and R. B. Corcino, "Asymptotic estimates for second kind generalized stirling numbers," *Journal of Applied Mathematics*, vol. 2013, Article ID 918513, 7 pages, 2013.

[35] C. B. Corcino and R. B. Corcino, "An asymptotic formula for r-Bell numbers with real arguments," *ISRN Discrete Mathematics*, vol. 2013, Article ID 274697, 7 pages, 2013.

[36] C. B. Corcino, R. B. Corcino, and N. Acala, "Asymptotic estimates for r-Whitney numbers of the second kind," *Journal of Applied Mathematics*, vol. 2014, Article ID 354053, 7 pages, 2014.

[37] C. B. Corcino, R. B. Corcino, and R. J. Gasparin, "Equivalent asymptotic formulas of second kind r-Whitney numbers," *Integral Transforms and Special Functions*, vol. 26, no. 3, pp. 192–202, 2015.

[38] C. B. Corcino, R. B. Corcino, J. M. Ontolan, C. M. Perez-Fernandez, and E. R. Cantallopez, "The Hankel transform of q-noncentral Bell numbers," *International Journal of Mathematics and Mathematical Sciences*, vol. 2015, Article ID 417327, 10 pages, 2015.

[39] A. de Médicis and P. Leroux, "Generalized Stirling numbers, convolution formulae and p; q-analogues," *Canadian Journal of Mathematics*, vol. 47, no. 3, pp. 474–499, 1995.

[40] B. S. El-Desouky, "The multiparameter noncentral Stirling numbers," *The Fibonacci Quarterly*, vol. 32, pp. 218–225, 1994.

[41] R. B. Corcino and M. M. Mangontarum, "On multiparameter q-noncentral Stirling and Bell numbers," *Ars Combinatoria*, vol. 118, pp. 201–220, 2015.

[42] M. M. Mangontarum and J. Katriel, "On q-boson operators and q-analogues of the r-Whitney and r-dowling numbers," *Journal of Integer Sequences*, vol. 18, article 15.9.8, 2015.

[43] M. Arik and D. D. Coon, "Hilbert spaces of analytic functions and generalized coherent states," *Journal of Mathematical Physics*, vol. 17, no. 4, pp. 524–527, 1976.

f_q-Derivations of G-Algebra

Deena Al-Kadi

Department of Mathematics and Statistic, Faculty of Science, Taif University, P.O. Box 888, Taif 21974, Saudi Arabia

Correspondence should be addressed to Deena Al-Kadi; dak12le@hotmail.co.uk

Academic Editor: Ilya M. Spitkovsky

We introduce the notion of f_q-derivation as a new derivation of G-algebra. For an endomorphism map f of any G-algebra X, we show that at least one f_q-derivation of X exists. Moreover, for such a map, we show that a self-map d_q^f of X is f_q-derivation of X if X is an associative medial G-algebra. For a medial G-algebra X, d_q^f is f_q-derivation of X if d_q^f is an outside f_q-derivation of X. Finally, we show that if f is the identity endomorphism of X then the composition of two f_q-derivations of X is a f_q-derivation. Moreover, we give a condition to get a commutative composition.

1. Introduction

Derivation is an important area of research in the theory of algebraic structure in mathematics. The theory of derivations of algebraic structures came from the development of Galois theory and the theory of invariants. Many researches have been done on derivations on different algebras (see [1–4]).

Several authors [5–9] have studied derivations in BCI-algebra after the work done in 2004 by Jun and Xin where the notion of derivation in ring and near-ring theory was applied to BCI-algebra [4]. As in [5], for a self-map d, for any algebra X, d is a left-right derivation (briefly (l,r)-derivation) of X if it satisfies the identity $d(x * y) = (d(x) * y) \wedge (x * d(y))$ for all $x, y \in X$. If d satisfies the identity $d(x * y) = (x * d(y)) \wedge (d(x) * y)$ for all $x, y \in X$, then d is a right-left derivation (briefly (r,l)-derivation) of X. If d is both (l,r)- and (r,l)-derivation, then d is a derivation of X.

Recently, in 2013, a new derivation named f_q-derivation of BCI-algebras was introduced. That is, in general, for any self-map d_q^f of an algebra X, f_q-derivation of X is defined by $d_q^f(x) = f(x) * q$ for all x and $q \in X$. The map d_q^f is called an outside f_q-derivation of X if it satisfies $d_q^f(x * y) = (f(x) * d_q^f(y)) \wedge (d_q^f(x) * f(y))$, $\forall x, y \in X$. If the map d_q^f satisfies the identity $d_q^f(x * y) = (d_q^f(x) * f(y)) \wedge (f(x) * d_q^f(y))$, $\forall x, y \in X$, then the map d_q^f is called an inside f_q-derivation of X. If d_q^f is both outside and inside f_q-derivation of X, then it is a f_q-derivation of X ([10]).

The notion of G-algebra was introduced in [11]. The aim of the paper is to complete the studies on G-algebra; in particular, we aim to apply the notion of f_q-derivation on G-algebra and obtain some related properties. We start with definitions and propositions on G-algebra taken from [11]. Then, we redefine the notion of f_q-derivation in G-algebra and prove that every self-map d_q^f of an associative, medial G-algebra is f_q-derivation, where f is an endomorphism of X. We also show that every self-map d_q^f of an associative, medial G-algebra is f_q-derivation. Then, we show that if f is the identity endomorphism of X, then, for a medial G-algebra, d_q^f is a f_q-derivation of X if d_q^f is an outside f_q-derivation of X. Further, we show that if f is the identity endomorphism of X and d_q^f, $d_q^{\prime f}$ are both outside (resp., inside) f_q-derivations of X, then the composition is an outside (resp., inside) f_q-derivation of X and consequently f_q-derivation. We conclude the section with a condition given on two f_q-derivations of X to get a commutative composition.

Definition 1. A G-algebra is a nonempty set X with a constant 0 and a binary operation $*$ satisfying the axioms:

(1) $x * x = 0$,

(2) $x * (x * y) = y$, for all x, y in X.

Proposition 2. *If $(X, *, 0)$ is a G-algebra, then the following conditions hold:*

(1) $x * 0 = x$,

(2) $0 * (0 * x) = x$, *for any* $x \in X$.

Proposition 3. *Let $(X, *, 0)$ be a G-algebra. Then, the following conditions hold for any $x, y \in X$:*

(1) $(x * (x * y)) * y = 0$,

(2) $x * y = 0 \Rightarrow x = y$,

(3) $0 * x = 0 * y \Rightarrow x = y$.

Definition 4. A G-algebra X satisfying $(x * y) * (z * u) = (x * z) * (y * u)$, for any x, y, z and $u \in X$, is called a medial G-algebra.

Lemma 5. *If X is a medial G-algebra, then, for any $x, y, z \in X$, the following axiom holds:*

$$(x * y) * z = (x * z) * y. \tag{1}$$

Theorem 6. *A G-algebra X is medial if and only if it satisfies the following conditions:*

(1) $y * x = 0 * (x * y)$ *for all* $x, y \in X$,

(2) $x * (y * z) = z * (y * x)$ *for all* $x, y, z \in X$.

2. Results

In this section we will introduce a new derivation of G-algebra motivated by [10, Definition 3.1]. We start by defining an endomorphism of G-algebra X.

Definition 7. Let X be a G-algebra and let f be a self-map of X. One says that f is an endomorphism if

$$f(x * y) = f(x) * f(y), \quad \forall x, y \in X. \tag{2}$$

Throughout the paper, d_q^f is a self-map of G-algebra X defined by $d_q^f(x) = f(x) * q$ for all $x \in X$, $q \in X$ and f is an endomorphism self-map of X unless otherwise mentioned.

For elements x and y of a G-algebra X, denote $x \wedge y$ by $y * (y * x)$. By considering that $x \wedge y = x$ in G-algebra, we redefine the notion of f_q-derivation in [10] to get the following definition.

Definition 8. A map d_q^f is called an outside f_q-derivation of X if

$$d_q^f(x * y) = f(x) * d_q^f(y), \quad \forall x, y \in X. \tag{3}$$

If the map d_q^f satisfies the following identity:

$$d_q^f(x * y) = d_q^f(x) * f(y), \quad \forall x, y \in X, \tag{4}$$

then the map is called an inside f_q-derivation of X. If d_q^f is both an outside and inside f_q-derivation of X, then d_q^f is a f_q-derivation of X.

TABLE 1

$*$	0	1	2
0	0	1	2
1	1	0	2
2	2	1	0

TABLE 2

x	0	0	0	1	1	1	2	2	2
y	0	1	2	0	1	2	0	1	2
$x * y$	0	1	2	1	0	2	2	1	0
$f(x * y)$	0	2	1	2	0	1	1	2	0
$d_0^f(x * y)$	0	2	1	2	0	1	1	2	0
$f(x)$	0	0	0	2	2	2	1	1	1
$d_0^f(x)$	0	0	0	2	2	2	1	1	1
$f(y)$	0	2	1	0	2	1	0	2	1
$d_0^f(y)$	0	2	1	0	2	1	0	2	1
$f(x) * d_0^f(y)$	0	2	1	2	0	1	1	2	0
$d_0^f(x) * f(y)$	0	2	1	2	0	1	1	2	0

TABLE 3

$*$	0	a	b	c
0	0	a	b	c
a	a	0	c	b
b	b	c	0	a
c	c	b	a	0

Remark 9. If d_q^f is f_q-derivation, then

$$d_q^f(x * y) = f(x) * d_q^f(y) = d_q^f(x) * f(y). \tag{5}$$

Example 10. Consider the G-algebra given by Cayley table (Table 1).

Define an endomorphism:

$$f : X \longrightarrow X, \quad \text{such that } x \longmapsto \begin{cases} 0 & \text{if } x = 0, \\ 2 & \text{if } x = 1, \\ 1 & \text{if } x = 2. \end{cases} \tag{6}$$

If $q = 0$, then Table 2 shows that d_0^f is an outside and an inside f_q-derivation of X. Hence, d_0^f is a f_q-derivation of X.

If we take $q = 2$, then d_2^f is not an outside f_q-derivation of X or an inside f_q-derivation of X since $d_2^f(1 * 2) = 2$ while $f(1) * d_2^f(2) = 0$ and $d_2^f(1) * f(2) = 1$.

Example 11. Let $X = \{0, a, b, c\}$. Consider the G-algebra given by Cayley table (Table 3).

Define an endomorphism:

$$f: X \longrightarrow X, \quad \text{such that } x \longmapsto \begin{cases} 0 & \text{if } x = 0, \\ b & \text{if } x = a, \\ a & \text{if } x = b, \\ c & \text{if } x = c. \end{cases} \tag{7}$$

It can be shown by direct calculation that d_q^f is f_q-derivation of X for all $q \in X$.

Proposition 12. *For any G-algebra X, there exists at least one f_q-derivation of X, that is, the map d_0^f.*

Proof. Let $q = 0$; then $d_0^f(x * y) = (f(x * y)) * 0 = (f(x) * f(y)) * 0 = f(x) * f(y)$ and $f(x) * d_0^f(y) = f(x) * (f(y) * 0) = f(x) * f(y)$. We also have $d_0^f(x) * f(y) = (f(x) * 0) * f(y) = f(x) * f(y)$. Hence, d_0^f is f_q-derivation of X. □

Proposition 13. *If X is an associative G-algebra, then d_q^f is an outside f_q-derivation of X, for all $q \in X$.*

Proof. We have $d_q^f(x * y) = (f(x * y)) * q = (f(x) * f(y)) * q$ and $f(x) * d_q^f(y) = f(x) * (f(y) * q) = (f(x) * f(y)) * q$, as X is associative. Hence, d_q^f is an outside f_q-derivation of X. □

Proposition 14. *If X is a medial G-algebra, then d_q^f is an inside f_q-derivation of X, for all $q \in X$.*

Proof. Since $d_q^f(x * y) = f(x * y) * q = (f(x) * f(y)) * q$ and $d_q^f(x) * f(y) = (f(x) * q) * f(y) = (f(x) * f(y)) * q$, as X is medial, therefore, d_q^f is an inside f_q-derivation of X. □

The next theorem follows from Propositions 13 and 14.

Theorem 15. *Let X be an associative medial G-algebra; then d_q^f is a f_q-derivation of X, for all $q \in X$.*

Next we provide an alternative proof of Theorem 15.

Theorem 16. *Let X be an associative medial G-algebra. Then, d_q^f is both an outside f_q-derivation of X and an inside f_q-derivation of X for any $q \in X$.*

Proof. Let $x, y, q \in X$. Then, on one hand, we have

$$\begin{aligned} d_q^f(x * y) &= f(x * y) * q \\ &= ((f(x) * f(y)) * q) * 0 \\ &= ((f(x) * q) * f(y)) * 0 \\ &= ((f(x) * q) * f(y)) \\ &\quad * (((f(x) * q) * f(y)) * ((f(x) * q) * f(y))) \\ &= (d_q^f(x) * f(y)) \\ &\quad * ((d_q^f(x) * f(y)) * ((f(x) * f(y)) * q)) \\ &= (d_q^f(x) * f(y)) \\ &\quad * ((d_q^f(x) * f(y)) * (f(x) * (f(y) * q))) \\ &= (d_q^f(x) * f(y)) \\ &\quad * ((d_q^f(x) * f(y)) * (f(x) * d_q^f(y))) \\ &= f(x) * d_q^f(y), \quad \text{as } y * (y * x) = x. \end{aligned} \tag{8}$$

By Definition 8, d_q^f is an outside f_q-derivation of X.

On the other hand,

$$\begin{aligned} d_q^f(x * y) &= f(x * y) * q = (f(x * y) * q) * 0 \\ &= (f(x * y) * q) \\ &\quad * ((f(x * y) * q) * (f(x * y) * q)) \\ &= ((f(x) * f(y)) * q) \\ &\quad * (((f(x) * f(y)) * q) * ((f(x) * f(y)) * q)) \\ &= (f(x) * (f(y) * q)) \\ &\quad * ((f(x) * (f(y) * q)) * ((f(x) * q) * f(y))) \\ &= (f(x) * d_q^f(y)) \\ &\quad * ((f(x) * d_q^f(y)) * (d_q^f(x) * f(y))) \\ &= d_q^f(x) * f(y). \end{aligned} \tag{9}$$

Therefore, d_q^f is an inside f_q-derivation of X. □

Using Proposition 3(2), we get the following.

Proposition 17. *If d_q^f is an outside (resp., inside) f_q-derivation of X, then $d_q^f(0) = f(x) * d_q^f(x)$, $\forall x \in X$ (resp., $d_q^f(0) = d_q^f(x) * f(x)$, $\forall x \in X$).*

Proof. It is obvious. □

Theorem 18. *Let X be a medial G-algebra. If d_q^f is an outside f_q-derivation of X, then d_q^f is a f_q-derivation of X.*

Proof. From Proposition 14, we know that d_q^f is an inside f_q-derivation of X. Thus, d_q^f is a f_q-derivation of X. □

Definition 19. A map d_q^f is said to be regular if $d_q^f(0) = 0$.

Proposition 20. *Let d_q^f be a f_q-derivation of X. If either $f(x) * d_q^f(y) = 0$ or $d_q^f(x) * f(y) = 0$, then d_q^f is a regular derivation.*

Proof. Since d_q^f is a f_q-derivation, we have $d_q^f(x * y)=f(x) * d_q^f(y) = d_q^f(x) * f(y)$. Consider that $f(x) * d_q^f(y) = 0$; then $d_q^f(0) = d_q^f(x * x) = f(x) * d_q^f(x) = 0$. Similarly, if $d_q^f(x) * f(y) = 0$, we have $d_q^f(0) = d_q^f(x * x) = d_q^f(x) * f(x) = 0$. This proves that d_q^f is a regular derivation. □

Proposition 21. *Let d_q^f be a regular f_q-derivation of X; then $d_q^f(x) = f(x)$, $\forall x \in X$.*

Proof. Since d_q^f is a regular f_q-derivation of X, then $d_q^f(0) = 0$. So $d_q^f(0) = d_q^f(x * x)=d_q^f(x) * f(x) = 0$. Therefore, $d_q^f(x) = f(x)$ from Proposition 3(2). □

Definition 22. Let X be a G-algebra and let d_q^f, $d_q^{\prime f}$ be two self-maps of X. Define $d_q^f \circ d_q^{\prime f} : X \to X$ by

$$\left(d_q^f \circ d_q^{\prime f}\right)(x) = d_q^f\left(d_q^{\prime f}(x)\right) \quad \forall x \in X. \tag{10}$$

Proposition 23. *Let X be a G-algebra and let f be the identity endomorphism of X. If d_q^f, $d_q^{\prime f}$ are outside f_q-derivations of X, then $d_q^f \circ d_q^{\prime f}$ is also an outside f_q-derivation.*

Proof. Consider the element $x * y$. Then $(d_q^f \circ d_q^{\prime f})(x * y) = d_q^f(d_q^{\prime f}(x * y))$. As $d_q^{\prime f}$ and d_q^f are outside f_q-derivations, we have $d_q^f(d_q^{\prime f}(x * y)) = d_q^f(f(x) * d_q^{\prime f}(y)) = f(x) * (d_q^f \circ d_q^{\prime f})(y)$. Thus, $d_q^f \circ d_q^{\prime f}$ is an outside f_q-derivation. □

Similarly, we can prove the following proposition.

Proposition 24. *For a G-algebra X, let f be the identity endomorphism of X. If d_q^f, $d_q^{\prime f}$ are inside f_q-derivations of X, then $d_q^f \circ d_q^{\prime f}$ is also an inside f_q-derivation.*

Combining Proposition 23 and Proposition 24 we have the following theorem.

Theorem 25. *Let X be a G-algebra and let f be the identity endomorphism of X. If d_q^f, $d_q^{\prime f}$ are both outside (resp., inside) f_q-derivations of X, then the composition is an outside (resp., inside) f_q-derivation of X.*

Proposition 26. *Let X be a G-algebra and let d_q^f, $d_q^{\prime f}$ be f_q-derivations of X such that $d_q^f \circ f = f \circ d_q^f$ and $d_q^{\prime f} \circ f = f \circ d_q^{\prime f}$; then $d_q^f \circ d_q^{\prime f} = d_q^{\prime f} \circ d_q^f$.*

Proof. Consider $d_q^{\prime f}$ as an outside f_q-derivation of X and d_q^f as an inside f_q-derivation of X; then for all $x, y \in X$ we have

$$\begin{aligned}\left(d_q^f \circ d_q^{\prime f}\right)(x * y) &= d_q^f\left(d_q^{\prime f}(x * y)\right) \\ &= d_q^f\left(f(x) * d_q^{\prime f}(y)\right) \\ &= d_q^f(f(x)) * f\left(d_q^{\prime f}(y)\right) \\ &= \left(d_q^f \circ f\right)(x) * \left(f \circ d_q^{\prime f}\right)(y).\end{aligned} \tag{11}$$

On the other hand,

$$\begin{aligned}\left(d_q^{\prime f} \circ d_q^f\right)(x * y) &= d_q^{\prime f}\left(d_q^f(x * y)\right) \\ &= d_q^{\prime f}\left(d_q^f(x) * f(y)\right) \\ &= f\left(d_q^f(x)\right) * d_q^{\prime f}(f(y)) \\ &= \left(f \circ d_q^f\right)(x) * \left(d_q^{\prime f} \circ f\right)(y) \\ &= \left(d_q^f \circ f\right)(x) * \left(f \circ d_q^{\prime f}\right)(y).\end{aligned} \tag{12}$$

From (11) and (12), we can see that $(d_q^f \circ d_q^{\prime f})(x * y) = (d_q^{\prime f} \circ d_q^f)(x * y)$. By putting $y = 0$, we get

$$\left(d_q^f \circ d_q^{\prime f}\right)(x) = \left(d_q^{\prime f} \circ d_q^f\right)(x). \tag{13}$$

Hence, $d_q^f \circ d_q^{\prime f} = d_q^{\prime f} \circ d_q^f$. □

3. Conclusion

In this paper, the notion of f_q-derivation of G-algebra is introduced and some related properties are investigated. The main results are Theorems 15 and 18 where we show that a self-map d_q^f of a G-algebra is f_q-derivation if G-algebra satisfies some properties. In Theorem 25, we show that in G-algebra the composition of two f_q-derivations of X is f_q-derivation if f is identity endomorphism of X. Moreover, we give in Proposition 26 a condition on two f_q-derivations of X to get a commutative composition.

Competing Interests

The author declares that there are no competing interests regarding the publication of this paper.

References

[1] H. A. Abujabal and N. O. Al-Shehri, "Some results on derivations of BCI-algebras," *The Journal of Natural Sciences and Mathematics*, vol. 46, no. 1-2, pp. 13–19, 2006.

[2] C. Prabpayak and U. Leerawat, "On derivations of BCC-algebras," *Kasetsart Journal*, vol. 43, no. 2, pp. 398–401, 2009.

[3] N. O. Al-Shehrie, "Derivations of B-algebras," *Journal of King Abdulaziz University-Science*, vol. 22, no. 1, pp. 71–83, 2010.

[4] Y. B. Jun and X. L. Xin, "On derivations of BCI-algebras," *Information Sciences*, vol. 159, no. 3-4, pp. 167–176, 2004.

[5] H. A. S. Abujabal and N. O. Al-Shehri, "On left derivations of BCI-algebras," *Soochow Journal of Mathematics*, vol. 33, no. 3, pp. 435–444, 2007.

[6] J. Zhan and Y. L. Liu, "On f-derivations of BCI-algebras," *International Journal of Mathematics and Mathematical Sciences*, no. 11, pp. 1675–1684, 2005.

[7] M. A. Javed and M. Aslam, "A note on f-derivations of BCI-algebras," *Communications of the Korean Mathematical Society*, vol. 24, no. 3, pp. 321–331, 2009.

[8] G. Muhiuddin and A. M. Al-Roqi, "On (α,β) -derivations in BCI-algebras," *Discrete Dynamics in Nature and Society*, vol. 2012, Article ID 403209, 11 pages, 2012.

[9] G. Muhiuddin and A. M. Al-Roqi, "On t-derivations of BCI-algebras," *Abstract and Applied Analysis*, vol. 2012, Article ID 872784, 12 pages, 2012.

[10] K. J. Lee, "A new kind of derivations in BCI-algebras," *Applied Mathematical Sciences*, vol. 7, no. 84, pp. 4185–4194, 2013.

[11] R. K. Bandru and N. Rafii, "On G-algebras," *Scientia Magna*, vol. 8, no. 3, pp. 1–7, 2012.

h-Adic Polynomials and Partial Fraction Decomposition of Proper Rational Functions over $\mathbb{R}$ or $\mathbb{C}$

Kwang Hyun Kim[1] **and Xin Zhang**[2]

[1] *Departments of Mathematics and Computer Science, Queensborough Community College, 222-05 56th Avenue, Bayside, NY 11364, USA*
[2] *Zicklin School of Business, Baruch College, One Bernard Baruch Way, New York, NY 10010, USA*

Correspondence should be addressed to Kwang Hyun Kim; kkim@qcc.cuny.edu

Academic Editor: Adolfo Ballester-Bolinches

The partial fraction decomposition technique is very useful in many areas including mathematics and engineering. In this paper we present a new and simple method on the partial fraction decomposition of proper rational functions which have completely factored denominators over $\mathbb{R}$ or $\mathbb{C}$. The method is based on a recursive computation of the h-adic polynomial in commutative algebra which is a generalization of the Taylor polynomial. Since its computation requires only simple algebraic operations, it does not require a computer algebra system to be programmed.

1. Introduction

Let F be $\mathbb{R}$ or $\mathbb{C}$ and $R = F[x]$ be a polynomial ring with the coefficients in F. We also assume the rational function to be proper (i.e., the degree of denominator is greater than the degree of numerator) with the denominator factored completely over F. Now, we will show how to apply our method to the following partial fraction decomposition over $F = \mathbb{C}$:

$$\frac{1}{x^3(x-1)^2} = \frac{p(x)}{x^3} + \frac{q(x)}{(x-1)^2}. \tag{1}$$

We multiply through by the least common denominator to clear the fractions:

$$\begin{aligned} 1 &= p(x)(x-1)^2 + q(x)x^3, \\ p(x) &\equiv \frac{1}{(x-1)^2} \mod x^3. \end{aligned} \tag{2}$$

Since $R/\langle x^3\rangle \cong F[[x]]/\langle x^3\rangle$, we can replace $p(x)$ with a power series in $F[[x]]$. From Proposition 6(a) which we will prove later, we get

$$\begin{aligned} \frac{1}{(x-1)^2} &= 1 + 2x + 3x^2 + \cdots \in F[[x]], \\ p(x) &\equiv 1 + 2x + 3x^2 \mod x^3. \end{aligned} \tag{3}$$

Since we have $\deg(p) < \deg(x^3)$, $p(x)$ is equal to $1+2x+3x^2$, which is the Taylor polynomial of order 2 for $1/(x-1)^2$ at $x = 0$. Similarly, $q(x)$ is the Taylor polynomial of order 1 for $1/x^3$ at $x = 1$ and we obtain

$$q(x) = 1 - 3(x-1). \tag{4}$$

Hence

$$\begin{aligned} \frac{1}{x^3(x-1)^2} &= \frac{1+2x+3x^2}{x^3} + \frac{1-3(x-1)}{(x-1)^2} \\ &= \frac{1}{x^3} + \frac{2}{x^2} + \frac{3}{x} + \frac{1}{(x-1)^2} + \frac{-3}{(x-1)}. \end{aligned} \tag{5}$$

In [1], Ma et al. explained several recent approaches of pfd (partial fraction decomposition) and introduced a fast recursive method of pfd over $\mathbb{C}$. Our method is more algebraic and works for pfd of any proper rational functions over $\mathbb{R}$ or $\mathbb{C}$ by extending the concept of Taylor polynomials into h-adic polynomials.

2. h-Adic Polynomials and Main Theorem

In this section we introduce the completion of a commutative ring which is useful in commutative algebra and algebraic geometry. We also define the h-adic polynomial of order n.

Definition 1 (ch. 10 in [3]). For any commutative ring R with unity and a proper ideal I of R, the completion of R with respect to I is

$$\widehat{R}^I := \varprojlim \left(\frac{R}{I^i}\right) = \left\{c = (c_0, c_1, \dots) \in \prod_{i \in \mathbb{N}} \frac{R}{I^i} \mid c_{i-1} = \pi_i(c_i)\right\} \quad (6)$$

with the quotient maps $\pi_i : R/I^{i+1} \xrightarrow{\pi_i} R/I^i$. We also use $\widehat{R}^h$ if $I = \langle h \rangle$ with $h \in R$.

If $R = F[x]$ with a maximal ideal $\mathfrak{m} = \langle x \rangle$, then $\widehat{R}^x \cong F[[x]]$ is the ring of formal power series. For $R = \mathbb{Z}$ with a prime ideal $\mathfrak{p} = \langle p \rangle$, $\widehat{R}^p$ is $\mathbb{Z}_p$, the ring of p-adic integers. Now we will define the h-adic polynomial of order n, which is crucial for our method.

Lemma 2 (h-adic expansion and h-adic polynomial of order n). *Let h be an irreducible polynomial in $R = F[x]$ with $\deg(h) \geqslant 1$ and $\langle h \rangle$ be the corresponding maximal ideal. Then, consider the following:*

(a) *The natural map $R \hookrightarrow \widehat{R}^h$ via $r \to \widehat{r} = (r, r, r, \dots)$ is injective and factors through the localization $R_{\langle h \rangle} = \{f/g \mid g \notin \langle h \rangle\}$ of R at the maximal ideal $\langle h \rangle$ to $\widehat{R}^h$.*

$$\begin{array}{ccc} R & \hookrightarrow & \widehat{R}^h \\ & \searrow \quad \nearrow & \\ & R_{\langle h \rangle} & \end{array} \quad (7)$$

(b) *$R \hookrightarrow \widehat{R}^h$ induces*

$$\frac{R}{h^n R} \cong \frac{R_{\langle h \rangle}}{h^n R_{\langle h \rangle}} \cong \frac{\widehat{R}^h}{h^n \widehat{R}^h}, \quad \text{where } n \in \mathbb{N}. \quad (8)$$

(c) *For $c \in \widehat{R}^h$, there exists the unique h-adic expansion $P(c, h) \in \widehat{R}^h$ such that*

$$c = P(c, h) := \left(a_0, a_0 + a_1 h, \dots, \sum_{i=0}^{n} a_i \cdot h^i, \dots\right) \in \widehat{R}^h, \quad (9)$$

where $a_i \in R$ is a polynomial with $\deg(a_i) < \deg(h)$.

Definition 3. For $c \in \widehat{R}^h$, $f/g \in R_{\langle h \rangle}$, and $r \in R$, we define the *h-adic polynomial of order n* for c, f/g, and r, respectively, as

$$P_n(c, h) := P(c, h)_n = \sum_{i=0}^{n} a_i \cdot h^i \in R,$$
$$P_n\left(\frac{f}{g}, h\right) := P_n\left(\widehat{\left(\frac{f}{g}\right)}, h\right), \quad (10)$$
$$P_n(r, h) := P_n(\widehat{r}, h).$$

Proof. (a) is true by the Krull intersection theorem for Noetherian domains (Corollary 10.18 in [3]) and (22.13) in [4]. (b) is true from Proposition 10.15 in [3].

Case (c). For $c = (c_0, c_1, \dots) \in \widehat{R}^h \subset (R/h) \times (R/h^2) \times \cdots$, let c_n' be a representative of c_n in R. Now we will use induction on n. For $n = 0$, let a_0 be the remainder of $c_0' \div h$. Then a_0 is independent of choice of c_0', $c_0 \equiv a_0 \bmod h$, and $\deg(a_0) < \deg(h)$. Assume that there exists $\sum_{i=0}^{n} a_i h^i$ with $\deg(a_i) < \deg(h)$ such that $c_n \equiv \sum_{i=0}^{n} a_i \cdot h^i \bmod h^{n+1}$. The condition $\pi_{n+1}(c_{n+1}) = c_n$ implies that

$$c_{n+1}' - \sum_{i=0}^{n} a_i h^i = e_{n+1} \cdot h^{n+1} \quad (11)$$

with $e_{n+1} \in R$. Let a_{n+1} and q_{n+1} be the remainder and the quotient of e_{n+1}/h. Then a_{n+1} does not depend on choice of c_{n+1}' and we have $\deg(a_{n+1}) < \deg(h)$. Since $e_{n+1} = a_{n+1} + h \cdot q_{n+1}$,

$$\begin{aligned} c_{n+1}' - \sum_{i=0}^{n} a_i h^i &= (a_{n+1} + h \cdot q_{n+1}) h^{n+1} \\ &= a_{n+1} \cdot h^{n+1} + q_{n+1} \cdot h^{n+2}. \end{aligned} \quad (12)$$

Hence

$$c_{n+1} \equiv \sum_{i=0}^{n+1} a_i h^i \mod h^{n+2}. \quad (13)$$

The uniqueness is clear by the construction. □

If $h = x - a$ with $a \in F$, $(x - a)$-adic polynomial of order n for f/g is the Taylor polynomial of order n for f/g at $x = a$. Here we present the following main theorem.

Theorem 4 (main theorem). *For $f, g, p, q \in R = F[x]$ and $n \in \mathbb{N}$, let h be irreducible in R with $\deg(h) \geqslant 1$ and $\gcd(h(x), g(x)) \in F$. Assume that*

$$\frac{f(x)}{g(x) h^n(x)} = \frac{p(x)}{h^n(x)} + \frac{q(x)}{g(x)} \quad (14)$$

with $\deg(p) < \deg(h^n)$. Then

$$p(x) = P_{n-1}\left(\frac{f}{g}, h\right). \quad (15)$$

Proof. By taking the modulus of h^n and applying Lemma 2(b), we have

$$p(x) \equiv \frac{f(x)}{g(x)} \equiv P_{n-1}\left(\frac{f}{g}, h\right) \mod h^n. \quad (16)$$

$\deg(p(x) - P_{n-1}(f/g, h)) < \deg(h^n)$ implies that

$$p(x) - P_{n-1}\left(\frac{f}{g}, h\right) = 0, \quad (17)$$

from which we complete the proof. □

In the next section we will explain how to compute h-adic polynomials with $\deg(h) \leqslant 2$.

3. Formulas of h-Adic Polynomials with $deg(h) \leqslant 2$ and the Product of h-Polynomials

For $F = \mathbb{R}$ or $\mathbb{C}$, an irreducible polynomial h has at most degree 2. Since we can convert $x-a$ and $(x-a)^2+c^2$ into t and t^2+c^2, respectively, by replacing $x-a$ with t, we only present recursive formulas of (x)-adic polynomials and (x^2+c^2)-adic polynomials.

Lemma 5. *For $p(\neq 0) \in R = F[x]$ with $\gcd(p,h) = 1$, let $y = p^{-m} \in R_{\langle h\rangle}$ with $m \in \mathbb{Z}$. Then*

$$m \cdot \frac{d}{dx}(p) \cdot y + p \cdot \frac{d}{dx}(y) = 0. \tag{18}$$

Proof. $p^m \cdot y = 1$ implies $m \cdot p^{m-1}(d/dx)(p) \cdot y + p^m \cdot (d/dx)(y) = p^{m-1}(m \cdot (d/dx)(p) \cdot y + p \cdot (d/dx)(y)) = 0.$ □

Proposition 6. *We present two formulas for h-adic polynomials with $h = x$:*

(a) *If $f(x) = (x-a)^{-m} = \sum_{i=0}^{\infty} c_i x^i$ with $m \in \mathbb{Z}$ and $a \neq 0$, then $c_0 = (-a)^{-m}$ and*

$$c_{k+1} = \frac{(m+k)}{a(k+1)} c_k \quad (k \geqslant 0). \tag{19}$$

(b) *If $f(x) = [(x-a)^2+b^2]^{-m} = \sum_{i=0}^{\infty} c_i x^i$ with $m \in \mathbb{Z}$ and $a^2+b^2 \neq 0$, then $c_0 = (a^2+b^2)^{-m}$, $c_{-1} = 0$, and*

$$c_{k+1} = \frac{2a(m+k)}{(a^2+b^2)(k+1)} c_k - \frac{(2m+k-1)}{(a^2+b^2)(k+1)} c_{k-1} \quad (k \geqslant 0). \tag{20}$$

Proof.

Case (a). From Lemma 5 with $p = x-a$ and $y = \sum_{i=0}^{\infty} c_i x^i$, we have

$$m \cdot 1 \cdot \sum_{i=0}^{\infty} c_i x^i + (x-a) \sum_{i=1}^{\infty} i \cdot c_i x^{i-1} = 0. \tag{21}$$

By comparing the coefficients of x^k, one can find

$$[x^k] : m \cdot c_k + k \cdot c_k - a \cdot (k+1) \cdot c_{k+1} = 0. \tag{22}$$

Thus $c_{k+1} = ((m+k)/a(k+1))c_k$ $(k \geqslant 0)$, where $c_0 = (-a)^m$.

Case (b). We can also prove part (b) by repeating the same argument with $p = (x-a)^2+b^2$. □

To compute the h-adic polynomial of a polynomial, we need formulas of the Taylor shift.

Definition 7 (coefficient vector of a h-adic polynomial of order n). For a given h-adic polynomial $f = \sum_{i=0}^{n} a_i h^i \in R$ of order $n \geqslant 1$, we define the corresponding coefficient vectors $v_f(h)$, $v_f^{\text{odd}}(h)$, and $v_f^{\text{even}}(h)$, respectively, as

$$\begin{aligned} v_f(h,n) &:= (a_0, a_1, \ldots, a_n)^t, \\ v_f^{\text{even}}(h,n) &:= (a_0, a_2, \ldots, a_{2\cdot\lfloor n/2\rfloor})^t, \\ v_f^{\text{odd}}(h,n) &:= (a_1, a_3, \ldots, a_{2\cdot\lfloor (n-1)/2\rfloor+1})^t, \end{aligned} \tag{23}$$

where v^t is the transpose of a row vector v and $\lfloor x\rfloor := \max\{k \in \mathbb{Z} : k \leqslant x\}$ (the floor function).

Definition 8 (the Taylor shift matrix). For $s \in F$ with $r, c \in \mathbb{N} \cup \{0\}$, $(r+1) \times (c+1)$ matrix $T_{r,c}(s)$ is

$$T_{r,c}(s) = (t_{i,j}(s))_{0\leqslant i\leqslant r, 0\leqslant j\leqslant c}$$

$$\text{where } t_{i,j}(s) = \begin{cases} \binom{j}{i} \cdot s^{j-i}, & i \leqslant j \\ 0, & i > j \end{cases}, \tag{24}$$

$$\binom{j}{i} = \frac{j!}{(j-i)! \cdot i!} \text{ with } j \geqslant i \geqslant 0.$$

From the binomial expansion, $\sum_{j=0}^{n} a_j (x+s)^j = \sum_{i=0}^{n} \left(\sum_{j=i}^{n} a_j \binom{j}{i} \cdot s^{j-i}\right) x^i$, we obtain the following Taylor shift formulas for the coefficient vectors.

Proposition 9 (Taylor shift). *For a polynomial $f \in R$ with $\deg(f) = d$,*

$$v_f(x-s,n) = T_{n,d}(s) \cdot v_f(x,d), \tag{25}$$

where $s \in F$ and $n \in \{0,1,2,\ldots\}$.

Remark 10. To compute the Taylor shift matrix which is a Toeplitz matrix, we may use recursive formulas given by

$$\begin{aligned} t_{0,0} &= 1, \\ t_{0,j} &= s \cdot t_{0,j-1} \quad \text{for } j \geqslant 0, \\ t_{i,j} &= \frac{(j-i+1)}{s \cdot i} \cdot t_{i-1,j} \quad \text{for } j \geqslant i \geqslant 1. \end{aligned} \tag{26}$$

Proposition 11. *We present three formulas for h-adic polynomials with $h = x^2 + c^2$:*

(a) (h-Adic shift for $\deg(h) = 2$) for a polynomial $f \in R = F[x]$ with $\deg(f) = d \geqslant 1$, the corresponding h-adic polynomial of order n, $P_n(f,h) = \sum_{i=0}^{n} (\beta_i + \alpha_i x) h^i$ with $\alpha_i, \beta_i \in F$, is determined by

$$\begin{aligned} v_\beta &= (\beta_0, \beta_1, \ldots, \beta_n)^t = T_{n,\lfloor d/2\rfloor}(-c^2) \cdot v_f^{even}(x,d), \\ v_\alpha &= (\alpha_0, \alpha_1, \ldots, \alpha_n)^t = T_{n,\lfloor (d-1)/2\rfloor}(-c^2) \cdot v_f^{odd}(x,d). \end{aligned} \tag{27}$$

If $d = 0$, then $f \in F$ and

$$P_n(f,h) = f + \sum_{i=1}^{n} 0 \cdot h^i. \tag{28}$$

(b) For $a \neq 0$ with $m \in \mathbb{N}$,

$$(x-a)^{-m} = \left[h - \left(a^2 + c^2\right)\right]^{-m} \cdot (x+a)^m. \quad (29)$$

For $a = 0$, let $q = \lfloor (m+1)/2 \rfloor$ and $\delta = 2q - m$. Then $m = 2q - \delta$ with $\delta \in \{0, 1\}$ and

$$(x)^{-m} = \left[h - c^2\right]^{-q} \cdot \left(x^{\delta}\right). \quad (30)$$

(c) For $a \neq 0$, $b \neq 0$, and $m \in \mathbb{N}$,

$$\left[(x-a)^2 + b^2\right]^{-m} = \left[(h-A)^2 + B^2\right]^{-m} \cdot \left[(x+a)^2 + b^2\right]^m \quad (31)$$

where $A = a^2 - b^2 + c^2$, $B = 2ab$.

For $a = 0$, $b \neq 0$, $b^2 \neq c^2$, and $m \in \mathbb{N}$,

$$\left[x^2 + b^2\right]^{-m} = \left[h - \left(-b^2 + c^2\right)\right]^{-m}. \quad (32)$$

Proof. Case (a). You can find a similar argument on p. 591 in [2]. For $d = 0$, it is clear. For $f(x) = \sum_{i=0}^{d} a_0 x^i \in R$ with $d \geqslant 1$, we have

$$\begin{aligned} f(x) &= \sum_{i=0}^{\lfloor d/2 \rfloor} a_{2i} x^{2i} + x \sum_{i=0}^{\lfloor (d-1)/2 \rfloor} a_{2i+1} x^{2i} \\ &= \sum_{i=0}^{\lfloor d/2 \rfloor} a_{2i} \left(h - c^2\right)^i + x \sum_{i=0}^{\lfloor (d-1)/2 \rfloor} a_{2i+1} \left(h - c^2\right)^i \quad (33) \\ &= \sum_{i=0}^{\lfloor d/2 \rfloor} \beta_i h^i + x \sum_{i=0}^{\lfloor (d-1)/2 \rfloor} \alpha_i h^i. \end{aligned}$$

Hence α_i and β_i can be computed by the Taylor shift (Proposition 9).

Cases (b) and (c). (b) and (c) are clear by direct computations. □

To get h-adic polynomials of arbitrary rational functions, we just multiply simple h-adic polynomials which are computed using Propositions 6, 9, and 11. In general, the product of two h-adic polynomials may carry a term like

$$\begin{aligned} &\left[(1+x) + (1-x)\left(1+x^2\right)\right] \\ &\cdot \left[(1-x) + (1+x)\left(1+x^2\right)\right] = \left(1-x^2\right) \\ &+ 2\left(1+x^2\right)^2 + \left(1-x^2\right)\left(1+x^2\right)^2 = 2 \quad (34) \\ &- 1\left(1+x^2\right) + 4\left(1+x^2\right)^2 - 1\left(1+x^2\right)^3. \end{aligned}$$

But if the coefficients of one of two h-adic polynomials are in F, then the product is carryless multiplication and the classic Cauchy product (convolution) formula still holds.

Proposition 12 (h-adic Cauchy product formula). *For $\widehat{r_1}, \widehat{r_2} \in \widehat{R}^h$, let $P_n(\widehat{r_1}, h) = \sum_{i=0}^{n} a_i h^i$, $P_n(\widehat{r_2}, h) = \sum_{i=0}^{n} b_i h^i$, and $P_n(\widehat{r_1} \cdot \widehat{r_2}, h) = \sum_{i=0}^{n} c_i h^i$ and assume $a_i \in F$. Then*

$$c_k = \sum_{i=0}^{k} a_i \cdot b_{k-i}, \quad (35)$$

where $0 \leqslant k \leqslant n$.

4. Example

Example 1. We compute the following partial fraction decomposition:

$$\frac{x^2 + x}{(x-1)^3 \left(x^2+4\right)^2} = \frac{p(x)}{(x-1)^3} + \frac{q(x)}{\left(x^2+4\right)^2}. \quad (36)$$

(i) $h = x - 1$: from Theorem 4, we get

$$\begin{aligned} p(x) &= P_2\left(\frac{x^2 + x}{\left(x^2 + 2^2\right)^2}, h\right) \\ &= P_2\left(P_2\left(x^2 + x, h\right) \cdot P_2\left(\frac{1}{\left(x^2+4\right)^2}, h\right), h\right). \end{aligned} \quad (37)$$

From the Taylor shift (Proposition 9), we have $P_2(x^2 + x, h) = 2 + 3h + h^2$.

From Proposition 6(b), we compute

$$\begin{aligned} P_2\left(\frac{1}{\left(x^2+4\right)^2}, h\right) &= P_2\left(\left[(h+1)^2 + 2^2\right]^{-2}, h\right) \\ &= \frac{1}{25} - \frac{4}{125} h + \frac{2}{625} h^2. \end{aligned} \quad (38)$$

Hence

$$\begin{aligned} &p(x) \\ &= P_2\left(\left(2 + 3h + h^2\right) \cdot \left(\frac{1}{25} - \frac{4}{125} h + \frac{2}{625} h^2\right), h\right) \\ &= \frac{2}{25} + \frac{7}{125}(x-1) - \frac{31}{625}(x-1)^2. \end{aligned} \quad (39)$$

(ii) $h = x^2 + 2^2$: using Proposition 11(b), we get $1/(x-1)^3 = (x+1)^3 \cdot (h-5)^{-3}$. From Theorem 4, we have

$$\begin{aligned} q(x) &= P_1\left(\frac{x^2 + x}{(x-1)^3}, h\right) \\ &= P_1\left((h-5)^{-3} \cdot \left(x^2 + x\right)(x+1)^3, h\right). \end{aligned} \quad (40)$$

From Proposition 6(a), we compute $P_1((h-5)^{-3}, h) = -1/125 - (3/625)h$.

Using Propositions 6(b) and 12, we get

$$\begin{aligned}\left(x^2+x\right)(x+1)^3 &= \left(0+x+x^2\right)\left(1+3x+3x^2+x^3\right)\\ &= \left(0+4x^2+4x^4\right)\\ &\quad+\left(1x+6x^3+1x^5\right).\end{aligned} \tag{41}$$

From Proposition 11(a), we compute

$$\begin{gathered}\begin{pmatrix}1 & -4 & 16\\ 0 & 1 & -8\end{pmatrix}\begin{pmatrix}0\\4\\4\end{pmatrix}=\begin{pmatrix}\mathbf{48}\\ \mathbf{-28}\end{pmatrix},\\ \begin{pmatrix}1 & -4 & 16\\ 0 & 1 & -8\end{pmatrix}\begin{pmatrix}1\\6\\1\end{pmatrix}=\begin{pmatrix}\mathbf{-7}\\ \mathbf{-2}\end{pmatrix}\Longrightarrow\end{gathered} \tag{42}$$

$$P_1\left(\left(x^2+x\right)(x+1)^3,h\right)=(48-7x)-(28+2x)h.$$

Using Proposition 12, we get

$$\begin{aligned}q(x) &= P_1\left(\left(-\frac{1}{125}-\frac{3}{625}h\right)\right.\\ &\quad\left.\cdot\left((48-7x)-(28+2x)h\right),h\right)\\ &= \frac{-48+7x}{125}+\frac{-4+31x}{625}h.\end{aligned} \tag{43}$$

Hence

$$\begin{aligned}\frac{x^2+x}{(x-1)^3\left(x^2+4\right)^2} &= \frac{2}{25(x-1)^3}+\frac{7}{125(x-1)^2}\\ &\quad-\frac{31}{625(x-1)}+\frac{-48+7x}{125\left(x^2+4\right)^2}\\ &\quad+\frac{-4+31x}{625\left(x^2+4\right)}.\end{aligned} \tag{44}$$

5. Main Algorithm (Algorithm 1)

For a given proper ($\deg(N)<\deg(D)$) fraction

$$\frac{N(x)}{D(x)}=\frac{N(x)}{\prod_{i=1}^{k_1}(x-a_i)^{m_i}\prod_{i=1}^{k_2}\left[(x-b_i)^2+c_i^2\right]^{n_i}}, \tag{45}$$

where $(x-a_i)$ and $[(x-b_i)^2+c_i^2]$ are relatively prime, respectively, we return $C_1,\ldots,C_{k_1}$ ($(x-a_i)$-adic polynomials of order (m_i-1)) and $D_1,\ldots,D_{k_2}$ ($[(x-b_i)^2+c_i^2]$-adic polynomial of order (n_i-1)) such that

$$\begin{aligned}&\frac{N(x)}{\prod_{i=1}^{k_1}(x-a_i)^{m_i}\prod_{i=1}^{k_2}\left[(x-b_i)^2+c_i^2\right]^{n_i}}\\ &\quad=\sum_{i=1}^{k_1}\frac{C_i(x)}{(x-a_i)^{m_i}}+\sum_{i=1}^{k_2}\frac{D_i(x)}{\left[(x-b_i)^2+c_i^2\right]^{n_i}}.\end{aligned} \tag{46}$$

From the computation of Theorems 3.1 and 6.1 in [2], the complexity of the main algorithm (Algorithm 1) with simple algebraic operations is given by $O(k\cdot n^2)$ with $M(n)=O(n^2)$, where $k=k_1+k_2$, $n=\deg(D(x))$, and $M(n)$ is any upper bound on the number of operations needed to multiply two n-th degree polynomials. Since $F=\mathbb{R}$ or $\mathbb{C}$, if an FFT (fast Fourier transform) is used, we can reduce the complexity to $O(k\cdot(n\log n))$ since $M(n)=O(n\log n)$ with FFT (Ch. 2 in [5]). For $k_2=0$, the best known complexity is $O((\log k)\cdot(n\log n))$ in [2]. If k is not big enough, our algorithm is still effective since it does not require computing

$$R(x)=\sum_{i=1}^{k}\left(\prod_{j=1,j\neq i}^{k}R_j(x)\right) \tag{47}$$

which is an overhead of the algorithm in [2], where $R_j(x)$ is $(x-a_j)^{m_j}$ or $[(x-b_j)^2+c_j^2]^{n_j}$.

6. Discussion

The main theorem and Propositions 6, 9, 11, and 12 still hold for any field with characteristic zero. Since an irreducible polynomial h over $\mathbb{R}$ or $\mathbb{C}$ has at most degree 2, our recursive formulas for $\deg(h)\leqslant 2$ cover all possible proper rational functions over $\mathbb{R}$ or $\mathbb{C}$. If $\deg(h)>2$, we need to replace Propositions 6, 11(b), and 11(c) with the Extended Euclidean algorithm (Ch. 2 in [5]) which requires an algebra system to be implemented. To compute a h-adic polynomial $P_n(f,n)=\sum_{i=0}^{n}r_ih^i$ of a polynomial f, we use the repeated division of f by h instead of Propositions 9 and 11(c). Then we obtain $\{r_i\}$ as the remainders.

$$\begin{aligned}f &= q_1h+r_0\\ q_1 &= q_2h+r_1\\ &\vdots\\ q_n &= q_{n+1}h+r_n.\end{aligned} \tag{48}$$

We can compute it quickly by applying the divide-and-conquer method (Problem 3.5 of [6]).

7. Conclusion

This paper shows that the partial fraction of a proper rational function with denominator factored completely over $\mathbb{R}$ or $\mathbb{C}$ can be found by computing suitable h-adic polynomials with $\deg(h)\leqslant 2$. We also present recursive formulas (Propositions 6, 9, 11, and 12) to compute h-adic polynomials with $\deg(h)\leqslant 2$. Since this algorithm only requires simple arithmetic operations, it can be implemented easily without a complex computer algebra system.

Conflicts of Interest

The authors declare that there are no conflicts of interest regarding the publication of this paper.

Output: C_i for $i = 1, \ldots, k_1$ and D_i for $i = 1, \ldots, k_2$.

(1) **for** $i \leftarrow 1$ **to** k_1 **do**

(2) Let $h = x - a_i$ and find $\widetilde{N}(h) = N(x)$ using the Taylor shift (Proposition 9). Compute C_i:

$$C_i = P_{m_i}\left(N(x)D(x)^{-1}(x-a_i)^{m_i}, x-a_i\right) = P_{m_i}\left(\widetilde{N}(h)\prod_{j=1,j\neq i}^{k_1}\left(h-(a_j-a_i)\right)^{-m_j}\prod_{j=1}^{k_2}\left[\left(h-(b_j-a_i)\right)^2+c_i^2\right]^{-n_i}, h\right)$$

using a binary splitting scheme (Lemma 4.1 on [2]), Proposition 6 and the h-adic Cauchy product formula (Proposition 12).

(3) **end**

(4) **for** $i \leftarrow 1$ **to** k_2 **do**

(5) Let $t = x - b_i$ and $h = t^2 + c_i^2$. Using the Taylor shift (Proposition 9), find $\widetilde{N}(t) = N(x)$. Using the h-adic shift (Proposition 11(a)), find

$$\widetilde{M}(h) = \widetilde{N}(t)\prod_{\substack{j=1,\\ a_j\neq b_i}}^{k_1}\left(t+(a_j-b_i)\right)^{m_j}\prod_{\substack{j=1,\\ a_j=b_i}}^{k_1}t^{\delta_j}\prod_{\substack{j=1,\\ b_j\neq b_i}}^{k_2}\left[\left(t+(b_j-b_i)\right)^2+c_j^2\right]^{n_i}$$

where $q_j = \lfloor (m_j+1)/2 \rfloor$ and $\delta_j = 2q_j - m_j$ in Proposition 11(b). Compute D_i:

$$D_i = P_{n_i}\left(N(x)D(x)^{-1}\left[(x-b_i)^2+c_i^2\right]^{n_i}, (x-b_i)^2+c_i^2\right) = P_{n_i}\left(\widetilde{M}(h)\cdot\prod_{\substack{j=1,\\ a_j\neq b_i}}^{k_1}\left(h-\left((a_j-b_i)^2+c_j^2\right)\right)^{-m_j}\prod_{\substack{j=1,\\ a_j=b_i}}^{k_1}\left(h-c_i^2\right)^{-q_j}\right.$$

$$\left.\cdot\prod_{\substack{j=1,\\ b_j\neq b_i}}^{k_2}\left[\left(h-\left((b_j-b_i)^2-c_j^2+c_i^2\right)\right)^2+\left(2(b_j-b_i)c_j\right)^2\right]^{-n_i}\prod_{\substack{j=1,\\ i\neq j,\\ b_j=b_i}}^{k_2}\left(h-\left(-c_j^2+c_i^2\right)\right)^{-n_i}, h\right)$$

using a binary splitting scheme (Lemma 4.1 on [2]), Proposition 6 and the h-adic Cauchy product formula (Proposition 12).

(6) **end**

ALGORITHM 1: Main algorithm.

Acknowledgments

The work of Xin Zhang has been partially supported through The CUNY Research Scholars Program (CRSP) and US Department of Education Queensborough MSEIP (P120A140057). The authors thank Dr. Jeehoon Park and Dr. Kostas Stroumbakis for their valuable comments.

References

[1] Y. Ma, J. Yu, and Y. Wang, "Efficient recursive methods for partial fraction expansion of general rational functions," *Journal of Applied Mathematics*, vol. 2014, Article ID 895036, 18 pages, 2014.

[2] H. T. Kung and D. M. Tong, "Fast algorithms for partial fraction decomposition," *SIAM Journal on Computing*, vol. 6, no. 3, pp. 582–593, 1977.

[3] M. F. Atiyah and I. G. Macdonald, *Introduction to Commutative Algebra. Addison-Wesley series in mathematics. Reading, Mass*, Addison-Wesley, London, UK, 1969.

[4] S. Kleiman and A. Altman, *A Term of Commutative Algebra*, Worldwide Center of Mathematics, LLC, 2013.

[5] D. E. Knuth, *The Art of Computer Programming*, vol. 2, Seminumerical Algorithms. Addison-Wesley Longman Publishing Co., Boston, MA, USA, 3rd edition, 1997.

[6] D. Bini and V. Y. Pan, *Polynomial and Matrix Computations*, vol. 1, Birkhäuser Boston, Boston, MA, USA, 1994.

Vector Spaces of New Special Magic Squares: Reflective Magic Squares, Corner Magic Squares, and Skew-Regular Magic Squares

Thitarie Rungratgasame, Pattharapham Amornpornthum, Phuwanat Boonmee, Busrun Cheko, and Nattaphon Fuangfung

Department of Mathematics, Faculty of Science, Srinakharinwirot University, Bangkok 10110, Thailand

Correspondence should be addressed to Thitarie Rungratgasame; thitarie@g.swu.ac.th

Academic Editor: Marianna A. Shubov

The definition of a regular magic square motivates us to introduce the new special magic squares, which are reflective magic squares, corner magic squares, and skew-regular magic squares. Combining the concepts of magic squares and linear algebra, we consider a magic square as a matrix and find the dimensions of the vector spaces of these magic squares under the standard addition and scalar multiplication of matrices by using the rank-nullity theorem.

1. Introduction

A classical magic square is an $n \times n$ square array of the distinct positive integers $1, 2, \ldots, n^2$ arranged in such a way that the summations of n entries along each row, each column, the main diagonal, and the cross diagonal are all equal to the same constant, called a *magic sum*. For example, the famous classical magic square appearing in Albrecht Dürer's engraving *Melancholia*, known as Yang Hui-Dürer magic square, is

$$\begin{bmatrix} 16 & 3 & 2 & 13 \\ 5 & 10 & 11 & 8 \\ 9 & 6 & 7 & 12 \\ 4 & 15 & 14 & 1 \end{bmatrix}. \quad (1)$$

In this article, a magic square is also defined in the same context as the classical one except that all of its entries can be any real numbers and need not be distinct. The idea to study a magic square as a matrix was initiated by Fox [1] in 1956 where he considered a magic square as a matrix of real numbers (each entry needs not to be distinct) and showed that the inverse of a 3×3 magic square with the magic sum $\mu \neq 0$ was also a magic square with the magic sum $1/\mu$. Certainly, many research studies of magic squares in the framework of linear algebra come afterwards.

The set of all $n \times n$ magic squares is well-known to be a vector space over $\mathbb{R}$ under the usual addition and scalar multiplication of matrices, denoted by $\mathrm{MS}(n)$. A magic square with the zero magic sum is called a *zero magic square*, and $0\mathrm{MS}(n)$ denotes the set of all $n \times n$ zero magic squares, which is a subspace of $\mathrm{MS}(n)$. In 1959, Ratliff [2] found the dimension of the vector space of all $n \times n$ magic squares whose entries are from any field F. Later in 1980, Ward [3] by applying the rank and nullity theorem derived that the dimension of the vector space of all $n \times n$ magic squares over $\mathbb{R}$ was equal to $n^2 - 2n$.

One of the popular types of the special magic squares that draw attentions from researchers is the regular magic squares; for example, see [4–9]. The property to define a regular magic square was noticed during the construction of magic squares (see [10] p. 202). We then challenge ourselves by constructing new types of magic squares which leads us to determine the following three types of magic squares.

2. Reflective Magic Squares

Before introducing the reflective magic squares, we shall recall the definition of the regular magic squares.

Definition 1. An $n \times n$ magic square $A = [a_{ij}]$ with a magic sum μ is said to be *regular* if

$$a_{ij} + a_{(n+1-i)(n+1-j)} = \frac{2\mu}{n} \quad \forall i, j = 1, 2, 3, \ldots, n. \tag{2}$$

The set of all $n \times n$ regular magic squares is denoted by RMS(n) and 0RMS(n) is the set of all $n \times n$ zero regular magic squares. The magic squares below are examples of 4×4 and 5×5 regular magic squares where the entries with the same symbol represent the pairs satisfying (2):

$$\begin{bmatrix} \spadesuit & \heartsuit & \clubsuit & \diamondsuit \\ \triangle & \blacksquare & \triangledown & \blacktriangle \\ \blacktriangle & \triangledown & \blacksquare & \triangle \\ \diamondsuit & \clubsuit & \heartsuit & \spadesuit \end{bmatrix},$$

$$\begin{bmatrix} \spadesuit & \heartsuit & \clubsuit & \diamondsuit & \bullet \\ \triangle & \blacksquare & \triangledown & \blacktriangle & \square \\ \bigstar & \triangleleft & \otimes & \triangleleft & \bigstar \\ \square & \blacktriangle & \triangledown & \blacksquare & \triangle \\ \bullet & \diamondsuit & \clubsuit & \heartsuit & \spadesuit \end{bmatrix}. \tag{3}$$

In the regular conditions, any two entries added together are located in the positions diametrically equidistant from the center of the square. This way of symmetry persuades us to add up a new pair of entries that are symmetric along the main diagonal line. In this fashion, we define the new magic squares.

$$\begin{bmatrix} \spadesuit & \heartsuit & \clubsuit & \diamondsuit \\ \heartsuit & \spadesuit & \square & \bullet \\ \clubsuit & \square & \spadesuit & \circ \\ \diamondsuit & \bullet & \circ & \spadesuit \end{bmatrix}. \tag{4}$$

Definition 2. An $n \times n$ magic square $A = [a_{ij}]$ with a magic sum μ is said to be *reflective* if

$$a_{ij} + a_{ji} = \frac{2\mu}{n} \quad \forall i, j = 1, 2, 3, 4, \ldots, n. \tag{5}$$

Nonetheless, condition (5) imposed here with the summation conditions of a magic square forces all entries on the main diagonal to be μ/n. In particular, all entries on the main diagonal of a zero magic square must be zero. Let FMS(n) and 0FMS(n) denote the set of all $n \times n$ reflective magic squares and zero reflective magic squares, respectively. Consequently, FMS(n) and 0FMS(n) are subspaces of MS(n).

Lee et al. [11] showed that the dimension of the vector space of all $n \times n$ regular magic square matrices (RMS(n)) is $(n-1)^2/2+1$ when n is odd and $n(n-2)/2+1$ when n is even. Their results lead us to pursue the dimension of FMS(n). We shall begin with the most basic approach.

Let $A \in$ 0FMS(4) such that

$$A = \begin{bmatrix} 0 & a_{12} & a_{13} & a_{14} \\ a_{21} & 0 & a_{23} & a_{24} \\ a_{31} & a_{32} & 0 & a_{34} \\ a_{41} & a_{42} & a_{43} & 0 \end{bmatrix}. \tag{6}$$

Then we derive the homogeneous system of linear equations of a_{ij}s' satisfying the row conditions, column conditions, and the reflective conditions, respectively. The main condition is omitted in this case because each entry on the main diagonal is zero. Moreover, the cross diagonal condition is replaced by the reflective conditions. Let $\overline{A}$ denote the coefficient matrix where the variables are $a_{12}, a_{13}, a_{14}, a_{21}, a_{23}, a_{24}, a_{31}, a_{32}, a_{34}, a_{41}, a_{42}, a_{43}$, as shown in the following example of the coefficient matrix $\overline{A}$ when $A \in$ 0FMS(4):

$$\begin{bmatrix} 1&1&1&0&0&0&0&0&0&0&0&0 \\ 0&0&0&1&1&1&0&0&0&0&0&0 \\ 0&0&0&0&0&0&1&1&1&0&0&0 \\ 0&0&0&0&0&0&0&0&0&1&1&1 \\ 0&0&0&1&0&0&1&0&0&1&0&0 \\ 1&0&0&0&0&0&0&1&0&0&1&0 \\ 0&1&0&0&1&0&0&0&0&0&0&1 \\ 0&0&1&0&0&1&0&0&1&0&0&0 \\ 1&0&0&1&0&0&0&0&0&0&0&0 \\ 0&1&0&0&0&0&1&0&0&0&0&0 \\ 0&0&1&0&0&0&0&0&0&1&0&0 \\ 0&0&0&0&1&0&0&1&0&0&0&0 \\ 0&0&0&0&0&1&0&0&0&0&1&0 \\ 0&0&0&0&0&0&0&0&1&0&0&1 \end{bmatrix}. \tag{7}$$

By using elementary row operations to $\overline{A}$ to get the row-reduced echelon matrix $\overline{A}_{\text{rr}}$ of $\overline{A}$, we derive a basis of 0FMS(4) to be $\{A_1, A_2, A_3\}$

$$A_1 = \begin{bmatrix} 0 & -1 & 1 & 0 \\ 1 & 0 & -1 & 0 \\ -1 & 1 & 0 & 0 \\ 0 & 0 & 0 & 0 \end{bmatrix},$$

$$A_2 = \begin{bmatrix} 0 & -1 & 0 & 1 \\ 1 & 0 & 0 & -1 \\ 0 & 0 & 0 & 0 \\ -1 & 1 & 0 & 0 \end{bmatrix},$$

$$A_3 = \begin{bmatrix} 0 & 0 & -1 & 1 \\ 0 & 0 & 0 & 0 \\ 1 & 0 & 0 & -1 \\ -1 & 0 & 1 & 0 \end{bmatrix}. \tag{8}$$

Moreover, the associate basis for FMS(4) is $\{A_1, A_2, A_3, U\}$, where

$$U = \begin{bmatrix} 1 & 1 & 1 & 1 \\ 1 & 1 & 1 & 1 \\ 1 & 1 & 1 & 1 \\ 1 & 1 & 1 & 1 \end{bmatrix}. \tag{9}$$

Therefore, dim(0FM(4)) = 3 and dim(FM(4)) = 4. Elements in 0FMS(4) and FMS(4) can be calculated by using these bases; for example,

$$2A_1 + 4A_3 = \begin{bmatrix} 0 & -2 & -2 & 4 \\ 2 & 0 & -2 & 0 \\ 2 & 2 & 0 & -4 \\ -4 & 0 & 4 & 0 \end{bmatrix} \in \text{0FMS}(4),$$
$$3A_2 + A_3 + 4U = \begin{bmatrix} 4 & 1 & 3 & 8 \\ 7 & 4 & 4 & 1 \\ 5 & 4 & 4 & 3 \\ 0 & 7 & 5 & 4 \end{bmatrix} \in \text{FMS}(4). \tag{10}$$

However, this basic method cannot be used to find the dimension of 0FMS(n) and FMS(n) in the general cases. The rank and nullity theorem plays an important role to obtain the results.

Theorem 3. *For $n \geq 4$, the dimension of* 0FMS(n) *is $(n^2 - 3n + 2)/2$ and the dimension of* FMS(n) *is $(n^2 - 3n + 4)/2$.*

Proof. Let $A = [a_{ij}] \in$ 0FMS(n), where $a_{ii} = 0$ and $a_{ij} \in \mathbb{R}$, for all $i, j \in \{1, 2, 3, \ldots, n\}$. By the summation of a_{ij}s' along each line to be zero, we then write the homogeneous system of linear equations of a_{ij}s' satisfying the row conditions, column conditions, and the reflective conditions, respectively. The cross diagonal condition is not included here because it is replaced by the reflective conditions. We denote R_i, the ith row of the coefficient matrix, and let $\overline{A}$ denote the coefficient matrix where the variables are all a_{ij} such that $i \neq j$ for all $i, j \in \{1, 2, 3, \ldots, n\}$. That is, there are $n^2 - n$ variables in the homogeneous system. By the rank-nullity theorem, dim(0FMS(n)) can be derived by finding the rank of the coefficient matrix $\overline{A}$ first. In this case, $\overline{A}$ is an $((n^2 + 3n)/2) \times (n^2 - n)$ matrix (see equation (7)) whose rows are ordered by n row conditions (starting from the first row), n column conditions (starting from the first column), and $n(n-1)/2$ reflective conditions (ordering pairs from left to right and then to the rows below). The rows of $\overline{A}$ are related in the following ways:

(1) R_{2n} is a linear combination of R_k for all $k = 1, 2, \ldots, 2n-1$.

(2) R_{3n-1} is a linear combination of R_1, R_{n+1}, and R_k for all $k = 2n+1, 2n+2, \ldots, 3n-2$.

(3) For each $2 \leq i \leq n-3$, $R_{2n+i(2n-1-i)/2}$ is a linear combination of R_i, R_{n+i}, and $n-2$ rows from the reflective conditions.

(4) $R_{(n^2+3n-4)/2}$ is a linear combination of R_k for all $k = n, n+1, \ldots, 2n-1$ and $(n^2 - 3n)/2$ rows from the reflective conditions.

(5) $R_{(n^2+3n-2)/2}$ is a linear combination of $R_{n-2}, R_n, R_{n+1}, R_{n+2}, \ldots, R_{2n-3}, R_{2n-1}$ and $(n^2 - 5n + 6)/2$ rows from the reflective conditions.

(6) $R_{(n^2+3n)/2}$ is a linear combination of R_k for all $k = n-1, n, n+1, \ldots, 2n-2$ and $(n^2 - 5n + 6)/2$ rows from the reflective conditions.

(7) The remaining $(n^2 + n - 2)/2$ rows, which are all rows except $R_{k(2n-1-k)/2}$ where $k = 0, 1, \ldots, n$, are linearly independent.

Therefore, dim(0FMS(n)) $= (n^2 - n) - (n^2 + n - 2)/2 = (n^2 - 3n + 2)/2$. Let U denote the $n \times n$ matrix with all entries of 1. We observe that, for any $A \in$ FMS(n) with a magic sum μ, there is the associate zero magic square $A_0 \in$ 0FMS(n), where $A_0 = A - (\mu/n)U$. It is easy to see that $\mathscr{B} \cup \{U\}$ forms a basis for FMS(n) for any basis $\mathscr{B}$ of 0FMS(n). Hence dim(FMS(n)) $=$ dim(0FMS(n)) $+ 1 = (n^2 - 3n + 4)/2$. □

The generalization of a regular magic square which we shall introduce next will be a magic square satisfying a condition of combining four entries instead of adding two entries.

3. Corner Magic Squares

The idea to construct this type of magic squares comes from our observation on any four entries in a magic square which form corners of a rectangle whose center is the same as the magic square and is symmetric horizontally and vertically. In particular, even the famous Yang Hui-Dürer magic square also satisfies the observed properties.

Definition 4. An $n \times n$ magic square $A = [a_{ij}]$ with a magic sum μ is said to be a *corner magic square* if

$$a_{ii} + a_{(n+1-i)(n+1-i)} + a_{i(n+1-i)} + a_{(n+1-i)i} = \frac{4\mu}{n} \tag{11}$$

for all $i = 1, 2, 3, \ldots, n$. The set of all $n \times n$ corner magic squares is denoted by CMS(n) and 0CMS(n) is the set of all $n \times n$ zero regular magic squares. The magic squares below are examples of 4×4 and 5×5 corner magic squares where the entries with the same symbol represent the entries satisfying (11):

$$\begin{bmatrix} \spadesuit & \heartsuit & \bigcirc & \spadesuit \\ \triangle & \blacksquare & \blacksquare & \triangledown \\ \triangleleft & \blacksquare & \blacksquare & \triangleright \\ \spadesuit & \diamond & \square & \spadesuit \end{bmatrix},$$
$$\begin{bmatrix} \spadesuit & \heartsuit & \clubsuit & \diamond & \spadesuit \\ \triangle & \blacksquare & \triangledown & \blacksquare & \triangleright \\ \bullet & \triangleleft & \bigstar & \bigcirc & \blacktriangle \\ \boxtimes & \blacksquare & \circ & \blacksquare & \circledcirc \\ \spadesuit & \ominus & \blacklozenge & \square & \spadesuit \end{bmatrix}. \tag{12}$$

By the linear property of (11), it is easy to check that both $0\mathrm{CMS}(n)$ and $\mathrm{CMS}(n)$ are subspaces of $\mathrm{MS}(n)$. Moreover, $\mathrm{RMS}(n) \subseteq \mathrm{CMS}(n)$ and $0\mathrm{RMS}(n) \subseteq 0\mathrm{CMS}(n)$, but the converse does not hold; for example,

$$\begin{bmatrix} 4 & 5 & 1 & 2 \\ 3 & 1 & 5 & 3 \\ 3 & 3 & 3 & 3 \\ 2 & 3 & 3 & 4 \end{bmatrix} \tag{13}$$

is an element in CMS(4), but not in RMS(4). Nevertheless, $\mathrm{CMS}(3) = \mathrm{RMS}(3)$ and $0\mathrm{CMS}(3) = 0\mathrm{RMS}(3)$.

To find a basis of 0CMS(4), we can use the Gauss-Jordan elimination method as before to retrieve a basis of 0CMS(4) to be $\{A_1, A_2, A_3, A_4, A_5, A_6, A_7\}$, where

$$A_1 = \begin{bmatrix} 1 & 1 & -1 & -1 \\ -1 & -1 & 1 & 1 \\ 0 & 0 & 0 & 0 \\ 0 & 0 & 0 & 0 \end{bmatrix},$$
$$A_2 = \begin{bmatrix} 0 & -1 & 1 & 0 \\ 1 & 0 & -1 & 0 \\ -1 & 1 & 0 & 0 \\ 0 & 0 & 0 & 0 \end{bmatrix},$$
$$A_3 = \begin{bmatrix} 0 & 1 & -1 & 0 \\ 1 & -1 & 0 & 0 \\ -1 & 0 & 1 & 0 \\ 0 & 0 & 0 & 0 \end{bmatrix},$$
$$A_4 = \begin{bmatrix} 1 & 1 & -1 & -1 \\ 0 & -1 & 1 & 0 \\ -1 & 0 & 0 & 1 \\ 0 & 0 & 0 & 0 \end{bmatrix},$$
$$A_5 = \begin{bmatrix} 1 & 0 & -1 & 0 \\ 0 & -1 & 1 & 0 \\ 0 & 0 & 0 & 0 \\ -1 & 1 & 0 & 0 \end{bmatrix},$$
$$A_6 = \begin{bmatrix} 1 & 1 & -2 & 0 \\ 0 & -1 & 1 & 0 \\ 0 & 0 & 0 & 0 \\ -1 & 0 & 1 & 0 \end{bmatrix},$$
$$A_7 = \begin{bmatrix} 1 & 2 & -2 & -1 \\ 0 & -2 & 2 & 0 \\ 0 & 0 & 0 & 0 \\ -1 & 0 & 0 & 1 \end{bmatrix}. \tag{14}$$

Moreover, the associate basis of CMS(4) is $\{A_1, \ldots, A_7, U\}$, where

$$U = \begin{bmatrix} 1 & 1 & 1 & 1 \\ 1 & 1 & 1 & 1 \\ 1 & 1 & 1 & 1 \\ 1 & 1 & 1 & 1 \end{bmatrix}; \tag{15}$$

that is, $\dim(0\mathrm{CMS}(4)) = 7$ and $\dim(\mathrm{CMS}(4)) = 8$. Apparently, $\dim(\mathrm{CMS}(4)) = \dim(\mathrm{MS}(4))$. This means that CMS(4) is the same space as MS(4). Furthermore, we can use the bases to find a magic square in 0MS(4) and MS(4); for example,

$$3A_2 + 5A_7 = \begin{bmatrix} 5 & 7 & -7 & -5 \\ 3 & -10 & 7 & 0 \\ -3 & 3 & 0 & 0 \\ -5 & 0 & 0 & 5 \end{bmatrix},$$
$$3A_2 + 5A_7 + 13U = \begin{bmatrix} 18 & 20 & 6 & 8 \\ 16 & 3 & 20 & 13 \\ 10 & 16 & 13 & 13 \\ 8 & 13 & 13 & 18 \end{bmatrix}. \tag{16}$$

Nevertheless, $\mathrm{CMS}(n)$ is not always the same set as $\mathrm{MS}(n)$. For the general case, we can apply the rank and nullity theorem to find the dimension of $0\mathrm{CMS}(n)$ and $\mathrm{CMS}(n)$.

Theorem 5. *For $n \geq 5$, the dimension of $0CMS(n)$ is $(2n^2-5n-1)/2$ when n is odd and $(2n^2-5n)/2$ when n is even. Moreover,* *the dimension of CMS(n) is* $(2n^2 - 5n + 1)/2$ *when n is odd and* $(2n^2 - 5n + 2)/2$ *when n is even.*

Proof. Let $A = [a_{ij}] \in 0\text{CMS}(n)$, where $a_{ij} \in \mathbb{R}$ for all $i, j \in \{1, 2, 3, \ldots, n\}$. By the summation of a_{ij}s' along each line to be zero, we then write the homogeneous system of linear equations of a_{ij}s' satisfying the row conditions, column conditions, the main diagonal condition, the cross diagonal condition, and the corner conditions, respectively. As before, let $\overline{A}$ denote the coefficient matrix.

Case 1 (n is odd). Including all conditions, there are $((5n + 3)/2)$ homogeneous equations in the system and $a_{((n+1)/2)((n+1)/2)} = 0$ by the corner condition at the center. Then all a_{ij} except $a_{((n+1)/2)((n+1)/2)}$ are variables in this system. We then arrange the equations so that the coefficient matrix $\overline{A}$ is an $((5n + 3)/2) \times (n^2 - 1)$ matrix of 0's and 1's where its rows are, respectively, ordered by n row conditions, n column conditions, 1 main diagonal condition, 1 cross diagonal condition, and $(n - 1)/2$ corner conditions. See the following matrix as an example of the coefficient matrix $\overline{A}$ when $A \in 0\text{CMS}(4)$:

$$\begin{bmatrix} 1&1&1&1&0&0&0&0&0&0&0&0&0&0&0&0\\ 0&0&0&0&1&1&1&1&0&0&0&0&0&0&0&0\\ 0&0&0&0&0&0&0&0&1&1&1&1&0&0&0&0\\ 0&0&0&0&0&0&0&0&0&0&0&0&1&1&1&1\\ 1&0&0&0&1&0&0&0&1&0&0&0&1&0&0&0\\ 0&1&0&0&0&1&0&0&0&1&0&0&0&1&0&0\\ 0&0&1&0&0&0&1&0&0&0&1&0&0&0&1&0\\ 0&0&0&1&0&0&0&1&0&0&0&1&0&0&0&1\\ 1&0&0&0&0&1&0&0&0&0&1&0&0&0&0&1\\ 0&0&0&1&0&0&1&0&0&1&0&0&1&0&0&0\\ 1&0&0&1&0&0&0&0&0&0&0&0&1&0&0&1\\ 0&0&0&0&0&1&1&0&0&1&1&0&0&0&0&0 \end{bmatrix}. \tag{17}$$

Observing the coefficient matrix $\overline{A}$, we derive the relationships of the rows of $\overline{A}$ as follows:

(1) R_{2n} is a linear combination of R_k for all $k = 1, 2, \ldots, 2n - 1$.

(2) R_{2n+2} is a linear combination of R_{2n+1} and R_k for all $k = 2n + 3, 2n + 4, \ldots, (5n + 3)/2$.

(3) The remaining rows are linearly independent.

Therefore, $\overline{A}$ has $(5n - 1)/2$ linearly independent rows. By the rank-nullity theorem, $\dim(0\text{CMS}(n)) = (n^2 - 1) - ((5n - 1)/2) = (2n^2 - 5n - 1)/2$.

Case 2 (n is even). In this case, the coefficient matrix $\overline{A}$ is an $((5n + 4)/2) \times n^2$ matrix of 0's and 1's such that its rows are, respectively, arranged by n row conditions, n column conditions, 1 main diagonal condition, 1 cross diagonal condition, and $n/2$ corner conditions (see (17) for example). Moreover, we derive the relationships of the rows of $\overline{A}$ as follows:

(1) R_{2n} is a linear combination of R_k for all $k = 1, 2, \ldots, 2n - 1$.

(2) R_{2n+2} is a linear combination of R_{2n+1} and R_k for all $k = 2n + 3, 2n + 4, \ldots, (5n + 4)/2$.

(3) The remaining rows are linearly independent.

Therefore, $\overline{A}$ has $5n/2$ linearly independent rows. By the rank-nullity theorem, $\dim(0\text{CMS}(n)) = n^2 - (5n/2) = (2n^2 - 5n)/2$.

Let U be the $n \times n$ matrix with all entries of 1. For any $A \in \text{CMS}(n)$ with a magic sum μ, we have the associate zero magic square $A_0 \in 0\text{CMS}(n)$ such that $A_0 = A - (\mu/n)U$. From this fact, $\mathcal{B} \cup \{U\}$ forms a basis for CMS(n) when $\mathcal{B}$ is a basis for 0CMS(n). The dimension of CMS(n) is then derived to be $\dim(0\text{CMS}(n)) + 1$. □

4. Skew-Regular Magic Squares

From the corner magic squares, we shift to observe any four entries in a magic square which form corners of a rectangle located in the position that its center is the same as the magic square and symmetrical axes are the main diagonal and cross diagonal.

Definition 6. An $n \times n$ magic square $A = [a_{ij}]$ with a magic sum μ is said to be a *skew-regular magic square* if

$$a_{ij} + a_{(n+1-i)(n+1-j)} + a_{ji} + a_{(n+1-j)(n+1-i)} = \frac{4\mu}{n}, \tag{18}$$

where $i \in \{1, 2, 3, \ldots, (n-1)/2\}$ and $j \in \{i+1, i+2, \ldots, n-i\}$ if n is odd and $i \in \{1, 2, 3, \ldots, (n-2)/2\}$ and $j \in \{i+1, i+2, \ldots, n-i\}$ if n is even. The set of all $n \times n$ skew-regular magic squares is denoted by SRMS(n) and 0SRMS(n) is the set of all $n \times n$ zero skew-regular magic squares. The matrices below are examples to indicate groups of four entries (not on the diagonals) in the same symbols satisfying (18) for skew-regular magic squares:

$$\begin{bmatrix} \spadesuit & \heartsuit & \blacksquare & \bigcirc \\ \heartsuit & \spadesuit & \bigcirc & \blacksquare \\ \blacksquare & \bigcirc & \spadesuit & \heartsuit \\ \bigcirc & \blacksquare & \heartsuit & \spadesuit \end{bmatrix},$$

$$\begin{bmatrix} \spadesuit & \heartsuit & \clubsuit & \diamondsuit & \spadesuit \\ \heartsuit & \spadesuit & \triangledown & \spadesuit & \diamondsuit \\ \clubsuit & \triangledown & \spadesuit & \triangledown & \clubsuit \\ \diamondsuit & \spadesuit & \triangledown & \spadesuit & \heartsuit \\ \spadesuit & \diamondsuit & \clubsuit & \heartsuit & \spadesuit \end{bmatrix}. \tag{19}$$

As before, the linear property of (18) implies that both 0SRMS(n) and SRMS(n) are subspaces of MS(n).

Theorem 7. *For $n \geq 6$,*

$$\dim 0\mathrm{SRMS}\,(n) = \begin{cases} \dfrac{3n^2 - 6n - 1}{4}, & \text{if } n \text{ is odd,} \\ \dfrac{3n^2 - 6n}{4}, & \text{if } n \text{ is even,} \end{cases}$$
$$\dim \mathrm{SRMS}\,(n) = \begin{cases} \dfrac{3n^2 - 6n + 3}{4}, & \text{if } n \text{ is odd,} \\ \dfrac{3n^2 - 6n + 4}{4}, & \text{if } n \text{ is even.} \end{cases} \tag{20}$$

Proof. Let $A = [a_{ij}] \in 0\mathrm{SRMS}(n)$, where $a_{ij} \in \mathbb{R}$ for all $i, j \in \{1, 2, \ldots, n\}$.

Case 1 (n is odd). All rows of the coefficient matrix $\overline{A}$ are ordered by n row conditions, n column conditions, the main diagonal condition, the cross diagonal condition, and $(n-1)^2/4$ skew-regular conditions, respectively. Including all conditions, there are $((n^2 + 6n + 9)/4)$ homogeneous equations in the system. By the sums along the middle column, the middle row, and all four corners on these two lines, we can conclude that $a_{((n+1)/2)((n+1)/2)} = 0$. Then we have the homogeneous system with the variables all a_{ij} except $a_{((n+1)/2)((n+1)/2)}$. Observing the $((n^2 + 6n + 9)/4) \times (n^2 - 1)$ coefficient matrix $\overline{A}$, we derive the relationships of the rows of $\overline{A}$ as follows:

(1) R_{2n} is a linear combination of R_k for all $k = 1, 2, \ldots, 2n - 1$.

(2) $R_{(n^2+6n+5)/4}$ is a linear combination of $R_1, R_2, \ldots, R_{(n-1)/2}, R_{(n+3)/2}, \ldots, R_n, R_{(3n+1)/2}, R_{2n+1}, R_{2n+2}$ and all rows from the skew-regular conditions except $R_{(n^2+6n+9)/4-k(k+1)}$ for all $k = 0, 1, \ldots, (n-3)/2$.

(3) $R_{(n^2+6n+9)/4}$ is a linear combination of $R_{(n+1)/2}$, $R_{(3n+1)/2}$ and $R_{(n^2+6n+9)/4-k(k+1)}$ for all $k = 1, \ldots, (n-3)/2$.

(4) The remaining rows are linearly independent.

By the rank and nullity theorem, $\dim 0\mathrm{SRMS}(n) = (n^2 - 1) - ((n^2 + 6n + 9)/4 - 3) = (3n^2 - 6n - 1)/4$.

Case 2 (n is even). The rows of $\overline{A}$ are arranged as in the previous case except that there are $(n^2 - 2n)/4$ skew-regular conditions; see the following matrix as an example of the coefficient matrix $\overline{A}$ when $A \in 0\mathrm{SRMS}(4)$:

$$\begin{bmatrix}
1&1&1&1&0&0&0&0&0&0&0&0&0&0&0&0\\
0&0&0&0&1&1&1&1&0&0&0&0&0&0&0&0\\
0&0&0&0&0&0&0&0&1&1&1&1&0&0&0&0\\
0&0&0&0&0&0&0&0&0&0&0&0&1&1&1&1\\
1&0&0&0&1&0&0&0&1&0&0&0&1&0&0&0\\
0&1&0&0&0&1&0&0&0&1&0&0&0&1&0&0\\
0&0&1&0&0&0&1&0&0&0&1&0&0&0&1&0\\
0&0&0&1&0&0&0&1&0&0&0&1&0&0&0&1\\
1&0&0&0&0&1&0&0&0&0&1&0&0&0&0&1\\
0&0&0&1&0&0&1&0&0&1&0&0&1&0&0&0\\
0&1&0&0&1&0&0&0&0&0&0&1&0&0&1&0\\
0&0&1&0&0&0&0&1&1&0&0&0&0&1&0&0
\end{bmatrix}, \tag{21}$$

and hence, $\overline{A}$ has $(n^2 + 6n + 8)/4$ rows and n^2 columns. Moreover, we derive the relationships of the rows of $\overline{A}$ as follows:

(1) R_{2n} is a linear combination of R_k for all $k = 1, 2, \ldots, 2n - 1$.

(2) $R_{(n^2+6n+8)/4}$ is a linear combination of $R_1, R_2, \ldots, R_n$ and $R_{2n+1}, R_{2n+2}, \ldots, R_{(n^2+6n+4)/4}$.

(3) The remaining rows are linearly independent.

Therefore, $\overline{A}$ has $(n^2 + 6n)/4$ linearly independent rows resulting in $\dim 0\mathrm{SRMS}(n) = n^2 - (n^2 + 6n)/4 = (3n^2 - 6n)/4$.

From the same reason as before, $\mathscr{B} \cup \{U\}$ is a basis for SRMS(n) for any basis $\mathscr{B}$ of 0SRMS(n) where U is the $n \times n$ matrix with all entries of 1. Hence dim SRMS(n) is derived afterward. □

5. Conclusion

According to the studies of regular magic squares, the reflective magic squares, corner magic squares, and skew-regular magic squares, as the generalization, can also lead to new research studies of their properties and applications.

Competing Interests

The authors declare that there is no conflict of interests regarding the publication of this paper.

References

[1] C. Fox, "Magic matrices," *The Mathematical Gazette*, vol. 40, pp. 209–2011, 1956.

[2] J. Ratliff, "The dimension of the magic square vector space," *The American Mathematical Monthly*, vol. 66, pp. 793–795, 1959.

[3] I. Ward, "Vector spaces of magic squares," *Mathematics Magazine*, vol. 53, no. 2, pp. 108–111, 1980.

[4] R. B. Mattingly, "Even order regular magic squares are singular," *The American Mathematical Monthly*, vol. 107, no. 9, pp. 777–782, 2000.

[5] P. Loly, I. Cameron, W. Trump, and D. Schindel, "Magic square spectra," *Linear Algebra and Its Applications*, vol. 430, no. 10, pp. 2659–2680, 2009.

[6] M. Z. Lee, E. Love, S. K. Narayan, E. Wascher, and J. D. Webster, "On nonsingular regular magic squares of odd order," *Linear Algebra and Its Applications*, vol. 437, no. 6, pp. 1346–1355, 2012.

[7] R. P. Nordgren, "On properties of special magic square matrices," *Linear Algebra and its Applications*, vol. 437, no. 8, pp. 2009–2025, 2012.

[8] C. J. Chan, M. G. Mainkar, S. K. Narayan, and J. D. Webster, "A construction of regular magic squares of odd order," *Linear Algebra and Its Applications*, vol. 457, pp. 293–302, 2014.

[9] L. Liu, Z. Gao, and W. Zhao, "On an open problem concerning regular magic squares of odd order," *Linear Algebra and Its Applications*, vol. 459, pp. 1–12, 2014.

[10] W. W. Rouse Ball and H. S. M. Coxeter, *Mathematical Recreations and Essays*, Courier Corporation, 1987.

[11] M. Lee, E. Love, and E. Wascher, *Linear Algebra of Magic Squares, Undergraduate Research*, Central Michigan University, Mount Pleasant, Mich, USA, 2006.

26

A Note on Primitivity of Ideals in Skew Polynomial Rings of Automorphism Type

Edilson Soares Miranda

Departamento de Ciências, Centro de Ciências Exatas, Universidade Estadual de Maringá, 87360-000 Goioerê, PR, Brazil

Correspondence should be addressed to Edilson Soares Miranda; esmiranda@uem.br

Academic Editor: Kaiming Zhao

We extend results about primitive ideals in polynomial rings over nil rings originally proved by Smoktunowicz (2005) for σ-primitive ideals in skew polynomial rings of automorphism type.

1. Introduction

Throughout this paper R denotes an associative ring but does not necessarily have an identity element and $\sigma : R \to R$ an automorphism of R, unless otherwise stated. We denote by $R[x;\sigma]$ the skew polynomial rings of automorphism type whose elements are polynomials $\sum_{i=0}^{n} a_i x^i$, $a_i \in R$, for every $i \geq 0$, with usual addition and the following multiplication: $xa = \sigma(a)x$ for all $a \in R$.

A ring R is said to be a Jacobson ring if every prime ideal of R is an intersection of (either left or right) primitive ideals of R. In [1], Smoktunowicz proved that if R is a nil ring and I an ideal of $R[x]$, then $R[x]/I$ is Jacobson radical if and only if $R[x]/I'[x]$ is Jacobson radical, where I' is the ideal of R generated by coefficients of polynomial from I. Also if R is a nil ring and I is a primitive ideal of $R[x]$, then $I = M[x]$ for some ideal M of R and affirmative answer to this question is equivalent to the Köthe conjecture. Our main results state that if R is a nil ring and I an ideal of $R[x;\sigma]$, then $R[x;\sigma]/I$ is σ-Jacobson radical if and only if $R[x;\sigma]/I'[x;\sigma]$ is σ-Jacobson radical, where I' is the ideal of R generated by coefficients of polynomial from I. Also if R is a nil ring and I is a σ-primitive ideal of $R[x;\sigma]$, then $I = M[x;\sigma]$ for some ideal M of R. This result includes, as particular cases, all the above results.

Now we recall some terminology and results; see [2–4]. A right ideal Q of a ring R is called modular in R if and only if there exists an element $b \in R$ such that $a - ba \in Q$ for every $a \in R$. An ideal I of a ring R is said to be a σ-invariant if and only if $\sigma(I) = I$. An ideal P of R is said to be a right σ-primitive in R if and only if there exists a modular maximal right ideal σ-invariant Q of R such that P is the maximal ideal contained in Q. For $f \in R[x;\sigma]$, $\deg(f)$ denotes the degree of f and $\mathrm{lc}(f)$ the leading coefficient of f.

2. Results

We begin with the following results that extend ([1, Lemma 1]) and the proof is also similar to the one in the paper.

Lemma 1. *Let R be a ring, J a right ideal of R, $f \in J[x;\sigma]$, Q a right ideal of $R[x;\sigma]$, and $b \in R[x;\sigma]$ such that $a - ba \in Q$ for every $a \in R[x;\sigma]$. If $b - fx \in Q$, then, for every $i \geq 1$, there are $f_i \in J[x;\sigma]$ such that $b - f_i x^i \in Q$ and $\deg(f_i) \leq \deg(f)$.*

Proof. We proceed by induction on n. If $n = 1$, we put $f_1 = f$. Suppose the lemma holds for some $n \geq 1$. Let

$$f_n = a_0 + a_1 x + \cdots + a_k x^k \in J[x;\sigma], \tag{1}$$

with $b - f_n x^n \in Q$ and $k \leq \deg(f)$. Consider

$$f_{n+1} = f\sigma(a_0) + a_1 + a_2 x + \cdots + a_k x^{k-1} \in J[x;\sigma]. \tag{2}$$

Since $b - fx \in Q$, then $fx = b + q$, $q \in Q$. Thus

$$b - f_{n+1} x^{n+1} = b - f_n x^n + (a_0 - ba_0) x^n - q a_0 x^n \in Q. \tag{3}$$

□

We denote by R^1 the usual extension of R to a ring with identity and by σ again the natural extension of σ to R^1.

The next lemma extends ([1, Lemma 2]).

Lemma 2. *Let I be an ideal of $R[x;\sigma]$ with $\sigma(I) = I$ and J a right ideal of R with $\sigma(J) = J$. Consider $p = a_0 + a_1x + \cdots + a_kx^k \in I$, $k > 0$, and*

$$U = \sum_{i\in\mathbf{Z}} J[x;\sigma]\,\sigma^i(a_k)\,R^1[x;\sigma]. \tag{4}$$

(i) *If $h \in U^l$, $l \geq 1$, and $\deg(h) \geq k$, then there exists $g \in U^{l-1}$ such that $h - g \in I$ and $\deg(g) < \deg(h)$.*

(ii) *Let Q be a right ideal of $R[x;\sigma]$, $b \in R[x;\sigma]$ such that $a - ba \in Q$ for every $a \in R[x;\sigma]$, and $I \subset Q$. If $b - fx \in Q$ with $f \in J[x;\sigma]$, $\deg(f) \geq 1$, and $b - g \in Q$, where $g \in U^{\deg(f)}$, then, for every $i > \deg(g)$, there exists $g_i \in J[x;\sigma]$ such that $b - g_ix^i \in Q$ and $\deg(g_i) < k$.*

Proof. (i) Let $h = c_0 + c_1x + \cdots + c_tx^t \in U^l$, $c_t \neq 0$, and $k \leq t$. We can write

$$c_t = \sum_{j=0}^{m'} \alpha_j\beta_j, \quad \alpha_j \in U,\ \beta_j \in U^{l-1}. \tag{5}$$

Then

$$\alpha_j = \sum_{i=0}^{n_j} t_{ji}\sigma^{q_{ji}}(a_k)\,u_{ji}, \quad t_{ji} \in J,\ u_{ji} \in R^1,\ q_{ij} \in \mathbb{Z}. \tag{6}$$

Hence

$$c_t = \sum_{j=0}^{m'}\sum_{i=0}^{n_j}\left(t_{ji}\sigma^{q_{ji}}(a_k)\,u_{ji}\beta_j\right) = \sum_{i=0}^{m} p_i\sigma^{l_i}(a_k)\,e_iq_i \tag{7}$$

with $p_i \in J$, $e_i, q_i \in R^1$, $q_i \in U^{l-1}$, and $l_i \in \mathbb{Z}$. Put

$$\begin{aligned} g &= h - c_tx^t \\ &\quad + \sum_{i=0}^{m} p_i\left(\sigma^{l_i}(p) - \sigma^{l_i}(a_k)\,x^k\right)\sigma^{-k}(e_iq_i)\,x^{t-k}. \end{aligned} \tag{8}$$

Therefore

$$g - h = \sum_{i=0}^{m} p_i\left(\sigma^{l_i}(p)\right)\sigma^{-k}(e_i)\,\sigma^{-k}(q_i)\,x^{t-k} \in I. \tag{9}$$

Since $J[x;\sigma]U^{l-1} \subset U^{l-1}$ and $h \in U^{l-1}$, then $g \in U^{l-1}$ and $\deg(g) < \deg(h)$.

(ii) By Lemma 1, for every $i \geq 1$, there exists $f_i \in J[x;\sigma]$ such that

$$b - f_ix^i \in Q, \quad \deg(f_i) \leq \deg(f). \tag{10}$$

Consider

$$g = \sum_{j=0}^{m} c_jx^j, \quad c_j \in U^{\deg(f)}. \tag{11}$$

For every $n > m$ denote

$$h_n = \sum_{j=0}^{m} f_{n-j}\sigma^{n-j}(c_j) \in J[x;\sigma] \cap U^{\deg(f)}. \tag{12}$$

Note that $\deg(h_n) \leq \deg(f_{n-j}) \leq \deg(f)$; thus for every $i \geq 1$

$$f_ix^i = b + q_i, \quad q_i \in Q. \tag{13}$$

Hence

$$b - h_nx^n = b - b\sum_{j=0}^{m} c_jx^j + \sum_{j=0}^{m} q_{n-j}c_jx^j \in Q. \tag{14}$$

Because $b - g \in Q$ and $g - bg \in Q$, then $b - bg \in Q$. We have that, for every $n > t$, there exists $h_n \in U^{\deg(f)} \subseteq J[x;\sigma]$ such that $b - h_nx^n \in Q$. If $\deg(h_n) < k$, then h_n is the g_n required. If $\deg(h_n) \geq k$, by first part of this lemma, there exists

$$\lambda_{n_1} \in U^{\deg(f)-1} \subseteq J[x;\sigma] \tag{15}$$

such that $h_n - \lambda_{n_1} \in I$ and $\deg(\lambda_{n_1}) < \deg(h_n)$. Thus $b - \lambda_{n_1}x^n \in Q$ for all $n > m$. If $\deg(\lambda_{n_1}) < k$, then λ_{n_1} is the g_n required. If $\deg(\lambda_{n_1}) \geq k$, using similar arguments as above, we can find $s \in \mathbb{N}$ such that

$$\lambda_{n_s} \in U^{\deg(f)-s} \subseteq J[x;\sigma] \tag{16}$$

with $b - \lambda_{n_s}x^n \in Q$ and $\deg(\lambda_{n_s}) \leq \deg(f) - s < k$ for every $n > m$. Hence λ_{n_s} is the g_n required. □

Let $r \in R$, Q a right ideal of $R[x;\sigma]$, $\sigma(Q) = Q$, and $b \in R[x;\sigma]$ such that $a - ba \in Q$ for all $a \in R[x;\sigma]$. Following [1] we have the following. We say that v is a "good number for r," if, for all sufficiently large n, there are $f_n \in R[x;\sigma]$ such that $b - rf_nx^n \in Q$ with $\deg(f_n) \leq v$. Let $A \subseteq R$; we denote

$$\widetilde{A} = \{a \in A - Q \mid \sigma(a) - a \in Q\}. \tag{17}$$

Lemma 3. *Let Q be a right ideal of $R[x;\sigma]$ maximal in the set of all right ideals σ-invariants with $b \in R[x;\sigma]$ such that $a - ba \in Q$ for all $a \in R[x;\sigma]$. Suppose $f \in R[x;\sigma]$ with $b - fx^j \in Q$ for some $j \geq 1$. If there is no right ideal J of R with $\sigma(J) = J$, $J \not\subseteq Q$, and $J \neq R$, then there exists a positive integer v and $r \in \widetilde{R}$ such that if $w \in rR[x;\sigma]$ with $b - wx^m \in Q$, $m \geq 0$, and $\deg(w) \leq v$, then $lc(w) \in \widetilde{rR}$, $lc(w)(Q \cap R) \subseteq Q$, and v is a good number for all $a \in \widetilde{rR}$.*

Proof. Let v be minimal positive integer such that there exists $w' = i_0 + i_1x + \cdots + rx^v \in R[x;\sigma]$ and $m \geq 1$ with $b - w'x^m \in Q$ and $\deg(w') = v$. It is clear that $r \notin Q$. If $c = \sigma(r) - r \notin Q$, put $g = \sigma(w') - w'$ and

$$A = \sum_{i\in\mathbf{Z}} \sigma^i(c)\,R^1. \tag{18}$$

Thus A is a right ideal of R with $\sigma(A) = A$ and $A \not\subseteq Q$. By assumption $A = R$, then $r = \sum_{i=0}^{s} \sigma^{q_i}(c)l_i$, where $l_i \in R^1$ and $q_i \in \mathbb{Z}$. Put

$$t = w' - \sum_{i=0}^{s} \sigma^{q_i}(g)\,\sigma^{-v}(l_i). \tag{19}$$

Comparing the leading coefficients of w' and $\sum_{i=0}^{s}\sigma^{q_i}(g)\sigma^{-v}(l_i)$, we have that

$$b - tx^m \in Q, \quad \deg(t) \le v - 1, \tag{20}$$

which contradicts the minimality of v. Therefore $\sigma(r)-r \in Q$; consequently $r \in \widetilde{R}$.

Suppose that $rq \notin Q$ for some $q \in R \cap Q$. Put $g' = w'\sigma^{-v}(q) \in Q$; using similar arguments as above we can have a contradiction. Hence $r(Q \cap R) \subseteq Q$.

If there exists $w \in rR[x;\sigma]$ with $b - wx^j \in Q$, $j \ge 0$, and $\deg(w) \le v$, then using similar arguments as above we can show that $\operatorname{lc}(w) \in \widetilde{rR}$ and $\operatorname{lc}(w)(Q \cap R) \subseteqq Q$. Moreover, if $a \in \widetilde{rR}$, put $B = aR + Q \cap R$; we have that B is a right ideal of R with $\sigma(B) = B$ and $B \nsubseteq Q$.

By assumption $B = aR + Q \cap R = R$. Thus $w' = aw'' + q'$, where $w'' \in R[x;\sigma]$, $\deg(w'') \le \deg(w')$, and $q' \in Q$. Therefore $b - aw''x^m \in Q$. Consequently v is a good number for all $a \in \widetilde{rR}$. □

Lemma 4. *Let J be a right ideal R with $\sigma(J) = J$, $J \nsubseteq Q$, and $J \ne R$ such that for all sufficiently large n there are $f_n \in J[x;\sigma]$ such that $b - f_nx^n \in Q$ and $\deg(f_n) \le k$, where Q is a right ideal of $R[x;\sigma]$ and $b \in R[x;\sigma]$ such that $a - ba \in Q$ for every $a \in R[x;\sigma]$. Then there exists a positive integer v and $r \in \widetilde{R}$ such that if $w \in rR[x;\sigma]$ with $b - wx^m \in Q$, $m \ge 0$, and $\deg(w) \le v$ one has that $lc(w) \in \widetilde{rR}$, $lc(w)(Q \cap R) \subseteq Q$, and v is a good number for all $a \in \widetilde{rR}$.*

Proof. Let v be minimal positive integer such that for all sufficiently large n there are $f_n \in J[x;\sigma]$ such that $b - f_nx^n \in Q$ and $\deg(f_n) \le v$. Put

$$w' = i_0 + i_1x + \cdots + i_{v-1}x^{v-1} + rx^v \in J[x;\sigma] \tag{21}$$

with $b - w'x^m \in Q$, $m \ge 0$, and $\deg(w') \le v$. By Lemma 1 and minimality of v we have that $r \notin Q$. Using the same ideas of Lemma 3, we have that $r \in \widetilde{R}$ and $r(Q \cap R) \subseteq Q$. Since $rR \subseteq J$, we have that the first part of lemma is satisfied.

Let $a \in \widetilde{rR} \subseteq \widetilde{J}$; we denote by B the right ideal of R:

$$B = \sum_{i\in\mathbb{Z}}\sigma^i(a)R^1, \quad \sigma(B) = B, \; B \subseteq J, \; B \nsubseteq Q. \tag{22}$$

For sufficiently large n there are $g_n \in B[x;\sigma] \subseteq J[x;\sigma]$ such that $b - g_nx^n \in Q$ and $\deg(g_n) \le v$. Put

$$g_n = c_{n_0} + c_{n_1}x + \cdots + c_{n_v}x^v \in B[x;\sigma]. \tag{23}$$

For every $0 \le j \le v$ we have that $c_{n_j} = \sum_{i=0}^{m_j}\sigma^{q_{n_i}}(a)l_{n_i}$, where $l_{n_i} \in R^1$ and $q_{n_i} \in \mathbb{Z}$. Consequently

$$c_{n_j} = \sum_{i=0}^{m_j}(\sigma^{q_{n_i}}(a) - a)l_{n_i} + a\sum_{i=0}^{m_j}l_{n_i}. \tag{24}$$

Since $a \in \widetilde{rR}$, we can write

$$c_{n_j} = s_{n_j} + r_{n_j}, \quad s_{n_j} \in Q \cap R, \; r_{n_j} \in R. \tag{25}$$

Put $h_n = r_{n_0} + r_{n_1}x + \cdots + r_{n_v}x^v$; thus $b - ah_nx^n \in Q$. Therefore v is a good number for all $a \in \widetilde{rR}$. □

Lemma 5. *Let Q be a right ideal of $R[x;\sigma]$, $b \in R[x;\sigma]$, such that $a - ba \in Q$ for all $a \in R[x;\sigma]$ and v is good number for all $a \in \widetilde{rR}$, where $r \in \widetilde{R}$. Assume that for every $w \in rR[x;\sigma]$ with $b - wx^m \in Q$, $m \ge 0$, and $\deg(w) \le v$ one has that $lc(w) \in \widetilde{rR}$ and $lc(w)(Q \cap R) \subseteq Q$. If there are p and $p' \in \widetilde{rR}$ with*

$$\left(\widetilde{pR} + Q \cap rR\right) \cap \left(\widetilde{p'R} + Q \cap rR\right) \subseteq Q, \tag{26}$$

then $v - 1$ is a good number for r.

Proof. Since v is a good number for p and p', then for every sufficiently large n there are $g_n \in pR[x;\sigma]$ and $g_n' \in p'R[x;\sigma]$ such that

$$b - g_nx^n \in Q, \tag{27}$$

$$b - g_n'x^n \in Q \tag{28}$$

with $\deg(g_n)$, $\deg(g_n') \le v$. Consider

$$\begin{aligned} g_n &= p_{n_0} + p_{n_1}x + \cdots + p_{n_v}x^v, \quad p_{n_v} \in \widetilde{pR}, \\ g_n' &= p_{n_0}' + p_{n_1}'x + \cdots + p_{n_v}'x^v, \quad p_{n_v}' \in \widetilde{p'R}. \end{aligned} \tag{29}$$

Since $p_{n_v} - p_{n_v}' \in Q$, then

$$p_{n_v} \in \left(\widetilde{pR} + Q \cap rR\right) \cap \left(\widetilde{p'R} + Q \cap rR\right) \subseteq Q, \tag{30}$$

a contradiction.

Thus there exists sufficiently large $i \in \mathbb{N}$ such that $c = p_{i_v} - p_{i_v}' \in \widetilde{rR}$; hence v is a good number for c. Then for all sufficiently large n there are $h_n \in R[x;\sigma]$ such that $b - ch_nx^n \in Q$ and $\deg(h_n) \le v$. We denote

$$h_n = r_{n_0} + r_{n_1}x + \cdots + r_{n_v}x^v. \tag{31}$$

Consider

$$k_n = ch_n + (g_i' - g_i)\sigma^{-v}(r_{n_v}) \in rR[x;\sigma]. \tag{32}$$

Since $g_i' - g_i \in Q$, then $b - k_nx^n \in Q$. Moreover

$$k_n = ch_n - cr_{n_v}x^v + \sum_{j=0}^{v-1}\left(p_{i_j}' - p_{i_j}\right)x^j\sigma^{-v}(r_{n_v}). \tag{33}$$

Consequently $v - 1$ is a good number for r. □

The following theorem extends ([1, Theorem 1]).

Theorem 6. *Let R be a nil ring and let I be a σ-primitive ideal in $R[x;\sigma]$. Then $I = I'[x;\sigma]$, where I' is an ideal σ-invariant of R.*

Proof. Assume by contradiction that there are $a_0, a_1, \ldots, a_k \in R$ with

$$a_0 + a_1x + \cdots + a_kx^k \in I, \quad a_k \notin I. \tag{34}$$

Since I is a σ-primitive ideal in $R[x;\sigma]$, there is a right ideal Q of $R[x;\sigma]$ with $\sigma(Q) = Q$ and $b \in R[x;\sigma]$ such that $a - ba \in Q$ for all $a \in R[x;\sigma]$. Moreover Q is a maximal in the set of right ideals σ-invariants and I is the maximal ideal contained in Q. We have that $R[x;\sigma]x \nsubseteq Q$; otherwise $b \in R$, which is impossible because R is a nil ring. By definition of Q it follows that $R[x;\sigma]x + Q = R[x;\sigma]$.

If $b - hx^i \in Q$ for some $i \geq 0$ with $h \in R[x;\sigma]$, then $\deg(h) \geq 1$. In fact, if $h \in R$, let $t \geq 1$ be the minimal positive integer with respect to $h^t \in Q$. Thus $(b - hx^i)\sigma^{-i}(h^{t-1}) \in Q$. Then $b\sigma^{-i}(h^{t-1}) \in Q$; hence $\sigma^{-i}(h^{t-1}) \in Q$. Consequently $h^{t-1} \in Q$, a contradiction.

Let J be a right ideal of R with $\sigma(J) = J$ and $J \nsubseteq Q$. We have that $J[x;\sigma]x + Q = R[x;\sigma]$. There exists $f \in J[x;\sigma]$ such that $b - fx \in Q$. Consider

$$U = \sum_{i\in\mathbb{Z}} J[x;\sigma]\,\sigma^i(a_k)\,R^1[x;\sigma]. \tag{35}$$

Since I is an ideal σ-prime and $a_k \notin I$, then $U \nsubseteq I$. Consequently $U \nsubseteq Q$, because I is the maximal ideal contained in Q. Then $U^{\deg(f)} + Q = R[x;\sigma]$. There exists $g' \in U^{\deg(f)}$ such that $b - g' \in Q$. By Lemma 2, for every $i \geq \deg(g')$, there are $g_i' \in J[x;\sigma]$ such that $b - g_i'x^i \in Q$ and $\deg(g_i') < k$. Lemmas 3 and 4 imply that there are $r' \in \tilde{R}$ and $v' \geq 1$ such that if $w \in r'R[x;\sigma]$ with $b - wx^m \in Q$, $m \geq 1$, and $\deg(w) \leq v'$, then $\mathrm{lc}(w) \in \widetilde{r'R}$ and $\mathrm{lc}\, w(Q \cap R) \subseteq Q$. Moreover v' is a good number for all $a \in \widetilde{r'R}$. Let v be minimal such that v is a good number for all $a \in \widetilde{r'R}$. We have that $v \leq v'$. Let $r \in \widetilde{r'R}$. Since v is a good number for r, then for sufficiently large n there are $h_n \in R[x;\sigma]$, such that

$$b - rh_nx^n \in Q, \quad \deg(h_n) \leq v. \tag{36}$$

Consider $f_n = rh_n$, then $b - f_nx^n \in Q$ and $\deg(f_n) \leq v$. For some $i \in \mathbb{N}$, there are $f_i, f_{i+1}, \ldots, f_{i+k} \in rR[x;\sigma]$, such that $b - f_jx^j \in Q$, $\deg(f_j) \leq v$, and $i \leq j \leq i+k$. Put

$$f_j = ra_{j_0} + ra_{j_1}x + \cdots + ra_{j_v}x^v = g_j + c_jx^v, \tag{37}$$

where $g_j = ra_{j_0} + ra_{j_1}x + \cdots + ra_{j_{v-1}}x^{v-1} \in rR[x;\sigma]$ and $c_j = ra_{j_v}$. Since $\deg(f_j) \leq v \leq v'$, then $c_j \notin Q$. Moreover,

$$\sigma(c_j) - c_j \in Q, \quad c_j(Q \cap R) \subseteq Q. \tag{38}$$

Since R is a nil ring, consider $e_j = c_j^{n_j}$, where n_j is a minimal with respect to the condition $c_j^{n_j} \notin Q$. Thus $\sigma(e_j) - e_j \in Q$ for all $i \geq 0$. We have that

$$\begin{aligned} f_j\sigma^j(e_j) &= g_j\sigma^j(e_j) + c_j\sigma^{j+v}(e_j)x^v \\ &= g_j\sigma^j(e_j) + c_j\left(\sigma^{j+v}(e_j) - e_j\right)x^v \\ &\quad + c_je_jx^v. \end{aligned} \tag{39}$$

Put $t_j = g_j\sigma^j(e_j) \in rR[x;\sigma]$. Thus,

$$f_j\sigma^j(e_j) - t_j \in Q \quad \deg(t_j) \leq v - 1 \tag{40}$$

for every $i \leq j \leq i+k$. Since $e_j \in \widetilde{rR} \subseteq \widetilde{r'R}$, if $v-1$ is not a good number for r, then Lemma 5 implies that

$$\bigcap_{j=1}^{i+k}\left(\widetilde{e_jR} + Q \cap rR\right) \nsubseteq Q. \tag{41}$$

In this case, there exists $s \in \bigcap_{j=1}^{i+k}(\widetilde{e_jR} + Q \cap rR)$ such that $s \notin Q$. Consequently $s - e_jd_j \in Q \cap rR$, $d_j \in R$, and $e_jd_j \in \widetilde{e_jR}$. Then $s \in \widetilde{rR} \subseteq \widetilde{r'R}$. Therefore v is a good number for s. Then for sufficiently large n there are $\overline{f}_n \in sR[x;\sigma]$, such that

$$b - \overline{f}_nx^n \in Q, \quad \deg(\overline{f}_n) \leq v. \tag{42}$$

Let

$$\overline{f}_n = \sum_{j=0}^{v} sb_{j_n}x^j. \tag{43}$$

Since $b - \overline{f}_jx^j \in Q$, $s - e_jd_j \in Q$, and $e_jd_j - be_jd_j \in Q$, then $(b - f_jx^j)e_jd_j \in Q$. Thus $be_jd_j - f_jx^je_jd_j \in Q$; hence $s - f_jx^je_jd_j \in Q$ for every $i \leq j \leq i+k$.

Let

$$\overline{g}_n = \sum_{j=0}^{v} f_{i+v-j}\sigma^{i+v-j}\left(e_{i+v-j}d_{i+v-j}b_{j_n}\right)x^{i+v} \in rR[x;\sigma]. \tag{44}$$

We have that $\overline{f}_n - \overline{g}_n \in Q$. Thus $b - \overline{g}_nx^n = (b - \overline{f}_nx^n) + (\overline{f}_n - \overline{g}_n)x^n \in Q$. Put

$$\overline{h}_n = \sum_{j=0}^{v} t_{i+v-j}\sigma^{i+v-j}\left(d_{i+v-j}b_{j_n}\right) \in rR[x;\sigma]. \tag{45}$$

We can write $b - \overline{h}_nx^{i+v+n}$ as

$$\begin{aligned} &b - \sum_{j=0}^{v}\left(t_{i+v-j} - f_{i+v-j}\sigma^{i+v-j}\left(e_{i+v-j}\right)\right)\sigma^{i+v-j}\left(d_{i+v-j}b_{j_n}\right) \\ &\cdot x^{i+v+n} - \overline{g}_nx^n. \end{aligned} \tag{46}$$

Thus for all sufficient large n

$$b - \overline{h}_nx^{i+v+n} \in Q, \quad \deg(\overline{h}_n) \leq v - 1. \tag{47}$$

Then $v-1$ is a good number for all $r \in \widetilde{r'R}$. This contradicts the minimality of v. □

Recall that the σ-Jacobson radical $J_\sigma(R)$ of a ring R is defined as the intersection of all σ-primitive ideals of R. A ring R is a σ-Jacobson radical if $J_\sigma(R) = R$.

Theorem 7. *Let R be a nil ring and let I be an ideal of $R[x;\sigma]$. Consider $\overline{I}$ the ideal of R generated by coefficients of polynomial from I. Then $R[x;\sigma]/\overline{I}[x;\sigma]$ is σ-Jacobson radical if and only if $R[x;\sigma]/I$ is σ-Jacobson radical.*

Proof. Assume by contradiction that $R[x;\sigma]/I$ is not σ-Jacobson radical. Then there is a σ-primitive ideal P of $R[x;\sigma]/I$ such that $P \neq R[x;\sigma]/I$. We have that there is an ideal K of $R[x;\sigma]$ such that $P = K/I$. Therefore K is a σ-primitive ideal of $R[x;\sigma]$. By Theorem 6, there is an ideal $\overline{P}$ of R such that $K = \overline{P}[x;\sigma]$. It is clear that $\overline{I} \subseteq \overline{P}$. Since

$$\frac{\left(R[x;\sigma]/\overline{I}[x;\sigma]\right)}{\left(\overline{P}[x;\sigma]/\overline{I}[x;\sigma]\right)} \simeq \frac{R[x;\sigma]}{K}, \tag{48}$$

then $\overline{P}[x;\sigma]/\overline{I}[x;\sigma]$ is a σ-primitive ideal, a contradiction. Using the fact that $I \subseteq \overline{I}[x;\sigma]$, the converse follows. □

Corollary 8. *If R is a nil ring, then the polynomial ring of type automorphism $R[x;\sigma]$ can not be homomorphically mapped onto a σ-simple σ-primitive ring.*

Competing Interests

The author declares that they have no competing interests.

References

[1] A. Smoktunowicz, "On primitive ideals in polynomial rings over nil rings," *Algebras and Representation Theory*, vol. 8, no. 1, pp. 69–73, 2005.

[2] E. Cisneros, M. Ferrero, and M. I. Conzles, "Prime ideals of skew polynomial rings and skew laurent polynomial rings," *Mathematical Journal of Okayama University*, vol. 32, pp. 61–72, 1990.

[3] N. Divinsky, *Rings and Radicals*, Allen and Unwin, London, UK, 1965.

[4] T. Y. Lam, *A First Course in Noncommutative Rings*, Graduate Texts in Mathematics, Springer, New York, NY, USA, 1991.

Permissions

All chapters in this book were first published in IJMMS, by Hindawi Publishing Corporation; hereby published with permission under the Creative Commons Attribution License or equivalent. Every chapter published in this book has been scrutinized by our experts. Their significance has been extensively debated. The topics covered herein carry significant findings which will fuel the growth of the discipline. They may even be implemented as practical applications or may be referred to as a beginning point for another development.

The contributors of this book come from diverse backgrounds, making this book a truly international effort. This book will bring forth new frontiers with its revolutionizing research information and detailed analysis of the nascent developments around the world.

We would like to thank all the contributing authors for lending their expertise to make the book truly unique. They have played a crucial role in the development of this book. Without their invaluable contributions this book wouldn't have been possible. They have made vital efforts to compile up to date information on the varied aspects of this subject to make this book a valuable addition to the collection of many professionals and students.

This book was conceptualized with the vision of imparting up-to-date information and advanced data in this field. To ensure the same, a matchless editorial board was set up. Every individual on the board went through rigorous rounds of assessment to prove their worth. After which they invested a large part of their time researching and compiling the most relevant data for our readers.

The editorial board has been involved in producing this book since its inception. They have spent rigorous hours researching and exploring the diverse topics which have resulted in the successful publishing of this book. They have passed on their knowledge of decades through this book. To expedite this challenging task, the publisher supported the team at every step. A small team of assistant editors was also appointed to further simplify the editing procedure and attain best results for the readers.

Apart from the editorial board, the designing team has also invested a significant amount of their time in understanding the subject and creating the most relevant covers. They scrutinized every image to scout for the most suitable representation of the subject and create an appropriate cover for the book.

The publishing team has been an ardent support to the editorial, designing and production team. Their endless efforts to recruit the best for this project, has resulted in the accomplishment of this book. They are a veteran in the field of academics and their pool of knowledge is as vast as their experience in printing. Their expertise and guidance has proved useful at every step. Their uncompromising quality standards have made this book an exceptional effort. Their encouragement from time to time has been an inspiration for everyone.

The publisher and the editorial board hope that this book will prove to be a valuable piece of knowledge for researchers, students, practitioners and scholars across the globe.

List of Contributors

Mina Ketan Mahanti
Department of Mathematics, College of Basic Science and Humanities, OUAT, Bhubaneswar, India

Amandeep Singh
DPS Kalinga, Bhubaneswar, India

Lokanath Sahoo
Gopabandhu Science College, Athagad, India

Sylvain Attan and A. Nourou Issa
Départemente de Mathématiques, Université d'Abomey-Calavi, 01 BP 4521 Cotonou, Benin

Refaat M. Salem
Mathematics Department, Faculty of Science, Al-Azhar University, Nasr City, Cairo, Egypt

Mohamed A. Farahat
Mathematics Department, Faculty of Science, Al-Azhar University, Nasr City, Cairo, Egypt
Department of Mathematics and Statistics, Faculty of Science, Taif University, Al-Hawiyah, Taif 21974, Saudi Arabia

Hanan Abd-Elmalk
Department of Mathematics, Faculty of Science, Ain Shams University, Abbasaya, Cairo, Egypt

Yanisa Chaiya, Preeyanuch Honyam and Jintana Sanwong
Department of Mathematics, Chiang Mai University, Chiang Mai 50200, Thailand

Lei Cao
Department of Mathematics, Georgian Court University, Lakewood, NJ 08701, USA

Selcuk Koyuncu
Department of Mathematics, University of North Georgia, Gainesville, GA 30566, USA

Shivani Dubey and Ajay Kumar
Department of Mathematics, University of Delhi, Delhi 110007, India

Mukund Madhav Mishra
Department of Mathematics, Hans Raj College, University of Delhi, Delhi 110007, India

Elif Ozel Ay, Gürsel Yesilot and Deniz Sonmez
Department of Mathematics, Yildiz Technical University, Davutpasa, Istanbul, Turkey

Vahagn Mikaelian
Yerevan State University, Alex Manoogian 1, 0025 Yerevan, Armenia
American University of Armenia, 40 Marshal Baghramyan Ave., 0019 Yerevan, Armenia

Samaher Adnan Abdul-Ghani, Shuker Mahmood Khalil, Mayadah Abd Ulrazaq and Abu Firas Muhammad Jawad Al-Musawi
Department of Mathematics, College of Science, Basrah University, Basrah 61004, Iraq

Yang-Hi Lee
Department of Mathematics Education, Gongju National University of Education, Gongju 32553, Republic of Korea

Soon-Mo Jung
Mathematics Section, College of Science and Technology, Hongik University, Sejong 30016, Republic of Korea

A. A. A. Agboola
Department of Mathematics, Federal University of Agriculture, Abeokuta, Nigeria

B. Davvaz
Department of Mathematics, Yazd University, Yazd, Iran

Le Gao
Department of Mechanical Engineering, Auburn University, Auburn, AL 36849, USA

N. K. Govil
Department of Mathematics and Statistics, Auburn University, Auburn, AL 36849, USA

Worachead Sommanee
Department of Mathematics and Statistics, Faculty of Science and Technology, Chiang Mai Rajabhat University, Chiang Mai 50300, Thailand

Yuriy V. Shablya
Tomsk State University of Control Systems and Radioelectronics, 40 Lenina Avenue, Tomsk 634050, Russia

Dmitry V. Kruchinin
Tomsk State University of Control Systems and Radioelectronics, 40 Lenina Avenue, Tomsk 634050, Russia
National Research Tomsk Polytechnic University, 30 Lenin Avenue, Tomsk 634050, Russia

Pakorn Palakawong na Ayutthaya and Bundit Pibaljommee
Department of Mathematics, Faculty of Science, Khon Kaen University, Khon Kaen 40002, Thailand
Centre of Excellence in Mathematics CHE, Si Ayutthaya Road, Bangkok 10400, Thailand

Abdelkarim Boua
Department of Mathematics, Faculty of Sciences of Agadir, Ibn Zohr University, Agadir, Morocco

A. Raji
Département de Mathématiques, Faculté des Sciences et Techniques, Université Moulay Ismaïl, Groupe d'Algèbre et Applications, BP 509, Boutalamine, Errachidia, Morocco

Asma Ali and Farhat Ali
Department of Mathematics, Aligarh Muslim University, Aligarh 202002, India

Somphong Jitman
Department of Mathematics, Faculty of Science, Silpakorn University, Nakhon Pathom 73000, Thailand

Aunyarut Bunyawat, SupanutMeesawat, Arithat Thanakulitthirat and Napat Thumwanit
Department of Mathematics, Mahidol Wittayanusorn School, Nakhon Pathom 73170, Thailand

Eunmi Choi
Department of Mathematics, HanNam Univ., Daejeon, Republic of Korea

David E. Dobbs
Department of Mathematics, University of Tennessee, Knoxville, TN 37996-1320, USA

Ruipu Bai
College of Mathematics and Information Science, Hebei University, Key Laboratory of Machine Learning and Computational Intelligence of Hebei Province, Baoding 071002, China

Shuai Hou and Yansha Gao
College of Mathematics and Information Science, Hebei University, Baoding 071002, China

Patricia L. Zoungrana and A. Nourou Issa
Département de Mathématiques de la Décision, Université Ouaga 2, 12 BP 412, Ouagadougou 12, Burkina Faso
Département de Mathématiques, Université d'Abomey-Calavi, 01 BP 4521, Cotonou, Benin

Mahid M.Mangontarum and Amila P. Macodi-Ringia
Department of Mathematics, Mindanao State University, Main Campus, 9700 Marawi City, Philippines

Omar I. Cauntongan
Department of Natural Sciences and Mathematics, Mindanao State University, Maigo School of Arts and Trades, 9206 Maigo, Philippines

Deena Al-Kadi
Department of Mathematics and Statistic, Faculty of Science, Taif University, Taif 21974, Saudi Arabia

Kwang Hyun Kim
Departments of Mathematics and Computer Science, Queensborough Community College, 222-05 56th Avenue, Bayside, NY 11364, USA

Xin Zhang
Zicklin School of Business, Baruch College, One Bernard Baruch Way, New York, NY 10010, USA

Thitarie Rungratgasame, Pattharapham Amornpornthum, Phuwanat Boonmee, Busrun Cheko and Nattaphon Fuangfung
Department of Mathematics, Faculty of Science, Srinakharinwirot University, Bangkok 10110,Thailand

Edilson Soares Miranda
Departamento de Ciências, Centro de Ciências Exatas, Universidade Estadual de Maringá, 87360-000 Goioerê, PR, Brazil

Index

A
Aberth-ehrlich Method, 85
Algebraic Polynomials, 1, 85
Auxiliary Primes, 50-51, 62

B
Bci-algebra, 79-81, 84
Bessel Polynomials, 102, 105
Big Prime Modular Gcd Algorithm, 50-53, 56-58
Binomial Coefficients, 88-89, 174
Boundary Value Problems, 39, 45

C
Classical Magic Square, 194
Coprime Polynomials, 50, 56, 60
Corner Magic Squares, 194, 196-199

D
Differential Equations, 45

E
Euclidean Algorithm, 50-53, 58

F
Factorization, 50-53, 121, 145
Fuzzy Sets Theory, 63

G
Gaussian Random Variables, 1
Good Punctured Polynomials, 121-122, 124, 126-127

H
Hilbert Spaces, 182
Hom-alternative Algebras, 9-10, 15, 19
Hom-bol Algebra, 8-9, 11, 15-17
Hom-lie Triple System, 8, 10
Hom-maltsev Algebra, 8-10, 14-15
Hom-maltsev Identity, 9-10
Hyers-ulam Stability, 70, 78
Hyperbolic Polynomials, 1
Hyperideal Expansion, 46-48

I
Inner Superderivation, 160
Integer Coefficients, 51
Interpolation, 101, 104
Intuitionistic Fuzzification, 63
Isomorphism, 84, 92, 136, 138-139, 141-142, 144-146

L
Lie Superalgebra, 159-160, 162-163, 165
Lie Supertriple System, 159, 162, 165
Lie-yamaguti Algebra, 158-159, 162, 165
Lie-yamaguti Superalgebra, 158-160, 163, 165
Linear Mappings, 160
Logic Algebras, 79

M
Malcev Superalgebra, 159
Maximal Regular Subsemigroups, 92
Meixner Polynomials, 101-105
Minimal Ring Extensions, 136, 138, 146, 148
Monoid Homomorphism, 20, 22

N
Neutrosophic Algebraic Structures, 79
Neutrosophic Bck-algebra, 80
Neutrosophic Set, 79-80, 84
Nonbinary Finite Fields, 124, 126
Noncentral Whitney Numbers, 168-170, 172, 174-175, 178, 180-181
Nonnegative Integer, 36, 38, 168, 170
Nonzero Polynomials, 50, 53-54, 58, 60
Nonzero Real Constants, 69, 71, 74
Number Theory, 33, 38, 101, 105, 148, 181

P
Partial Differential Equations, 45
Poisson Kernel, 45
Polynomial, 1, 21, 33-35, 50-54, 57-58, 62, 85, 90-91, 104, 121-122, 125, 127, 169, 173, 176, 201, 205
Polynomial Rings, 21, 50-52, 201, 205
Polynomials, 1, 21, 24, 45, 50-54, 56-58, 60-62, 85, 90-91, 101-105, 121-122, 124-127, 168, 173-182, 201
Positive Integers, 145, 194
Prime Near Ring, 114-120

Q
Quadratic Mapping, 69-70, 72
Quadratic-additive Mapping, 69-72, 74

R
Random Trigonometric Polynomial, 1
Rational Numbers, 50, 70-72, 127
Reflective Magic Squares, 194-195, 199
Regular Ordered Semirings, 106, 109-110

Right Hom-alternative Algebra, 15-17

S

Skew Generalized Power Series Extension, 20
Skew Generalized Power Series Rings, 20
Skew-regular Magic Squares, 194, 198
Symmetric Integer Matrices, 33

T

Ternary Hom-nambu Algebra, 10
Topology, 39
Trivial Good Punctured Polynomials, 122

W

Whitney Numbers, 167-170, 172, 174-175, 178, 180-182

Z

Zeros Of A Polynomial, 85, 90-91

www.ingramcontent.com/pod-product-compliance
Lightning Source LLC
Chambersburg PA
CBHW080646200326
41458CB00013B/4748

* 9 7 8 1 6 4 7 2 8 3 4 6 9 *